T0075688

Lecture Notes
in Computational Science
and Engineering

68

Editors

Timothy J. Barth
Michael Griebel
David E. Keyes
Risto M. Nieminen
Dirk Roose
Tamar Schlick

Sidney Yip · Tomás Diaz de la Rubia
Editors

Scientific Modeling and Simulations

Previously published in *Scientific Modeling and Simulations*, Volume 15, Issues 1–3, 2008

 Springer

Editors

Sidney Yip
Department of Nuclear Engineering
Massachusetts Institute of Technology
77 Massachusetts Avenue
Cambridge, MA 02139-4307
USA
syip@mit.edu

Tomás Diaz de la Rubia
Lawrence Livermore National Laboratory
University of California
7000 East Avenue
P.O. Box 808
Livermore, CA 94551
USA
delarubia@llnl.gov

ISBN: 978-1-4020-9740-9   e-ISBN: 978-1-4020-9741-6

Lecture Notes in Computational Science and Engineering ISSN 1439-7358

Library of Congress Control Number: 2009922358

Mathematics Subject Classification (2000): 65-01, 65P99, 35B27, 65R20, 70-08, 42C40, 76S05, 76T20

Contents

Scientific Modeling and Simulations

Advocacy of computational science (Editors' preface)

Tomas Diaz de la Rubia · Sidney Yip

Originally published in the journal Sci Model Simul, Volume 15, Nos 1–3, 1–2.
DOI: 10.1007/s10820-008-9099-8 © Springer Science+Business Media B.V. 2008

In its cover letter to President G. W. Bush on May 27, 2005, the President's Information Technology Advisory Committee wrote, "Computational Science—the use of advanced computing capabilities to understand and solve complex problems—has become critical to scientific leadership, economic competitiveness, and national security." This single sentence explains why this particular scientific endeavor has such a broad impact. It also conveys the essence of our vision for Scientific Modeling and Simulations.

There exist a number of studies at the federal level which are unanimous in recognizing that computational methods and tools can help to solve large-scale problems in our society, most recently problems concerning energy and the environment. On the other hand, there has been relatively little said about the fundamental attributes of Computational Science that justify the mobilization of a new science and technology community and the generations that follow. Along with funding support and the increased expectations of advanced computations, what is also needed is a sustained environment to continuously assess the intellectually stimulating and scientifically useful aspects of this endeavor, and promote active exchanges between different parts of the growing community. In our opinion it is only through this kind of advocacy and stewardship that the full potentials and rewards of Computational Science can be realized.

We see the conceptualization of a problem (modeling) and the computational solution of this problem (simulation), as the foundation of Computational Science. This coupled endeavor is unique in several respects. It allows practically any complex system to be analyzed with predictive capability by invoking the multiscale paradigm—linking unit-process models at lower length (or time) scales where fundamental principles have been established to calculations at the system level. It allows the understanding and visualization of cause-effect through simulations where initial and boundary conditions are prescribed specifically to gain insight. Furthermore, it can complement experiment and theory by providing the details that

T. D. de la Rubia
Lawrence Livermore National Security, Livermore, CA 64550, USA

S. Yip (✉)
Massachusetts Institute of Technology, Cambridge, MA 02139, USA
e-mail: syip@mit.edu

cannot be measured nor described through equations. When these conceptual advantages in modeling are coupled to unprecedented computing power through simulation, one has a vital and enduring scientific approach destined to play a central role in solving the formidable problems of our society. Yet, to translate these ideals into successful applications requires the participation of all members of the science and technology community. In this spirit Scientific Modeling and Simulations advocates the scientific virtues of modeling and simulation, and also encourages discussions on cross fertilization between communities, exploitations of high-performance computing, and experiment-simulation synergies.

The community of multiscale materials modeling has evolved into a multidisciplinary group with a number of identified problem areas of interest. (See the Retrospective for a series of meetings held by this community). Collectively the group is familiar with the intrinsic advantages of modeling and simulation and the potentials to enable significant advances in materials research. We believe that this and other similar groups can play a more expanded role in the advocacy of Computational Science, by highlighting the insights derived from a materials-focus study that can impact other problems involving basically the same physical system. Consider a well-known materials behavior as an example. The strength variation of the cement slurry from mixing the powder with water to hardened setting is a phenomenon of which multiscale modeling and simulation may be the only way to achieve some degree of molecular-level understanding. This is the kind of complex problem that a breakthrough can have very wide-spread impact; in this case, there would be significant consequences in communities, ranging from structural materials to colloidal science (paint, coating) to architecture. The possibilities for cross fertilization between a scientific community which understands materials and their innovation at the fundamental level and many others where materials play a critical role is virtually limitless. We recognize of course that the transfer of insight from one problem to another does not usually follow an obvious path; also what is insight can be different when seen in different context. Nonetheless, the potential benefit of a forum for open discussions is clear.

We know of no way to build the community we have in mind other than to engage a broad-based participation of scientists who share in the belief that this is a vital and promising method of scientific inquiry beneficial to all disciplines. The contributors to Scientific Modeling and Simulations have started the process of articulating the usefulness of modeling and simulation, each in its own context. From this one can begin to get a sense of the many different areas of applications worthy of investigation.

A retrospective on the journal of computer-aided materials design (JCAD), 1993–2007

Sidney Yip

Originally published in the journal Sci Model Simul, Volume 15, Nos 1–3, 3–4.
DOI: 10.1007/s10820-008-9098-9 © Springer Science+Business Media B.V. 2008

In 1993 The Journal of Computer-Aided Materials Design (JCAD) was launched by a small publishing company in the Netherlands, ESCOM. The Publisher, Dr. Elizabeth Schram, had invited four scientists with common interests in computer simulations of materials to serve as editors. The motivation then was that atomistic and molecular simulations were being recognized as emerging computational tools for materials research, so it seemed reasonable to anticipate that they could be integrated into the process of design. "Since in both analysis and design one can intrinsically exploit the remarkable (and ever-increasing) capabilities of computer hardware and software technology, the intersection of computation, in the broad sense, and materials design, with its holistic nature, naturally delineates a range of issues which need to be addressed at the present time." As for the name of the Journal, the decision was at least partially influenced by the success of another ESCOM journal, the Journal of Computer-Aided Molecular Design (JCAMD).

The four editors were Anthony K. Cheetham (University of California Santa Barbara) Ulrich W. Suter (ETH Zurich), Erich Wimmer (then with Biosym in Paris), and me. After about 3 years it was decided that it would be more efficient if one person would act as the Principal Editor to oversee all the operational affairs, with the other three assisting primarily in soliciting manuscripts. The understanding was that this arrangement would rotate about every three years. History shows that I was chosen to go first in this capacity, and subsequent handoffs never occurred.

How did things work out? If I were asked, my answer would be the journey was up and down. JCAD witnessed the period when multiscale materials modeling and simulation began to flourish from modest topical workshops to sizable international conferences. One of the most satisfying experiences for me is that the Journal has played a sustained role in building this community.

In the late 1990s the DOE launched the initiative ASCI (Accelerated Strategic Computing Initiative) as part of its stockpile stewardship program, which provided major impetus for

S. Yip (✉)
Departments of Nuclear Science and Engineering and Materials Science and Engineering, Massachusetts Institute of Technology, Cambridge, MA 02139, USA
e-mail: syip@mit.edu

large-scale simulation of materials behavior in extreme environments. A large workshop on the Dynamics of Metals was conducted in Bodega Bay by the Lawrence Livermore National Laboratory; the follow up work was reported in another JCAD special issue.

One of the early workshops, considered a significant "kindling event", took place at the Institute for Theoretical Physics, UC Santa Barbara, in 1995; the proceedings, *Modeling of Industrial Materials: Connecting Atomistic and Continuum Scales*, were published as a special issue in JCAD. This was followed by two international workshops, first in Beijing, then in Hong Kong, both with Hanchen Huang as the lead organizer. Together these gatherings paved the way for the MMM Conferences.

The multiscale materials modeling (MMM) series began in London in 2002 under the leadership of Xiao Guo, continued in UCLA in 2004 under Nasr Ghoniem, and in Freiburg in 2006 under Peter Gumbsch. Each meeting was larger than the previous one. The fourth gathering will take place in Florida State Univ. in October this year, organized by Anter El Azab.

Tomas Diaz de la Rubia, who organized and chaired the Bodega Bay workshop and was present at the Santa Barbara workshop, played a key role in keeping the multiscale materials modeling community engaged, along with other laboratory scientists such as John Moriarty and Elaine Chandler. It is perhaps appropriate that the last issue of JCAD contains a set of the invited papers presented at the third MMM Conference in Freiburg.

Looking ahead I can see how the lessons learned can be put to good use. One observation is that community building is not a natural phenomenon—it does not happen by itself, at least not for very long. Another is that the design of materials components or systems, in the context of multiscale modeling and simulation, is still very much in the future. Materials design poses different challenges from molecular (drug) design, which is not to say the latter is any less difficult. After 14 years, we still do not have many success stories to showcase the rewards of materials design. With hindsight I can now say the choice of the journal title was premature, more of a wish than a reality. JCAD is now relaunched as *Scientific Modeling and Simulation* which describes much more closely the interest of the community that JCAD had served. In closing this brief historical account, I thank the many individuals who have contributed to the existence of the journal, especially several members of the Editorial Board for being very helpful at critical times. To everyone in the community I would like to say that the spirit of JCAD will live on one way or another.

Extrapolative procedures in modelling and simulations: the role of instabilities

Göran Grimvall

Originally published in the journal Sci Model Simul, Volume 15, Nos 1–3, 5–20.
DOI: 10.1007/s10820-008-9093-1 © Springer Science+Business Media B.V. 2008

Abstract   In modelling and simulations there is a risk that one extrapolates into a region where the model is not valid. In this context instabilities are of particular interest, since they can arise without any precursors. This paper discusses instabilities encountered in the field of materials science, with emphasis on effects related to the vibrations of atoms. Examples deal with, i.a., common lattice structures being either metastable or mechanically unstable, negative elastic constants that imply an instability, unexpected variations in the composition dependence of elastic constants in alloys, and mechanisms governing the ultimate strength of perfect crystals.

Keywords   Modelling · Simulation · Extrapolation · Instability · Lattice dynamics · Ultimate strength

1 Introduction

"It is a common error which young physicists are apt to fall into to obtain a law, a curve, or a mathematical expression for given experimental limits and then apply it to points outside those limits. This is sometimes called extrapolation. Such a process, unless carefully guarded, ceases to be a reasoning process and becomes one of pure imagination specially liable to error when the distance is too great.

In temperature our knowledge extends from near the absolute zero to that of the sun, but exact knowledge is far more limited. In pressures we go from the Crookes vacuum still containing myriads of flying atoms, to pressures limited by the strength of steel, but still very minute compared with the pressure at the center of the earth and the sun, where the hardest steel would flow like the most limpid water. In velocities we are limited to a few miles per second."

G. Grimvall (✉)
Theoretical Physics, Royal Institute of Technology, AlbaNova University Center,
106 91 Stockholm, Sweden
e-mail: grimvall@kth.se

The two preceding paragraphs are taken from the presidential address 1899 by Henry A Rowland, the first president of the American Physical Society (reprinted in APS News, January 2008, p. 8). His words were well chosen, in particular in view of the surprises that were very soon to come. Two examples of that will now be given; one experimental and one theoretical, although in both cases with an interplay between experimental data and a mathematical description. They show unexpected phenomena, which prevent a simple extrapolation. A third example illustrates how a simple conceptual model may fail, in a counterintuitive way.

1.1 Superconductivity

At low temperatures, the electrical resistivity $\rho(T)$ of a pure metallic specimen varies rapidly with the temperature T, but in a very complicated way, which is still difficult to accurately account for. Lacking further understanding, one may try to describe the temperature dependence by a simple power law,

$$\rho(T) = aT^n$$

where a and n are fitted to experimental data. But no material is completely free of defects. They give rise to a temperature independent impurity resistivity ρ_{imp}. The total resistivity then takes the form

$$\rho(T) = \rho_{imp} + aT^n$$

This is in reasonable agreement with experiments, at least in a restricted temperature interval, as exemplified in Fig. 1 for Ag, a metal that does not become superconducting. Thus one expects that if resistivity data are extrapolated down to $T = 0\,\text{K}$, one would measure ρ_{imp}. It was a complete surprise when Kamerlingh Onnes [1] in 1911 found that the resistance of Hg did not follow this expected behavior, but suddenly dropped to such low values that no resistance could be measured. He called the new phenomenon superconductivity.

Fig. 1 The resistivity $\rho(T)$ of silver, as a function of temperature T, for a specimen with residual resistivity $\rho_{imp} \approx 2,98613$ pΩm. Figure based on data from Barnard and Caplin [2]

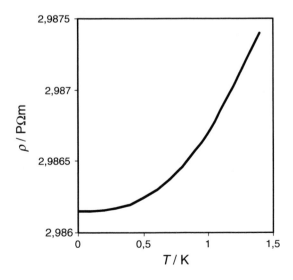

Table 1 The experimental heat capacity per atom at 25 °C, normalised to Boltzmann's constant k_B	Substance	$(C_P/\text{atom})/k_B$
	Cu	2.94
	NaCl	3.04
	PbS	2.97
	CaBr$_2$	3.01
Data from Barin [4]	C (diamond)	0.74

1.2 Low-temperature heat capacity

Almost two centuries ago, Petit and Dulong [3] noted that the heat capacity C_P of solids showed a striking regularity. In modern language, the Dulong-Petit rule gives

$$C_P = 3k_B/\text{atom}$$

where k_B is Boltzmann's constant. In the early 1900's it was well established that this rule is approximately obeyed, with the striking exception of diamond (Table 1). Furthermore, the value $3k_B/\text{atom}$ is exactly what the equipartition theorem in classical statistical physics gives for an ensemble of atoms vibrating as harmonic oscillators. Therefore it was puzzling why not all substances had a heat capacity C_P that was at least as high as about $3k_B/\text{atom}$. (There could be other contributions to the heat capacity, but they would always be positive and thus increase C_P.)

In a famous paper from 1906, Einstein [5] explained the exceptional behavior of diamond. In his model of lattice vibrations, the heat capacity has the form

$$C(T) = 3k_B \left(\frac{\theta_E}{T}\right)^2 \frac{\exp(\theta_E/T)}{\left[\exp(\theta_E/T) - 1\right]^2}$$

Here θ_E is a characteristic temperature (the Einstein temperature). For a harmonic oscillator, in which a force with force constant k acts on a particle of mass m, one has

$$\theta_E = \frac{\hbar}{k_B}\sqrt{\frac{k}{m}}$$

where $\hbar$ is Planck's constant. This extremely simple model explains the deviation from the Dulong-Petit rule, and its failure at low temperatures, Fig. 2.

It is now easy to understand why diamond differs from the other substances in Table 1. The force between the atoms, represented by the force constant k, is among the largest found for solids. Furthermore, the atomic mass m of diamond (carbon) is among the lightest in the Periodic Table. As a consequence the ratio k/m, and hence the characteristic temperature θ_E, is much larger than for any other solid. The heat capacity in the Einstein model is a universal function of the ratio T/θ_E. At room temperature, $T/\theta_E \approx 1/5$ for diamond, while the corresponding ratio for Pb is about 4. One could therefore loosely say that room temperature for diamond is equivalent to 15 K for Pb. It should be added that if we had included in Table 1 other substances with small atomic masses and strong interatomic forces, for instance B, TiC and SiO$_2$, we would also get $C_P/\text{atom} < 3k_B$, but not nearly as small as for diamond.

The Einstein model represents the beginning of modern descriptions of materials, based on forces between discrete and vibrating atoms and invoking quantum physics. It is remarkable that the model was formulated about six years before the technique of X-ray scattering

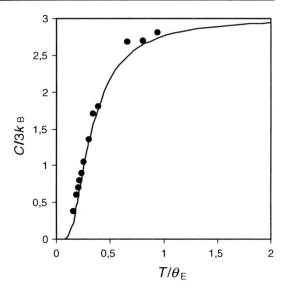

Fig. 2 The Einstein model of the heat capacity C of a solid, plotted as a function of the reduced temperature T/θ_E. Symbols refer to the data points for diamond used by Einstein in his original paper [5]. He fitted $\theta_E = 1320\,K$ to the data

from crystals had been developed, i.e. before one knew the crystal structure of diamond [6] or of any other solid [7].

1.3 Crystals modelled as stacking of atomic spheres

It is customary, and often very useful, to represent atoms in a solid with spheres stacked in a certain configuration. Among the elements in the Periodic Table, 17 have the face centered cubic structure (fcc) and 15 have the body centered cubic structure (bcc) as the equilibrium phase at ambient conditions. Some of the fcc elements are stable also in the bcc structure (or some other structure), and vice versa, at other temperatures and pressures. The most important example is iron, which transforms from bcc to fcc at 1184 K, and back to bcc at 1665 K, before it melts at 1809 K. At room temperature and under high pressure, Fe takes the hexagonal closed packed (hcp) structure. In the light of this fact, consider for instance tungsten. It has only been observed in the bcc structure, but one can imagine that if W atoms were stacked in a fcc lattice it would represent a possible but metastable configuration; a state with so high energy that it could not be observed before the lattice melts. This idea of a variety of hypothetical metastable structures has been extensively used in materials science, both in accounts of experimental thermodynamic data and in ab initio calculations, where the energies of different configurations are obtained from strictly theoretical quantum mechanical calculations.

At the end of the 1980's there was a growing concern that the experimental and the theoretical ways to find the energy (enthalpy) difference between a stable and an assumed metastable structure disagreed strongly in some cases, for instance for tungsten. It was not clear if the semi-empirical fitting to thermodynamic data, or the ab initio theory, was to be blamed. Skriver [8], in one of the most ambitious ab initio works at the time, wrote: "In a comparison between the calculated structural energy differences for the 4d transition metals and the enthalpy differences derived from studies of phase diagrams, we find that, although the crystal structures are correctly predicted by the theory, the theoretical energy differences are up to a factor of 5 larger than their 'experimental' counterparts. The reasons for this discrepancy may lie in the local-density approximation or in the neglect of the non-spherical part of the charge distribution. Furthermore, the derived enthalpy differences are certainly model dependent and may change as the model is improved."

The origin of the discrepancy later turned out to be caused not by large errors in the ab initio work but related to the fact that, for instance, fcc W does not represent a metastable configuration; see a review by Grimvall [9]. The semi-empirical and the ab initio approaches are essentially correct, but in some cases they give quantities that should not be compared. The fcc W lattice structure is dynamically (mechanically) unstable at $T = 0$ K under small shear deformations. For an unstable lattice one may *assume* a rigid lattice and calculate the total energy U as a solution to the Schrödinger equation, but the concept of a vibrational entropy S cannot be defined. Then also the Gibbs energy $G = U - TS$, which determines the thermodynamic properties, has no well-defined meaning. The reason why it took so long for this insight to be part of the general knowledge may in retrospect seem a bit puzzling. Born [10] had written in 1940 on the elastic instability of some simple structures, and the subject is thoroughly treated in a well-known book by Born and Huang [11] from 1954. On the other hand, the instability is counterintuitive. For instance, when spheres are stacked in the densest possible packing configuration (fcc), it is natural to assume that a small disturbance of that structure would give a state of higher energy and not one of a lower energy as is implied in a mechanical instability. However, there are still open questions regarding the effect of the lattice instabilities on the thermodynamic functions, in particular the possible stabilization of the lattice at high temperatures when the electron band structure is changed due to the thermal vibrational "disorder" [12].

2 Elastic shear instability and melting

How crystals melt has been a long-standing problem. Thermodynamics requires that melting takes place when the Gibbs energy $G(T)$ of the liquid phase becomes smaller than that of the solid. However, it has also been assumed that the shear modulus G_{sh} in the solid phase gradually goes to zero, as the melting temperature T_m is approached. Born [13] wrote in 1939: "In actual fact there can be no ambiguity in the definition of, or the criterion for, melting. The difference between a solid and a liquid is that the solid has elastic resistance against shearing stress while the liquid has not. Therefore a theory of melting should consist of an investigation of the stability of a lattice under shearing stress." This view was advocated by Sutherland already in 1891. Sutherland's plot of experimental data of the isotropically averaged shear modulus G_{sh} as a function of the reduced temperature T/T_m is shown in a paper by Brillouin [14]; cf. Fig. 3. The experimental points seemed to fit a parabola,

$$\frac{G_{sh}(T)}{G_{sh}(0)} = 1 - \left(\frac{T}{T_m}\right)^2$$

thus suggesting that G_{sh} extrapolates to zero at $T = T_m$. Brillouin wrote: "As the melting point is approached we may *guess by extrapolation* (italics added here) that the macroscopic rigidity will tend to zero while the microscopic rigidity remains finite." Brillouin further remarked that with a vanishing shear modulus in the solid phase, the latent heat of melting would be zero, which is not in agreement with experiment, but that the latent heat in any case is much smaller than the heat of vaporisation and therefore the model could give a reasonable approximate description. (See also a comment by Born [13] on the latent heat.)

With modern experimental techniques, the single crystal elastic constants C_{ij} can often be measured up to close to the melting point. In a crystal of cubic lattice structure there are

Fig. 3 The normalised elastic
shear modulus $G_{sh}(T)/G_{sh}(0)$
of polycrystalline materials as a
function of the reduced
temperature T/T_m, with some of
the data points used by
Sutherland and referring to Ag,
Pb and Sn. In the original graph
by Sutherland, published 1891 in
Phil. Mag. vol. 32, p. 42 and
reproduced 1938 in a paper by
Brillouin [14], there are also
some points for Fe, Al, Cu, Au
and Mg, with $T/T_m < 0.5$. The
dashed curve is Sutherland's
parabolic fit

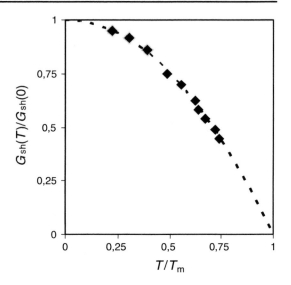

only two independent elastic shear constants, $C' = (C_{11} - C_{12})/2$ and C_{44}. For lead they
have been measured to within 1 K below the melting point $T_m = 601$ K. Both C' and C_{44}
show a very regular behavior as T_m is approached, and none of them extrapolates to zero at
T_m (Fig. 4).

3 Shear constants in alloys

The previous section discussed how the elastic shear constants extrapolate as a function of
temperature. We now consider how they vary as a function of the composition in an alloy
$A_{1-c} B_c$ with concentration $1 - c$ of the element A and c of the element B. In most real such

Fig. 4 The experimental elastic
shear moduli,
$C' = (C_{11} - C_{12})/2$ (lower
curve) and C_{44} (upper curve) for
Pb, with data from Vold et al.
[15], do not even approximately
extrapolate to zero at the melting
point $T_m = 601$ K

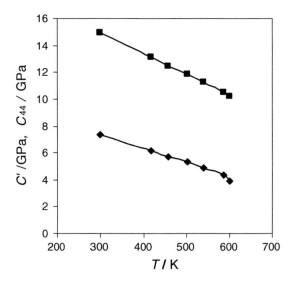

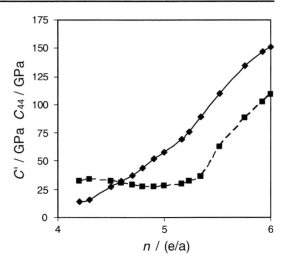

Fig. 5 The measured elastic shear constants C' (solid curve, diamonds) and C_{44} (dashed curve, squares) of binary bcc Zr-Nb-Mo alloys at room temperature, plotted versus the number n of d-electrons per atom ($n_{Zr} = 4$, $n_{Nb} = 5$, $n_{Mo} = 6$). Data from the compilation by Every and McCurdy [16]

systems one does not have a continuous solid solution from $c = 0$ to $c = 1$. This is obvious if A and B have different crystal structures, but normally there will also be regions with intermediate phases, for instance with approximate simple compositions such as AB_2, or much more complex phases. The zirconium-niobium-molybdenum system is a rare exception. There is complete solid solubility in bcc Nb-Mo at all concentrations and room temperature. In Zr-Nb it is only at high temperatures that there is complete solubility in the bcc phase at all concentrations, but the bcc phase can be retained by quenching to low temperatures when the Nb content is larger than about 20 at.%. The experimentally determined shear constants C' and C_{44} at room temperature vary in a regular way, Fig. 5.

Inspection of Fig. 5, and taking into account that the experimental uncertainty in C' and C_{44} is at least as large as the height of the symbols, suggests that Zr in a hypothetical bcc structure is close to being either stable or unstable under shear, $C' \approx 0$. However, there is a general decrease in the interatomic forces, as exemplified by the bulk modulus B, towards either end of the transition metal series in the Periodic Table. Therefore one might get further insight if C' and C_{44} are normalised as C'/B and C_{44}/B. From Fig. 6 we then see that C_{44} tends to stiffen significantly as one approaches bcc Zr ($n = 4$), while the near instability related to C' prevails. Above 1138 K, the thermodynamically stable phase of Zr has the bcc structure, and below that temperature it has the hcp structure. The bcc structure cannot be retained in Zr at room temperature by quenching from high temperatures. This has been taken as evidence that the bcc phase is dynamically unstable at low temperatures. Several theoretical phonon calculations corroborate this idea. A recent ab initio calculation (Kristin Persson, private communication 2008) gives $C' = -3\,\text{GPa}$ and $C_{44} = 28\,\text{GPa}$ at $T = 0\,\text{K}$, in good agreement with Fig. 5. The stabilisation of the bcc phase at *high* temperatures has been ascribed to the effect of the thermal motion of atoms on the electronic structure, i.e. on the effective forces between the atoms [12].

If a discussion of the dynamical stability of bcc Zr was based exclusively on an extrapolation to pure Zr of the elastic constants in the experimentally accessible range in bcc Zr-Nb alloys, one could not draw any definite conclusion. For instance, the constant C' in Fig. 5 could have increased again as $n = 4$ was approached, implying that bcc Zr is a metastable phase. With Ag-Zn as an example we shall now see how dangerous simple extrapolations can be.

Fig. 6 Data for C' and C_{44} as in Fig. 5 but now normalised with respect to the experimental bulk modulus, C'/B and C_{44}/B, with $B = (C_{11} + 2C_{12})/3$

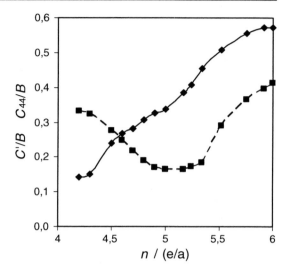

The thermodynamically stable phase of Ag has the fcc structure, and pure Zn the hcp structure. In a binary Ag-Zn system at room temperature, the fcc structure is present only from 0 to about 30 at.% Zn, and the bcc structure occurs only in a small concentration interval around 50 at.% Zn and at high temperatures. Experiments therefore give very little information about the possibility for fcc and bcc Ag-Zn solid solutions to be metastable, rather than dynamically (mechanically) unstable, states. But in ab initio calculations one can easily prescribe that the lattice structure is fcc or bcc, for any assumed solid solution concentration, and calculate C' and C_{44} from the energy change when the crystal is subject to a static shear. The result, from Magyari-Köpe et al. [17], is shown in Figs. 7 and 8. These graphs illustrate how dangerous an extrapolation can be, even if it is based on knowledge over a wide range of concentrations. For instance, on going from 0 to about 70 at.% of Zn in the bcc lattice C_{44} remains almost constant and C' increases slowly but steadily, after which they decrease rapidly, eventually making the lattice unstable ($C' < 0$), Fig. 7. A similar behavior is seen for C_{44} of the fcc structure (Fig. 8). In the fcc structure C' decreases between 0 and 50 at.% of Zn in Ag, and the curve seems to extrapolate to zero (i.e. a lattice instability) at about 60 at.% Zn (Fig. 8). But the theoretical calculation shows a very different behavior, with a reversal of the trend and C' becoming large as the Zn content is further increased.

4 Melting of superheated solids

Supercooling is a well-known phenomenon, where a liquid can be cooled to temperature below the freezing temperature, before a phase transformation takes place through the nucleation and growth of the solid phase. Superheating is much more difficult to achieve. The geometrical constraints on the formation of a nucleus of the new phase are much less severe when going from the ordered solid structure to the atomic disorder in the liquid, than in the reverse direction. It is an interesting scientific question to ask when and how a lattice eventually will transform to the liquid through so called homogeneous melting, i.e. in the absence of free surfaces or phase boundaries where nucleation of the liquid phase normally takes place. Many researchers have addressed the problem, relying on simple conceptual

Fig. 7 The elastic shear constants C' (solid curve, diamonds) and C_{44} (dashed curve, squares) as calculated by Magyari-Köpe et al. [17] with ab initio methods for hypothetical random Ag-Zn solid solutions in a bcc lattice structure

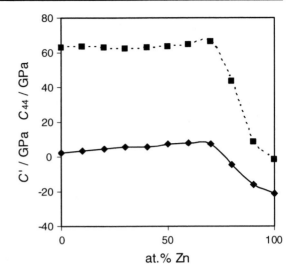

Fig. 8 As in Fig. 7 but for the fcc lattice structure

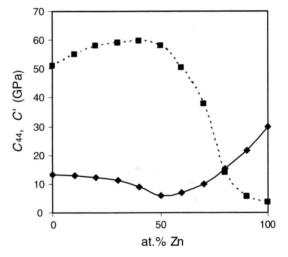

models, correlations of experimental data and numerical simulations; see Mei and Lu [18] for a review. Modern work is almost exclusively based on computer simulations, usually in the form of molecular dynamics calculations with an assumed interaction between the atoms. The possible mechanisms can be divided into two main types—a sudden instability such as the vanishing of a shear constant, and a thermal fluctuation that leads to a nucleation-and-growth process, respectively.

One possible mechanism has already been treated in Sect. 2; that the lattice becomes dynamically unstable under shear. Figure 9 shows such an instability arising in a molecular dynamics simulation of superheated Au when the lattice undergoes thermal expansion at $P = 0$, while the formation of lattice defects is suppressed [19]. Herzfeld and Goeppert Mayer [20] suggested the possibility of a vanishing bulk modulus (spinodal decomposition), rather than vanishing shear resistance. Naively, one might expect a simple mechanism in which the lattice will "shake apart" when the thermal vibrational displacement of the

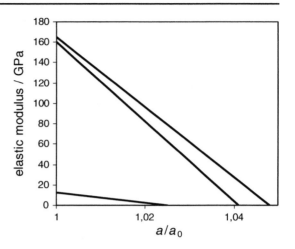

Fig. 9 The elastic shear constant C' (lowest curve) becomes zero as the lattice parameter a increases due to thermal expansion, while C_{44} (middle curve) and the bulk modulus B (upper curve) still are finite,. The straight lines are extrapolations of data within the stability range. Figure based on results from Wang et al. [19] obtained in a molecular dynamics simulation for fcc Au at $P = 0$

atoms becomes large enough. This is often referred to as the Lindemann [21] melting criterion from 1910, although it was not given its usual formulation until 1956 by Gilvarry [22]. Lennard-Jones and Devonshire [23] used an order-disorder description of the Bragg–Williams type [24] to model melting. In the 1960's it was suggested by Mizushima [25], Kuhlmann-Wilsdorf [26] and several others that at a certain critical temperature there is a spontaneous formation of dislocations. More recently Gómez et al. [27] also argued for a dislocation-mediated melting mechanism. The vibrational entropy of atoms at the dislocation core was assumed in Refs 25 and 26 to play an important role, but realistic simulations [28] show that the relative effect on the free energy is small. Gorecki [29] attributed correlations between various melting parameters to the formation of vacancies in metals. Jin and Lu [30], in a molecular dynamics study of Al, advocated a classical nucleation-and-growth mechanism. Jin et al. [31] suggested a mechanism that combines the Lindemann argument and a shear instability. In a molecular dynamics simulation for fcc Au, Wang et al. [19] found that in defect-free lattice, the volume increase due to thermal expansion would eventually cause a shear instability, see Fig. 9. Sorkin et al. [32], in a similar study for bcc V, found that not only thermal expansion but also expansion due to defect would lead to a shear instability. Forsblom and Grimvall [33,34], in a molecular dynamics study relevant for Al, found that melting takes place though a multi-stage process in which vacancy-interstitial pairs are thermally created. The relatively mobile interstitials then form an aggregate of a few interstitials, which serves as a nucleation site for further point-defect creation and the eventual melting. No shear instability was found in this simulation. A mechanism different from those just discussed is the entropy catastrophe suggested by Fecht and Johnson [35]. They noted that because the heat capacity of the superheated solid is expected to be larger than that of the liquid (cf. Forsblom et al. [36]) the entropy of the ordered solid will be larger than that of the disordered liquid above a critical temperature. This paradoxical behavior was identified with the onset of instability in a superheated solid phase.

Thus one can envisage many different melting mechanisms, and it is conceivable that several of them can occur if other mechanisms are suppressed. For instance, in the molecular dynamics simulation of Wang et al. [19] at constant pressure $P = 0$, there were no point defects created, and melting took place through the loss of rigidity. The somewhat similar molecular dynamics simulation of Forsblom and Grimvall [33,34] allowed for the thermal generation of vacancies and interstitials. Melting was initiated through a coalescence of such

Table 2 Some proposed melting mechanisms

Mechanism	Reference and year of publication
Large thermal vibrational amplitude	Lindemann [21] 1910, Gilvarry [22] 1956
Vanishing bulk modulus on expansion	Herzfeld and Goeppert Mayer [20] 1934
Bragg–Williams type disorder	Lennard-Jones and Devonshire [23] 1939
Rigidity loss at $T = T_m$	Sutherland 1890 (quoted in [14]), Brillouin [14] 1938, Born [13] 1939
Rigidity loss due to expansion	Wang et al. [19] 1997, Sorkin et al. [32] 2003
Entropy catastrophe	Fecht and Johnson [35] 1988
Proliferation of dislocations	Mizushima [25] 1960, Kuhlmann-Wilsdorf [26] 1965, Cotterill et al. [38] 1973, Gómez et al. [27] 2003
Proliferation of vacancies	Gorecki [29] 1974
Fluctuation, nucleation and growth	Jin and Lu [30] 1998, Jin et al. [31] 2001
Migrating interstitials coalesce	Forsblom and Grimvall [33,34] 2005

defects, thus precluding a study of the loss of rigidity at an even higher temperature. Table 2 summarises some of the mechanisms that have been suggested, with an incomplete list of references to such works. In a real system, with a hierarchy of possible melting mechanisms, the question is which of them that is encountered first when a solid is superheated [37].

5 Theoretical strength of solids

The actual strength of metallic engineering materials can be several orders of magnitude lower than what one expects from knowledge of the direct forces between the atoms. This fact, which for some time seemed very puzzling, gradually got an explanation in the 1930's and 1940's. Real materials are never free from defects. In particular they contain dislocations; linear defects that can move more or less easily through the lattice and thus deform a specimen. But it is of considerable theoretical and practical importance to study how an initially *perfect* crystal yields under an external load, i.e. to find its theoretical, or ultimate, strength. In comparison with the study of homogeneous melting, there are both similarities and essential differences. The ultimate strength is a much more complex phenomenon. The load can be applied in many ways; pure tension or shear, or a combination thereof. Seemingly similar materials can show different failure mechanisms even when they are subject to the same load conditions. The strained lattice may develop phonon instabilities analogous to those discussed above for the shear constants C' and C_{44} but now for short-wavelength phonons, as demonstrated in work by Clatterbuck et al. [39]. Furthermore, temperature is an important parameter in the determination of the ultimate strength. All these complications make a search for the onset of failure a formidable task. A brief review has been given by Šob et al. [40].

Like in the melting problem, there is a hierarchy of possible failure mechanisms, and again the question is which one of them that comes first and thus determines the ultimate strength. This is illustrated in Fig. 10, which gives the energy $U(\varepsilon)$ as a function of the strain ε for Nb. With no other instabilities present, the energy would increase with ε along the so called orthorhombic deformation path (upper curve) until one reaches the inflexion point in $U(\varepsilon)$. There the tension $\sigma = dU/d\varepsilon$ has its highest value, thus defining a theoretical tensile strength. But before that, the energy can be lowered through a "branching-off" along a tetragonal deformation path (lower curve), which means that the traditional approach of indentifying the theoretical strength with the inflexion point in $U(\varepsilon)$ is not correct in this case.

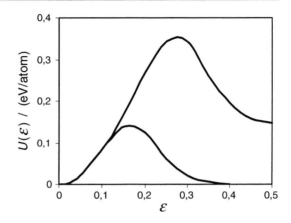

Fig. 10 The energy $U(\varepsilon)$ per atom as function of the relaxed $\langle 100 \rangle$ tensile strain ε, calculated for Nb along a tetragonal deformation path (upper curve), and along an orthorhombic path, which is "branching off" (lower curve). Figure based on data from Luo et al. [41]

6 Another model—the linear chain

This paper started with a discussion of one of the most celebrated models in condensed matter physics, the Einstein model of lattice vibrations. We end with another model in the same field, which is found in almost all textbooks; the linear chain introduced in 1912 by Born and von Kármán [42] along with a similar approach to three dimensions. Consider an infinite row of mass points with masses m, separated a distance a and connected by a spring with force constant k. Propagating waves (phonons) in the chain have wavenumber $q (= 2\pi/\lambda$ where λ is the wavelength) and frequency ω, with

$$\omega(q) = 2\sqrt{\frac{k}{m}} \sin\left(\frac{qa}{2}\right)$$

The model is used in textbooks to introduce and illustrate several very important concepts. Thus q has a maximum value π/a at the boundary of the first Brillouin zone. From $\omega(q)$ one can define a normalized density of states $D(\omega)$, which is a constant for small ω in a linear chain. The thermal energy and the heat capacity can also be calculated. Furthermore, the model is easily extended to a chain of alternating different masses, with a qualitative application to infrared properties of ionic solids. However, if we consider the root-mean-square $\langle u^2 \rangle^{1/2}$ of the atomic vibrational amplitude and use the standard expression

$$\langle u^2 \rangle = \int\limits_0^{\omega_{\max}} \frac{\hbar}{M\omega} \left[\frac{1}{2} + \frac{1}{\exp(\hbar\omega/k_B T) - 1}\right] D(\omega) d\omega$$

we find that $\langle u^2 \rangle^{1/2}$ diverges at all T, which might suggest a dynamical instability. The origin of the divergence lies in the fact that the integral is only an approximation to a discrete sum of amplitudes from all the individual vibrational quantum states labelled by a discrete set of wavenumbers q. The smallest possible q is not zero but $\pi/(Na)$, where N is the number of atoms in the chain. The divergence in $\langle u^2 \rangle^{1/2}$ varies as $\log N$ at $T = 0$ K. Since the singularity occurs in the limit of long wavelengths, and $\langle u^2 \rangle^{1/2}$ refers to displacements relative to an assumed coordinate system that is fixed in space, the distance between two neighboring atoms is not much affected. In d dimensions, the same integral expression for $\langle u^2 \rangle^{1/2}$ holds, when multiplied by the factor d. With $d = 3$, $D(\omega) \sim \omega^2$ at small ω and the exact definition of the lower integration limit is not important. The divergence of $\langle u^2 \rangle^{1/2}$ with increasing chain length is normally of no concern, and not mentioned in textbooks, but there are cases requiring detailed attention to the problem [43].

7 Four types of extrapolative procedures

The topic of this paper is the role of extrapolations in modelling and simulations. With reference to the examples we have treated, the following categorization may be helpful.

- An entirely new phenomenon suddenly arises beyond a critical value of a parameter.

Such a case is rare, with superconductivity being a prime example. It would take almost half a century from its discovery in 1911 to the theoretical explanation on a microscopic basis through the BCS theory (Bardeen et al. [44]) in 1957. Before that one just had to accept that the extrapolation of the resistivity to low T, which worked so well for many metals, was not valid below a critical temperature T_c for certain other metals. Starting from the BCS model, one may today calculate the critical temperature and other quantities of interest in the superconducting state. The obstacle is mainly one of having enough computer power. It is then necessary to remark that there is not yet any corresponding theory for the so-called high-temperature superconductors.

- A new theoretical insight explains a gradual change in trend of a physical quantity.

With changing physical parameters one may enter a regime where new effects must be taken into consideration, thus making a simple extrapolation gradually less relevant until it finally may fail completely. Arguably the best example of this in condensed matter physics is the Einstein model of lattice vibrations. On the basis of a single parameter (the Einstein temperature, θ_E) it explains how far down in temperature one can go and still ignore quantum effects. This view holds not only for the heat capacity, but for many other properties of solids involving the thermal vibrations of atoms, for instance the temperature dependence of the thermal expansion, the thermal conductivity of insulators, and the normal electrical conductivity of metals. Furthermore, the Einstein model and its various refinements may give confidence in the extrapolation of certain properties outside the range where there is good experimental knowledge, if one knows from the model that quantum effects will not be important.

- An established theoretical basis is used to obtain unknown quantities.

The elastic constants C_{ij} of metals with cubic crystal structures provide a good example to illustrate this point. The compilation by Every and McCurdy [16] in the Landolt-Börnstein tables gives an almost complete account of what was experimentally known about C_{ij} in the 1980's. There are data for 31 metallic elements in their pure state, but for only as few as 52 binary metallic alloys and then often for only a few compositions. In a need of data of alloys, it would then be tempting to interpolate and extrapolate, relying on the available meagre information. Until recently, this was the only practical alternative. But we also know that properties like the elastic constants are determined by the electronic structure of the solid, i.e. the solution to the Schrödinger equation. The steady progress in the computing power has in many cases made ab initio calculations a fast and cheap alternative to experiments. Such calculations may also reveal variations in the calculated quantity, which could not be anticipated on the basis of experimental data but are possible to understand with reference to the calculated electronic structure.

- An established theoretical basis is used in simulations to identify physical mechanisms.

If we can account for forces between the atoms, we can study many phenomena in detail, which would be difficult or impossible to study experimentally. Two examples have been discussed here; homogeneous melting mechanisms in superheated solids and failure mechanisms when a specimen is loaded to its theoretical strength. One could go further and model very complex phenomena, for instance plastic deformation on a nanoscale [45]. Such studies can give a detailed insight into the role of competing mechanisms, which are conceptually simple but strongly dependent on details. The simulations require substantial computing power but no new theory per se.

8 Conclusions

The danger of extrapolations has been illustrated through several examples from materials science. An extrapolation may be questionable because there is an unanticipated change in trend outside the range of previous knowledge, and caused by the fact that one enters a new physical regime of effects. An extrapolation can also be completely wrong or irrelevant because of the presence of an instability. In traditional mathematical modelling, based on analytical expressions, instabilities may be overlooked if they do not appear as natural extensions of the model, and one has to be explicitly aware of their existence. A ubiquitous model can contain features that are not generally known, because they are of importance only in special applications, as was the case in the infinite linear chain. On the other hand, one may avoid many such difficulties in good numerical simulations. If the basic theory underlying the simulations is general enough (for instance solving the Schrödinger equation and minimising the total energy with respect to *all* parameters of interest) one would expect a reliable model result. However, as was alluded to in the section on the theoretical strength, this may require an enormous computational effort. It is then tempting to make simplifications by introducing constraints, and thereby exclude certain crucial outcomes in the simulation. A thorough insight, or "intuition", may be needed to avoid unwarranted simplifications. It has sometimes been argued that intuition is nothing but having seen many examples of results, which may have direct or indirect relevance for the problem at hand. The discussion of instabilities and of dubious extrapolations in this paper could then add to such an intuition. Two other aspects of modelling and simulation will be treated in separate papers to follow in this journal; characteristic quantities and numerical accuracy, respectively.

Acknowledgements   I want to thank Mattias Forsblom, Blanka Magyari-Köpe, Vivuds Ozoliņs and Kristin Persson (KP) for numerous discussions on the topics covered in this paper, and KP for calculating C' and C_{44} of bcc Zr. This work has received funding from the Swedish Foundation for Strategic Research (SSF) under the project ATOMICS.

References

1. Kamerlingh Onnes, H.: Further experiments with liquid helium. On the change of the electrical resistance of pure metal at very low temperature. Leiden Commun. **122b**, (1911)
2. Barnard, B., Caplin, A.D.: 'Simple' behaviour of the low temperature electrical resistivity of silver? Commun. Phys. **2**, 223–227 (1977)
3. Petit, A.T., Dulong, P.L.: Sur quelques points importans de la theorie de la chaleur. Ann. Chim. Phys. **10**, 395–413 (1819)
4. Barin, I.: Thermochemical Data of Pure Substances. VCH Verlag (1989)
5. Einstein, A.: Planck's theory of radiation and the theory of specific heat. Ann. Physik **22**, 180–190 (1906)

6. Bragg, W.H., Bragg, W.L.: Structure of diamond. Nature (UK) **91**, 557 (1913)
7. Bragg, W.L.: Structure of some crystals as indicated by their diffraction of X-rays. Proc. R. Soc. Lond. A, Math. Phys. Sci. **89**, 248–277 (1913)
8. Skriver, H.L.: Crystal structure from one-electron theory. Phys. Rev. B **31**, 1909–1923 (1985)
9. Grimvall, G.: Reconciling ab initio and semiempirical approaches to lattice stabilities. Ber. Bunsenges. Phys. Chem. **102**, 1083–1087 (1998)
10. Born, M.: On the stability of crystal lattices. I. Proc. Camb. Philos. Soc. **36**, 160–172 (1940)
11. Born, M., Huang, K.: Dynamical Theory of Crystal Lattices. Oxford University Press (1954)
12. Souvatzis, P., Eriksson, O., Katsnelson, M.I., Rudin, S.P.: Entropy driven stabilization of energetically unstable crystal structures explained from first principles theory. Phys. Rev. Lett. **100**, 09590/1–4 (2008)
13. Born, M.: Thermodynamics of crystals and melting. J. Chem. Phys. **7**, 591–603 (1939)
14. Brillouin, L.: On thermal dependence of elasticity in solids. Phys. Rev. **54**, 916–917 (1938)
15. Vold, C.L., Glicksman, M.E., Kammer, E.W., Cardinal, L.C.: The elastic constants for single-crystal lead and indium from room temperature to the melting point. J. Phys. Chem. Solids **38**, 157–160 (1977)
16. Every, A.G., McCurdy, A.K.: Second and higher order elastic constants. In: Nelson, D.F. (ed.) Landolt-Börnstein Numerical Data and Functional Relationships in Science and Technology, New Series III/29a. Springer-Verlag (1992)
17. Magyari-Köpe, B., Grimvall. G., Vitos, L.: Elastic anomalies in Ag-Zn alloys. Phys. Rev. B **66**, 064210/1–7 (2002)
18. Mei, Q.S., Lu, K.: Melting and superheating of crystalline solids: from bulk to nanocrystals. Prog. Mater. Sci. **52**, 1175–1262 (2007)
19. Wang, J., Li, J., Yip, S., Wolf, D., Phillpot, S.: Unifying two criteria of Born: elastic instability and melting of homogeneous crystals. Physica A **240**, 396–403 (1997)
20. Herzfeld, K.F., Goeppert Mayer, M.: On the theory of fusion. Phys. Rev. **46**, 995–1001 (1934)
21. Lindemann, F.A.: Molecular frequencies. Phys. Zeits. **11**, 609–612 (1910)
22. Gilvarry, J.J.: The Lindemann and Grüneisen laws. Phys. Rev. **102**, 308–316 (1956)
23. Lennard-Jones, J.E., Devonshire, A.F.: Critical and co-operative phenomena IV. A theory of disorder in solids and liquids and the process of melting. Proc. R. Soc. Lond. A **170**, 464–484 (1939)
24. Bragg, W.L., Williams, E.J.: Effect of thermal agitation on atomic arrangement in alloys. Proc. R. Soc. Lond. A **145**, 699–730 (1934)
25. Mizushima, S.: Dislocation model of liquid structure. J. Phys. Soc. Jpn. **15**, 70–77 (1960)
26. Kuhlmann-Wilsdorf, D.: Theory of melting. Phys. Rev. **140**, A1599–A1610 (1965)
27. Gómez, L., Dobry, A., Geuting, Ch., Diep, H.T., Burakovsky, L.: Dislocation lines as the precursor of the melting of crystalline solids observed in Monte Carlo simulations. Phys. Rev. Lett. **90**, 095701/1–4 (2003)
28. Forsblom, M., Sandberg, N., Grimvall, G.: Vibrational entropy of dislocations in Al. Philos. Mag. **84**, 521–532 (2004)
29. Gorecki, T.: Vacancies and changes of physical properties of metals at the melting point. Z. Metallkd. **65**, 426–431 (1974)
30. Jin, Z.H., Lu, K.: Melting of surface-free bulk single crystals. Philos. Mag. Lett. **78**, 29–35 (1998)
31. Jin, Z.H., Gumbsch, P., Lu, K., Ma, E.: Melting mechanisms at the limit of superheating. Phys. Rev. Lett. **87**, 055703/1–4 (2001)
32. Sorkin, V., Polturak, E., Adler, J.: Molecular dynamics study of melting of the bcc metal vanadium. 1. Mechanical melting. Phys. Rev. B **68**, 174102/1–7 (2003)
33. Forsblom, M., Grimvall, G.: How superheated crystals melt. Nat. Mater. **4**, 388–390 (2005)
34. Forsblom, M., Grimvall, G.: Homogeneous melting of superheated crystals: molecular dynamics simulations. Phys. Rev. B **72**, 054107/1–10 (2005)
35. Fecht, H.J., Johnson, W.L.: Entropy and enthalpy catastrophe as a stability limit for crystalline material. Nature **334**, 50–51 (1988)
36. Forsblom, M., Sandberg, N., Grimvall, G.: Anharmonic effects in the heat capacity of Al. Phys. Rev. B **69**, 165106/1–6 (2004)
37. Tallon, J.L.: A hierarchy of catastrophes as a succession of stability limits for the crystalline state. Nature **342**, 658–660 (1989)
38. Cotterill, R.M.J., Jensen, E.J., Kristensen, W.D.: A molecular dynamics study of the melting of a three-dimensional crystal. Phys. Lett. **44A**, 127–128 (1973)
39. Clatterbuck, D.M., Krenn, C.R., Cohen, M.L., Morris, J.W., Jr.: Phonon instabilities and the ideal strength of aluminum. Phys. Rev. Lett. **91**, 135501/1–4 (2003)
40. Šob, M., Pokluda, J., Černý, M., Šandera, P., Vitek, V.: Theoretical strength of metals and intermetallics from first principles. Mater. Sci. Forum **482**, 33–38 (2005)
41. Luo, W., Roundy, D., Cohen, M.L., Morris, J.W., Jr.: Ideal strength of bcc molybdenum and niobium. Phys. Rev. B **66**, 094110/1–7 (2002)

42. Born, M., von Kármán, Th.: Über Schwingungen in Raumgittern. Physik. Zeits. **13**, 297–309 (1912)
43. Florencio, J. Jr., Lee, M.H.: Exact time evolution of a classical harmonic-oscillator chain. Phys. Rev. A **31**, 3231–3236 (1985)
44. Bardeen, J., Cooper, L.N., Schrieffer, J.R.: Theory of superconductivity. Phys. Rev. **108**, 1175–1204 (1957)
45. Van Vliet, K.J., Li, J., Zhu, T., Yip, S., Suresh, S.: Quantifying the early stages of plasticity through nanoscale experiments and simulations. Phys. Rev. B **67**, 104105/1–15 (2003)

Characteristic quantities and dimensional analysis

Göran Grimvall

Originally published in the journal Sci Model Simul, Volume 15, Nos 1–3, 21–39.
DOI: 10.1007/s10820-008-9102-4 © Springer Science+Business Media B.V. 2008

Abstract   Phenomena in the physical sciences are described with quantities that have a numerical value and a dimension, i.e., a physical unit. Dimensional analysis is a powerful aspect of modeling and simulation. Characteristic quantities formed by a combination of model parameters can give new insights without detailed analytic or numerical calculations. Dimensional requirements lead to Buckingham's Π theorem—a general mathematical structure of all models in physics. These aspects are illustrated with many examples of modeling, e.g., an elastic beam on supports, wave propagation on a liquid surface, the Lennard-Jones potential for the interaction between atoms, the Lindemann melting rule, and saturation phenomena in electrical and thermal conduction.

Keywords   Dimensional analysis · Characteristic quantity · Scaling ·
Buckingham's theorem · Lennard-Jones interaction

1 Introduction

Phenomena in the physical sciences are described with quantities that have a numerical value and a dimension, i.e., a physical unit. There are only seven independent such units: the SI base units for length (metre, m), mass (kilogram, kg), time (second, s), electric current (ampere, A), thermodynamic temperature (kelvin, K), amount of substance (mole, mol) and luminous intensity (candela, cd). Numerous other named units are used in science and technology, both within SI (e.g., joule, J) and outside SI (e.g., horsepower, hp). However, they can all be expressed in the seven base units, sometimes in a rather complicated form. For instance, 1 Btu (British thermal unit) $= 788.169\,\text{ft·lbf}$ (foot·poundforce) $= 1055.056\,\text{J}$ $= 1055.056\,\text{kg m}^2\,\text{s}^{-2}$. Molar entropy requires five SI base units: $\text{m}^2\text{kg s}^{-2}\,\text{K}^{-1}\,\text{mol}^{-1}$.

G. Grimvall (✉)
Theoretical Physics, Royal Institute of Technology, AlbaNova University Center,
106 91 Stockholm, Sweden
e-mail: grimvall@kth.se

The fact that seven units suffice to give a quantitative account of any physical phenomenon has significant consequences for the mathematical form that a physical modeling can take. Some of the consequences are trivial, for instance that the model parameters must enter so that all the members in a sum are expressed in the same SI base units, or that the argument of trigonometric, exponential and logarithmic functions is a number, i.e. dimensionless. Checking for violations of such rules is one of the most powerful methods to detect misprints and other errors in the derivation of physical relations. Dimensional arguments are then used posteriorly, as a routine. But they can also be used in a creative way, for instance to identify characteristic quantities, which should be included in the modeling of a complex phenomenon, or to give a deeper understanding of a phenomenon through an insightful interpretation of a characteristic quantity. This review deals with such non-trivial aspects. Through several examples we first illustrate the concept of characteristic quantities. That is used in an account of Buckingham's Π theorem, which gives a general mathematical framework for all physical models. As an application we then discuss the almost century-old Lennard-Jones potential, which is still frequently used in modeling. The popular Lindemann rule for the melting temperature of solids is shown to be a simple consequence of Buckingham's Π theorem, rather than a melting criterion. Finally, the "saturation" of electrical and thermal conductivities provides an example where dimensional arguments give significant insight, even though a fundamental and closed-form algebraic formulation is lacking.

Many of our examples start from well known relations found in textbooks from various fields of physics and engineering. Since the purpose is to illustrate general features, rather than solving a specific problem, they are often presented without further motivation.

2 Four examples of characteristic quantities

2.1 Waves at sea

A nautical chart gives information about depths. Consider a certain location where the depth is indicated to be $d = 20$ m. The acceleration of gravity is $g = 9.8$ m/s^2. These two quantities can be combined as

$$\sqrt{gd} = 14 \text{ m/s}$$

Is there a physical interpretation of this velocity? Something that propagates with about 14 m/s? The answer is the phase velocity of gravitational waves on a water surface, i.e., the familiar waves whose properties are governed by gravity forces (as opposed to capillary waves to be considered later). But the velocity in our example is not what we would normally observe, if we look at waves on a sea, because it refers to the special case of shallow water. What is then meant by shallow? In a physical statement like fast or slow, large or small, heavy or light, we must compare two quantities of the same physical dimension (unit). In our case the depth $d = 20$ m must be in some sense small compared with another relevant length. The wavelength λ of the wave seems to be an obvious choice of such a second length. Indeed, as we shall see in Sect. 3, the condition of shallow water can be formulated as $d \ll \lambda$. (The wavelength of the tidal waves is so long that even the oceans can be regarded as shallow.)

This interpretation illustrates important principles, which govern models of physical phenomena. Three more examples in Sect. 2, and a mathematical account of waves on liquid surfaces in Sect. 3, will provide further background, before we summarize these principles in Sect. 4.

2.2 Temperature profile in the ground

The thermal diffusivity of granite is $a = 1.1 \times 10^{-6}\,\mathrm{m^2/s}$. If this quantity is multiplied by a certain time t, and we form $\sqrt{at}$, the result is a distance L. What could be a relevant time t, and what is then the interpretation of the distance L? Consider a granite rock on a clear day. The surface layer of the rock is heated up during the day and cooled during the night, giving an approximately sinusoidal variation of the temperature with time. But far below the surface, the temperature variation will be negligible. If the time t is chosen to be the length of a day, $t_d = 24 \times 3600\,\mathrm{s}$, we get

$$L_d = \sqrt{at_d} \approx 0.31\,\mathrm{m}$$

We can interpret this as a *characteristic distance* below which the diurnal variations in the temperature are in some sense small. With t being the length of a year, $t_y = 3.15 \times 10^7\,\mathrm{s}$, we get the analogous result for the characteristic penetration depth of the annual temperature variations:

$$L_y = \sqrt{at_y} \approx 5.9\,\mathrm{m}$$

This is in good agreement with the familiar fact that the temperature in an underground cave is approximately constant throughout the year.

Given the thermal diffusivity a, we could also combine it with a certain length L_i and get a time t_i:

$$t_i = \frac{L_i^2}{a}$$

The temperature at a depth L_i cannot respond immediately to a change in the temperature at the surface. The quantity t_i can be viewed as a *characteristic time* lag before the temperature change occurs at the depth L_i.

Our interpretations of the distances L_d and L_y, and the time t_i, can be put on firm mathematical footing, if we solve the corresponding model problem of heat conduction into a semi-infinite medium. The result is given in Appendix 1.1.

2.3 Terminal velocity

An object is dropped in air from a large height. As it accelerates, the air resistance gradually becomes increasingly important, and the velocity asymptotically approaches the so called terminal velocity (speed) v_t. It is obvious that if the object is heavy, it will take a longer time, and require a longer distance of fall, before the terminal velocity is reached. What does heavy mean in this context? The actual mass m of the object should be compared with a quantity that also has the physical dimension of mass and is expressed in model parameters describing the fall.

In a standard model of the force F due to air resistance one writes [9]

$$F = \frac{1}{2}C_d A \rho v^2$$

A is the object's cross section area perpendicular to the direction of motion, ρ the density of air, v the velocity of the object and C_d the dimensionless so called drag coefficient. C_d depends on several additional parameters, but it will be considered as constant in our discussion.

Let us now form the quantity

$$M = AL\rho$$

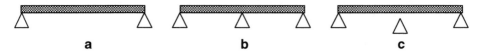

Fig. 1 A beam resting on supports

where L is the distance the object has fallen. M has the dimension of mass. An interpretation of M is that if $m \gg M$, the air resistance is still small when the object has fallen the distance L. In other words, the mass m of the object is to be compared with the total mass of the air inside that "tube" of cross section area A and length L, which is cut out by the trajectory of the falling object. The relevance of this visually appealing interpretation is confirmed, if one compares with the analytic result in Appendix 1.2 for the motion of an object under the influence of gravity and air resistance.

As an alternative approach, we can form a quantity L^* with the dimension of length from the quantities m, A and ρ:

$$L^* = \frac{m}{A\rho}$$

The interpretation of this length is that effects due to air resistance are small as long as the distance L of fall obeys $L \ll L^*$.

2.4 Engineering science *versus* school physics

Consider a uniform beam of mass m and length L, which rests either on two or on three supports, as in Fig. 1a and b. What are the forces on the supports?

In ordinary school physics, the case of two supports immediately gives the forces $mg/2$ on each side, but with three supports the problem "cannot be solved". Nevertheless, textbooks in elasticity [26] give the result $3mg/16$ for the two outer supports and $10mg/16$ for the middle support. But that is for a special (although very common) case, namely that the supports are stiff, and the beam is flexible.

As another extreme case, consider an absolutely rigid beam and represent the supports by springs, which can easily be compressed. The beam will sink down a bit but remain in a horizontal position, when placed on the supports. Thus all three springs are compressed an equal amount, and therefore exert the same force $mg/3$ on the beam.

There are several other special cases, leading to other results for the forces. Suppose that the beam is rigid, and the supports are not in the same horizontal plane. For instance, in Fig. 1c the middle support is below the beam, i.e., the forces are the same as in Fig. 1a. Since, in practice, it is not possible to have three supports at exactly the same height, the standard solid mechanics result quoted above must assume that the beam is deflected so much under its own weight, that the variation in height of the supports is negligible. For a beam in geometry 1a, the middle of the beam sinks down a distance d, with [26]

$$d = \frac{5}{384} \cdot \frac{mgL^3}{EI}$$

Here L is the length of the beam, I the second moment of area ("moment of inertia"; unit m^4) and E the elastic constant (Young's modulus, unit N/m^2 = kg/(m s^2)). A theoretical analysis [26] shows that when the (stiff) middle support lays the distance δ below the level of the outer supports, the force from the middle support is reduced to

$$F = \frac{5}{8}mg - \frac{48\delta EI}{L^3}$$

The force $F = 0$ when $\delta = d$, as expected.

3 Waves revisited

The propagation of waves in shallow water was briefly considered in Sect. 2.1. We shall now discuss general expressions for wave propagation on the surface of liquids, in the conventional model [17]. A wave is described by its wavelength λ, its frequency f and its propagation speed. One always has

$$v = \lambda f$$

Therefore we can choose either λ or f as the independent model parameter. However, for v we must distinguish between the group velocity v_g (which is what we usually refer to when we watch waves at sea) and the phase velocity v_p. The driving force for an oscillatory vertical motion of the surface may be gravity, given by the acceleration g, and surface tension (surface energy) given by γ. Furthermore, the modeling should include the density ρ of the liquid and the (constant) depth d from the liquid surface to the bottom.

When gravitational and capillary effects, and the depth, are simultaneously included, the general expression for the phase velocity is [17]

$$c_p = \sqrt{\left(\frac{\lambda g}{2\pi} + \frac{2\pi\gamma}{\lambda\rho}\right)\tanh(2\pi d/\lambda)}$$

It can be rewritten in three equivalent forms, which will be used in Sect. 5 to illustrate Buckingham's Π theorem. The only difference between the following three representations is that they emphasize gravitational waves in deep water, capillary waves in deep water, and waves in shallow water, respectively:

$$\frac{c_p}{\sqrt{\lambda g}} = \sqrt{\left(\frac{1}{2\pi} + 2\pi \cdot \frac{\gamma}{\lambda^2 g\rho}\right)\tanh\left(2\pi \cdot \frac{d}{\lambda}\right)}$$

$$\frac{c_p\sqrt{\lambda\rho}}{\sqrt{\gamma}} = \sqrt{\left(\frac{1}{2\pi} \cdot \frac{\lambda^2 g\rho}{\gamma} + 2\pi\right)\tanh\left(2\pi \cdot \frac{d}{\lambda}\right)}$$

$$\frac{c_p}{\sqrt{gd}} = \sqrt{\left(\frac{1}{2\pi} \cdot \frac{\lambda}{d} + 2\pi \cdot \frac{\gamma}{\lambda g\rho d}\right)\tanh\left(2\pi \cdot \frac{d}{\lambda}\right)}$$

Table 1 gives these wave velocities in the three limits of gravity-induced waves in deep water, capillary waves in deep water, and waves in shallow water, respectively. It is interesting to note that the velocity of the gravitational wave only depends on g, λ and d and not on the type of liquid. One may compare this with the well known result that the frequency of a mathematical pendulum does not depend on the mass or material of the pendulum bob.

4 Characteristic quantities

In Sect. 2 we encountered a number of characteristic quantities. They set a length scale, a time scale, a mass scale and so on, leading to a deeper insight into physical phenomena, when they are compared with other quantities of the same physical dimension. The comparison may be straight forward, such as comparing the depth with the wavelength to decide if a wave is propagating in shallow or deep water. A more interesting, and sometimes quite subtle, case is when a characteristic quantity is formed through a combination of quantities of different dimensions. In the example of heat penetration into the ground, we saw how the thermal diffusivity could be combined with the duration of a day or a year to set a length scale characteristic of how far diurnal or annual variations in the surface temperature penetrate into the ground. In the example of the falling sphere, we defined a mass scale M and a length scale L^*, which were relevant to decide whether air resistance effects were important. In the example of a beam resting on supports, we found a length scale, which gave us insight into how level the three supports must be, if they are regarded as lying in the same horizontal plane.

In all these cases the argumentation was only qualitative. The water was considered as shallow if the depth $d \ll \lambda$. The annual temperature variations in the ground were considered as small at depths $z \gg \sqrt{a t_y}$. The effect of air resistance on a falling sphere with mass m was considered as small if $m \gg A L \rho$. These arguments rest on a comparison of two quantities with the same dimension, but they do not specify to what extent the inequalities should be obeyed. Of course it depends on the accuracy we require in the description. But there is also another aspect. A wave can be described by its wavelength λ and, equally justified, by its wave number $k = 2\pi/\lambda$. A frequency be may expressed either as f or as its corresponding angular frequency $\omega = 2\pi f$. In a circular or cylindrical geometry, one may choose the radius, the diameter or the circumference as the characteristic length. (An example involving the circumference is given in Appendix 2.1.) The velocity of a wave can refer to its phase velocity v_p or its group velocity v_g, with $v_p = \omega/k$ and $v_g = d\omega/dk$. Depending on which choice we make, an associated characteristic quantity is changed by a numerical factor. The numerical pre-factors in the wave velocities in Table 1 lie between the smallest value $1/(2\sqrt{2\pi}) \approx 0.20$ and the largest value $3\sqrt{2\pi}/2 \approx 3.76$.

Our examples may suggest that the simplest algebraic expression involving the model quantities also gives a characteristic quantity, which is numerically relevant to better than an order of magnitude. This is indeed the case in almost all modeling, but there are exceptions. The familiar Poiseuille's law for laminar flow through a tube expresses the flow q (volume per time) as

$$q = \frac{\pi}{8} \cdot \frac{R^4 \Delta P}{\eta L}$$

L is the length and R the radius of the tube, ΔP the pressure difference between the tube ends and η the dynamic viscosity of the liquid. If we had chosen the tube diameter D instead of the radius R, the numerical pre-factor would change from $\pi/8 \approx 0.39$ to $\pi/128 \approx 0.025$. In the example with a beam resting on two supports, the maximum deflection d had a numerical pre-factor as small as $5/384$. Another noteworthy case is the onset of Rayleigh-Bénard

Table 1 Wave velocities	Wave type	Phase velocity	Group velocity
	Deep water, gravitational wave	$\frac{1}{\sqrt{2\pi}}\sqrt{g\lambda}$	$\frac{1}{2\sqrt{2\pi}}\sqrt{g\lambda}$
	Deep water, capillary wave	$\sqrt{2\pi}\sqrt{\frac{\gamma}{\lambda\rho}}$	$\frac{3\sqrt{2\pi}}{2}\sqrt{\frac{\gamma}{\lambda\rho}}$
	Shallow water	$\sqrt{gd}$	$\sqrt{gd}$

convection instability in a bounded horizontal liquid layer subject to a vertical temperature gradient. Then the numerical pre-factor is as large as 1708. However, both in this example and in that of Poiseuille's law, the simple algebraic form of the characteristic quantity contains a high power of one of the model parameters (h^4 and R^4, respectively). In the deflection d of the middle of a beam on two supports, the length L of the beam enters as L^3. If L had instead been introduced as the length from the end point to the middle of the beam, the pre-factor would be $5/48 \approx 0.10$ instead of $5/384 \approx 0.013$. In section 6 on scaling we will discuss the dimensionless Reynolds number Re which describes the behavior of motion in fluids. For air, a critical value of Re is about 2×10^5. These examples show that there are exceptions to the common feature of physical models that numerical factors are of the order of one, when the model is dimensionally consistent in SI units.

A very important aspect of the characteristic quantities is that they indicate the border line or transition from one physical behavior (regime) to another. Such a transition may be sudden, as in the Rayleigh instability in the heated liquid layer, but usually it is gradual, as in our example of waves propagating into increasingly shallow water. (Such a change of the wave velocity is the reason why waves come in parallel to a shoreline.)

The fact that a characteristic quantity marks the limit of applicability of a certain theory is an important insight, even when the understanding of the physics beyond that limit is poor. In Sect. 10 we will exemplify this aspect with conductivity saturation.

The Planck length $l_P = 1.62 \times 10^{-35}$ m, and the Planck time $t_P = 5.39 \times 10^{-44}$ s are two characteristic quantities formed from Planck's constant $\hbar(= h/2\pi)$, the constant of gravity G, and the velocity of light c as

$$l_P = \sqrt{\frac{\hbar G}{c^3}}, \qquad t_P = \sqrt{\frac{\hbar G}{c^5}}$$

In this case there is no understanding at all of the physics that describes phenomena at distances smaller than the Planck length or at times shorter than the Planck time.

5 Buckingham's Π theorem

In Sect. 3 we gave three equivalent mathematical expressions of the phase velocity. They can all be written in the mathematical form of a relation between dimensionless quantities Π_i;

$$\Pi_1 = \Psi(\Pi_2, \; \Pi_3)$$

where Ψ is an appropriate function. The quantities Π_i are given in Table 2 for the three considered cases.

Table 2 Dimensionless groups Π_i used in the mathematical expressions for the phase velocity in three special cases of wave propagation on a liquid surface

Wave type, phase velocity	Dimensionless groups		
	Π_1	Π_2	Π_3
Deep water, gravitational wave	$\dfrac{c_p}{\sqrt{\lambda g}}$	$\dfrac{\gamma}{\lambda^2 g \rho}$	$\dfrac{d}{\lambda}$
Deep water, capillary wave	$\dfrac{c_p \sqrt{\lambda \rho}}{\sqrt{\gamma}}$	$\dfrac{\lambda^2 g \rho}{\gamma}$	$\dfrac{d}{\lambda}$
Shallow water	$\dfrac{c_p}{\sqrt{g d}}$	$\dfrac{\gamma}{\lambda g \rho d}$	$\dfrac{d}{\lambda}$

In Appendix 1.1 we give an explicit expression for a temperature profile $T(z, t)$ in the ground. Appendix 1.2 gives the distance $L(t)$ an object has fallen in the case of air resistance. In both these cases the results can be rewritten as relations for dimensionless Π_i. Our examples suggest that such a mathematical representation is a generic feature of physical phenomena. In fact, Buckingham's Π theorem (π theorem, pi theorem) says that all physical relations can be given this form. It is named after the symbol chosen for the dimensionless groups in a paper by Buckingham [3], but other scientists have independently arrived at equivalent results.

In a slightly restricted version of the theorem, it can be formulated as follows. Let a physical phenomenon be described by n linearly independent quantities a_i $(i = 1, \ldots, n)$, whose physical dimensions can be expressed in m independent units. From a_i one can then form $n - m$ dimensionless quantities $\Pi_1, \ldots, \Pi_{n-m}$, which are related as

$$\Pi_1 = \Psi (\Pi_2, ..., \Pi_{n-m})$$

We check that this is consistent with the general equation for the phase velocity of waves on a liquid surface. There are $n = 6$ quantities in that equation ($c_p, g, \rho, \gamma, \lambda, d$), which can be expressed in $m = 3$ independent units (m, s, kg). This leaves us with $n - m = 3$ independent dimensionless quantities Π_i, cf. Table 2.

A simple example will illustrate two other, and important, points. Suppose that we seek the period T of a mathematical pendulum with length L and mass m, swinging under the force of gravity described by the acceleration g. There are $n = 4$ model parameters (T, L, m, g), which require $m = 3$ independent units (s, m, kg), suggesting that we can form $n - m = 1$ ratio Π. However, the mass unit appears in only one of the model parameters. Thus it can not be included in a dimensionless Π, unless there is another model parameter with a physical dimension that also includes mass. (If that had been the case, we would have 5 model parameters, expressed in 3 independent units, which would give $n - m = 2$ quantities Π_i. The fact that the pendulum period does not contain the mass is of course well known, but for the purpose of illustration we did not want to use that knowledge from the outset.) The true relation consistent with Buckingham's theorem involves 3 model parameters (T, L, g) expressed in 2 independent units (s, m), which gives us

$$\frac{T^2 g}{L} = \Psi = \text{constant}$$

The value of the dimensionless constant on the right-hand side can not be obtained from dimensional analysis, but we recognize that the mathematical form is the same as in the familiar result

$$T = 2\pi \sqrt{\frac{L}{g}}$$

The Π_i to be used in Buckingham's approach are not unique. For instance, we could have formed the inverted combination $\Pi = L/(gT^2)$ in the example above and still obtained the same physical relation. A product of two Π_i will also give a dimensionless quantity. However, the useful Π_i are often simple and natural combinations of some of the quantities that describe the phenomenon. Buckingham's theorem can be given a more rigorous mathematical formulation, but that falls outside the scope of this short review.

An in-depth treatment of dimensional analysis is given in many books, for instance by Taylor [23], Barenblatt [2] and Szirtes [22]. Here we shall mainly focus on its use in the systematic description of negligibly small effects, but first we make some comments on scaling.

6 Scaling

The most important consequence of dimensional analysis and Buckingham's theorem deals with scaling and similarity. In the example of a falling sphere with air resistance included, the distance $L(t)$ covered after a time t is (Appendix 1.2)

$$L = a \ln \left[\cosh \left(t \sqrt{\frac{g}{a}} \right) \right]$$

Here $a = 2m/(C_d A \rho)$ is a length parameter. The mass m of a homogeneous sphere of radius R, varies as R^3, while its cross section area A varies as R^2. Hence two spheres of the same material but different size will not fall equally fast. But if their densities scale as $1/R$, the ratio m/A, and therefore also the parameter a, will be constant. Then all data points for falling spheres of varying sizes will "collapse" onto the same curve $L(t)$.

From the structure of Buckingham's theorem, it is obvious that scaling onto a universal "master curve" is very common. It has extremely important applications in fluid mechanics. For instance, the drag coefficient C_d (which we approximated by a constant) is a function of the dimensionless Reynolds number Re [9]:

$$Re = \frac{v d \rho}{\eta}$$

Here v is the relative velocity of the body through the unperturbed fluid, d is a characteristic length of the body perpendicular to the direction of motion, ρ is the density of the fluid and η is the dynamic viscosity. The drag coefficient varies slowly with the Reynolds number in wide ranges of Re values. Then C_d can be regarded as a constant, and we have the familiar result that the air resistance increases with the velocity as v^2. However, there is also a critical region where Re varies rapidly; in air it happens at about $Re = 3 \times 10^5$. If an experiment is performed on a scale model which is similar in geometry to the actual object, the experimental conditions can be arranged so that, e.g., the Reynolds number, and therefore the drag coefficient C_d, is unchanged.

7 Systematics in the neglect of certain effects

7.1 Gravity waves in deep water

In our discussion of characteristic quantities, we noted that they may be associated with the transition from one physical regime to another. On either side of this transition region, certain effects may be neglected. We now illustrate this with the help of Buckingham's Π theorem. From the first row in Table 2 we see that the phase velocity of gravity waves can be written

$$\Pi_1 = \Psi_{\text{wave}}(\Pi_2, \Pi_3)$$

The special case of wave propagation in deep water, and with negligibly small surface tension, corresponds to the mathematical limit

$$\Pi_1 = \Psi_{\text{wave}}(0, \infty) = \text{constant}$$

If we had chosen $\Pi_g = \lambda/d$ instead of d/λ, the equivalent relation would be

$$\Pi_1 = \Psi_{\text{wave}}(0, 0) = \text{constant}$$

Fig. 2 Schematic geometry of a
sheet with a hole

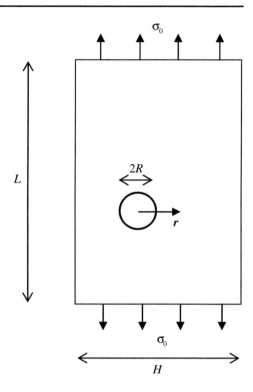

Ignoring the depth and the surface tension thus means that we take the limit $\Psi_{\text{wave}}(0, 0)$ (or $\Psi_{\text{wave}}(0,\infty)$), of course under the assumption that Ψ has no singularity in these limits.

7.2 Small hole in a large sheet under stress

Consider a sheet (Fig. 2) of width H, length L and thickness D, in which there is a hole with radius R. A uniform stress σ_0 is applied along the edge of the sheet. The sheet material is characterized by the Young's modulus E and the shear modulus G. We seek the stress $\sigma(r)$ at the point r shown in the figure. There are $n = 9$ parameters in this model. Five of them (H, L, D, R, r) have the dimension of length and the remaining four (σ_0, σ, E, G) can all be expressed in the unit of pressure, i.e., there are altogether $m = 2$ independent units. (Note that we should consider σ_0 etc. to be expressed in only one unit, pascal, in spite of the fact that pascal contains the three SI units m, s, kg.) According to Buckingham's theorem, we should now form $n - m = 7$ dimensionless and independent ratios. In the spirit of Buckingham's approach, we choose them as in the relation

$$\frac{\sigma(r)}{\sigma_0} = \Psi_{\text{elast}}\left(\frac{r}{R}, \frac{R}{H}, \frac{R}{L}, \frac{D}{H}, \frac{\sigma_0}{E}, \frac{\sigma_0}{G}\right)$$

Next, we specialize to the limit of a *small* hole ($R \ll H, L$) in a thin sheet ($D \ll H, L$) and for low stresses (i.e., small elastic deformations) $\sigma_0 \ll E, G$. That gives

$$\frac{\sigma(r)}{\sigma_0} = \Psi_{\text{elast}}\left(\frac{r}{R}, 0, 0, 0, 0, 0\right)$$

Finally we take the special case that $r = R$, i.e., we seek the stress at the edge of the hole. Then

$$\frac{\sigma(r)}{\sigma_0} = \Psi_{\text{elast}}(1, 0, 0, 0, 0, 0)$$

The value of $\Psi_{\text{elast}}(1,0,0,0,0,0)$ does not depend on the magnitude of R, as long as the inequalities above are obeyed. We obtained this result without any specific knowledge about the function Ψ_{elast}. A detailed mathematical analysis [25] shows that

$$\sigma(r, R) = \sigma_0 \left[1 + \frac{R^2}{2r^2} + \frac{3R^4}{2r^4} \right]$$

which is the explicit form of $\sigma(r)/\sigma_0 = \Psi_{\text{elast}}(r/R, 0, 0, 0, 0, 0)$.

Our approach assumed a continuum mechanics treatment, and therefore we did not discuss a possible *minimum* size of the hole radius. A more complete treatment should include the dimensionless ratio a/R, where a is the size of an atom, with the requirement that $a/R \ll 1$. Further, one may ask what is the meaning of the condition $D \ll H, L$, since the thickness D does not enter in a planar problem. However, we assumed that a constant stress σ_0 was applied to the sheet, and we could have added a requirement for the corresponding total force $\sigma_0 HD$.

8 The Lennard-Jones model

Many thousand works on the modeling and simulation of properties of solids and liquids assume that the atoms or molecules interact through a specific interaction potential, which has been given a simple form. Arguably the most popular such interaction is the Lennard-Jones potential V_{LJ} [13]. (The author of Ref. [13] is John Edward Jones. He married Kathleen Lennard in 1926, adding his wife's surname to his own to become Lennard-Jones.) The potential has been used to model phenomena ranging from materials failure in systems with billions of atoms, to protein folding and even simulations of machine learning. A particularly important area of application is physical chemistry, as exemplified by a systematic study of transport properties in fluids by Dyer et al. [4].

A common mathematical form of the Lennard-Jones potential is

$$V_{\text{LJ}}(r) = 4\varepsilon \left[\left(\frac{\sigma}{r} \right)^{12} - \left(\frac{\sigma}{r} \right)^6 \right]$$

where ε is a parameter determining the strength of the interaction, and σ is a parameter that sets a characteristic length scale. The exponent 6 was originally motivated by the form of dipole interactions. The exponent $12(= 2 \times 6)$ has been chosen mainly for mathematical convenience. Many works have generalized the potential to have other exponents, (m, n). Here it suffices to discuss the (12,6) form given above.

The fact that there are only two independent parameters in the Lennard-Jones model, one with the dimension of length and one with the dimension of energy (in SI units $J = kg\, m^2/s^2$), has far reaching consequences for the interrelation of properties modeled with the potential. Let the atoms in a monatomic solid be arranged in different crystal structures, for instance the face centered cubic (fcc), the hexagonal closed packed (hcp), and the body centred cubic (bcc) structures. For simplicity we exemplify with a cubic structure, which has a single lattice parameter a. In equilibrium at 0 K, and in classical physics (i.e., ignoring quantum effects) it is straight forward to calculate the cohesive energy E_{coh} in these structures. The

Table 3 The cohesive energy per atom (ignoring quantum-mechanical zero point vibrations) expressed in the natural unit of a Lennard-Jones model. Data from [1,14,21]

Crystal structure	fcc	hcp	bcc
Cohesive energy/ε	8.610	8.611	8.24

problem involves four quantities, E_{coh}, ε, a and σ. They can be combined in the dimensionless quantities $\Pi_1 = E_{\text{coh}}/\varepsilon$ and $\Pi_2 = a/\sigma$. It follows from Buckingham's theorem that E_{coh} will be proportional to ε (Table 3). Likewise, the lattice parameters of all assumed crystal structures, and the static atomic displacements near defects, scale as σ. Furthermore, all formation energies associated with static lattice defects, for instance vacancies, dislocations and grain boundaries, scale as ε.

An atom displaced from its equilibrium position will be subject to a restoring force, expressed as a force constant k. The magnitude of k depends on the character of the displacement, and on the position of all other atoms in the system, but k always has the physical dimension of force per length, or N/m = J/m^2 = kg/s^2. Such a combination in the Lennard-Jones interaction is ε/σ^2, and this is how force constants scale. Furthermore, elastic constants have the unit Pa = N/m^2 = J/m^3. Hence, elastic constants calculated from V_{LJ} scale as ε/σ^3. For instance, the bulk modulus of the fcc structure is 75.2 ε/σ^3 [1].

We next turn to thermal properties, for instance the Gibbs energy G, and do not exclude quantum effects. The atoms will vibrate, and we must include their mass M. Three more quantities are required; the temperature T (SI unit K), Boltzmann's constant k_B (SI unit J/K = kg m^2/(s^2 K)), and Planck's constant h (unit J s = kg m^2/s). In the discussion to follow it is convenient to replace the lattice parameter a with the volume per atom, V. We now have 8 quantities (ε, G, σ, V, M, k_B, T, h), expressed in 4 independent units (m, s, kg, K). Following the philosophy of Buckingham's theorem, it is natural to form $8 - 4 = 4$ dimensionless quantities Π_i as in

$$\frac{G}{\varepsilon} = \Psi_G \left(\frac{k_B T}{\varepsilon}, \frac{V}{\sigma^3}, \frac{h^2}{M\varepsilon\sigma^2} \right)$$

The subscript G on Ψ_G denotes the mathematical function Ψ, which yields the Gibbs energy. The equilibrium state is that which minimizes the Gibbs energy. Thus, for every value of $k_B T/\varepsilon$ we get the equilibrium value of V/σ^3.

At the melting point $T = T_m$, the Gibbs energies of the solid and the liquid phases are equal. Then we can write

$$\Psi_{G,\text{sol}} \left(\frac{k_B T_m}{\varepsilon}, \frac{V_{\text{sol}}}{\sigma^3}, \frac{h^2}{M\varepsilon\sigma^2} \right) = \Psi_{G,\text{liq}} \left(\frac{k_B T_m}{\varepsilon}, \frac{V_{\text{liq}}}{\sigma^3}, \frac{h^2}{M\varepsilon\sigma^2} \right)$$

V_{sol} and V_{liq} have values at T_m determined by the equilibrium condition, which gives V/σ^3 for each phase a definite numerical value. It follows from the equation above that if substances are described by different values of σ and ε in the Lennard-Jones model, but have the same equilibrium crystal structure, their T_m scales as ε. The presence of the quantity $h^2/(M\varepsilon\sigma^2)$ means that the numerical value of the ratio $k_B T_m/\varepsilon$ depends on whether quantum effects are included or not.

We next turn to the root-mean-square vibrational displacement u of an atom in a solid, relative to its equilibrium position. The model quantities are the same as for the Gibbs energy,

Property	SI unit	Scaling expression
Length	m	σ
Energy	$J = kg\, m^2\, s^{-2}$	ε
Mass	kg	m
Planck's constant	$Js = kg\, m^2\, s^{-1}$	$\hbar = h/(2\pi)$
Boltzmann's constant	$J/K = kg\, m^2\, s^{-2}\, K^{-1}$	k_B
Force	$N = kg\, m\, s^{-2}$	ε/σ
Pressure	$Pa = kg\, m^{-1}\, s^{-2}$	ε/σ^3
Elastic constant	$N/m^2 = kg\, m^{-1}\, s^{-2}$	ε/σ^3
Density	$kg\, m^{-3}$	m/σ^3
Time	s	$\sqrt{m\sigma^2/\varepsilon}$ or $\hbar/\varepsilon$
Frequency	s^{-1}	$\sqrt{\varepsilon/(m\sigma^2)}$ or $\varepsilon/\hbar$
Viscosity	$Pa\, s = kg\ m^{-1}\, s^{-1}$	$\sqrt{m\varepsilon/\sigma^4}$
Surface energy (tension)	$J/m^2 = N/m = kg\ s^{-2}$	ε/σ^2
Temperature	K	ε/k_B

Table 4 Combinations of quantities which show how certain physical properties scale with the model parameters in the Lennard-Jones potential

except that G is replaced by u. We can write

$$\frac{u}{\sigma} = \Psi_u \left(\frac{k_B T}{\varepsilon}, \frac{V}{\sigma^3}, \frac{h^2}{M\varepsilon\sigma^2} \right)$$

For a given crystal structure at $T = T_m$, all the dimensionless ratios in the argument of Ψ_u have the same value. This is used in Section 9 for a discussion of the Lindemann melting rule.

We have seen how various physical properties get a very simple scaling behavior. Table 4 summarizes the natural units for various properties, when they are modeled with a Lennard-Jones potential.

9 The Lindemann melting criterion

At the melting temperature, the root-mean-square u of the thermal atomic vibrational displacement from the equilibrium position in a lattice is found to be an almost constant fraction of the nearest-neighbor atomic distance, for several elements (Table 5). This has then been taken as a general aspect of melting. The lattice would "shake apart" if the vibrational amplitude is larger. The rule is named after Lindemann, who wrote a seminal paper on lattice dynamics in 1910 [18], although the formulation as a melting criterion was not given until 1956 by Gilvarry [7]. In the paper on instabilities (Grimvall, this volume) it was argued that melting is not caused by an instability of the Lindemann type. Still, the regularity in Table 5 is striking, and requires an explanation.

We noted in Sect. 8 that u/σ has a unique value at $T = T_m$ for a given crystal structure modeled with the Lennard-Jones potential. In other words, the Lindemann melting criterion is fulfilled for all fcc solids modeled with the Lennard-Jones potential. Likewise, it is fulfilled

Table 5 The ratio $u/(2a)$ at the melting temperature, where u is the root-mean-square of the thermal atomic vibrational displacement, and $2a$ is the nearest-neighbor atomic distance in the lattice (from Grimvall [11])

Na (bcc)	Cu (fcc)	Pb (fcc)	Al (fcc)	Mg (hcp)	Tl (hcp)
0.13	0.11	0.10	0.10	0.11	0.12

for all bcc solids, and so on, but with a ratio u/σ that is not quite the same for all structures. Furthermore, we would get the same regular behavior of u/σ if all the solids were modeled with another two-parameter interaction of the type

$$V(r) = \varepsilon\phi\left(\frac{r}{\sigma}\right)$$

where ε sets an energy scale, σ a length scale and ϕ is a "shape function" that determines how the interaction between atoms depends on their mutual distance.

The elemental metals is such a restricted class of solids that the interaction may be reasonably represented by a potential with a common shape function ϕ, but different bonding strengths ε and different atomic sizes σ. Then the Lindemann criterion would be fulfilled within a certain crystal structure, and it would be no surprise if, e.g., the two closed packed structures (fcc and hcp) have very similar results for u/σ at $T = T_m$.

We conclude that the regularities for metallic elements expressed by the Lindemann melting rule is nothing but a consequence of their similar effective binding potential. Our argument would not be valid, for instance, in a binary compound like NaCl, where the interaction is more complex than what can be modeled with a single two-parameter potential. Indeed, it turns out that the Lindemann criterion fails to account for the pressure dependence of the melting temperature in alkali halides [28].

10 Saturating conductivities – a still unsolved problem

In electrical conduction in metals, electrons can be viewed as moving in straight lines between points where they are scattered (Fig. 3). Thermal conduction mediated by lattice vibrations (phonons) has an analogous description. The mean free path λ is the average distance between such scattering events. A simple model for the electrical conductivity σ in metals gives [1,11,15]

$$\sigma = \frac{ne^2\tau}{m} = \frac{ne^2\ell}{mv_F}$$

Here n is the number of conduction electrons per volume, e the electron charge, m the electron mass, τ the average time between scattering events, and v_F the electron Fermi velocity, with

$$\ell = v_F\tau$$

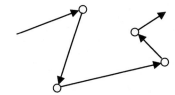

Fig. 3 Schematic illustration of electrical or thermal conduction in solids. A carrier of charge (electron) or heat (electron, phonon) travels in straight lines between scattering events, with a mean free path ℓ between each scattering site

The mean free path decreases with increased scattering rate, e.g., caused by the addition of impurity atoms or the increased thermal motion of the atoms at high temperature. However, the mean free path can not be shorter than the distance between neighboring "scattering centers". That distance is, at least, equal to the distance a between nearest atomic neighbors in a lattice. Thus we now have two length scales, λ and a. In the spirit of Buckingham's Π theorem, we expect that the schematic description in Fig. 3 breaks down when the ratio a/ℓ is no longer small. In fact, there is a somewhat more restrictive condition, since electrons and phonons can be represented by waves with a wavelength λ. It is meaningless to describe the scattering of a wave as in Fig. 3, if the wavelength is not much shorter than the distance between the scattering centers. However, the electrons and most of the phonons of interest here have a wavelength that is of the order of the distance between atoms (cf. electrons or phonons with wave vectors near the first Brillouin zone boundary).

A consequence of the arguments above is that the standard descriptions of electrical conduction in metals, and of thermal conduction through phonons, will fail when the scattering is strong. It should not matter what is the cause of the scattering, be it thermal disorder or various types of lattice defects. Experiments confirm this picture, for instance work by Mooij [19] and Fisk and Webb [5] for the electrical conductivity σ, and by Zeller and Pohl [29] for the thermal conductivity κ.

Gunnarsson et al. [12] reviewed theories for the electrical conduction in saturating systems, and Slack [20] reviewed thermal conduction. There is no simple formulation of σ or κ expressed as an accurate and simple algebraic expression, but there are many approximate formulas. For instance, Wiesmann et al. [27] noted that the shape of the electrical resistivity $\rho(T)$ is well approximated by the so called shunt resistor model:

$$\frac{1}{\rho(T)} = \frac{1}{\rho_{\text{ideal}}(T)} + \frac{1}{\rho_{\text{sat}}}$$

Here $\rho_{\text{ideal}}(T)$ is the "normal" resistivity one would obtain if there were no saturation effects, and ρ_{sat} is the value at which the actual resistivity saturates when ρ_{ideal} becomes very large (e.g., because T is high or because of strong scattering by static lattice defects).

In a nearly-free electron model, the expression $1/\rho = \sigma = ne^2\ell/(mv_{\text{F}})$ can be rewritten as [11]

$$\rho = k\left(\frac{a_0}{\ell}\right)\left(\frac{a_0\hbar}{e^2}\right)$$

Here k is a dimensionless constant, which varies with the electron number density n, and typically is of the order of 40. The quantity $a_0\hbar/e^2 = 0.22\,\mu\Omega\,\text{m}$ is a fundamental unit of resistivity, expressed in the Bohr radius a_0. If λ is of the order of a diameter of an atom, ρ becomes of the order of 1 $\mu\Omega\,\text{m}$, while typical values of ρ_{sat} are of the order of 0.1 $\mu\Omega\,\text{m}$. Thus, an extrapolation of the analytic model for free-electron-like metals only approximately gives the saturation value of the resistivity, when the mean free path approaches the distance between nearest neighbor atoms.

The thermal conductivity κ_{el} of metals is largely due to electron transport. In an elementary model, one has the Wiedemann-Franz-Lorenz law [1, 11, 15]

$$\kappa_{\text{el}} = LT\sigma$$

where L is the Lorenz number. Experimental data suggest that this relation remains valid in the saturation regime; see Grimvall [11, 10]

We next turn to the thermal conductivity of insulators. In the simplest theory, with the thermal conductivity κ_{ph} mediated by phonons, one has [1,11,15]

$$\kappa_{ph} = \frac{1}{3} n c_V C \ell$$

Here n is the number of energy carriers (phonons) per volume, c_V the heat capacity per particle (phonon), C the average velocity of the phonons and ℓ the phonon mean free path. Let $\ell \approx a$, where a is a typical nearest neighbor distance. With a Debye frequency ω_D, we get

$$\kappa_{ph} \sim \frac{k_B \omega_D}{a}$$

This value typically is ~ 1 W/(m K). Within an order of magnitude it is equal to typical saturation values of the thermal conductivity [20].

We have already remarked that there seems to be no fundamental theory leading to a simple closed-form and algebraic result for the electrical resistivity, or the phonon thermal conductivity, in the saturation regime. Instead one may have to resort to numerical modeling [6]. In fact, the situation is even more complex and troublesome. There are systems where the electrical resistivity does not saturate at the expected universal value, for instance the high-T_c cuprate superconductors [12].

11 Conclusions

Dimensional analysis is a powerful aspect of modeling and simulation, which can give new insights without detailed analytic or numerical calculations. Dimensionless combinations of parameters that are assumed to describe a phenomenon may lead to the identification of characteristic quantities, which have the dimension of length, time, mass, speed, and so on. They are helpful in a qualitative understanding of the problem, for instance to what extent certain effects may be neglected. They may also indicate the breakdown of a physical mechanism, when one goes beyond the values of such characteristic quantities – an insight of importance even when one does not have a theory to handle the new regime. A particularly useful approach in modeling and in the analysis of experimental data is to represent the results obtained under different conditions in such a way that, after a rescaling of quantities, they fall on a common graph or give the same number. Buckingham's Π theorem provides the theoretical framework for such an approach.

Appendix 1

1.1 Thermal conduction in a semi-infinite medium

Let the temperature at the surface of a semi-infinite medium (e.g., the surface of the ground) vary periodically as $T_s(t) = T_0 + \Delta T \sin(\omega t)$. The differential equation for the heat flow in a medium with the thermal diffusivity a yields the temperature $T(z,t)$ at depth z and time t:

$$T(z,t) = T_0 + \Delta T \exp\left[-z\sqrt{\omega/(2a)}\right] \sin\left[\omega\left(t - z/\sqrt{2a\omega}\right)\right]$$

The temperature variation decreases exponentially with the depth z. It is reduced by a factor $1/e$ at $z = \sqrt{2a/\omega}$. The temperature variation at the depth z is delayed relative to the temperature at the surface by the time $z/\sqrt{2a\omega}$.

1.2 Falling sphere with air resistance

An object with mass m falls under the influence of the gravitational force mg and the retarding force $F = \frac{1}{2}C_{\mathrm{d}}\rho A v^2$ due to air resistance. If the motion starts from rest, the distance L covered after the time t is given by solving Newton's equation of motion:

$$L = a \ln\left[\cosh\left(t\sqrt{\frac{g}{a}}\right)\right]$$

Here a is a parameter with the dimension of length,

$$a = \frac{2m}{C_{\mathrm{d}}A\rho}$$

Solving for t, and in a series expansion that keeps only the first term in the small quantity L/a, gives

$$t \approx \sqrt{\frac{2L}{g}}\left(1 + \frac{L}{6a}\right) = \sqrt{\frac{2L}{g}}\left(1 + \frac{C_{\mathrm{d}}}{12}\frac{M}{m}\right)$$

where $M = AL\rho$. In many realistic applications, C_{d} is of the order of $1/2$. (The expression for $L(t)$ is easily verified if we consider the acceleration $\mathrm{d}^2L/\mathrm{d}t^2 = mg - F$.)

Appendix 2

2.1 Spider silk

Consider a liquid cylinder, where the energy of the system is determined by the surface energy. If the cylinder is long enough, it is unstable under variations in shape such as indicated in Fig. 4. Suppose that the radius $R(z)$ is modulated with a wavelength λ and a small amplitude δ along the axial coordinate z:

$$R(z) = R_1(\delta) + \delta \sin(2\pi z/\lambda)$$

The radius $R_1(\delta)$ is adjusted so that the total volume is conserved, i.e., the material is incompressible. It is not difficult to show that the surface area of the cylinder is decreased upon modulation, provided that the wavelength λ is larger than the circumference $2\pi R$. The corresponding decrease in the surface energy of the system (the Rayleigh-Plateau instability) is the key to the formation of sticky spheres on threads in a cobweb, where the spider can can produce more than 10^4 small spheres in just one hour [24]. In the idealized model, the system is unstable if its length L is larger than the circumference, so that a wavelike perturbation with $\lambda > 2\pi R$ can be accommodated. In practice, this model is too simplified, for instance because it does not allow for the viscosity of the fluid, but it still shows the qualitative relationship between the wavelength of the disturbance and the radius of the cylinder. See Goren [8] for other examples of the instability of a liquid cylinder.

Fig. 4 Geometry of a
deformation that gives the
Rayleigh-Plateau instability

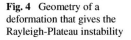

Fig. 5 Geometry leading to the
Rayleigh-Bénard instability

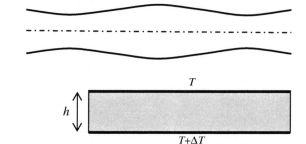

2.2 Rayleigh instability and Bénard cells

In a horizontal liquid layer bounded by rigid plates, the upper surface has the temperature T and the lower surface has a higher temperature, $T + \Delta T$ (Fig. 5). For small ΔT, thermal conduction gives a vertical transport of thermal energy, but at a certain critical value, convection currents arise [16]. The pertinent physical quantities include properties of the liquid, viz. the thermal diffusivity a (SI unit m²/s), the kinematic viscosity ν m²/s), and the thermal volume expansion coefficient β (1/K). Further, a model must contain the acceleration of gravity g (m/s²) and the liquid layer thickness h (m). Given these quantities, we can form a quantity with the physical dimension of the vertical temperature gradient dT/dz in the liquid layer:

$$\frac{a\nu}{g\beta h^4}$$

A mathematical analysis shows that the instability arises when the temperature gradient exceeds

$$\frac{dT}{dz} = Ra_c \cdot \frac{a\nu}{g\beta h^4}$$

where $Ra_c \approx 1708$ is the critical value of the Rayleigh number

$$Ra = \left(\frac{dT}{dz}\right) \frac{g\beta h^4}{a\nu}$$

Physically, the instability occurs when the energy liberated through the motion due to Archimedes' buoyancy equals the energy associated with internal friction in the liquid. The convection pattern is referred to as Bénard cells.

References

1. Ashcroft, N.W., Mermin, N.D.: Solid State Physics. Holt, Rinehart and Winston, Austin (1976)
2. Barenblatt, G.I.: Dimensional Analysis. Gordon and Breach, New York (1987)
3. Buckingham, E.: On physically similar systems: illustrations of the use of dimensional equations. Phys. Rev. **4**, 345–376 (1914)
4. Dyer, K.M., Pettitt, B.M., Stell, G.: Systematic investigation of theories of transport in the Lennard-Jones fluid. J. Chem. Phys. **126**, 034502/1–034502/9 (2007)
5. Fisk, Z., Webb, G.W.: Saturation of the high-temperature normal-state electrical resistivity of superconductors. Phys. Rev. Lett. **36**, 1084–1086 (1976)
6. Gilman, Y., Allen, P.B., Tahir-Kheli, J., Goddard, W.A., III.: Numerical resistivity calculations for disordered three-dimensional metal models using tight-binding Hamiltonians. Phys. Rev. B **70**, 224201/1–224201/3 (2004)

7. Gilvarry, J.J.: The Lindemann and Grüneisen laws. Phys. Rev. **102**, 308–316 (1956)
8. Goren, S.L.: The instability of an annular thread of fluid. J. Fluid Mech. **12**, 309–319 (1962)
9. Granger, R.A.: Fluid Mechanics. Dover, New York (1985)
10. Grimvall, G.: Transport properties of metals and alloys. Physica **127B**, 165–169 (1984)
11. Grimvall, G.: Thermophysical Properties of Materials. Enlarged and revised edition. North-Holland (1999)
12. Gunnarsson, O., Calandra, M., Han, J.E.: Colloquium: saturation of electrical resistivity. Rev. Mod. Phys. **75**, 1085–1099 (2003)
13. Jones, J.E.: On the determination of molecular fields—II. From the equation of state of a gas. Proc. R. Soc. Lond. A **106**, 463–477 (1924)
14. Kihara, T., Koba, S.: Crystal structures and intermolecular forces of rare gases. J. Phys. Soc. Jpn. **7**, 348–354 (1952)
15. Kittel, C.: Introduction to Solid State Physics, 8th edn. Wiley, New York (2005)
16. Landau, L.D., Lifshitz, E.M.: Fluid Mechanics, 2nd edn. Elsevier (1987)
17. Lighthill, J.: Waves in Fluids. Cambridge University Press, Cambridge (1978)
18. Lindemann, F.A.: Molecular frequencies. Physik. Zeits. **11**, 609–612 (1910)
19. Mooij, J.H.: Electrical conduction in concentrated disordered transition metal alloys. Phys. Stat. Solidi A **17**, 521–530 (1973)
20. Slack, G.A.: The thermal conduction of a non-metallic crystal. In: Ehrenreich, H., Seitz, F., Turnbull, D. (eds.) Solid State Physics, vol. 34, pp. 1–71. Academic Press, New York (1979)
21. Stillinger, F.H.: Lattice sums and their phase diagram implications for the classical Lennard-Jones model. J. Chem. Phys. **115**, 5208–5212 (2001)
22. Szirtes, T.: Applied Dimensional Analysis and Modeling. McGraw-Hill, New York (1998)
23. Taylor, E.S.: Dimensional Analysis for Engineers. Clarendon Press, Oxford (1974)
24. Thompson, D.W. In: Bonner, J.T. (ed.) On Growth and Form, Abridged edn. Cambridge University Press, Cambridge (1961)
25. Timoshenko, S.: Strength of Materials. Part II. Advanced Theory and Problems. McMillan, New York (1930)
26. Timoshenko, S.: Strength of Materials. Part I. Elementary Theory and Problems. McMillan, New York (1930)
27. Wiesmann, H., Gurvitch, M., Lutz, H., Ghosh, A., Schwarz, B., Strongin, M., Allen, P. B., Halley, J.W.: Simple model for characterizing the electrical resistivity in A-15 superconductors. Phys. Rev. Lett. **38**, 782–785 (1977)
28. Wolf, G.H., Jeanloz, R.: Lindemann melting law: anharmonic correction and test of its validity for minerals. J. Geophys. Res. **89**, 7821–7835 (1984)
29. Zeller, R.C., Pohl, R.O.: Thermal conductivity and specific heat of non-crystalline solids. Phys. Rev. B **4**, 2029–2041 (1971)

Accuracy of models

Göran Grimvall

Originally published in the journal Sci Model Simul, Volume 15, Nos 1–3, 41–57.
DOI: 10.1007/s10820-008-9103-3 © Springer Science+Business Media B.V. 2008

Abstract   The ability of a mathematical model to accurately describe a phenomenon depends on how well the model incorporates all relevant aspects, how robust the model is with respect to its mathematical form, and with respect to the numerical values of input parameters. Some models are primarily intended to reproduce known data. In other cases the purpose is to make a prediction outside the range of knowledge, or to establish details in a physical mechanism. These aspects of modelling are discussed, with examples mainly taken from the field of materials science.

Keywords   Modeling · Robustness · Data fitting · Einstein model · CALPHAD · Anharmonic effects

1 Introduction

There is a rich literature on the numerical and mathematical aspects of modeling, with applications in a wide spectrum of fields. Much less has been written on the interplay between experiment, simulation and the analytic form of models in science and engineering. That will be in the focus of this paper, which complements the two preceding papers by the author in this volume.

Several specific examples will be taken from the field of materials science, but the message is of a general character. Some of the examples deal with fundamental formulas found in textbooks, but now with remarks on aspects that would often go unnoticed even for the most ambitious student. Other examples are taken from the current research frontier, but deal with problems of a long-standing interest.

We are primarily interested in models that are given the mathematical form of an algebraic formula. Their purpose can be manifold. It can be to explain experimental results, and in this

G. Grimvall (✉)
Theoretical Physics, Royal Institute of Technology, AlbaNova University Center, 106 91 Stockholm, Sweden
e-mail: grimvall@kth.se

way provide a theoretical account of a phenomenon. Another purpose can be to reproduce data in a form that is suitable for further use, or to extrapolate known data into regions where the information is meager, uncertain or absent. In some cases, we want to apply a model to a set of data, in order to extract a specific physical quantity. Furthermore, it is of increasing importance to use models in a simulation of properties that are inaccessible to direct experiments.

Whatever the purpose is, we would like to know how accurately the model can account for the real situation. That leads to the following questions:

- Does the model contain all the relevant mechanisms or effects in the studied phenomenon?
- How robust is the model result with respect to variations in the mathematical form of the model?
- How robust is the model result with respect to variations in the numerical values of model parameters?
- How accurately are the model parameters known?

2 Robustness

2.1 The oil peak problem

Half a century ago, Hubbert [1,2] introduced a simple model to predict when the U.S. oil production would peak. Since then, the modeling of the world's oil production has become a central issue. Hubbert's approach is briefly as follows. The total amount of oil, Q_∞, is assumed to be fixed. It includes presently known resources, as well as those not yet discovered. Since the annual oil production $P(t)$ cannot increase indefinitely, it is assumed to have some bell-shaped form when plotted versus time. The total area under this curve is Q_∞. The curve is further described by two characteristic quantities: the width W of the "bell" (defined in some reasonable way), and the time t_p where the curve peaks. Data of the oil production are assumed to be available up to a time t_0. With given shape $P(t)$ and total amount of oil Q_∞, the parameters W and t_p are adjusted to give the least root mean square deviation (RMSD) between $P(W, t_p)$ and the actual data in the time interval $(-\infty, t_0)$.

One attractive mathematical form of the bell shape is the Gaussian curve;

$$P(t) = \frac{Q_\infty}{W\sqrt{2\pi}} \exp \left[\frac{-(t - t_p)^2}{2W^2} \right]$$

Another common form, and used by Hubbert, is the Logistic curve;

$$P(t) = \frac{Q_\infty}{w} \cdot \frac{\exp \left[(t - t_p)/w \right]}{\left[1 + \exp \left[(t - t_p)/w \right] \right]^2}$$

If we take

$$w = \frac{W\sqrt{\ln(4)}}{\ln \left(3 + \sqrt{8} \right)}$$

the two curves are normalized to have the same width at half peak height; Fig. 1

Bartlett [3] has discussed the robustness of predictions based on Hubbert-type modeling, for U.S. as well as world oil production. It was first established that the difference between

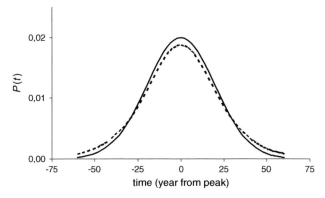

Fig. 1 The Gaussian model (solid curve) and the Logistic model (dashed curve), normalized to enclose the same area (1 unit), and adjusted to peak at the same time $t = 0$ and have the same width at half peak height

an analysis based on the Gaussian and on the Logistic curve was insignificant; cf. Fig. 1. In a sensitivity analysis of the modeling of the U.S. oil production, Bartlett found that fixing W and t_p at their primary values (i.e., those which give the smallest RMSD), and increasing Q_∞ by 8.1 %, increased the RMSD of the fit by 70 %. Starting from the primary fit and increasing t_p by 4.4 years increased the RMSD by about 80 %. Similarly, if W was increased by 5.7 years, the RMSD increased by about 52 %. Further, with the world resource $Q_\infty = 2.0 \times 10^{12}$ barrels of oil (Hubbert's original assumption), the best fit of the Gaussian curve puts the peak in the world oil production at the year $t_p = 2004$. Increasing Q_∞ to 3.0×10^{12} and to 4.0×10^{12} barrels moves the peak to the years 2019 and 2030, respectively.

The work of Bartlett is chosen here because it has a good discussion of the robustness of the model prediction. Hubert's original work does not address this aspect in detail, but Bartlett refers to several other studies, which come to essentially the same result. It should be added that the purpose of the present discussion is to present an example of a type of modeling, and not to treat the oil peak problem *per se*.

2.2 The entropy of TiC

The heat capacity $C(T)$ of titanium carbide (TiC) is known experimentally [4], from low to high temperatures. Therefore, also the entropy $S(T)$ is known;

$$S(T) = \int_0^T \frac{C(T)}{T} \, dT$$

In an attempt to account for these data in a simple mathematical form, one may model $S(T)$ with the Einstein model for lattice vibrations (cf. Grimvall, this volume). Then the entropy per formula unit of TiC has the form [5]

$$S_E(T, \theta_E) = 6k_B [(1 + n) \ln(1 + n) - n \ln(n)]$$

with the Bose-Einstein statistical factor

$$n = \frac{1}{\exp(\theta_E / T) - 1}$$

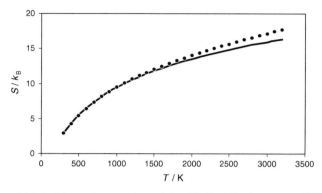

Fig. 2 Einstein model fit (solid curve) to experimental data [4] (dots) for the entropy of TiC. The model has been fitted to the experiments at $T = 1000$ K

Here k_B is Boltzmann's constant and θ_E is the Einstein temperature. We can fit θ_E by requiring that $S_E(T)$ agrees with the experimental value $S_{exp}(T)$ at a particular temperature T_{fit};

$$S_E (T_{fit}, \theta_E) = S_{exp} (T_{fit})$$

Figure 2 shows that the Einstein model gives a good account of the experimental entropy at intermediate and high temperatures. (It is well known that the model fails at very low temperature, where C and S should be proportional to T^3.) Note that it is sufficient to fit at a single temperature T_{fit} rather than, for instance, minimizing the RMSD of $S_E(T) - S_{exp}(T)$.

The Einstein temperature θ_E is a *characteristic quantity* that sets a temperature scale for many thermal properties. Furthermore, the combination $\hbar/(k_B\theta_E)$ gives a characteristic time, and the combination $M(k_B\theta_E/\hbar)^2$ gives a characteristic force constant for the interaction between atoms with typical mass M. Therefore, the aim could be to reverse the modeling process and use the experimental data to extract the characteristic quantity θ_E, rather than attempting to reproduce the entropy over a wide range of temperatures. The result will depend on the chosen temperature T_{fit}. An analogous fit of data to the Einstein model can be made for the heat capacity $C(T)$;

$$C_E (T_{fit}, \theta_E) = C_{exp} (T_{fit})$$

In physics, both $S(T)$ and $C(T)$ are assigned absolute values; the entropy because $S(0) = 0$ according to the third law of thermodynamics. On the contrary, the Gibbs energy $G(T) = U - TS + PV$ does not have an absolute value, because the internal energy $U(T)$, like all energies in physics, is defined only relative to a reference level. But we can use the experimental data for a difference in G, for instance $G(T) - G(298.15 \text{ K})$, in a fit to an Einstein model;

$$G_E (T_{fit}, \theta_E) - G_E (298.15 \text{ K}, \theta_E) = G_{exp} (T_{fit}) - G_{exp} (298.15 \text{ K})$$

The values obtained for θ_E when C_E, S_E and $G_E(T) - G_E(298.15 \text{ K})$ are fitted to experimental data for TiC are given in Table 1. We see that the results for θ_E are quite different in these three cases, in spite of the fact C, S and G are closely related through exact thermodynamic relations. This is not only because C, S and G give different weights to the individual phonon frequencies ω; cf. Appendix 1. The failure of the fit to the heat capacity reflects that $C(T)$ of harmonic vibrations never exceeds the classical value of $3k_B$/atom, while

Table 1 The Einstein temperature θ_E derived from a fit at the temperature T_{fit} to the recommended [4] experimental heat capacity, entropy and Gibbs energy, respectively

T_{fit} / K	300	400	500	600	700	800	1000	1500	2000	2500	3000
θ_E /K, from C	658	632	549	446	319	–	–	–	–	–	–
θ_E /K, from S	575	585	586	583	580	576	570	553	531	502	471
θ_E /K, from G		555	560	564	566	567	569	566	560	551	540

Fig. 3 The gross features of the phonon density of states $F(\omega)$ of TiC [5] (solid curve) and the Einstein model (represented by thick vertical line). The Einstein peak is a delta function enclosing the same area as $F(\omega)$

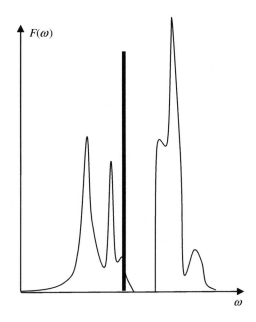

anharmonicity can make $C_{exp}(T)$ larger than this value. In that case there is no solution to the equation $C_E(T, \theta_E) = C_{exp}(T)$. Further comments on Table 1 are made in Sect. 3.3

2.3 Discussion

Figure 3 shows the gross features of the phonon density of states $F(\omega)$ of TiC, together with the Einstein model peak that reproduces the experimental entropy at $T_{fit} = 1000$ K. It may seem remarkable that this drastic simplification of the true $F(\omega)$ can give such a good account of the experimental entropy data. There are two main reasons for that. Firstly, both the actual $F(\omega)$ and the model $F(\omega)$ are *normalized*, so that the area under $F(\omega)$ (i.e., the number of phonon modes) is the same. We may refer to this as a sum rule for $F(\omega)$. Secondly, the entropy is obtained from an *integration* over the density of states $F(\omega)$. An integrated property is often insensitive to features in the integrand. In fact, the entropy of harmonic vibrations in the limit of high T depends on a single parameter $\omega(0)$, which is obtained as an average over $F(\omega)$, see Appendix 1. Finally we note that the Einstein model, like any model assuming harmonic lattice vibrations, leaves out anharmonic effects. That is the reason why it underestimates the entropy at high T. On the other hand, anharmonicity is indirectly accounted for in the varying θ_E-values in Table 1. This is further discussed in Sect. 3.3.

The oil peak model also contains a sum rule (the value of Q_∞) as well as an integration of $P(t)$. Moderate fluctuations in $P(t)$ do not much affect the predicted t_p of the model.

Further, the fitting procedure itself invokes a kind of integration, since the RMSD is a sum (i.e., an integral) of contributions over the fitting interval $(-\infty, t_0)$. Moreover, the optimum fit represents a minimum of the RMSD value. As will be further discussed in Sect. 5, the error in the modeling is expected to be "small to the second order".

In Sect. 1 we asked if the model contains all the relevant mechanisms or effects. It is obvious that a modeling of the oil peak should allow for political and technological changes, which could be drastic. With only a gradual change in such external conditions, our oil-peak modeling is quite robust. In the case of the entropy of TiC, the most serious deficiency is the absence of anharmonicity. Some further, and small, effects are discussed in Sect. 3. Considering that it is a one-parameter model, the robustness is remarkable.

3 Fitting of data—two different objectives

3.1 The CALPHAD method

In many complex modeling situations one needs an explicit mathematical expression, which accurately reproduces input data over a wide range. An outstanding example in this respect is the CALPHAD approach to the calculation of phase diagrams. Keeping the old acronym, it now refers to Computer Coupling of Phase Diagrams and Thermochemistry. As an illustration of key concepts, consider steel. It is an alloy based on iron, with the addition of carbon and several other alloying elements, for instance Cr, Ni, Mn and Si. Pure iron has the body centered cubic crystal structure (bcc) up to 1184 K, followed by the face centered cubic (fcc) structure at higher temperatures. Carbon atoms have a higher solubility in the fcc phase than in the bcc phase. A common hardening process of steel is to heat it into the fcc region, so that the carbon content can be increased. In a subsequent quenching, the specimen returns to the bcc phase, where the concentration of carbon now becomes higher than what corresponds to equilibrium. This forces the material to undergo structural changes that lead to a hardened state.

The point of interest to us is that the transition temperature between the bcc and the fcc phases, which lies at 1184 K for pure iron, varies significantly with the content of carbon and other alloying elements. Mapping out such phase region boundaries, i.e., establishing the phase diagram, is fundamental in materials science. It has been done through painstaking and expensive experiments, but the experimental information is very incomplete. For instance, assume that a chemical element is alloyed with five other elements, each of which can be present in five different concentrations. The number of compositionally different alloys is $5^5 \approx 3000$. Furthermore, one would like to investigate each of them at, say, five different temperatures. This example shows the impossibility of a broad experimental investigation.

The CALPHAD method offers a practical solution, through computer modeling of phase diagrams. It is based on the fact that the equilibrium structure of a material, at any composition c and temperature T, is that which has the lowest Gibbs free energy $G(c, T)$. Experimental thermodynamic and thermophysical data, like the enthalpy, heat capacity, bulk modulus and thermal expansion coefficient, are used to generate $G(T)$ for pure elements and simple substances. Mathematical modeling then yields $G(c, T)$ for alloys of different compositions and structures, through interpolation and extrapolation procedures. Finally, a computer program searches for that state, which has the lowest Gibbs energy at a given composition and temperature.

In the procedure sketched above, one should have the best possible $G(T)$ for the pure substances. It is the reproduction of the experimental data that matters, and an understanding of the physical origin of contributions to $G(T)$ is of secondary importance. (Sect. 3.2 addresses the question of the physical origin.) The CALPHAD research community typically expresses $G(T)$ of an element as in the following example for bcc molybdenum at zero pressure, in the temperature interval 298.15 K < T < 2896 K [6]:

$$G - H^{\text{SER}} = -7747.257 + 131.9197T - 23.56414T \ln(T) - 0.003443396T^2$$
$$+ 5.662834 \times 10^{-7}T^3 - 1.309265 \times 10^{-10}T^4 + 65812.39T^{-1}$$

Here G is given in joule/mole, T is the numerical value of the temperature expressed in kelvin, and H^{SER} is the stable element reference enthalpy. The form of the T-dependence in this interval may be understood from the discussion in Sect. 3.2 and in Appendix 1.

The generation of the explicit function $G(T)$ for a substance requires a deep competence and a substantial amount of work. Each piece of experimental information must be carefully evaluated and used in an optimization procedure, which has similarities to a RMSD fit but also involves a subjective judgment of the quality of the data. The final results for the recommended thermodynamic and thermophysical quantities must be consistent with the exact mathematical relations that exist between various such quantities.

3.2 Separation of contributions to the heat capacity

While the previous section focused on the reproduction of thermodynamic data, we shall now study the underlying physical mechanisms. The heat capacity $C_P(T)$ of a solid can be accounted for as a sum of almost independent contributions, some of which are so small that they can be ignored in certain applications, but large enough to be included in a more detailed analysis. Fundamental research on the heat capacity has usually dealt with the limit of low temperature, because interesting physical phenomena can be seen in that region. In this example, however, we will discuss intermediate and high temperatures. The purpose is to illustrate a fitting procedure, rather than physical rigor, and some simplifications are made in the following discussion. A detailed discussion of thermodynamic quantities in solids can be found in, e.g., Ref. [5].

The dominating contribution to the heat capacity comes from phonons in the harmonic approximation, C_{har}. In a crystal with N atoms there are $3N$ different phonon modes, with frequencies ω described by a frequency distribution function (density of states) $F(\omega)$. Each individual phonon frequency gives a contribution to the heat capacity, which has the mathematical form of the Einstein model. Typically, N is of the order of Avogadro's constant, 6×10^{23} mol^{-1}, but in the high-temperature expansion of C_{har} only a few averages $\omega(n)$ of $F(\omega)$ are important; see Appendix 1. We can write

$$C_{\text{har}} = 3Nk_B \left[1 - \frac{1}{12}\left(\frac{\hbar\omega(2)}{k_B T}\right)^2 + \frac{1}{240}\left(\frac{\hbar\omega(4)}{k_B T}\right)^4 - \dots \right]$$

Another contribution to the heat capacity comes from the anharmonic correction to the expression C_{har} above. That correction has the high-temperature expansion

$$C_{\text{anh}} = aT + bT^2 + \dots$$

For the electronic contribution, C_{el}, we use the form (see Sect. 6.2)

$$C_{\text{el}} = \gamma(T)\, T \approx \gamma_{\text{band}}(0)\, T$$

At very high temperature, the formation of lattice vacancies may add a term

$$C_{vac} = c \exp\left(-E_{vac}/k_B T\right)$$

to the heat capacity. Here E_{vac} is the vacancy formation energy. The pre-factor c has a temperature dependence, which we ignore in comparison with the strong temperature dependence in the exponential term.

The sum of all these contributions to the heat capacity gives an expression of the general mathematical form

$$C_{total} = A + BT^{-4} + CT^{-2} + DT + ET^2 + F \exp\left(-G/T\right)$$

Even though several approximations have been made, there are no less than seven independent constants $A - G$. If C_{total} is fitted to experimental C_P data, one may get a good numerical fit, but it would be useless as the basis for a physical interpretation of the individual contributions. Very likely, part of an effect with known temperature dependence is, incorrectly, included in a combination of other numerically fitted terms.

Instead a performing an unrestricted fit, some of the constants A to G may be given values derived from other information than C_P, as follows. The constants A, B and C can be identified with terms in the high-temperature expansion of C_{har}, where $\omega(2)$ and $\omega(4)$ are calculated from, e.g., neutron scattering experiments (see Appendix 1). The constants F and G may be best obtained through independent information on E_{vac} combined with a theoretical estimation of the pre-factor c, perhaps including its temperature dependence. In a direct fit, it could be extremely difficult to distinguish between a polynomial and an exponential term when the fitting interval is not wide enough (see Sect. 4.3), so that is not a good procedure to obtain F and G.

Two coefficients then remain; D and E. The term linear in T contains contributions from electrons and from anharmonicity. When the electronic density of states $N(E)$ does not have significant variations close to the Fermi level E_F, one can calculate γ_{el} from theoretical band structure result $N(E_F)$, but in transition metals this is often too crude an approximation (see Sect. 6.2). Finally, the anharmonic term C_{anh} is very hard to calculate. It cannot be reduced to an expression that is both accurate and algebraically simple. Numerical simulations offer the only realistic theoretical approach. Then one assumes some parameterized effective interaction between the atoms. But to verify if the numerical simulation yields a good account, one should also try to extract C_{anh}, starting from the experimental C_P and subtracting C_{har}, C_{el} and C_{vac} as outlined above. Such an analysis has been done for aluminum [7]. In a similar, but less ambitious, analysis the heat capacity of ThO_2 and UO_2 was analyzed with the purpose of extracting non-vibrational contributions to C_P [8].

3.3 Discussion

The three examples above, where we considered the Einstein model description of the entropy of TiC, the CALPHAD approach to the Gibbs energy, and the individual contributions to the heat capacity, deal with the thermodynamic quantities S, G and C. They are connected through exact thermodynamic relations. Nevertheless, our examples are quite different from a modeling point of view. The two equations

$$S = -\left(\frac{\partial G}{\partial T}\right)$$

and

$$C = -T \left(\frac{\partial^2 G}{\partial T^2} \right)$$

relate S and C to the first and the second derivative of G, respectively. Conversely, S could be regarded as a quantity that is obtained after one integration of C, and G as obtained after two integrations. Features in $C(T)$ will therefore be much less pronounced in $S(T)$, and smaller still in $G(T)$. With this background we now return to the θ_E-values for TiC in Table 1.

It has already been remarked that anharmonicity is the reason why a fit of the heat capacity $C_E(T, \theta_E)$ to $C_{exp}(T)$ fails even at a moderate temperature (800 K). This is in contrast to the integrated properties $S(T, \theta_E)$ and $G(T, \theta_E)$, which allow θ_E to be extracted from a fit to experiments at all temperatures T_{fit}. But there is a fundamental difference between the entropy and the Gibbs energy. The θ_E-value obtained from the entropy can be given a direct interpretation in terms of the shifts in the phonon frequencies with temperature (including thermal expansion) [5]. In θ_E obtained from G there is no such simple interpretation, because of a partial double counting of phonon frequency shifts. We see in Table 1 that θ_E representing G peaks at a higher temperature, and does not fall off as rapidly with T, when compared with θ_E derived from S. This is explained as follows. The model enthalpy H_E cannot increase above the classical value $3k_B T$ per atom, but H_{exp} has no such restriction. The only way to accommodate the anharmonic effect present in H_{exp} when we fit G_{exp} to $G_E = H_E - T S_E$ is to let S_E have a smaller value, i.e., be represented by a larger θ_E. This effect is superimposed on the general decrease of θ_E that is seen in the data derived from the entropy.

Two further comments should be made. The θ_E-values in Table 1 derived from S and G show a broad maximum as a function of T. On the high temperature side, the decrease in θ_E is due to anharmonicity, and on the low temperature side it is connected with the fact that a single parameter is insufficient to describe $S(T)$ and $G(T)$; cf. the terms containing $\omega(n)$ in Appendix 1. Finally, there is a small electronic contribution $C_{el}(T)$ to the heat capacity. We have ignored that, with the consequence that the anharmonic effect is slightly exaggerated.

We conclude that the unique properties of the vibrational entropy make S particularly suitable for thermodynamic analysis, a feature not always recognized, for instance when experimental data are used to deduce θ_E (or, more common, the Debye temperature θ_D). The smooth decrease in θ_E obtained from the entropy of TiC (Table 1) is very characteristic of anharmonicity in solids. Deviations from such a regular behavior have been used to detect errors in published thermodynamic data for Ca [9] and K_2O [10].

Returning to the CALPHAD representation of the Gibbs energy, we note that the temperature dependence of most of the individual terms can be motivated by the expected theoretical forms of $C(T)$ discussed in Sect. 3.2. However, it would not be possible to identify the different contributions to the heat capacity with terms of a certain T-dependence in the expression $-T(\partial^2 G_{CALPHAD}/\partial T^2)$. Suppose now that $G(T)$ based on experimental information is not very well known. It can then be argued that one should not use more than three fitting parameters, A, B and C. The following two forms might be tried;

$$G(T) = A - BT - CT^2$$

and

$$G(T) = A - BT - CT \ln T$$

The first expression would lead to the linear heat capacity $C_P(T) = 2CT$ and the second expression to the constant heat capacity $C_P(T) = C$. These examples show the difficulty, or the danger, encountered when one takes the derivative of a fitted quantity. For instance, the tabulated [4] heat capacity of elements in the liquid state is usually a constant, i.e., independent of T. In a few cases for which there is more accurate information, $C_P(T)$ shows a shallow minimum at roughly twice the melting temperature [5, 11]. The liquid heat capacity may be determined from measurements of the change in the enthalpy $H(T)$, when a specimen is quenched from high temperature. Since $C_P = \partial H / \partial T$, a four-parameter expression $H = A + BT + CT^2 + DT^3$ is needed to generate a heat capacity that displays a minimum. This may be beyond the accuracy of the experiment.

4 Fitting in log-log plots

4.1 Introduction

It is characteristic of the algebraic formulas in physical models that the model parameters appear in powers, with the exponent usually being an integer but sometimes is a simple fraction. For instance, the emitted power according to the Stefan-Boltzmann radiation law varies with the temperature T as T^4, the Debye theory says that the heat capacity of a solid at low temperatures is proportional to T^3, and the average speed of a particle in a classical gas varies as $T^{1/2}$. These "regular" values of the exponents are well understood theoretically.

A more general case appears frequently in practical engineering. We have a quantity Q whose q-dependence is not known, and we make the *Ansatz*

$$Q = a \, q^n$$

The exponent n is obtained from the slope of the straight line, when data are plotted as $\ln Q$ versus $\ln q$;

$$\ln Q = \ln a + n \ln q$$

Although there are examples of very complex phenomena, where the exponent nevertheless gets a regular value (e.g., in certain so-called critical phenomena), one normally finds that when complex phenomena can be approximately described by a straight line in a log-log plot, the exponent has an "irregular" value. We will illustrate that with the thermal conductivity of insulators. Appendix 2 and 3 give further aspects of data analyzed in log-log plots.

4.2 Thermal conduction in insulators

The thermal energy transport in a crystalline solid insulator is mediated by phonons. They can be scattered by, e.g., lattice defects and other phonons. The theoretical account of the thermal conductivity κ is extremely complicated, although the basic physics is well understood. In the simplest textbook model for κ at high temperature, and in a pure specimen, one has

$$\kappa \propto \frac{1}{T}$$

However, when actual data of $\ln \kappa$ are plotted versus $\ln T$, one finds that a better approximation is

$$\kappa \propto \frac{1}{T^x}$$

The exponent typically is $x \approx 1.3$, but with large variations between different materials [5, 12]. In order to understand this fact, we discuss a more elaborate model [5], where

$$\kappa = k \frac{\theta_E^3}{T}$$

The approximate dependence of κ on the phonon frequencies ω is here represented in an average way by the Einstein temperature θ_E. The quantity k contains several other characteristics of the substance, which also vary somewhat with T, but we will regard k as a constant. Then,

$$x = -\frac{d \ln \kappa}{d \ln T} = 1 - 3 \frac{T}{\theta_E} \cdot \frac{d\theta_E}{dT}$$

We noted in Table 1 that $d\theta_E/dT < 0$, which explains why $x > 1$. It is now easy to understand why the exponent x does not have a simple and universal value, but can vary significantly with the substance.

5 Second-order effects

5.1 Introduction

In modeling, one often neglects an effect with the motivation that it is of "second order". Here we are interested in a property Q that varies with the parameter q, and has a maximum or a minimum for variations of q around $q = q_0$. Since the extremum implies that $dQ/dq = 0$ for $q = q_0$, we have to lowest order in $q - q_0$:

$$Q(q) \approx Q(q_0) + \alpha (q - q_0)^2$$

$Q(q_0)$ may be easy to calculate, while the constant α is more difficult to obtain. But with an approximate estimate of α, one may find that the second-order correction is exceedingly small in the application of interest. Then $Q(q_0)$ gives a satisfactory result even in the non-ideal situation. This powerful argument will be illustrated with the thermal conductivity κ, but identical mathematical results hold for the electrical conductivity σ, the dielectric permittivity ε and the magnetic permeability μ.

5.2 Weakly inhomogeneous materials

Consider a one-phase material where the composition shows fluctuations on a length scale such that one can assign a varying conductivity κ to each point in the material. Let $\langle \ldots \rangle$ denote a spatial average over the entire specimen. Then the effective conductivity of the material is [5]

$$\kappa_{eff} \approx \langle \kappa \rangle \left\{ 1 - \frac{1}{3} \frac{\langle (\kappa - \langle \kappa \rangle)^2 \rangle}{\langle \kappa \rangle^2} \right\}$$

As an example, let the local conductivity vary along the z-direction as

$$\kappa(z) = \kappa_0 + \Delta \sin(kz)$$

Then

$$\kappa \approx \kappa_0 \left[1 - \frac{1}{6} \cdot \left(\frac{\Delta}{\kappa_0} \right)^2 \right]$$

With $\Delta/\kappa_0 = 0.2$, the correction, relative to the case of a uniform conductivity, is less than 1 %, which would be completely negligible in most practical cases.

6 Mislead by simple textbook results

Fundamental formulas found in textbooks in solid state physics are sometimes presented in such a way that they give the impression of a very high accuracy. However, the high accuracy may apply to only the most common situations. Many advanced computer simulations deal with conditions where the textbook formula may be wrong by a factor of two or more. We will illustrate that with examples from the thermodynamics of solids at high temperature.

6.1 The vibrational heat capacity

Experimental data for the heat capacity usually refer to C_P, i.e., the heat capacity at constant pressure. On the other hand, theoretical simulations are often performed under the condition of constant volume. There is an exact relation between the heat capacity C_V at constant volume, and C_P:

$$C_V(T) = C_P(T) - VT\beta^2 K_T$$

Here V is the specimen volume, β the cubic expansion coefficient and K_T the isothermal bulk modulus. Anharmonic effects can make $C_P(T)$ larger than the classical result, $3k_B$/atom, at high temperature. The dominant anharmonic effect usually comes from changes in the effective interatomic forces caused by thermal expansion. In fact, graphs in many textbooks show that $C_V(T)$ for an insulator, derived from $C_P(T)$ as above, seems to approach precisely $3k_B$/atom. This has given rise to a widespread belief that there are no anharmonic effects left in $C_V(T)$. However, the harmonic approximation requires that the vibrational amplitude of the atoms is small; a condition that is not fulfilled at high temperature. Although the anharmonic contributions to C_V are often negligible, they are very significant in, e.g., solid Mo and W [13] and in molecular dynamics simulations reaching into the superheated solid state of a solid [7].

There is one further complication when simulations are compared with experimental data referring to "constant volume". The operational definition of $C_V(T)$ at the temperature T_1 is

$$C_V(T_1) = \frac{\Delta Q}{\Delta T}$$

where ΔT is the temperature increase when the amount of heat ΔQ is added to the specimen, under the condition of unchanged specimen volume $V(T_1)$. But if this procedure is carried out at another temperature $T_2 \neq T_1$, the specimen volume to be kept constant is now $V(T_2)$. Thus the heat capacity at "constant volume" V is not the same as the heat capacity at "fixed volume" V_0. It is the former condition that is most easily obtained from experiments, while simulations usually refer to the latter case. To lowest order in the small quantity $(V - V_0)/V_0$ one has [5, 13, 14]

$$C_V(T) - C_{V_0}(T) = (V - V_0)T \left[K_T(\partial\beta/\partial T)_p + 2\beta(\partial K_T/\partial T)_p + \beta^2 K_T(\partial K_T/\partial P)_T\right]$$

There is a tendency of cancellation between the terms on the right hand side, making the correction small in some cases, e.g., KCl [14] and solid Al [7], but it cannot be neglected in, e.g., liquid Al [11] and solid Mo and W [13].

Table 2 The Fermi temperature T_F of some metals [15]

Element	Na	K	Al	Pb
T_F / K	37 500	24 600	134 900	108 700

6.2 The electronic heat capacity

The electronic heat capacity of metals is often presented as

$$C_{el} = \gamma_{band} T = \frac{2\pi^2}{3} N (E_F) k_B^2 T$$

where $N(E_F)$ is the electron density of states at the Fermi energy E_F. Some textbooks recognize that there is also an enhancement factor $(1 + \lambda_{el-ph})$ due to interaction between electrons and phonons [5]. Typically λ_{el-ph} is 0.4–0.6, but it can be as large as about 1.5 (Pb, Hg). We have

$$C_{el,exp} = \gamma_{exp} T = \gamma_{band} (1 + \lambda_{el-ph}) T$$

This relation gives a very accurate description of low temperature experiments. One may ask if such data can be accurately extrapolated to high temperature. The answer is no, and for several reasons, as we will now see.

The enhancement factor $(1 + \lambda_{el-ph})$ is temperature dependent. Above the Debye temperature it is almost negligible [5]. In practice, this means that only the term $\gamma_{band} T$ may remain above room temperature. It is then tempting to assume that C_{el} continues to be linear in T, but with a reduced slope compared to the low-T limit. This assumption seems to be strongly supported by the following result obtained for a free electron gas [5]:

$$\gamma_{el} = \frac{2\pi^2}{3} N (E_F) k_B^2 \left[1 - \frac{3\pi^2}{10} \left(\frac{T}{T_F} \right)^2 \right]$$

Here $T_F = E_F / k_B$ is the Fermi temperature. Some characteristic values of T_F are given in Table 2. With those numbers, the correction to a linear T-dependence is exceedingly small, less than 0.1 % at the melting temperature. But this argument is very misleading. The heat capacity C_{el} probes $N(E)$ in a "window" of approximate width $\pm 5 k_B T$ around E_F. It follows that C_{el} is linear in T only to the extent that $N(E)$ can be regarded as a constant within that window. In the free electron model, $N(E)$ varies slowly over energies of the order of $k_B T_F$, which leads to the mathematical expression above. In a real system $N(E)$ may vary significantly over a smaller energy distance $k_B T^*$ from the Fermi level (Fig. 4). Then $C_{el} \sim T$ only as long as $T \ll T^*$.

Modern electron structure calculations can give $N(E)$ with high accuracy. Such information could be used in a straight forward and numerically accurate calculation of C_{el}. But even that result may be misleading. At high temperature, the sharp features of $N(E)$ present in the calculated $N(E)$ are significantly broadened, due to the vibrations of the lattice [16,17]. A calculation of C_{el} then becomes a demanding task.

Our statement about the width of the energy "window" deserves a comment. Usually, one says that the Fermi-Dirac function is constant (1 or 0) except in the approximate energy interval $E_F \pm k_B T$. However, the electronic heat capacity gets its largest contribution from states at energies $E_F \pm 2.4 k_B T$, and gives zero weight to states at $E = E_F$ (due to particle-hole symmetry). $E_F \pm 5 k_B T$ is a better measure of the total range of the energy window probed by

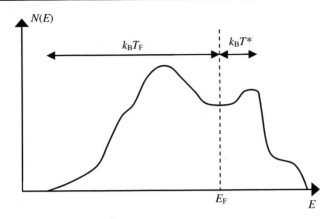

Fig. 4 A schematic picture of the electron density of states $N(E)$. T_F and T^* are two characteristic quantities that describe $N(E)$. See the text for a discussion

C_{el}. This should be compared with the electronic entropy, which is an integral of $C_{el}(T)/T$ (cf. Sect. 3.3) and therefore probes a narrower window, approximately $E_F \pm 3k_B T$.

7 Conclusions

All models involve constraints and approximations. Constraints that have the form of sum rules or normalization conditions can make a model robust to fine structure in the model input. Robustness is also common in properties that depend on the integration of model quantities. For instance, we have seen that an analysis of the entropy may have significant advantages compared to a direct analysis of the heat capacity. A third cause of robustness arises when the simplest model represents an extremum, and more elaborate modeling gives corrections that are small to the second order. Often the existence of an extremum is obvious from, e.g., symmetry arguments, and no detailed treatment of the second order effect is necessary.

A "brute force" numerical fitting to data may be well motivated in some cases, for instance in the CALPHAD approach to phase diagrams, but it does not give insight into the detailed physical origin of model quantities. In particular, although the conventional wisdom that "everything gives a straight line in a log-log plot" contains a certain truth, such a direct fitting of data may lead to completely erroneous physical interpretations, as we have exemplified with the difficulties in separating polynomial terms due to anharmonicity from an exponential term due to vacancy formation, in the heat capacity of solids..

Finally, we have discussed examples where the common text book treatment of a phenomenon is presented in such a way that the corresponding mathematical relation may be taken as literally true, or at least as an extremely good approximations. But those relations may be merely building blocks in a model that deals with a different, or more complex, property. One has to be aware of the fact that although the simple textbook relations can be extremely accurate in a certain limit, they may be quite inaccurate away from that limit. This is particularly important when simulations make it possible to study systems under conditions that are very far from those accessible to experiments or used as illustrations in the textbooks.

Appendix 1

Moment frequencies and thermodynamic functions

Let $F(\omega)$ be the phonon density of states. A complete mathematical representation of $F(\omega)$ is obtained in terms of moment frequencies $\omega(n)$ defined as [5]

$$\omega(n) = \left[\int \omega^n F(\omega) \, d\omega \right]^{1/n} \quad n \neq 0$$

$$\ln [\omega(0)] = \int \ln(\omega) \, F(\omega) \, d\omega$$

Here $F(\omega)$ is normalized as

$$\int F(\omega) \, d\omega = 1$$

The entropy per atom, at high temperature and expressed in moment frequencies $\omega(n)$, is

$$S(T) = 3k_B \{ 1 + \ln [k_B T / \hbar \omega(0)] \\ + (1/24) [\hbar \omega(2) / k_B T]^2 - (1/960) [\hbar \omega(4) / k_B T]^4 + \dots \}$$

This expression converges very rapidly at high T, where only a single parameter, the logarithmically averaged phonon frequency $\omega(0)$, suffices to give a very accurate account of the harmonic entropy S. The corresponding expression for the heat capacity of harmonic vibrations is

$$C(T) = 3k_B \{ 1 - (1/12) [\hbar \omega(2) / k_B T]^2 + (1/240) [\hbar \omega(4) / k_B T]^4 + \dots \}$$

The Gibbs energy $G = U - TS + PV$ has a constant part in U, which does not follow from fundamental physics but must be fixed through a convention, for instance so that $U + PV = 0$ at $T = 298.15 \, \text{K}$ for the elements in their equilibrium state. We can write (for $P = 0$)

$$G(T) = U_0 - 3k_B T \{ \ln [k_B T / \hbar \omega(0)] - (1/24) [\hbar \omega(2) / k_B T]^2 \\ - (1/2880) [\hbar \omega(4) / k_B T]^4 + \dots \}$$

All of the expressions above contain moment frequencies $\omega(n)$. It can be a very good approximation to calculate them from the density of states $F(\omega)$ determined, for instance, in neutron scattering experiments.

Appendix 2.

Sum of power laws in a log-log plot

Assume that a physical phenomenon has two contributions to a quantity Q. Each of them can be exactly described by a power law, as in the expression

$$Q = Q_1 \left(\frac{q}{q_0} \right)^m + Q_2 \left(\frac{q}{q_0} \right)^n$$

The quantities Q_1 and Q_2 have the same physical dimension as Q, with $Q_2 \ll Q_1$. Data for Q are available in a region of q near q_0, and plotted as $\ln Q$ versus $\ln q$. Neglecting terms of order smaller than Q_2/Q_1, we find after some elementary mathematics that

$$\frac{Q(q)}{Q(q_0)} \approx \left(\frac{q}{q_0}\right)^{m+\varepsilon}$$

with $\varepsilon \ll 1$ given by

$$\varepsilon = \left(\frac{Q_2}{Q_1}\right)(n-m)$$

We conclude that when a quantity Q has contributions as above, with exponents m and n, a single power law with an "irregular" approximate exponent $m + \varepsilon$ may give a good description.

Appendix 3.

Power law versus exponential behavior

Suppose that the correct temperature dependence of a certain quantity is

$$Q_{\text{true}} = aT + bT^2$$

but we don't know this and try to fit it to an expression of the form

$$Q_{\text{fit}} = cT + d\exp\left(-E/k_B T\right)$$

A good fit can be obtained if $f(T) = Q_{\text{true}} - cT = aT + bT^2 - cT$ is well represented in a certain temperature interval around $T = T_1$ by $d\exp(-E/k_B T)$, i.e., if $\ln f(T)$ plotted versus $1/T$ is approximately a straight line with slope $-E/k_B$. This is the case if $\partial^2 \ln f/\partial(1/T)^2 = 0$. Then $E = -k_B[\partial \ln f/\partial(1/T)]$, evaluated at $T = T_1$. One always obtains

$$E = \left(2 + \sqrt{2}\right)k_B T_1 \approx 3.4 k_B T_1$$

for any a, b and c. We conclude that it may be very difficult to distinguish between polynomial and exponential representations when one is fitting to data in a narrow interval. If, however, one finds a very good fit with an exponential term, that result might be ruled out on physical grounds, when one considers the numerical values of the fitted parameters [18].

References

1. Hubbert, M.K.: Nuclear energy and the fossil fuels. Publication No. 95, Shell Development Company (1956)
2. Hubbert, M.K.: Technique of prediction as applied to the production of oil & gas. NBS special publication **631**, 16–141. U.S. Department of Commerce/National Bureau of Standards (1982)
3. Bartlett, A.A.: An analysis of U.S. and world oil production patterns using Hubbert-style curves. Math. Geol **32**, 1–17 (2000)
4. Barin, I.: Thermochemical Data of Pure Substances. VCH Verlag (1989)
5. Grimvall, G.: Thermophysical Properties of Materials. Enlarged and Revised Edition. North-Holland (1999)

6. Fernández Guillermet, A.: Critical evaluation of the thermodynamic properties of molybdenum. Int. J. Thermophys. **6**, 367–393 (1985)
7. Forsblom, M., Sandberg, N., Grimvall, G.: Anharmonic effects in the heat capacity of Al. Phys. Rev. B **69**, 165106/1–165106/6 (2004)
8. Peng, S., Grimvall, G.: Heat capacity of actinide dioxides. J. Nucl. Mater. **210**, 115–122 (1982)
9. Grimvall, G., Rosén, J.: Heat capacity of fcc calcium. Int. J. Thermophys. **3**, 251–257 (1982)
10. Peng, S., Grimvall, G.: A method to estimate entropies and heat capacities applied to alkali oxides. Int. J. Thermophys. **15**, 973–981 (1994)
11. Forsblom, M., Grimvall, G.: Heat capacity of liquid Al: molecular dynamics simulations. Phys. Rev. B **72**, 132204/1–132204/4 (2005)
12. Ashcroft, N.W., Mermin, N.D.: Solid State Physics. Holt, Rinehart and Winston, NY (1976)
13. Fernández Guillermet, A., Grimvall, G.: Analysis of thermodynamic properties of molybdenum and tungsten at high temperatures. Phys. Rev. B **44**, 4332–4340 (1991)
14. Wallace, D.C.: Thermodynamics of Crystals. Wiley, London (1972)
15. Kittel, C.: Introduction to Solid State Physics, 8th edn. Wiley, London (2005)
16. Souvatzis, P., Eriksson, O., Katsnelson, M.I., Rudin, S.P.: Entropy driven stabilization of energetically unstable crystal structures explained from first principles theory. Phys. Rev. Lett. **100**, 09590/1–09590/4 (2008)
17. Asker, C., Belenoshko, A.B., Mikhaylushkin, A.S., Abrikosov, I.A.: First-principles solution to the problem of Mo lattice stability. Phys. Rev. B **77**, 220102/1–220102/4 (2008)
18. Grimvall, G.: Separation of anharmonic and vacancy effects in thermophysical properties. High Temp.–High Press **15**, 245–247 (1983)

Multiscale simulations of complex systems: computation meets reality

Efthimios Kaxiras · Sauro Succi

Originally published in the journal Sci Model Simul, Volume 15, Nos 1–3, 59–65.
DOI: 10.1007/s10820-008-9096-y © Springer Science+Business Media B.V. 2008

Abstract   Multiscale simulations are evolving into a powerful tool for exploring the nature of complex physical phenomena. We discuss two representative examples of such phenomena, stress corrosion cracking and ultrafast DNA sequencing during translocation through nanopores, which are relevant to practical applications. Multiscale methods that are able to exploit the potential of massively parallel computer architectures, will offer unique insight into such complex phenomena. This insight can guide the design of novel devices and processes based on a fundamental understanding of the link between atomistic-scale processes and macroscopic behavior.

Keywords   Multiscale simulations · Biomolecules · Corrosion

1 Introduction

Most physical phenomena of interest to humankind involve a large variety of temporal and spatial scales. This generic statement applies to systems as diverse as the brittle fracture of solids under external forces, which can lead to the failure of large structures such as bridges or ships starting at nanometer-scale cracks, to tidal currents in bays extending over many miles, whose behavior is dictated by the water viscosity determined from the molecular-scale interactions of water molecules.

Analytical models are typically formulated to capture the behavior of simple, homogeneous systems, or small deviations from such idealized situations, which are expressed as

E. Kaxiras (✉)
Department of Physics and School of Engineering and Applied Sciences, Harvard University,
Cambridge, MA 02138, USA
e-mail: kaxiras@physics.harvard.edu

S. Succi
Istituto Applicazioni Calcolo, CNR, Viale del Policlinico 137, 00161 Roma, Italy

S. Succi
Initiative in Innovative Computing, Harvard University, Cambridge, MA 02138, USA

linear terms of a variable that describes the departure from uniformity. Such theories cannot cope with situations far from equilibrium, or involving very large deviations away from homogeneity, as is the case for many interesting phenomena in the spirit of the examples mentioned above. By necessity, a realistic description of such complex phenomena must rely on computational models, and more specifically on the use of multiscale approaches. This last requirement is dictated by the need to use the available computational resources in the most effective manner, allocating them judiciously to resolve the finer scales of the physical systems (which are usually the most computationally demanding) only where and when absolutely necessary. We feel that this notion is so central to computational approaches that it deserves the status of a "principle", and we will refer to it as the *Principle of Least Computation* (PLC). In other words, multiscale approaches aim at using the least possible computer power to describe satisfactorily the behavior of a complex physical system. The fact that these approaches often employ the latest and most sophisticated computer technology that is available, is a testament to the complexity of the problem at hand.

In the present article we wish to motivate the need for multiscale approaches by discussing the essential features of two specific examples of complex physical phenomena from very different domains. Through these examples, we hope to demonstrate how multiscale approaches can be formulated to satisfy the PLC, and what their prospects are for answering important scientific questions and addressing specific technological needs. We do not attempt to provide extensive references to the relevant literature, as this is beyond the scope of a perspective article and has already been done elsewhere [1]. We will only point to certain publications which have contributed key ideas on the way toward a comprehensive framework of multiscale modeling.

2 Two representative examples

Our first example has to do with Stress Corrosion Cracking (SCC): in this phenomenon, solids that are normally tough (ductile) become weak (brittle) and crack under the influence of external stresses due to the exposure to a hostile environment in which chemical impurities affect the structure at the microscopic, atomistic scale (corrosion). SCC is relevant to the stability or failure of systems in many practical applications [2]. The length scale at which this phenomenon is observed in everyday-life situations is roughly millimeters and above. At this scale cracks are detectable by naked eye, and if not arrested they can lead to the failure (occasionally catastrophic) of large structures such as ships, airplanes, bridges, etc. Evidently, the presence of chemical impurities changes the nature of bonds between the atoms, which is crucial in the region near the tip of the crack, where bonds are being stressed to the point of breaking. These atomic scale changes are ultimately responsible for the macroscopic scale change in the nature of the solid and can turn a normally ductile material into a brittle one leading to cracking. A breakdown of the different length and scale regimes relevant to this phenomenon is illustrated in Fig. 1.

A portion of a solid of linear dimension 1 mm contains of order 10^{20} atoms. Treating this number of atoms with chemical accuracy is beyond the range of any present or foreseeable computational model. Moreover, such a treatment would be not only extremely computationally demanding but terribly wasteful: most of the atoms are far away from the crack tip, they are not exposed to the effect of chemical impurities, and have little to contribute to the cracking phenomenon, until the crack tip reaches them. Thus, a sensible treatment would describe the vast majority of atoms in the solid by a simple continuum model, such as continuum elasticity theory, which is well suited for the regions far from the crack tip. Closer to the

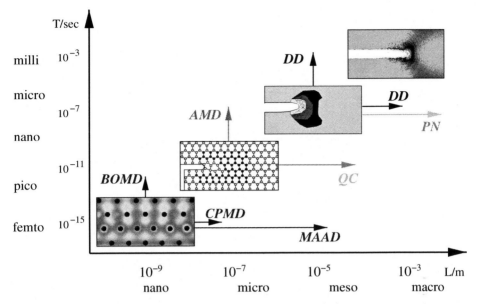

Fig. 1 Illustration of the length and time scales in Stress Corrosion Cracking, with the four different length scales, macro- (a continuum description of the crack in an otherwise elastic solid), meso- (a description of the plastic zone near the crack tip in terms of dislocations), micro- (a resolution of the dislocation structure at the atomistic level with classical interatomic interactions) and nano-scale (with the ions and corresponding valence electrons, treated quantum mechanically). The various acronyms refer to methodologies for extending the reach of conventional approaches, or coupling approaches across several scales (see text for details)

tip, several scales become increasingly important. First, the plastic zone, a scale on the order of μm, contains a large number of dislocation defects the presence of which determines the behavior of the system. The nature and interaction of these defects is adequately described by continuum theories at large scales [3,4]. However, atomistic scale simulations are required to derive effective interactions between these defects when they directly intersect each other, or in any other way come into contact at length scales equal to interatomic separation, when the continuum level of description breaks down. This introduces the need to couple atomic-scale structure and motion to dislocation dynamics [5], a coupling that has already been demonstrated and provided great insight to the origin of crystal plasticity [6]. Finally, in order to include the effects of chemical impurities on the crack tip, where the important processes that determine the brittle or ductile nature of the material take place, it is necessary to turn on a quantum mechanical description of the system, including ions and electrons.

These four levels of description, continuum at the macro-scale (mm), defect dynamics at the meso-scale (μm), atomistic scale dynamics at the micro-scale (100 nm) and quantum-mechanical calculations for the nano-scale (1–10 nm), are shown schematically in Fig. 1, together with the corresponding time scales which range from femto-seconds for the coupled motion of ions and electrons to milli-seconds and beyond for the macroscopic scale. This separation of spatial and temporal scales is only intended to show the conceptual division of the problem into regions that can be successfully handled by well developed computational methodologies. Such methodologies may include Density Functional Theory at the quantum-mechanical level, molecular dynamics with forces derived from effective interatomic potentials at the the scale of 100 nm, dislocation dynamics at the scale of several microns, and continuum elasticity (using, for instance, finite elements to represent a solid of arbitrary shape) in the macroscopic region. The goal of mutliscale simulations is to couple the various regimes, as appropriate for a particular problem, in order to achieve a satisfactory description of this complex phenomenon.

A number of computational schemes have already been developed to address this type of phenomenon. The work of Car and Parrinello [7] from over two decades ago, is a general method for extending the reach of quantum-mechanical calculations to systems larger than a few atoms, and can be viewed as a pioneering attempt to couple the purely quantum-mechanical regime to the classical atomistic. More recently, Parrinello and coworkers [8] have developed a method that extends the reach of this type of approach to significantly longer time-scales. The multiscale-atomistics-*ab initio*-dynamics (MAAD) [9] approach was an attempt to couple seamlessly within the same method three scales, the quantum-mechanical one, the classical atomistic, and the continuum, that enabled the first realistic simulation of brittle fracture in silicon. A direct link between the atomistic scale and the scale of defect (dislocation, domain boundaries, etc.) dynamics was accomplished by the quasi-continuum (QC) method [10], that enabled realistic simulations of large deformation of materials in situations like nano-indentation [11] and polarization switching in piezoelectrics [12]. The methods originally developed by Voter [13] for accelerating molecular dynamics (AMD) can extend the time-scales of classical atomistic approaches by a few orders of magnitude, addressing an important bottleneck in multiscale simulations. Finally, dislocation dynamics (DD) and continuum formulations like the Peierls-Nabarro (PN) theory, can make the connection to truly marcoscopic scales; however, these approaches depend crucially on input from finer scales to determine the values of important parameters, so they cannot by themselves have predictive power unless a connection is made to the finer scales.

We turn next to our second example. This has to do with ultrafast sequencing of DNA through electronic means, a concept that is being vigorously pursued by several experimental groups [14, 15]. The idea is to form a small pore, roughly of the same diameter as the DNA double helix, and detect the DNA sequence of bases by measuring the tunneling current across two electrodes at the edges of the pore during the translocation of the biopolymer from one side of the pore to the other. Estimates of the rate at which this sequencing can be done are of order 10 kbp/s, which translates to sequencing the entire human genome in a time frame of a couple of days. While this goal has not been achieved yet, the successful demonstration of this concept has the potential to produce long-lasting changes in the way medicine is practiced.

This system is also a very complex one, involving several scales that are shown schematically in Fig. 2. At the coarsest scale, the biopolymer in solution needs to be directed toward the nanopore. The scale of the biopolymer in solution is of order μm, and the time scale involved in finding the pore is of order fractions of a second. At these scales, the system can be reasonably modeled as consisting of a continuous polymeric chain in a uniform solvent, possibly under the influence of an external non-uniform but continuous field that drives it toward the pore. In experiments, this can be achieved either by a concentration gradient on the two sides of the pore, or by an electric field acting on a polymer with uniform charge distribution. Understanding the translocation process itself involves modeling the polymer at scales set by the persistence length ($\sim$100 nm for DNA), while including the effect of the solvent. This is already a demanding proposition, because the number of molecules that constitute the solvent is macroscopic (10^{24}) and in fact it makes no sense to monitor their dynamics at an individual basis, since it is only the cumulative effect that is felt by the polymer.

We have recently succeeded in producing an efficient coupling between the solvent motion, described by the Lattice Boltzmann approach to model fluid dynamics [16], and the polymer motion, described by beads representing motion at the scale of the persistence length and

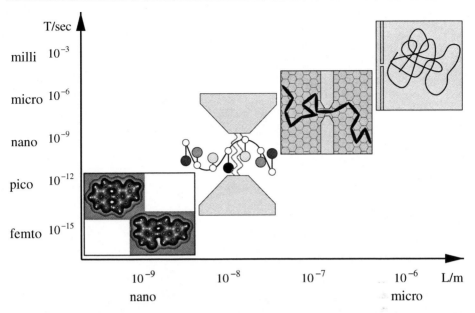

Fig. 2 Illustration of the length and time scales in ultrafast DNA sequencing by electronic signals during translocation through a nanopore. At the micro-scale, the biopolymer is a continuous line moving in a homogeneous solvent under external fields, at the next level (100 nm) the biopolymer is resolved at the persistence length and the fluid is modeled by a transport theory on a grid (the Lattice Boltzmann Equation), at the 10 nm scale the biopolymer structure is resolved at the level of individual bases (color coded) and electron current across the leads can be modeled by quantum transport equations, and at the nm scale the individual bases are modeled explicitly in terms of their atomic and electronic structure

following newtonian dynamics under the influence of local friction from the fluid [17]. This lattice-Boltzmann plus molecular-dynamics (LBMD) description was very successful in reproducing details of the observed translocation process in a realistic manner, as observed in experiments for both single-file and multi-file translocation [18]. Nevertheless, the approach is still far from capturing the full problem of electronic DNA sequencing. What is missing is the information at the 10 nm scale, with the biopolymer described as a sequence of individual DNA bases, the electronic properties of which will determine the electron current across the two leads on either side of the pore. The electron current can be calculated from quantum-transport theories, once the relevant electronic states are known. To this end, detailed calculations of the electronic properties of DNA bases need to be carried out, at the level of quantum chemistry, including all the details of the base-pair atomistic structure. Assembling the entire picture in a coherent model is a serious challenge to current computational capabilities and is likely to require profound conceptual and mathematical innovations in the way we handle the phase-space of complex systems [19,20].

3 Problems and prospects

The cases described above are representative of complex systems which by their nature demand multiscale approaches. There are many subtle issues on how multiscale approaches are formulated, both at the practical level, with each approach requiring great attention to computational details, as well as at the fundamental level, where questions of basic notions arise when degrees of freedom are eliminated in the process of coarsening [19], but these are well beyond the scope of the present article. Here, we wish to address only two general points. The first is the bottleneck of time-scale integration; the second is the bottleneck of efficient allocation of computational resources.

As the two examples mentioned earlier make clear, it is usually feasible to produce multiscale methods that successfully integrate several spatial scales, spanning many orders of magnitude in length. However, it is much more difficult to integrate temporal scales. Most often, the time-step of the simulation is dictated by the shortest time scale present. Even if ingenious approaches can be devised to accelerate the time evolution, they can produce a speedup of a couple orders of magnitude at best, which is rather restricted compared to the vast difference in time scales for different parts of the process (see Figs. 1, 2). At present, we are not aware of coupling schemes that can overcome this bottleneck in a satisfactory way, integrating over many orders of the time variable [21].

Second, we suggest that it may be necessary to rethink traditional algorithmic approaches in view of recent developments in computer architectures. Specifically, taking advantage of massively parallel architectures or streamlining in graphics-processing units, can produce very significant gains, at the cost of writing codes that are specifically suited for these architectures. We have found that in our LBMD scheme, speedups of order 10^4 in computation time could be achieved by using a multi-processor architecture, the IBM BluGene/L, because when properly formulated, the problem scales linearly with the number of available processors with excellent parallel efficiency. The mapping of the physical system to the computer architecture is not trivial and required specialist's skills. We feel that this aspect of the computational implementation should not be overlooked in trying to construct useful multiscale models.

As closing remarks, we express our conviction that multi-scale approaches are essential in faithfully capturing the nature of physical reality in a wide range of interesting phenomena. We are hopeful that carefully planned and executed computational efforts will indeed come very close to an accurate and useful representation of complex phenomena. This challenge has not been met yet, in other words, computation is still shy of full success in meeting physical reality. But the gap is closing fast.

Acknowledgements We are indebted to our coworkers involved in the work mentioned here, and especially to Simone Melchionna, Maria Fyta, Massimo Bernaschi and Jayantha Sircar, for their contributions to the LBMD project.

References

1. Lu, G., Kaxiras, E.: Overview of multiscale simulations of materials. In: Rieth, M., Schommers, W. (eds.) Handbook of Theoretical and Computational Nanothechnology, vol. X, pp. 1–33. American Scientific Publishers (2005)
2. Winzer, N., Artens, A., Song, G., Ghali, E., Dietzel, W., Ulrich, K., Hortand, N., Blawert, C.: A critical review of the stress corrosion cracking (SCC) of Magnesium alloys. Adv. Eng. Mater. **7**, 659 (2005)
3. Hirth, J.P., Lothe, J.: Theory of Dislocations. Krieger Publshing Co., Malabar, Florida (1992)
4. Peierls, R.: The size of a dislocation. Proc. Phys. Soc. London **52**, 34–37 (1940); Nabarro, F.R.N.: Dislocations in a simple cubic lattice. Proc. Phys. Soc. London **59**, 256–272 (1947)
5. Devincre, B., Kubin, L.: Mesoscopic simulations of dislocations and plasticity. Mater. Sci. Eng. A **234–236**, 8–14 (1994)
6. Bulatov, V., Abraham, F.F., Kubin, L., Yip, S.: Connecting atomistic and mesoscopic simulations of crystal plasticity. Nature **391**, 669–681 (1998)

7. Car, R., Parrinello, M.: Unified approach for molecular-dynamics and density-functional theory. Phys. Rev. Lett. **55**, 2471–2474 (1985)
8. Kuhne, T.D., Krack, M., Mohamed, F.R., Parrinello, M.: Efficient and accurate Car-Parrinello-like approach to Born-Oppenheimer molecular dynamics. Phys. Rev. Lett. **98**, 066401 (2007)
9. Abraham, F., Broughton, J., Bernstein, N., Kaxiras, E.: Spanning the length scales in dynamic simulation. Comput. Phys. B. **12**, 538–546 (1998); Broughton, J., Abraham, F., Bernstein, N., Kaxiras, E.: Concurrent coupling of length scales: methodology and application. Phys. Rev. B. **60**, 2391–2403 (1999)
10. Tadmor, E.B., Ortiz, M., Phillips, R.: Quasicontinuum analysis of defects in solids. Philos. Mag. A. **73**, 1529–1563 (1996)
11. Smith, G.S., Tadmor, E.B., Bernstein, N., Kaxiras, E.: Multiscale simulation of silicon nanoindentation. Acta Mater. **49**, 4089 (2001)
12. Tadmor, E.B., Waghmare, U.V., Smith, G.S., Kaxiras, E.: Polarization switching in PbTiO3: an ab initio finite element simulation. Acta Mater. **50**, 2989 (2002)
13. Voter, A.F.: A method for accelerating the molecular dynamics simulation of infrequent events. J. Chem. Phys. **106**, 4665–4677 (1997); Voter, A.F.: Hyperdynamics: accelerated molecular dynamics of infrequent events. Phys. Rev. Lett. **78**, 3908–3911 (1997)
14. Kasianowicz, J.J., Brandin, E., Branton, D., Deamer, D.: Characterization of individual polynucleotide molecules using a membrane channel. Proc. Natl. Acad. Sci. USA **93**, 13770–13773 (1996); Meller, A., Nivon, L., Brandin, E., Golovchenko, J., Branton, D.: Rapid nanopore discrimination between single polynucleotide molecules. Proc. Natl. Acad. Sci. **97**, 1079–1084 (2000); Li, J., Gershow, M., Stein, D., Brandin, E., Golovchenko, J.A.: DNA molecules and configurations in a solid-state nanopore microscope. Nat. Mater. **2**, 611–615 (2003); Storm, A.J., Storm, C., Chen, J., Zandbergen, H., Joanny, J.-F., Dekker, C.: Fast DNA translocation through a solid-state nanopore. Nanoletter **5**, 1193–1197 (2005)
15. Dekker, C.: Solid state nanopores. Nat. Nanotech. **2**, 209–215 (2007)
16. Succi, S.: The Lattice Boltzmann Equation for Fluid Dynamics and Beyond. Clarendon Press, Oxford (2001); Benzi, R., Succi, S., Vergassola, M.: The lattice Boltzmann equation: theory and applications. Phys. Rep. **222**, 145–197 (1992)
17. Fyta, M.G., Melchionna, S., Kaxiras, E., Succi, S.: Multiscale coupling of molecular dynamics and hydrodynamics: application to DNA translocation through a nanopore. Multiscale Model. Sim. **5**, 1156–1173 (2006); Fyta, M.G., Melchionna, S., Kaxiras, E., Succi, S.: Exploring DNA translocation through a nanopore via a multiscale Lattice-Boltzmann molecular-dynamics methodology. Int. J. Mod. Phys. C. **18**, 685–692 (2007); Fyta, M.G., Kaxiras, E., Melchionna, S., Succi, S.: Multiscale modeling of biological flows. Comput. Sci. Eng. **20** (2008) to appear
18. Bernaschi, M., Melchionna, S., Succi, S., Fyta, M., Kaxiras, E.: Quantized current blockade and hydrodynamic correlations in biopolymer translocation through nanopores: evidence from multiscale simulations. Nano Lett. **8**, 1115–1119 (2008)
19. Öttinger, H.C.: Stochastic Processes in Polymeric Fluids: Tools and Examples for Developing Simulation Algorithms. Springer, Berlin (1996); Beyond Equilibrium Thermodynamics. Wiley, New York (2005)
20. Praprotnik, M., Kremer, K., Delle Site, L.: Adaptive molecular resolution via a continuous change of the phase space dimensionality. Phys. Rev. E. **75**, 017701 (2007)
21. E,W., Engquist, B., Li, X.T.: Heterogeneous multiscale methods: a review. Commun. Comp. Phys. **2**, 367–450 (2007); and in Kevrekidis, I.G., Gear, C.W., Hummer, G.: Equation-free: the computer-aided analysis of complex multiscale systems. AiChe J. **50**, 1346–1355 (2004)

Chemomechanics of complex materials: challenges and opportunities in predictive kinetic timescales

Krystyn J. Van Vliet

Originally published in the journal Sci Model Simul, Volume 15, Nos 1–3, 67–80.
DOI: 10.1007/s10820-008-9111-3 © Springer Science+Business Media B.V. 2008

Abstract   What do nanoscopic biomolecular complexes between the cells that line our blood vessels have in common with the microscopic silicate glass fiber optics that line our communication highways, or with the macroscopic steel rails that line our bridges? To be sure, these are diverse materials which have been developed and studied for years by distinct experimental and computational research communities. However, the macroscopic functional properties of each of these structurally complex materials pivots on a strong yet poorly understood interplay between applied mechanical states and local chemical reaction kinetics. As is the case for many multiscale material phenomena, this chemomechanical coupling can be abstracted through computational modeling and simulation to identify key unit processes of mechanically altered chemical reactions. In the modeling community, challenges in predicting the kinetics of such structurally complex materials are often attributed to the so-called rough energy landscape, though rigorous connection between this simple picture and observable properties is possible for only the simplest of structures and transition states. By recognizing the common effects of mechanical force on rare atomistic events ranging from molecular unbinding to hydrolytic atomic bond rupture, we can develop perspectives and tools to address the challenges of predicting macroscopic kinetic consequences in complex materials characterized by rough energy landscapes. Here, we discuss the effects of mechanical force on chemical reactivity for specific complex materials of interest, and indicate how such validated computational analysis can enable predictive design of complex materials in reactive environments.

Keywords   Chemomechanics · Rough energy landscape · Computational materials science · Kinetic barriers

K. J. Van Vliet (✉)
Department of Materials Science and Engineering, Massachusetts Institute of Technology, Cambridge, MA 02139, USA
e-mail: krystyn@MIT.EDU

1 Introduction

Although computational modeling and simulation of material deformation was initiated with the study of structurally simple materials and inert environments, there is an increasing demand for predictive simulation of more realistic material structure and physical conditions. In particular, it is recognized that applied mechanical force can plausibly alter chemical reactions inside materials or at material interfaces, though the fundamental reasons for this chemomechanical coupling are studied in a material-specific manner. Atomistic-level simulations can provide insight into the unit processes that facilitate kinetic reactions within complex materials, but the typical nanosecond timescales of such simulations are in contrast to the second-scale to hour-scale timescales of experimentally accessible or technologically relevant timescales. Further, in complex materials these key unit processes are "rare events" due to the high energy barriers associated with those processes. Examples of such rare events include unbinding between two proteins that tether biological cells to extracellular materials [1], unfolding of complex polymers, stiffness and bond breaking in amorphous glass fibers and gels [2], and diffusive hops of point defects within crystalline alloys [3].

Why should we consider ways for computational modeling to bridge this gap between microscopic rare events and macroscopic reality? The answer lies chiefly in the power of computational modeling to abstract general physical concepts that transcend compositional or microstructural details: accurate incorporation of mechanically altered rare events can help to predict the macroscopic kinetics that govern phenomena as diverse as creep in metal alloys, hydrolytic fracture of glass nanofibers, and pharmacological drug binding to cell surface receptors (Fig. 1). Modeling of this chemomechanical coupling is especially important and challenging in materials of limited long-range order and/or significant entropic contributions to the overall system energy. Here, we explore the concepts of rare events and rough energy landscapes common to several such materials, and show how mechanical environment defined by material stiffness and applied force can alter kinetic processes in the most complex of these systems: solvated biomolecules. Finally, we point to the potential of such computational analyses to address other complex materials for which breakthroughs in materials design will offer significant technological and societal impact.

2 Putting rare events and rough energy landscapes in context of real materials

2.1 Rare events and rough energy landscapes

Energy landscapes or energy surfaces are graphical representations of the energy of a system as a function of reaction coordinates, $E(\chi)$. These reaction coordinates can represent the physical distance between atomic or molecular coordinates of a material (e.g., distance between atomic nuclei), but more generally represents any relevant order parameter in the material phase space. The energetic basins or minima represent thermodynamically favored configurations, separated by many intermediate states during transitions between these minima. The utility of such free energy landscapes in predictions of material dynamics is that these multidimensional surfaces convey the pathways between local and global energetic minima. The energy basins associated with these minima define the thermodynamics of the system, and the connectivity among these basins defines the chemical reaction kinetics. There exist infinite reaction paths between any two states A and B, but the path of highest activation barrier traverses the saddle point on this $E(\chi)$ surface and is called the transition state. This reaction proceeds at a rate r:

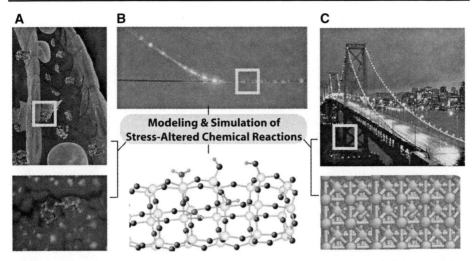

Fig. 1 Chemomechanics of complex materials requires abstraction of molecular-scale reactions under defined mechanical states. This level of abstraction (boxes) is indicated for several disparate macroscale applications, which are all well described as molecular reactions that are altered by the stiffness and/or applied stress state of the material interface. (**a**) In blood vessels, stress altered kinetics of reversible binding of soluble molecules (ligand, red) to cell surface bound molecules (receptor, blue) can be simulated with steered molecular dynamics; (**b**) In fiber optics, kinetics of hydrolytic cleavage (orange) in amorphous silicate glasses (gray and red) can be identified via nudged elastic band methods; (**c**) In structural steel alloys, migration barriers governing diffusion kinetics of carbon (red) in iron crystals (gray) under creep stress can be predicted via ab initio methods

$$r = k\,[\text{A}]\,[\text{B}] \tag{1}$$

where k is the rate constant of the transition from A to B:

$$k = \nu \ \exp\left(-E_b/RT\right) \tag{2}$$

and E_b is the minimum activation barrier the system must overcome to transition from A to B, and ν is the attempt frequency (e.g., atomic collision frequency). The existence of local and global minima of defined connectivity in this phase space demonstrates why certain systems may be kinetically trapped in local minima for hours, whereas others are able to achieve states predicted by thermodynamic or global minima within seconds. Rare events or transitions are those with relatively high activation barriers, which consequently occur at exponentially lower rates. Physical examples of rare events include nucleation and migration of many-atom defects in crystals or amorphous solids, and large conformational changes in solvated proteins [4–7].

The challenge in actually using the concept of energy landscapes to predict such material transitions is that the free energy landscape of real materials is typically not known *a priori*. Figure 1a illustrates a very simple one-dimensional energy landscape between two states or configurations, showing that the transition to a configuration of lower energy (local minimum) requires sufficient input energy to overcome a single transition barrier. Although the energy differences between configurations A and B can be calculated directly through various methods (e.g., ab initio approximations of the Schrodinger equation describing material configurations at the level of electronic structure [8–10]), it is computationally costly to use such approaches to map out an entire landscape for a material comprising hundreds to thousands of atoms. Thus, simplified descriptions of one-dimensional paths between conformations of interest (Fig. 2a) are often constructed using empirical force fields, identifying

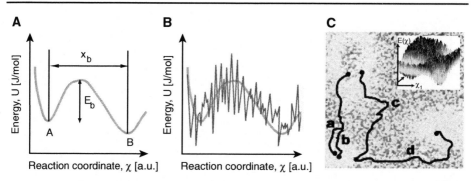

Fig. 2 Transition states and rough energy landscapes. (**a**) Kinetics for a material reaction described by a simple one-dimensional energy landscape with well-separated transition states A and B can be described via transition state theory, with energy barriers E_b and energetic distances x_b that define reaction rates k; (**b**) Structurally complex materials often exhibit so-called rough energy landscapes, with access to many energetically competitive minima between states A and B; (**c**) More realistically, these landscapes are three-dimensional (inset), such that two similar configurations/energy states can diverge to attain different final states (compare *a* and *b*) and two dissimilar configurations/energy states can converge to attain similar final states (compare *c* and *d*) Adapted from Ref. [50]

minimum energy paths between the two states of interest via nudged elastic band (NEB)-type approaches [11, 12], or even by drawing from experimental measurements of reaction times that set relative barrier heights according to Eqs. (1) and (2) [13–15]. Energy landscapes described as "rough" exhibit many local minima of comparable barrier heights (Fig. 2b), such that neighboring conformations of lower barriers are rapidly sampled whereas neighboring conformations of higher barriers are rarely accessed.

Often, the concept of energy landscapes is not invoked to directly predict properties from quantitatively accurate $E(\chi)$ for complex materials, but to understand that systems with multiple energetic minima will exhibit multiple timescales of relaxation to the local and global energetic minima. For example, thermal fluctuations in conformations within glasses, colloids, polymers, and functional states of proteins have long been conceptualized as sampling of neighboring conformations in a high dimensional energy landscape [16]. Thus, a decrease in temperature forces the material system to sample conformations of lower energy barriers; this trapping within many local minima effectively freezes out sampling of larger conformational changes of higher barriers. The most physically relevant and experimentally accessible transitions in complex materials, however, are typically these high-barrier or "rare events" such as crack nucleation and protein unfolding. As Fig. 2c illustrates qualitatively, the roughness of the three-dimensional energy landscape causes the transition states and reaction rates that are sampled to depend quite strongly on the initial configuration of the system: small differences in the initial trajectory can lead to large differences in the minima that are sampled, and likewise two very different paths in phase space can require the same activation energy. For complex materials, then, the computational challenge is to adequately sample the ensemble of initial configurations and the frequency of rare events, in order to predict mechanisms of chemical reactions or macroscopic kinetic consequences of those reactions. Next, we will consider the increasing complexity of rough energy landscapes in highly alloyed crystals, glasses, and solvated proteins.

2.2 Diving in to rough energy landscapes of alloys, glasses, and biomolecules

Elemental metallic crystals such as body-centered cubic iron exhibit long range translational and rotational order, and as such have well-defined energy minima corresponding to the underlying atomic lattice. The energy landscape of such Fe crystals is thus relatively smooth, notwithstanding the magnetic spin state of the atoms, and the activation barriers for nucleation of vacancies within the lattice can be determined directly via density functional theory (DFT) [3,17–21]. The diffusion of single vacancies within the lattice must overcome an activation barrier, but in this case the energetic barrier of the diffusive unit process—atomic hopping to adjacent lattice sites—is defined by transition pathways that are essentially limited to < 100 > and < 111 > crystallographic directions. Such barriers can then be calculated readily through NEB approaches that identify a minimum energy path consisting of intermediate configurations between state A (vacancy in one defined lattice site) and state B (vacancy in adjacent lattice site), where energies of these intermediate states can be calculated from ab initio methods such as DFT [3,22,23].

Now contrast this case of simple bcc Fe with the simplest approximation of high-carbon ferritic steel. As ubiquitous as this alloy is in failure-critical industrial applications, the additional chemical complexity created by a thermodynamic supersaturation of Fe self-vacancies and interstitial carbon immediately impedes such straightforward calculations of self-diffusivity. Ab initio formation energies of point defect clusters indicate that vacancies will be predominantly sequestered as point defect clusters such as divacancy-carbon clusters [3]. There exist many possible paths and unit processes of self-diffusion for even this triatomic defect cluster: a single vacancy could dissociate from the cluster along specific directions, the divacancy pair could dissociate from the carbon, etc. The minimum energy paths and associated activation barriers for each possible final state could still be computed via NEB, albeit at considerable computational expense for realistic carbon concentrations of up to 1 wt%C in body-centered cubic iron to form hardened structural steels.

Next, consider a glass such as amorphous silica. Silica is a naturally occurring material, comprising silicate tetrahedra (SiO_4) in the crystalline form of quartz; it is the most abundant material in the Earth's crust and the key component of fiberoptic and dielectric-thin—film communication platforms. It is well known that water reduces the strength of silica through several mechanisms, including hydrolytic weakening of quartz due to interstitial water [24–26] and stress corrosion cracking of amorphous silica due to surface water [17,19–21]. This interaction is representative of a broader class of chemomechanical degradation of material strength, including stress corrosion and hydrogen embrittlement in metals, and enzymatic biomolecular reactions. Owing to the large system size required to recapitulate such amorphous structures, atomistic reactions and deformation of such materials are typically studied via molecular dynamics (MD) simulations based on approximate empirical potentials. To study the unit process of this degradation in this glass, here the process of a single water molecule attacking the siloxane bonds of amorphous silica [25], the lack of long-range order confers an immediate challenge, as compared to crystalline analogues: where will this failure occur? That is, simulations of hydrolytic fracture must identify which of the thousands of distinct silica bond configurations, which are strained yet thermodynamically stable just prior to fracture, will be most susceptible to failure and thus should be simulated in detail. In direct simulation, either via molecular dynamics or quasi-static deformation by energy minimization, no information is available *a priori* to determine how far from stability the system is located and what failure mechanism will be activated. The system must be driven beyond the instability point and then allowed to explore the energy landscape in order to find a lower energy failed configuration. In the amorphous case, there are numerous instability points and a matching number of failed configurations. Therefore, the failure behavior will be highly dependent on the time available to explore the energy landscape (in molecular dynamics) or on the number of search directions attempted during minimization (in NEB identification of minimum energy paths). Here, efficient searching of these strained regions of interest, or failure kernels, can be identified by recourse to computational modeling and simulation methods we have developed for crystalline materials: namely, the identification of an unstable localized vibration mode in the material [27]. In the context of dislocation nucleation in metallic crystals, we have previously described this unstable mode as the λ-criterion [28–30] which can be computed from eigenvectors and stresses calculable for each atom (for MD simulations) or material region under affine strain (for continuum simulations); we located the failure kernel of homogeneous dislocation nucleation at crystallographic locations where $\lambda_{min} = 0$. This unstable mode identification thus isolates independent regions of the many-atom amorphous material that are most strained and thus most susceptible to bond failure. For silicate glass nanofibers under applied stress (Fig. 1b), we have found that the size of this failure kernel depends on the cutoff chosen for this vibrational amplitude but can be as small as 200 atoms within the very complex amorphous glass; this system size discourages use of ab initio methods to study the failure processes in detail, but makes efficient use of classical and reactive empirical potentials that are currently in development for silicate glasses, water, and organic molecules.

Proteins and other biomacromolecules can also be idealized as complex materials. The chemical reactions among such proteins and protein subunits define fundamental processes such as protein folding (the prediction of three-dimensional protein structures from knowledge of lower-order structural details) and reversible binding of drug molecules to cell surface-bound molecules. From a computational modeling and simulation perspective that recapitulates *in vitro* or *in vivo* environments, accurate prediction of mechanically altered chemical reaction kinetics between biomolecules must consider several challenges. For molecular dynamics (MD) or steered MD (SMD) simulations, these challenges include protein solvation in aqueous media, definitions of solution pH via protonation of specific amino acid residues comprising the protein, and the many configurations that will be accessed due to enhanced entropic contributions of long organic molecules. The structure-function paradigm of biology states that small changes in protein structure can confer ample changes in function of that protein. Although the structure of many proteins has been experimentally determined and shared on the Protein Data Bank (PDB) public database, it is important to note that these "solved structures" are ensemble or time-averaged snapshots of the many configurations a protein will access even when exploring a global energetic basin at body temperature. Recent experiments on photoreceptor proteins have demonstrated that very small-scale variations in configurations of a protein structure (root-mean-square deviation $<< 0.1$ nm) can have large functional consequences in the activation response of such proteins to light [31]. The effects of such small configurational changes in the initial simulated structure are typically neglected in modern simulation of protein dynamics, and instead the community has focused on simulations of increasing duration to access rare events such as unfolding over timescales approaching 100 ms or to access forced unbinding at increasingly realistic (slow) velocities < 10 m/s [32–34].

These increased timescales of observation are an important and reasonable goal for predictive protein simulations. However, in light of the rough or high-dimensional energy landscape of proteins (see Fig. 2c), the strong dependence of energetic trajectories on the initial configuration, and thus on the inferred reaction kinetics, warrants additional consideration. For example, the equilibrium dissociation constant K_D is a key parameter that represents the ratio between the unbinding rate or off-rate k_{off} and binding rate or on-rate k_{on} of a ligand-receptor complex. The off-rate k_{off} is expressed in units of $[s^{-1}]$, and is the inverse of the lifetime of the complex τ. In particular, integrated computational and experimental studies of forced unbinding of complementary macromolecules (e.g., ligand-receptor complexes such as Fig. 1a) can reveal key energetic and kinetic details that control the lifetime of the bound complex. The reversible binding of such complexes is a rare event that can be altered by two distinct mechanical cues: applied mechanical forces that may result from fluid shear flow or other far-field stresses, or the mechanical stiffness of materials to which the molecules are tethered.

In the following section, we will consider the effects of both initial configuration and effective mechanical stiffness of molecules to which ligands are tethered, as these relate to the inferred energetic and kinetic properties of a particularly strong and useful biomolecular complex, biotin-streptavidin. Steered molecular dynamics simulations of this forced unbinding show why the mechanical stiffness effect was noted as a possibility but not calculated explicitly in the pioneering work of Evans. These simulations also show that tether stiffness and initial configuration should be carefully considered both in the interpretation of biophysics experiments and in the design of ligand-functionalized drug delivery materials.

3 Forced unbinding of biomolecular complexes

Three decades ago, Bell established that the force required to rupture the adhesion between two biological cell surfaces F_R should increase with the rate of loading $dF/dt = F'$:

$$F_R = k_B T/x_b \ln \left(\left[F' x_b \right] / \left[k_B T k_{off} \right] \right) \qquad (3)$$

where k_B is Boltzmann's constant, T is absolute temperature, x_b is the energetic distance between the bound and unbound states in units of χ, and is sometimes termed the unbinding width. This model has since been applied to the rupture between individual molecules such as adhesive ligands to cell-surface receptors, and Evans and Ritchie have attributed this rate-dependent rupture force of the molecular complex as a tilt in the one-dimensional energy landscape [35,36]. They and many others, including our group, have demonstrated that the dynamic strength of this bond can be measured experimentally by several methods that quantify the rupture force at defined loading rates $F_R \left(F' \right)$ [36–39].

Although such experiments are far from equilibrium, in that the dissociation of the molecular pair is forced by the applied loading rate, Eq. (3) shows that the equilibrium bond parameters k_{off} and x_b should be obtainable via extrapolation and slope of acquired F_R vs. ln F', respectively. Here, it is presumed that the loading rates are sufficiently slow that at least some details of the rough energy landscape $E(\chi)$ are accessed. One of the most commonly employed experimental methods used to obtain these so-called dynamic force spectra is molecular force spectroscopy or dynamic force spectroscopy, an atomic force microscopy (AFM)-based approach that measures the force-displacement response between a ligand tethered to an AFM cantilevered probe and its receptor presented at an opposing surface (such as a rigid, flat mica surface or a compliant living cell surface) [40]. In such experiments, the loading rate F' is expected to alter the observed rupture force according to Eq. (3), and could be defined as the product of the velocity of the ligand v and the stiffness of the force transducer to which the ligand is attached k_s, or $F' = k_s v$.

3.1 Does stiffness matter? Why k_s perturbs the accessible molecular rupture forces

Upon compiling experimental results from several groups acquired over several years for a single molecular complex, biotin-streptavidin, we noticed that Bell's prediction was qualitatively accurate but that various researchers reported significant differences in F_R at the same ostensible loading rate F'. Biotin-streptavidin is a well studied biomolecular complex because, under equilibrium conditions and in the absence of applied load, the binding lifetime

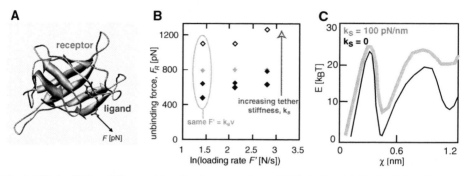

Fig. 3 Effects of tether stiffness on biomolecular rupture force. (**a**) Biotin ligand is displaced from the strep-tavidin receptor pocket via a transducer of stiffness ks moving at constant velocity v; (**b**) the observed rupture forces depend on loading rate as expected from Bell's model, but also on k_S; (**c**) This stiffness-dependence can be attributed to direct alteration of the energy landscape of the complex, even before load is applied to the biotin ligand. Adapted from Ref. [1]

of the biotin ligand to the streptavidin receptor is long (as compared to other ligand-receptor and antibody-antigen complexes. Thus, this complex has become a tool to bioengineers who use the complex to effectively "glue" other molecules together with nearly covalent bond strength, and to biophysists who use the complex as a benchmark for long binding lifetimes τ.

However, when we observed that experimentalists using the same AFM approach and biotin-streptavidin preparations measured the same rupture forces for 100-fold differences in loading rate, or 150 pN differences in rupture force for the same applied loading rate [1], we developed a matrix of steered molecular dynamics simulations to identify possible sources of this discrepancy. These simulations are described fully in [1], in which we acquired the structure of this complex from the PDB, solvated the protein in simple-point charge (SPC) water with ions to mimic charge neutrality of physiological buffers, and then selected a single configuration of this complex from the so-called "equilibration trajectory" of an unconstrained MD simulation of 100 ns duration. This configuration was not strictly equilibrated in terms of representing the global minimum of the protein structure, of course; instead, it was chosen at a timepoint in the trajectory which we deemed to be sufficient to enter an energetic minimum in the admittedly rough energy landscape (20 ns), according to our protocol for objectively choosing an initial configuration of proteins as we outlined in Ref. [41]. We then conducted SMD simulations (Fig. 3a), applying a simulated loading rate via a Hookean spring connected to the ligand, and achieved the same magnitude of $F' = k_s v$ for pairs of low k_s/ high v, and high k_s/ low v(Fig. 3b).

We were initially surprised that we obtained very different rupture forces F_R for pairs of simulations conducted for different spring constants and velocities but the same resulting F', differing as much as 60 pN for F' on the order of 1000 pN/s; this difference was well in excess of our standard deviation of replicate simulations for the same configuration under the same loading conditions, differing only in the initial velocities of atoms within the complex. We also confirmed this finding with AFM force spectroscopy experiments on biotin-streptavidin, at admittedly lower loading rates, and observed the same result: the velocity of two AFM cantilevers of differing stiffness k_s could not be altered to attain the same rupture force at a given loading rate defined as $F' = k_s v$. However, more careful consideration of our findings showed a clear and independent effect of this tether stiffness k_s on the observed rupture force. We conceptualized this effect as a direct perturbation of the energy landscape, superposing an additional energetic term $E = k_s \chi^2$ to the existing potential of the complex. Importantly, this energy landscape is perturbed *even before any external force is applied*, and effectively changes the accessible energy states and trajectories accessible to the complex.

Although Evans had previously noted the possibility of stiffness directly altering the nature of the energy landscape [42], he had not included this term explicitly in his model of molecular unbinding kinetics without any loss of accuracy in the interpretation of his own experiments. Why did this tether stiffness contribute so strongly to our simulations and AFM force spectroscopy experiments, but so little to Evans et al.'s experiments on the same complex? Figure 3c shows that the magnitude of this stiffness effect naturally scales directly with the stiffness of the force transducer stiffness, as superposed on a one-dimensional energy landscape that has been based on several measurements and simulations of the bound states of this particular complex. In Evans et al.'s biomembrane force probe-based experiments on this complex, the stiffness of the biomembranes was so low ($k_s < 10\,pN/nm$) that the energy landscape was not noticeably perturbed. In contrast, in our AFM-based experiments with cantilevers of stiffness $k_s \sim 100\,pN/nm$, there is noticeable alteration of the initial landscape, such that the barriers to dissociation of the complex are increased with increasing ks. Naturally, SMD-based simulations of the complex employ springs of even larger stiffness ($k_s \sim 1000\,pN/nm$) in order to force the dissociation in accessible simulation timescales, and perturb the initial landscape of the complex to an even greater extent.

Note that the actual rupture force of the complex is unchanged by this tether stiffness, but rather the observed rupture force F_{obs} increases with increasing tether stiffness. For biophysical applications, the actual rupture force of the complex F_R can thus be computed from this summed effect:

$$F_{obs} = F_R - k_s x_b/2 = k_B T/x_b \ln \left(\left[F' x_b \right] / [k_B T k_{off}] \right) \tag{4}$$

However, for practical and biological applications, this direct contribution of tether stiffness to the applied force required to dissociate the complex presents the following implication. The force required to dissociate a ligand from its target (such as a cell surface receptor), and thus the binding lifetime of that complex, will depend directly on the stiffness of the material to which the ligand is tethered (e.g., the conjugating biomolecules between the drug ligand and a solid support) and, by extension, to the stiffness of the material to which the receptor is tethered (e.g., the mechanical compliance of the cell surface region that presents the receptor protein).

3.2 Enough is enough: practical requirements of rare event sampling in MD

As noted, MD simulations have been utilized to model protein behavior and function for over 30 years, in fact as long as Bell's model has been known [43,44]. The timescale limitation of such atomistic simulations has been addressed by the development of Steered Molecular Dynamics (SMD), a variation on classical MD in which rare events such as ligand-receptor unbinding are induced by mechanical force. One key goal of simulations that faithfully recapitulate experiments is to consider replicate simulations that explore the variation as well as the magnitude of simulated parameters and extracted properties of individual and complexed molecules. However, here we must ask for the sake of tractability, when have we simulated "enough" of the ensemble, given that we will never full access the entire ensemble via direct simulation?

MD studies typically utilize a single, long-timescale trajectory to draw conclusions about biomolecular properties such as complex stabilization, binding pathways, and ligand-receptor unbinding [45–47]. However, proteins *in vitro* and *in vivo* show considerable variation in measured properties; protein properties including rupture forces and photoactivity are typically distributions and cannot be adequately characterized by a single measurement, but can depend critically on small configurational differences that are accessible for a given protein structure [31]. Therefore, it is reasonable that simulations should aim to sample the property *distribution* if they are to be representative of natural or experimental systems, especially if the goal is to compare simulation results at two different conditions. Encouragingly, a few simulation-focused researchers have begun to report MD simulated results for multiple initial configurations [48,49]. For more in the community to consider this initially daunting approach, however, it is very helpful to know how many initial configurations are necessary to capture sufficient variation without incurring excessive computational expense.

We have considered this sufficient number for the biotin-streptavidin complex, seeking to identify whether there were configurational subsets that gave rise to higher or lower rupture forces under identical applied loading conditions [50]. We first created configurational subsets of the solvated complexes that differed by less than some cutoff root-mean-square deviation among atoms (e.g., RMSD < 10 Å), and then applied the same loading rate details to each configurational member of the cluster; to the centroid configuration within each cluster (of median RMSD); and to one of the cluster centroids while varying only the initial velocity of the atoms as set by the simulated temperature and a random number generator. As Fig. 4 shows, the variation in rupture force observed for these three different cases was quite similar, demonstrating that at least for this particular case, configuration is not a strong determinant of observed rupture force. More generally, though, we created randomized subsets of size N = 4 to N = 40, to identify the sufficient number of simulations required to reproduce this distribution of observed rupture forces. In retrospect, statistical analysis of epidemiology, economic forecasts, and even polymer rheology would have predicted what we observed: in order to capture the full distribution of rupture forces and correctly identify the mean value, we needed to simulate at least 30 configurations. In fact, a statistical rule of thumb is that approximately 30 samples are needed to adequately characterize the mean of an underlying distribution, because for this sampling number the sampling distribution approaches a normal distribution [51]. Thus, we can conclude that, in the absence of a correlation between configuration and observed molecular property, at least 30 replicate simulations will be necessary to ensure adequate characterization of the observed property. We find that the relative error in our simulation output F_R is equal to the relative error of the calculated final parameter k_{off}, and so 30 replicate simulations at each loading rate would be appropriate. If this number of initial configurations is intractable due to system size, however, the confidence interval of the simulated predictions could be calculated and reported. Consideration of such qualifications by the atomistic simulation community, particularly for predictive modeling of amorphous and biomolecular systems, should help to inform our interpretation of brute-force simulated dynamics.

4 Potential advances for chemomechanical analysis of other complex materials

Several other complex materials can benefit from computational analyses that establish links between chemical kinetics and mechanics. Consider the example of cement, a material so inexpensive and commonly used (1.7 billion tons/year, with 5 billion tons/year estimated by 2050 [52]) that few would initially consider its study as high-impact, and fewer still would even know the composition and structure of this ubiquitous material. Despite its status as the most widely used material on Earth, cement remains an enigma in the fundamental science of materials and a key international target of environmentally sustainable development. Cement is the solid composite material that forms at room temperature from mixing a grey

Fig. 4 Configurational variations may not be predictive of property variations. (**a**) Rupture force distributions are quite similar when obtained via steered molecular dynamics of configurationally distinct groups (clusters) of biotin-streptavidin; a single cluster for which every member was simulated; or a single configuration for which initial atomic velocities were randomly varied. This indicates an absence of configuration-property correlation for this particular complex and property, so 30 randomly selected configurations will capture the actual distribution of this property. Adapted from Ref. [49]

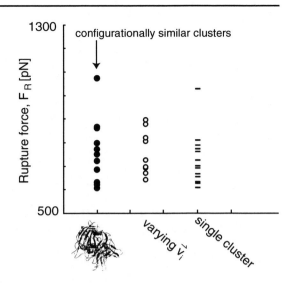

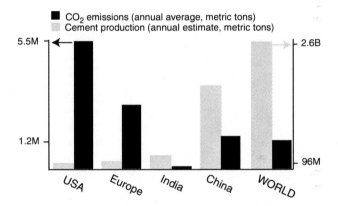

Fig. 5 National production of cement and CO_2 in metric tons are inversely related as of 2007, but are projected to increase rapidly in industrializing nations such as India and China. Predictive modeling and simulation of cement chemomechanics can contribute to new compositions and/or processing temperatures and pressures that lower the CO_2 production associated with cement production. Data source for cement production and carbon dioxide: Cement Industry Environmental Impact Assessment (2005) and US Geological Survey (2007), respectively

Portland cement powder with water, is the most widely used material on Earth at $1\,m^3$/person/year, typically in the form of concrete comprising cement and stone aggregates. Each ton of cement produced results in an emission of 750 to 850 kg of carbon dioxide, amounting to a per capita emission of some 250 kg of CO_2 per cubic meter of concrete consumed. As shown in Fig. 5, cement production for infrastructural applications is currently most prevalent in industrializing nations such as India and China.

The composition and structure of Calcium-Silica-Hydrate (C-S-H), the basic molecular unit of cementitious materials, continues to defy quantitative characterization. Cement is a nanocomposite material, and is called "liquid stone" because it gels by a poorly understood, exothermic and confined aqueous reaction between calcium/silicate powders and water. As the processing of the currently used cement compositions requires reduction of calcium carbonate ($CaCo_3$, limestone) to CaO prior to reaction with water, every produced ton of cement contributes approximately 1 ton of CO_2 to the atmosphere, representing $> 5\%$ of the global CO_2 emission on earth and an estimated "environmental burden cost" of $100 ton [52].

Experimental efforts to alter cement chemistry and processing without altering the kinetic rates of gelation and final stiffness and strength of cementitious composites are ongoing, but have been largely iterative: the structure of the C-S-H constituents and the details of the gelation over hour-scale gelation reactions are unknown. Importantly, the mechanical properties of the C-S-H nanoparticles have recently been identified through nanoscale indentation experiments and differ in packing density [53–55]. Here, computational modeling and simulation can offer valuable insight into structural predictions and dynamics of hydrated C-S-H layers and nanoparticles that impact curing times in defined ambient environments, susceptibility to hydrolytic fracture, and ultimate compressive strength of such complex materials.

What would the benefit of such fundamental modeling and simulation efforts in such an established material be? The chief advances would focus on reduced energy required to create CaO, achieved through either new cement compositions requiring less Ca or more likely through new processing routes to synthesize or reduce CaO. If materials research efforts could predict cementitious materials that reduced CaO concentrations (e.g., by identifying alternative divalent ions) and/or processing temperature/time (while maintaining critical gelation times and compressive strength), the CO_2 savings could be sufficient for the world to meet Kyoto Protocol targets. Alternatively, if cement compositions could be modified to include noxious gas-getting nanoparticles such as TiO_2, again with predictive alteration of the kinetics of gelation and the compressive strength of the fully cured gel, the very surface of the buildings rising rapidly in India and China could be used to meliorate air quality in congested cities. These are not blue-sky promises, but the foreseeable result of computational studies focused on complex nanocomposites critical to physical infrastructure. The above considerations of stress-altered reactions in other complex materials such as alloys, glasses, and biomolecules will be important examples and case studies to guide accurate benchmarking of such models.

5 Summary and outlook

Here we have discussed the challenge of predictive modeling and simulation for materials of increasing structural complexity. Although the materials are dissimilar in both structure and application, the key dobstacles to informative computational studies of stress-altered reaction kinetics are strikingly similar. Prediction of transition states and associated kinetics in materials with high-dimensional or rough energy landscapes require careful consideration of sufficient system size, identification of strained regions of interest, effects of conjugate variables such as force transducer stiffness, and sufficient sampling of initial configurations. From alloys to glasses to biomolecules, these requirements of predictive modeling and simulation serve to connect the rich and largely unexplored terrain of chemomechanically coupled behavior in and between materials. Just as these concepts are interesting within the material application of interest, they are also inspirational for new materials and material processes that can potentially improve material performance while reducing associated environmental burdens.

References

1. Walton, E.B., Lee, S., Van Vliet, K.J.: Extending Bell's model: how force transducer stiffness alters measured unbinding forces and kinetics of molecular complexes. Biophys. J. **94**, 2621–2630 (2007)
2. Silva, E.C., Tong, L., Yip, S., Van Vliet, K.J.: Size effects on the stiffness of silica nanowires. Small **2**(2), 239–243 (2006)
3. Forst, C.J., Slycke, J., Van Vliet, K.J., Yip, S.: Point defect concentrations in metastable Fe–C alloys. Phys. Rev. Lett. **96**(17) (2006)
4. Andrews, B.T., Schoenfish, A.R., Roy, M., Waldo, G., Jennings, P.A.: The rough energy landscape of superfolder GFP is linked to the chromophore. J. Mol. Biol. **373**(2), 476–490 (2007)
5. Gruebele, M.: Rough energy landscape of folded and unfolded proteins. Abstr. Pap. Am. Chem. Soc. **227**, U267–U267 (2004)
6. Wang, J., Huang, W.M., Lu, H.Y., Wang, E.K.: Downhill kinetics of biomolecular interface binding: Globally connected scenario. Biophys. J. **87**(4), 2187–2194 (2004)
7. Onuchic, J.N., Wolynes, P.G.: Energy landscapes, glass transitions, and chemical-reaction dynamics in biomolecular or solvent environment. J. Chem. Phys. **98**(3), 2218–2224 (1993)
8. Simka, H., Willis, B.G., Lengyel, I., Jensen, K.F.: chemistry predictions of reaction processes in organometallic vapor phase epitaxy. Progress in Crystal Growth and Characterization of Materials **35**(2–4), 117–149 (1997)
9. Wesolowski, T., Muller, R.P., Warshel, A.: Ab initio frozen density functional calculations of proton transfer reactions in solution. J. Phys. Chem. **100**(38), 15444–15449 (1996)
10. Kim, K., Jordan, K.D.: Comparison of density-functional and Mp2 calculations on the water monomer and dimer. J. Phys. Chem. **98**(40), 10089–10094 (1994)
11. Henkelman, G., Uberuaga, B.P., Jonsson, H.: A climbing image nudged elastic band method for finding saddle points and minimum energy paths. J. Chem. Phys. **113**(22), 9901–9904 (2000)
12. Henkelman, G., Jonsson, H.: Improved tangent estimate in the nudged elastic band method for finding minimum energy paths and saddle points. J. Chem. Phys. **113**(22), 9978–9985 (2000)
13. Sorensen, M.R., Brandbyge, M., Jacobsen, K.W.: Mechanical deformation of atomic-scale metallic contacts: Structure and mechanisms. Phys. Rev. B **57**(6), 3283–3294 (1998)
14. Pincet, F., Husson, J.: The solution to the streptavidin-biotin paradox: the influence of history on the strength of single molecular bonds. Biophys. J. **89**, 4374–4381 (2005)
15. Merkel, R., Nassoy, P., Leung, A., Ritchie, K., Evans, E.: Energy landscapes of receptor–ligand bonds explored with dynamic force spectroscopy. Nature **397**, 50–53 (1999)
16. Coureux, P.-D., Fan, X., Sojanoff, V., Genick, U.K.: Picometer-scale conformational heterogeneity separates functional from nonfunctional states of a photoreceptor protein. Structure **16**, 863–872 (2008)
17. Kellou, A., Grosdidier, T., Aourag, H.: An ab initio study of the effects and stability of vacancies, antisites and small radius atoms (B, C, N, and O) in the B2–FeAl structure. In: Fisher, D.J. (ed.) Defects and Diffusion in Metals—An Annual Retrospective VII, vol. 233–234, pp. 87–95. Defect and Diffusion Forum (2004)
18. Lau, T.T., Foerst, C.J., Lin ,X., Gale, J.D., Yip, S., Van Vliet, K.J.: Many-body potential for point defect clusters in Fe–C alloys. Phys. Rev. Lett. **98**(21):215501 (2007)
19. Legris, A.: Recent advances in point defect studies driven by density functional theory. In: Fisher, D.J. (ed.) Defects and Diffusion in Metals—An Annual Retrospective VII, vol. 233–234, pp. 77–86. Defect and Diffusion Forum (2004)
20. Mizuno, T., Asato, M., Hoshino, T., Kawakami, K.: First-principles calculations for vacancy formation energies in Ni and Fe: non-local effect beyond the LSDA and magnetism. J. Magn. Magn. Mater. **226**, 386–387 (2001)
21. Simonetti, S., Pronsato, M.E., Brizuela, G., Juan, A.: The C–C pair in the vicinity of a bcc Fe bulk vacancy: electronic structure and bonding. Phys. Status Solidi B Basic Solid State Phys. **244**(2), 610–618 (2007)
22. Becquart, C.S., Souidi, A., Domain, C., Hou, M., Malerba, L., Stoller, R.E.: Effect of displacement cascade structure and defect mobility on the growth of point defect clusters under irradiation. J. Nucl. Mater. **351**(1–3), 39–46 (2006)
23. Domain, C.: Ab initio modelling of defect properties with substitutional and interstitials elements in steels and Zr alloys. J. Nucl. Mater. **351**(1–3), 1–19 (2006)
24. Cao, C., He, Y., Torras, J., Deumens, E., Trickey, S.B., Cheng, H.P.: Fracture, water dissociation, and proton conduction in SiO2 nanochains. J. Chem. Phys. **126**(21):211101 (2007)
25. Zhu, T., Li, J., Yip, S., Bartlett, R.J., Trickey, S.B., De Leeuw, N.H.: Deformation and fracture of a SiO2 nanorod. Mol. Simul. **29**(10–11), 671–676 (2003)
26. McConnell, J.D.C.: Ab initio studies on water related species in quartz and their role in weakening under stress. Phase Transit. **61**(1–4), 19–39 (1997)

27. Silva, E., Van Vliet, K.J., Yip, S.: Effects of water on chemomechanical instabilities in amorphous silica: nanoscale experiments and molecular simulation. PhD Thesis, Silva, MIT, unpublished results (2007)
28. Van Vliet, K.J., Li, J., Zhu, T., Yip, S., Suresh, S.: Quantifying the early stages of plasticity through nanoscale experiments and simulations. Phys. Rev. B **67**(10)(2003)
29. Li, J., Zhu, T., Yip, S., Van Vliet, K.J., Suresh, S.: Elastic criterion for dislocation nucleation. Mater. Sci. Eng. A Struct. Mater. Prop. Microstruct. Process. **365**(1–2), 25–30 (2004)
30. Li, J., Van Vliet, K.J., Zhu, T., Yip, S., Suresh, S.: Atomistic mechanisms governing elastic limit and incipient plasticity in crystals. Nature **418**(6895), 307–310 (2002)
31. Coureux P.-D., Fan, Z., Stojanoff, V., and Genick,U.K.: Picometer-scale conformational heterogeneity separates functional and nonfunctional states of a photoreceptor protein. Structure **6**:863–872 (2008)
32. Puklin-Faucher, E., Gao, M., Schulten, K., Vogel, V.: How the headpiece hinge angle is opened: new insights into the dynamics of integrin activation. J. Cell Biol. **175**(2), 349–360 (2006)
33. Wriggers, W., Mehler, E., Pitici, F., Weinstein, H., Schulten, K.: Structure and dynamics of calmodulin in solution. Biophys. J. **74**(4), 1622–1639 (1998)
34. Sheng, Q., Schulten, K., Pidgeon, C.: Molecular-dynamics simulation of immobilized artificial membranes. J. Phys. Chem. **99**(27), 11018–11027 (1995)
35. Evans, E., Ritchie, K.: Dynamic strength of molecular adhesion bonds. Biophys. J. **72**(4), 1541–1555 (1997)
36. Merkel, R., Nassoy, P., Leung, A., Ritchie, K., Evans, E.: Energy landscapes of receptor–ligand bonds explored with dynamic force spectroscopy. Nature **397**(6714), 50–53 (1999)
37. Lee, S., Mandic, J., Van Vliet, K.J.: Chemomechanical mapping of ligand–receptor binding kinetics on cells. Proc. Natl. Acad. Sci. USA **104**(23), 9609–9614 (2007)
38. Hinterdorfer, P., Baumgartner, W., Gruber, H.J., Schilcher, K., Schindler, H.: Detection and localization of individual antibody-antigen recognition events by atomic force microscopy. Proc. Natl. Acad. Sci. USA **93**(8), 3477–3481 (1996)
39. Stroh, C.M., Ebner, A., Geretschlager, M., Freudenthaler, G., Kienberger, F., Kamruzzahan, A.S.M., Smith-Gil, S.J., Gruber, H.J., Hinterdorfer, P.: Simultaneous topography and recognition Imaging using force microscopy. Biophys. J. **87**(3), 1981–1990 (2004)
40. Van Vliet, K.J., Hinterdorfer, P.: Probing drug–cell interactions. Nano Today **1**(3), 18–25 (2006)
41. Walton, E.B., Van Vliet, K.J.: Equilibration of experimentally determined protein structures for molecular dynamics simulation. Phys. Rev. E **74**(6):061901 (2006)
42. Evans, E.: Looking inside molecular bonds at biological interfaces with dynamic force spectroscopy. Biophys. Chem. **82**, 83–87 (1999)
43. McCammon, J.A., Gelin, B.R., Karplus, M.: Dynamics of folded proteins. Nature **267**, 585–590 (1977)
44. McCammon, J.A., Karplus, M.: Simulation of protein dynamics. Annu. Rev. Phys. Chem. **31**:29–45 (1980)
45. Craig, D., Gao, M., Schulten, K., Vogel, V.: Structural insights into how the MIDAS ion stabilizes integrin binding to an RGD peptide under force. Structure **12**(11), 2049–2058 (2004)
46. Isralewitz, B., Izrailev, S., Schulten, K.: Binding pathway of retinal to bacterio-opsin: a prediction by molecular dynamics simulations. Biophys. J. **73**(6), 2972–2979 (1997)
47. Kosztin, D., Izrailev, S., Schulten, K.: Unbinding of retinoic acid from its receptor studied by steered molecular dynamics. Biophys. J. **76**(1), 188–197 (1999)
48. Curcio, R., Caflisch, A., Paci, E.: Change of the unbinding mechanism upon a mutation: a molecular dynamics study of an antibody–hapten complex. Protein Sci. **14**(10), 2499–2514 (2005)
49. Paci, E., Caflisch, A., Pluckthun, A., Karplus, M.: Forces and energetics of hapten-antibody dissociation: a biased molecular dynamics simulation study. J. Mol. Biol. **314**(3), 589–605 (2001)
50. Krishnan, R., Walton, E.B., Van Vliet, K.J.: In review (2008)
51. Mendenhall, W., Beaver, R.J., Beaver, B.M.: Introduction to Probability and Statistics. 12th ed., Thomson Higher Education, Belmont, CA (2006)
52. Chaturvedi, S., Ocxhendorf, J.: Global environmental impacts due to concrete and steel. Struct. Eng. Int. **14**(3), 198–200 (2004)
53. Ulm, F.J., Constantinides, G., Heukamp, F.H.: Is concrete a poromechanics material?—a multiscale investigation of poroelastic properties. Mater. Struct. **37**(265), 43–58 (2004)
54. Jennings, H.M., Thomas, J.J., Gevrenov, J.S., Constantinides, G., Ulm, F.J.: A multi-technique investigation of the nanoporosity of cement paste. Cem. Concr. Res. **37**(3), 329–336 (2007)
55. Constantinides, G., Ulm, F.J.: The effect of two types of C-S-H on the elasticity of cement-based materials: Results from nanoindentation and micromechanical modeling. Cem. Concr. Res. **34**(1), 67–80 (2004)

Tight-binding Hamiltonian from first-principles calculations

Cai-Zhuang Wang · Wen-Cai Lu · Yong-Xin Yao ·
Ju Li · Sidney Yip · Kai-Ming Ho

Originally published in the journal Sci Model Simul, Volume 15, Nos 1–3, 81–95.
DOI: 10.1007/s10820-008-9108-y © Springer Science+Business Media B.V. 2008

1 Introduction

The tight-binding method attempts to represent the electronic structure of condensed matter using a minimal atomic-orbital like basis set. To compute tight-binding overlap and Hamiltonian matrices directly from first-principles calculations is a subject of continuous interest. Usually, first-principles calculations are done using a large basis set or long-ranged basis set (e.g. muffin-tin orbitals (MTOs)) in order to get convergent results, while tight-binding overlap and Hamiltonian matrices are based on a short-ranged minimal basis representation. In this regard, a transformation that can carry the electronic Hamiltonian matrix from a large or long-ranged basis representation onto a short-ranged minimal basis representation is necessary to obtain an accurate tight-binding Hamiltonian from first principles.

The idea of calculating tight-binding matrix elements directly from a first-principles method was proposed by Andersen and Jepsen in 1984 [1]. They developed a scheme which transforms the electronic band structures of a crystal calculated using a long-ranged

C.-Z. Wang (✉) · Y.-X. Yao · K.-M. Ho
US Department of Energy, Ames Laboratory, Ames, IA 50011, USA
e-mail: wangcz@ameslab.gov

C.-Z. Wang · Y.-X. Yao · K.-M. Ho
Department of Physics and Astronomy, Iowa State University, Ames, IA 50011, USA

W.-C. Lu
State Key Laboratory of Theoretical and Computational Chemistry, Jilin University, Changchun 130021,
People's Republic of China

J. Li
Department of Materials Science and Engineering, University of Pennsylvania, Philadelphia,
PA 19104, USA

S. Yip
Department of Nuclear Science and Engineering, Massachusetts Institute of Technology, Cambridge,
MA 02139, USA

S. Yip
Department of Materials Science and Engineering, Massachusetts Institute of Technology, Cambridge,
MA 02139, USA

basis set of muffin-tin orbitals (MTO's) into a much shorter-ranged tight-binding representation. In the framework of this transformation, tight-binding matrix elements can be calculated by first-principles LMTO method and casted into an effective two-center tight-binding Hamiltonian called LMTO-TB Hamiltonian [1]. More recently, an improved version of such a "downfolding" LMTO method, namely the order-N MTO [2], has also been developed which allows the LMTO-TB Hamiltonian matrix elements to be extracted more accurately from a full LMTO calculation [3].

Another approach to determine the tight-binding Hamiltonian matrix elements by first-principles calculations was developed by Sankey and Niklewski [4] and by Porezag et al. [5]. In their approach, the matrix elements are calculated directly by applying an effective one-electron Hamiltonian of the Kohn-Sham type onto a set of pre-constructed atomic-like orbitals. The accuracy of the tight-binding Hamiltonian constructed in this way depends on the choice of atomic-like basis orbitals. More recently, McMahan and Klepeis [6] have developed a method to calculate the two-center Slater-Koster hopping parameters and effective on-site energies from minimal basis functions optimized for each crystal structure, in terms of $\mathbf{k}$-dependent matrix elements of one-electron Hamiltonian obtained from first-principles calculations.

All of the above mentioned work was derived from a description of electronic structures using a fixed minimal basis set, except the work of McMahan and Klepeis [6]. It should be noted that while a fixed minimal basis set can give a qualitative description of electronic structures, it is too sparse to give an accurate description of the energetics of systems in varying bonding environments. A much larger basis set would be required in the first-principles calculations in order to get accurate and convergent results if the basis set is going to be kept fixed for various structures. Thus, it is clear that in order for a minimal basis set to have good transferability, it is important to focus our attention on the changes that the basis must adopt in different bonding environments.

In the past several years, we have developed a method for projecting a set of chemically deformed atomic minimal basis set orbitals from accurate *ab initio* wave functions [7–12]. We call such orbitals "quasi-atomic minimal-basis orbitals" (QUAMBOs) because they are dependent on the bonding environments but deviate very little from free-atom minimal-basis orbitals. While highly localized on atoms and exhibiting shapes close to orbitals of the isolated atoms, the QUAMBOs span exactly the same occupied subspace as the wavefunctions determined by the first-principles calculations with a large basis set. The tight-binding overlap and Hamiltonian matrices in the QUAMBO representation give exactly the same energy levels and wavefunctions of the occupied electronic states as those obtained by the fully converged first-principles calculations using a large basis set. Therefore, the tight-binding Hamiltonian matrix elements derived directly from *ab initio* calculations through the construction of QUAMBOs are highly accurate.

In this article, we will review the concept and the formalism used in generating the QUAMBOs from first-principles wavefunctions. Then we show that tight-binding Hamiltonian and overlap matrix elements can be calculated accurately by the first-principles methods through the QUAMBO representation. By further decomposing the matrix elements into the hopping and overlap parameters through the Slater-Koster scheme [13], the transferability of the commonly used two-center approximation in the tight-binding parameterization can be examined in detail. Such an analysis will provide very useful insights and guidance for the development of accurate and transferable tight-binding models. Finally, we will also discuss a scheme for large scale electronic structure calculation of complex systems using the QUAMBO-based first-principles tight-binding method.

2 Quasi-atomic minimal-basis-sets orbitals

The method to project the QUAMBOs from the first-principles wave functions has been described in detail in our previous publications [7–12]. Some of the essential features of the method will be reviewed here using Si as an example. If the Si crystal structure contains N silicon atoms and hence $4N$ valence electrons in a unit cell, the total number of minimal sp^3 basis orbitals per unit cell will be $4N$. In our method, the $4N$ QUAMBOs (A_α) are spanned by $2N$ occupied valence orbitals which are chosen to be the same as those from the first-principles calculations, and by another $2N$ unoccupied orbitals which are linear combinations of a much larger number of unoccupied orbitals from first-principles calculations. The condition for picking such $2N$ unoccupied orbitals is the requirement that the resulting QUAMBOs deviate as little as possible from the corresponding $3s$ and $3p$ orbitals of a free Si atom (A_α^0). The key step in constructing the above mentioned QUAMBOs is the selection of a small subset of unoccupied orbitals, from the entire virtual space, that are maximally overlapped with the atomic orbitals of the free atom A_α^0.

Suppose that a set of occupied Bloch orbitals $\phi_\mu(\mathbf{k},\mathbf{r})$ ($\mu = 1, 2, \ldots, n_{occ}(\mathbf{k})$) and *virtual* orbitals $\phi_v(\mathbf{k},\mathbf{r})$ ($v = n_{occ}(\mathbf{k}) + 1, n_{occ}(\mathbf{k}) + 2, \ldots, n_{occ}(\mathbf{k}) + n_{vir}(\mathbf{k})$), labeled by band μ or v, and wave vector $\mathbf{k}$, have been obtained from first-principles calculations using a large basis set, our objective is to construct a set of quasi-atomic orbitals $A_\alpha(\mathbf{r} - \mathbf{R}_i)$ spanned by the occupied Bloch orbitals $\phi_\mu(\mathbf{k},\mathbf{r})$ and an optimal subset of orthogonal virtual Bloch orbitals $\varphi_p(\mathbf{k},\mathbf{r})$

$$A_\alpha(\mathbf{r} - \mathbf{R}_i) = \sum_{\mathbf{k},u} a_{\mu\alpha}(\mathbf{k},\mathbf{R}_i)\phi_\mu(\mathbf{k},\mathbf{r}) + \sum_{\mathbf{k},p} b_{p\alpha}(\mathbf{k},\mathbf{R}_i)\varphi_p(\mathbf{k},\mathbf{r}) \tag{1}$$

where

$$\varphi_p(\mathbf{k},\mathbf{r}) = \sum_v T_{vp}(\mathbf{k})\phi_v(\mathbf{k},\mathbf{r}), \ (p = 1, 2, \ldots, n_p(\mathbf{k}) < n_{vir}(\mathbf{k})) \tag{2}$$

The orthogonal character of $\varphi_p(\mathbf{k},\mathbf{r})$ gives $\sum_v T_{vp}^*(\mathbf{k})T_{vq}(\mathbf{k}) = \delta_{pq}$, in which T is a rectangular matrix which will be determined later.

The requirement is that A_α should be as close as possible to the corresponding free atom orbitals A_α^0. Mathematically, this is a problem of minimizing $\langle A_\alpha - A_\alpha^0 | A_\alpha - A_\alpha^0 \rangle$ under the side condition $\langle A_\alpha | A_\alpha \rangle = 1$. Therefore the Lagrangian for this minimization problem is

$$L = \langle A_\alpha - A_\alpha^0 | A_\alpha - A_\alpha^0 \rangle - \lambda \left(\langle A_\alpha | A_\alpha \rangle - 1 \right) \tag{3}$$

The Lagrangian minimization leads to

$$A_\alpha(\mathbf{r} - \mathbf{R}_i) = D_{i\alpha}^{-1/2} \left[\sum_{\mathbf{k},\mu} \langle \phi_\mu(\mathbf{k},\mathbf{r}) \mid A_\alpha^0(\mathbf{r} - \mathbf{R}_i) \rangle \phi_\mu(\mathbf{k},\mathbf{r}) \right.$$

$$\left. + \sum_{\mathbf{k},p} \langle \varphi_p(\mathbf{k},\mathbf{r}) \mid A_\alpha^0(\mathbf{r} - \mathbf{R}_i) \rangle \varphi_p(\mathbf{k},\mathbf{r}) \right] \tag{4}$$

where

$$D_{i\alpha} = \sum_{\mathbf{k},\mu} \left| \langle \phi_\mu(\mathbf{k},\mathbf{r}) \mid A_\alpha^0(\mathbf{r} - \mathbf{R}_i) \rangle \right|^2 + \sum_{\mathbf{k},p} \left| \langle \varphi_p(\mathbf{k},\mathbf{r}) \mid A_\alpha^0(\mathbf{r} - \mathbf{R}_i) \rangle \right|^2 \tag{5}$$

For this optimized A_α, the mean-square deviation from A_α^0 is

$$\Delta_{i\alpha} = \langle A_\alpha - A_\alpha^0 | A_\alpha - A_\alpha^0 \rangle^{1/2} = [2(1 - D_{i\alpha}^{1/2})]^{1/2} \tag{6}$$

It is clear from Eqs. (5) and (6) that the key step to get quasi-atomic minimal-basis-set orbitals is to select a subset of virtual orbitals $\varphi_p(\mathbf{k},\mathbf{r})$ which can maximize the matrix trace

$$S = \sum_{i,\alpha,\mathbf{k},p} \langle \varphi_p(\mathbf{k},\mathbf{r}) | A_\alpha^0(\mathbf{r} - \mathbf{R}_i) \rangle \langle A_\alpha^0(\mathbf{r} - \mathbf{R}_i) | \varphi_p(\mathbf{k},\mathbf{r}) \rangle \tag{7}$$

The maximization can be achieved by first diagonalizing the matrix

$$B_{\nu\nu'}^{\mathbf{k}} = \sum_{i,\alpha} \langle \phi_\nu(\mathbf{k},\mathbf{r}) \, | \, A_\alpha^0(\mathbf{r} - \mathbf{R}_i) \rangle \langle A_\alpha^0(\mathbf{r} - \mathbf{R}_i) \, | \, \phi_{\nu'}(\mathbf{k},\mathbf{r}) \rangle \tag{8}$$

for each $\mathbf{k}$-point, where ν and ν' run over all unoccupied states up to a converged upper cutoff. The transformation matrix T which defines the optimal subset of virtual Bloch orbitals $\varphi_p(\mathbf{k},\mathbf{r})$ ($p = 1, 2, \ldots, n_p(\mathbf{k})$) by Eq. (2) is then constructed using the $\sum_k n_p(\mathbf{k})$ eigenvectors *with the largest eigenvalues* of the matrixes $B^{\mathbf{k}}$, each of such eigenvectors will be a column of the transformation matrix T. Given $\varphi_p(\mathbf{k},\mathbf{r})$, the localized QUAMBOs are then constructed by Eqs. (4) and (5). As one can see from the above formalism development that the key concept in this QUAMBO construction is to keep the bonding states (occupied state) intact and at the same time searching for the minimal number of anti-bonding states (which are usually not the lowest unoccupied states) from the entire unoccupied subspace. The bonding states that kept unchanged and the anti-bonding states constructed from the unoccupied states can form the desirable localized QUAMBOs.

Figure 1 shows the s- and p- like QUAMBOs of Si in diamond structure with different bond lengths of 1.95 Å, 2.35 Å and 2.75 Å, and in fcc structure with bond lengths of 2.34 Å, 2.74 Å, and 3.14 Å, respectively. The QUAMBOs are in general non-orthogonal by our construction as discussed above. One can see that the QUAMBOs constructed by our scheme are indeed atomic-like and well localized on the atoms. These QUAMBOs are different from the atomic orbitals of the free atoms because they are deformed according to the bonding environment. It is clear that the deformations of QUAMBOs are larger with shorter interaction distances. When the bond length increases to be 2.75 Å, the QUAMBOs are very close to the orbitals of a free atom.

As we discussed above, the effective one-electron Hamiltonian matrix in the QUAMBO representation by our construction preserves the occupied valence subspace from the first-principles calculations so that it should give the exact energy levels and wavefunctions for the occupied states as those from first-principles calculations. This property can be seen from Fig. 2 where the electronic density-of-states (DOS) of Si in the diamond structure calculated using QUAMBOs are compared with that from the original first-principles calculations. It is clearly shown that the electronic states below the energy gap are exactly reproduced by the

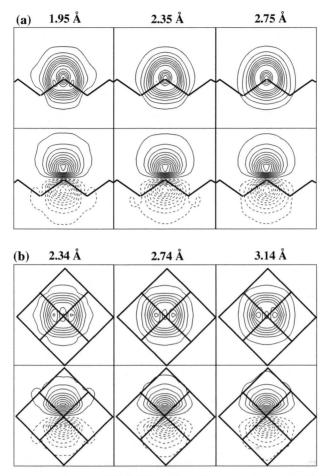

Fig. 1 Non-orthogonal s- and p- like QUAMBOs in Si (a) diamond structure in the (110) plane for three bond lengths 1.95 Å, 2.35 Å and 2.75 Å, and (b) fcc structure in the (100) plane for three bond lengths 2.34 Å, 2.74 Å, and 3.14 Å

QUAMBOs, while the unoccupied states have been shifted upwards so that the energy gap between the valence and conduction states increases from $\sim$0.7 eV to $\sim$1.8 eV. This shift is expected because the QUAMBOs contain admixtures of eigenstates from the higher energy spectrum.

It should be noted that the formalism for the QUAMBOs construction discussed in the section is based on the wavefunctions from first-principles calculations using all-electrons or norm-conserving pseudopotentials [14]. The formalism for constructing the QUAMBOs from first-principles calculations using ultra-soft pseudopotential (USPP) [15] or projector augmented-wave (PAW) [16], as implemented in the widely used VASP code [17,18], is similar and has been recently worked out by Qian et al. [12]. Moreover, Qian et al. also adopt a projected atomic orbital scheme [19–21] which replaces the unoccupied subspace from the first-principles calculations in the above formula with a projection of the unoccupied part of the atomic orbitals, and improve the efficiency and stability of the QUAMBO construction procedure [12].

Fig. 2 Electronic density of
states of diamond Si obtained by
using the QUAMBOs as basis
set, compared with those from the
corresponding LDA calculations
using the PW basis set

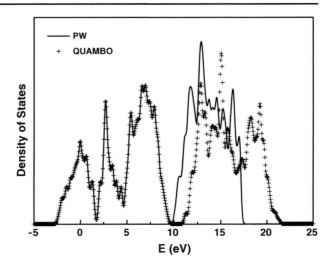

3 Tight-binding matrix elements in terms of QUAMBOs

Once the QUAMBOs have been constructed, overlap and effective one-electron Hamiltonian
matrices in representation of QUAMBOs are readily calculated from first-principles.

$$S_{i\alpha,j\beta} = \langle A_\alpha(\mathbf{r} - \mathbf{R}_i) | A_\beta(\mathbf{r} - \mathbf{R}_j) \rangle \tag{9}$$

$$H_{i\alpha,j\beta} = \langle A_\alpha(\mathbf{r} - \mathbf{R}_i) | H | A_\beta(\mathbf{r} - \mathbf{R}_j) \rangle \tag{10}$$

H in Eq. 10 can then be expressed by using the corresponding eigenvalues ε_n and eigenfunc-
tions ϕ_n from original DFT calculations, i.e., $H = \sum_n \varepsilon_n |\phi_n\rangle \langle\phi_n|$, and thus the matrix
elements $H_{i,\alpha,j\beta}$ can be calculated easily.

Note that in our approach the electronic eigenvalues and eigenfunctions of the occupied
states from first-principles calculations are exactly reproduced by the QUAMBO representation. Although the overlap and effective one-electron Hamiltonian matrices in terms of the
QUAMBOs are in a minimal basis representation, the matrices obtained from our method
go beyond the traditional two-center approximation. Therefore, the Slater-Koster tight-binding parameters [13] obtained by inverting such first-principles matrices are expected to be
environment-dependent.

In order to examine how the overlap and hopping integrals are dependent on the environment and to see how serious the error the two-center approximation will make in traditional
tight-binding approaches, we have performed calculations for 3 types (i.e, diamond, simple
cubic (sc), and face-centered cubic (fcc)) of crystal structures of Si with several different bond
lengths for each type of structures in order to study the tight-binding parameters in different
bonding environments. Based on the overlap and effective one-electron Hamiltonian matrix
elements from our QUAMBO scheme, the Slater-Koster overlap integrals $s_{ss\sigma}$, $s_{sp\sigma}$, $s_{pp\sigma}$,
and $s_{pp\pi}$, and hopping integrals $h_{ss\sigma}$, $h_{sp\sigma}$, $h_{pp\sigma}$, and $h_{pp\pi}$ are then extracted using the
Slater-Koster geometrical factors [13]. The results for the overlap and hopping integrals as a
function of interatomic distance in the three different crystal structures are plotted in Figs. 3
and 4, respectively.

Figure 3 shows the overlap parameters $s_{ss\sigma}$, $s_{sp\sigma}$, $s_{pp\sigma}$, and $s_{pp\pi}$ from different structures
and different pairs of atoms, plotted as a function of interatomic distance. Note that the two-
center nature of overlap integrals for fixed atomic minimal basis orbitals may not necessarily

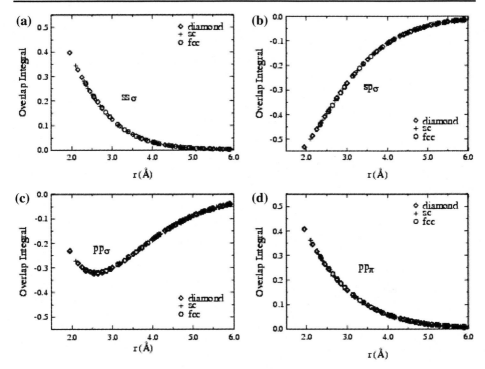

Fig. 3 Overlap integrals as a function of interatomic distance for Si in the diamond, sc, and fcc structures

hold for the QUAMBOs because QUAMBOs are deformed according to the bonding environments of the atoms. Nevertheless, the overlap parameters obtained from our calculations as plotted in Fig. 3 fall into smooth scaling curves nicely. These results suggest that the two-center approximation is adequate for overlap integrals.

By contrast, the hopping parameters as plotted in Fig. 4 are far from being transferable, especially for $h_{pp\sigma}$. Even for the best case of $h_{ss\sigma}$, the spread in the first neighbor interaction is about 1 eV. For a given pair of atoms, the hopping parameters $h_{pp\sigma}$ and $h_{pp\pi}$ obtained from the decompositions of different matrix elements can exhibit slightly different values, especially for the sc and fcc structures. The hopping parameters from different structures do not follow the same scaling curve. For a given crystal structure, although the bond-length dependence of hopping parameters for the first and second neighbor interactions can be fitted to separate smooth scaling curves respectively, these two scaling curves cannot be joined together to define an unique transferable scaling function for the structure. These results suggest that under the two-center approximation, it is not possible to describe the scaling of the tight-binding hopping parameters accurately.

It is interesting to note from Fig. 4 that the structure which has larger coordination number tends to have larger hopping parameter (in magnitude) as compared to the lower-coordinated structure at the same interatomic separation. It is also interesting to note that the scaling curve of the second neighbor interactions tends to be above that of the first neighbors at the same interatomic distance. These behaviors are indications of significant contributions from three-center integrals, because more contribution from the three-center integrals is expected for pair of atoms that have more neighbors which enhance the effective hopping between the two atoms.

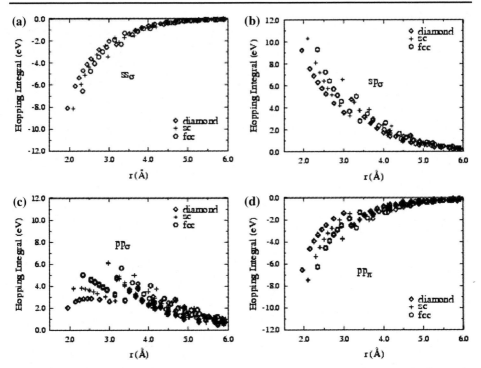

Fig. 4 Non-orthogonal tight-binding hopping integrals for Si as a function of interatomic distance in the diamond, sc, and fcc structures obtained by decomposing the QUAMBO-based effective one-electron Hamiltonian according to the Slater-Koster tight-binding scheme

To express the tight-binding Hamiltonian matrix in terms of QUAMBOs also allows us to address the issue of the effects of orthogonality on the transferability of tight-binding models from the first-principles perspective. We can construct orthogonal QUAMBOs from non-orthogonal ones using the symmetrical orthogonalization method of Löwdin [22]. Starting from the Bloch sum of non-orthogonal QUAMBOs

$$\tilde{A}_\alpha(\mathbf{k}, \mathbf{r}) = \frac{1}{\sqrt{N}} \sum_n \exp(i\mathbf{k} \cdot \mathbf{r}_n) A_\alpha(\mathbf{r} - \mathbf{r}_n) \tag{11}$$

the overlap matrix of $\tilde{A}_\alpha(\mathbf{k}, \mathbf{r})$ can be defined as

$$S_{\alpha\beta}(\mathbf{k}) = \sum_n \exp(i\mathbf{k} \cdot \mathbf{r}_n) \langle A_\alpha(\mathbf{r}) \mid A_\beta(\mathbf{r} - \mathbf{r}_n) \rangle \tag{12}$$

For each k-point, we perform the symmetrical orthogonalization method of Löwdin [22],

$$\tilde{A}_\alpha^{orthog}(\mathbf{k}, \mathbf{r}) = \sum_\beta S_{\beta\alpha}(\mathbf{k})^{-1/2} \tilde{A}_\beta(\mathbf{k}, \mathbf{r}) = \frac{1}{\sqrt{N}} \sum_{n,\beta} \exp(i\mathbf{k} \cdot \mathbf{r}_n) S_{\beta\alpha}(\mathbf{k})^{-1/2} A_\beta(\mathbf{r} - \mathbf{r}_n)$$

$$= \frac{1}{\sqrt{N}} \sum_n \exp(i\mathbf{k} \cdot \mathbf{r}_n) \sum_{\mathbf{k}',\beta} S_{\beta\alpha}(\mathbf{k}')^{-1/2} A_\beta^{\mathbf{k}'}(\mathbf{r} - \mathbf{r}_n) \tag{13}$$

Then the orthogonal QUAMBOs can be expressed by

$$\tilde{A}_{\alpha}^{orthog}(\mathbf{r} - \mathbf{r}_n) = \sum_{\mathbf{k},\beta} S_{\beta\alpha}(\mathbf{k})^{-1/2} A_{\beta}^{\mathbf{k}}(\mathbf{r} - \mathbf{r}_n) \tag{14}$$

since

$$\left\langle A_{\alpha}^{orthog}(\mathbf{r} - \mathbf{r}_n) \,\middle|\, A_{\beta}^{orthog}(\mathbf{r} - \mathbf{r}_{n'}) \right\rangle$$

$$= \frac{1}{N} \sum_{\mathbf{k},\mathbf{k}'} \left\langle \tilde{A}_{\alpha}^{orthog}(\mathbf{k},\mathbf{r}) \,\middle|\, \tilde{A}_{\beta}^{orthog}(\mathbf{k}', \mathbf{r}) \right\rangle \exp(i\mathbf{k} \cdot \mathbf{r}_n - i\mathbf{k}' \cdot \mathbf{r}_{n'})$$

$$= \frac{1}{N} \sum_{\mathbf{k}} \exp[i\mathbf{k} \cdot (\mathbf{r}_n - \mathbf{r}_{n'})] \delta_{\alpha\beta}$$

$$= \delta_{nn'} \delta_{\alpha\beta} \tag{15}$$

Figure 5 shows the orthogonal s- and p- like QUAMBOs in Si diamond and fcc structures with three different bond lengths, respectively. In comparison with the non-orthogonal QUAMBOs as shown in Fig. 1, the orthogonal QUAMBOs are much more tightly localized on the center atoms, but some wavefuction components have been pushed out to the neighboring atoms in order to satisfy the orthogonal requirement. Using the orthogonal QUAMBOs, the effective one-electron Hamiltonian matrix in the orthogonal QUAMBO representation can be calculated and the orthogonal Slater-Koster hopping integrals can be extracted following the decomposition procedures discussed in the non-orthogonal tight-binding case. The results are plotted in Fig. 6. It is interesting to note that the orthogonal hopping parameters as a function of interatomic distance decay much faster than their non-orthogonal counterparts. Therefore, the interactions in the orthogonal tight-binding scheme are essentially dominated by the first neighbor interactions which depend not only on the interatomic separations but also on the coordination of the structures. In contrast to the non-orthogonal model, the magnitudes of the orthogonal hopping parameters decrease as the coordination number of the structure increases. These coordination-dependence of the hopping parameters and the short-range nature of the interactions are qualitatively similar to the environment-dependent tight-binding model of Wang et al. [23,24]. In their model, the coordination dependence of the hopping parameters is considered through a bond-length scaling function, and the short-ranged interactions is guaranteed by the screening function. However, though small, the contributions from the second and higher neighbor hopping parameters are not entirely negligible. In particular, some hopping parameters in the orthogonal TB scheme are found to change sign at the second and higher neighbors. The sign changes in the second and higher neighbor interactions can be attributed to the effects of the orthogonality which push some orbital wavefunctions to the nearby atomic sites in order to satisfy the orthogonal condition as one can see from Fig. 5. Such effects have not been noticed in previous tight-binding models.

4 Large-scale electronic calculations using the QUAMBO scheme

The above development in QUAMBO construction and "exact" tight-binding matrix elements calculation enables us to perform tight-binding electronic-structure calculations for large systems directly from the first-principles approach, without a fitting procedure to generate tight-binding parameters. A scheme based on this idea has been developed by Yao et al. [25]. In this scheme, an overlap or tight-binding Hamiltonian matrix of a big system is built

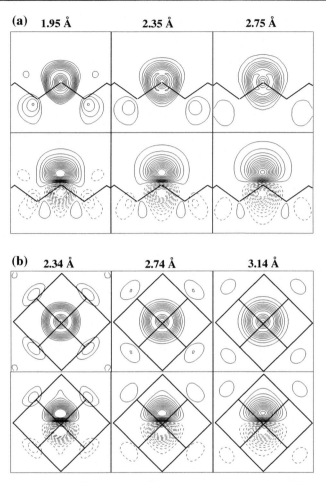

Fig. 5 Orthogonal s- and p- like QUAMBOs in Si (a) diamond structure in the (110) plane for three bond lengths 1.95 Å, 2.35 Å and 2.75 Å, and (b) fcc structure in the (100) plane for three bond lengths 2.34 Å, 2.74 Å, and 3.14 Å

by filling in the $n \times n$ "exact" sub-matrices (where n is the number of minimal basis orbitals for each atom) for every pair of atoms in the system. Note that the QUAMBOs and hence the $n \times n$ sub-matrices of tight-binding are dependent on the environment around the pair of atoms, the $n \times n$ "exact" sub-matrices has to be calculated for every pair of atoms in the system. This can be done by first performing first-principles calculations for a relatively small system with the same environment around the pair of atoms as if they are in the big systems, then the $n \times n$ tight-binding matrix for this pair of atoms can be constructed following the QUAMBO scheme. This approach will break the first-principles calculations of a big system into calculations for many much smaller sub-systems. In many cases of our interest (e.g., defects in crystals), the bonding environment of many different atom pairs in the big system may be essentially the same, therefore, first-principles calculations are needed only for a limited number of smaller systems and an accurate tight-binding overlap and Hamiltonian matrices for the big system can be constructed.

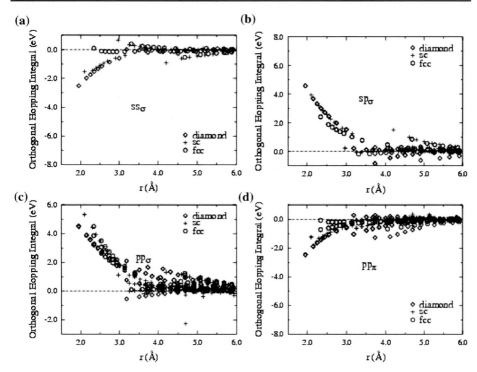

Fig. 6 Orthogonal tight-binding hopping integrals for Si as a function of interatomic distances in the diamond, sc, and fcc structures obtained by decomposing the QUAMBO-based effective one-electron Hamiltonian according to the Slater-Koster tight-binding scheme

The method was recently applied to studies the electronic structure of graphene nanoribbons [25]. For calculating the electronic structure of perfect armchair-grapheme nano-ribbons (A-GNRs) of different width, three different types of atoms in the nano-ribbons have been identified as illustrated in Fig. 7 where atom **a** represents a carbon atom inside the ribbon, atom **b** represents a carbon atom at the edges, and atom **c** is a hydrogen atom for passivation. Only one training sample of a $N_a = 7$ A-GNRs as shown in Fig. 7 and a single first-principles calculation are needed to extract all the necessary "exact" 4×4 or 4×1 tight-binding matrices for each pair of **a-a**, **a-b**, **b-b**, and **b-c** atoms from these three type of non-equivalent atoms, respectively. Fig. 8 shows the band structures and electronic density of states (DOS) for A-GNR with the width $N_a = 7$ and 13 (solid lines) from the QUAMBO-tight-binding scheme using the small 4×4 and 4×1 tight-binding matrices generated from the $N_a = 7$ training cell as described above. The results from full first-principles calculations (circle) were also shown for comparison. One can see that the QUAMBO-TB band structures agree very well with the full first-principles results up to 1eV above the Fermi-level. The electronic band gap variation of a perfect A-GNR as a function of the width of the nanoribbon has also been studied. Fig. 9 shows the oscillating behavior of band gap with a period of $N_a = 3$ obtained from our QUAMBO-TB scheme agree very well the results from first-principles calculations [26–28]. The efficiency of the QUAMBO-TB scheme enable us to calculate the electronic structure of much wider grapheme nano-ribbon, as one can also see from Fig. 9 where the band gap of a nanoribbon up to 100 Å in width has been calculated by our QUAMBO-TB method.

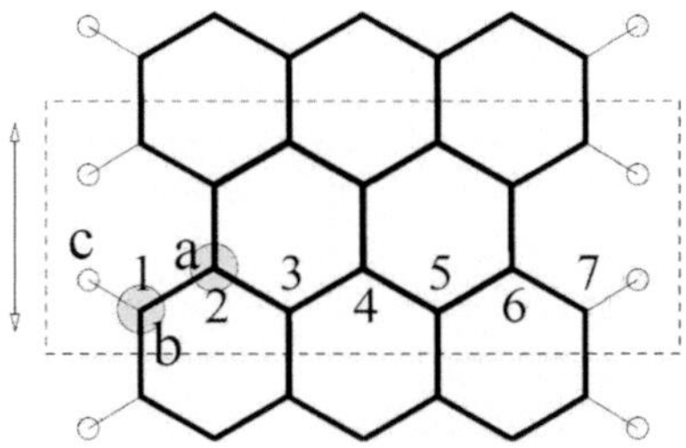

Fig. 7 (Color online) A-GNR
with $N_a = 7$ was chosen to be a
training cell. Dotted rectangle
indicates the unit cell. The left
arrow gives the periodical
direction. Atom **a** and **b** are
treated to be different due to
different local environment

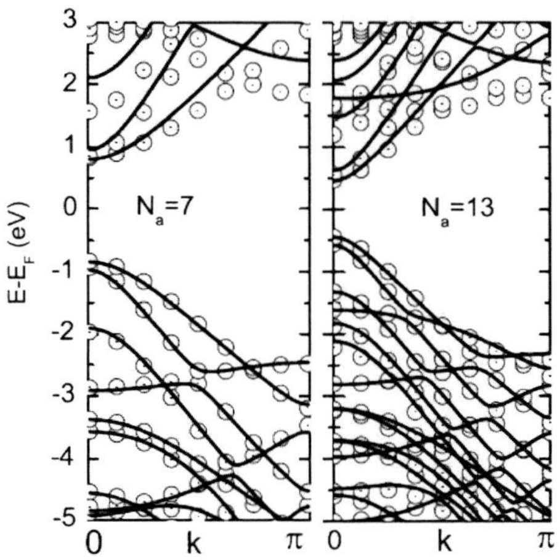

Fig. 8 TB band structures based
on the QUAMBO-generated TB
parameters (solid line) compared with first-principles DFT
results (circle) for A-GNR with
$N_a = 7$ and 13 respectively

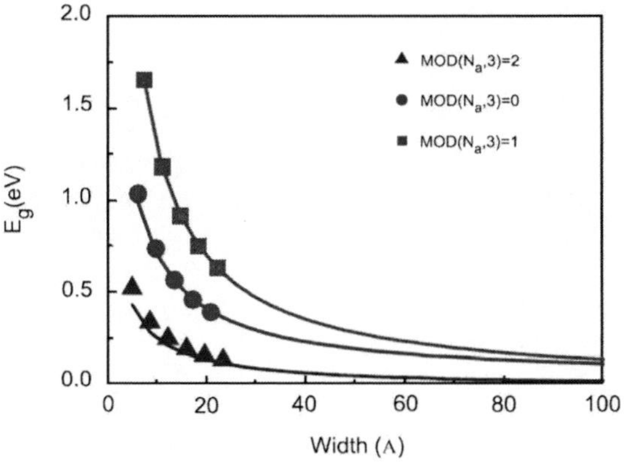

Fig. 9 (Color online) TB band
gap (solid lines) of A-GNR with
different size compared with
first-principles DFT results
(symbols)

    The efficiency of the QUAMBO-TB scheme also enable us to study the electronic structure
of grapheme nano-ribbons with defects, which usually require a much large unit cell and it
is not easy to calculate using straightforward first-principles calculations. Yao et al. have
studied the electronic structures of a $N_a = 6$ A-GNR with random edge defects on one edge
of the ribbon at different defect ratio [25]. The supercell used in the calculation contains 1200
carbon atoms and about 200 hydrogen atoms. The edge defects were generated by randomly
removing pairs of carbon atoms at one side of A-GNR as shown on Fig. 10(a). The carbon
atoms at the defected edge were again passivated by hydrogen atoms. For this defect system,
only some additional QUAMBO-TB matrix elements around the edge defects are needed to

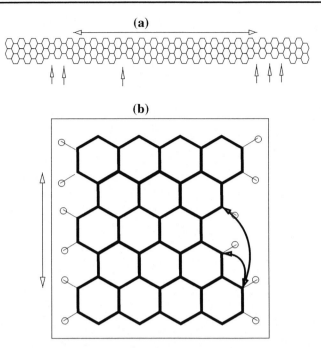

Fig. 10 (Color online) (a) Schematic view of a part of a supercell of $N_a = 6$ A-GNR containing more than one thousand atoms with edge defects randomly distributed on one side.. (b) The training cell used to generate the additional TB parameters for the A-GNR with edge defects

be generated using a training cell as shown in Fig. 10(b), where the curved arrows indicate the new matrix elements between these sites to be added to the existing QUAMBO-TB matrix elements database from the $N_a = 7$ training cell as discussed above. Based on this set of QUAMBO-TB matrix elements from first-principles calculations performed on two small unit cells, actuate tight-binding overlap and Hamiltonian matrices for the defected graphene nano-ribbons of various defect concentration can be constructed, and the electronic structure of A-GNRs with random edge defects can be studied. The results of band gap as the function of defect ratio in the $N_a = 6$ A-GNR are shown in Fig. 11. The random distribution of the edge defects gives some variation of the band gap at each defect concentration; however, there exists a general trend of the band gap with increasing defect concentration. The band gap reaches its minimum (which is quite small) at the edge defect ratio of 70%. This implies that edge defects have a significant effect on electronic structures of A-GNRs, which is consistent with the indications from experiments [29].

5 Concluding remarks

Using the recently developed quasi-atomic minimal-basis-set orbitals, we show that accurate tight-binding Hamiltonian and overlap matrix elements can be extracted from first-principles calculations. Based on the information from the QUAMBO-TB matrix elements, the transferability of two-center tight-binding models can be examined from a first-principles perspective. Our studies show that tight-binding models with two-center approximation are not adequate for describing the effective one-electron Hamiltonian matrix elements under

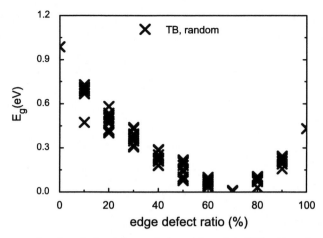

Fig. 11 Band gap as a function of edge defects ratio in an $N_a = 6$ A-GNR with random edge defects. A perfect $N_a = 6$ and $N_a = 5$ A-GNRs corresponds to 0% and 100% defect ratio in this plot respectively

different bonding environments. While we discuss about Si and C systems in this article, similar analyses have been carried out for other systems such as Al, Mo, Fe, SiC etc. [9, 11, 12]. Such analyses provide useful insights and guidance for generating accurate and transferable tight-binding models. In particular, we show that environment-dependence of the tight-binding parameters need to be adequately described, and it may also be necessary to include three-center integrals explicitly if we want to describe accurately the electronic structures of complex systems by tight-binding approach. Although the QUAMBO-TB scheme can help us gaining insight into how the tight-binding interactions are dependent on the environment, how to model and parameterize such environment-dependence of the tight-binding interaction still remains an open question and need much further investigation.

Another route to utilize the QUAMBO-TB scheme for calculating the electronic structure of a large system is to use a divide-and-conquered strategy which divides the Hamiltonian and overlap matrices of a big system into a set of much smaller $n \times n$ QUAMBO-TB matrices of pair of atoms with different bonding environment. First-principles calculations are needed for small number of atoms around the pairs, yet a QUAMBO-TB matrix for the whole large system can be constructed accurately. We have shown that such "QUAMBO-on-demand" approach has been quite successful for the studies of electronic structure in grapheme nanoribbons. One could construct a variety of training cells to generate a database of Hamiltonian parameters for a catalogue of local bonding environments. This opens a promising avenue to do electronic-structure simulations and total energy calculations for big systems directly from first principles. The computational savings thus achieved is analogous to savings obtained using Green's function boundary condition near infinite half space, but it can handle complex geometric arrangements. A sophisticated computational technology needs to be developed in the future to automate this process.

Acknowledgements Ames Laboratory is operated for the U.S. Department of Energy by Iowa State University under Contract No. DE-AC02-07CH11358. This work was supported by the Director for Energy Research, Office of Basic Energy Sciences including a grant of computer time at the National Energy Research Supercomputing Center (NERSC) in Berkeley. Work of JL is supported by NSF CMMI-0728069, AFOSR, ONR N00014-05-1-0504, and Ohio Supercomputer Center.

References

1. Andersen, O.K., Jepsen, O.: Phys. Rev Lett. **53**, 2571 (1984)
2. Andersen, O.K., Sha-Dasgupta, T.: Phys.Rev. B **62**, R16219 (2000)
3. Andersen, O.K., et al.: Electronic structure and physical properties of solid. The uses of the LMTO method. In: Dreysse, H. (ed.) Springer Lecture Notes in Physics. Springer, New York (2000)
4. Sankey, O.F., Niklewski, D.J.: Phys. Rev. B **40**, 3979 (1989)
5. Porezag, D., Frauenheim, Th., Köhler, Th., Seifert, G., Kaschner, R.: Phys. Rev. B **51**, 12947 (1995)
6. McMahan, A.K., Klepeis, J.E.: . Phys. Rev. B **56**, 12250 (1997)
7. Lu, W.C., Wang, C.Z., Schmidt, M.W., Bytautas, L., Ho, K.M., Ruedenberg, K.: J. Chem. Phys. **120**, 2629 (2004)
8. Lu, W.C., Wang, C.Z., Schmidt, M.W., Bytautas, L., Ho, K.M., Ruedenberg, K.: J. Chem. Phys. **120**, 2638 (2004)
9. Lu, W.C., Wang, C.Z., Chan, Z.L., Ruedenberg, K., Ho, K.M.: Phys Rev. B **70**, 041101 (R) (2004)
10. Lu, W.C., Wang, C.Z., Ruedenberg, K., Ho, K.M.: Phys. Rev. B **72**, 205123 (2005)
11. Chan, T.-L., Yao, Y.X., Wang, C.Z., Lu, W.C., Li, J., Qian, X.F., Yip, S., Ho, K.M.: Phys. Rev. B **76**, 205119 (2007)
12. Qian, X.-F., Li, J., Wang, C.-Z., Qi, L., Chan, T.-L., Yao, Y.-X., Ho, K.-M., Yip, S.: Phys. Rev. B, (to be published)
13. Slater, J.C., Koster, G.F.: Phys. Rev. **94**, 1498 (1954)
14. Hamann, D.R., Schluter, M., Chiang, C.: Phys. Rev. Lett. **43**, 1494 (1979)
15. Vanderbilt, D.: Phys. Rev. B **41**, 7892 (1990)
16. Blöchl, P.E.: Phys. Rev. B **50**, 17953 (1994)
17. Kresse, G., Hafner, J.: Phys. Rev. B **47**, 558 (1993)
18. Kresse, G., Furthmller, J.: Phys. Rev. B **54**, 169 (1996)
19. Pulay, P.: Chem. Phys. Lett. **100**, 151 (1983)
20. Sæbø, S., Pulay, P.: Chem. Phys. Lett. **113**, 13 (1985)
21. Sæbø, S., Pulay, P.: Annu. Rev. Phys. Chem. **44**, 213 (1993)
22. Löwdin, P.: J. Chem. Phys. **18**, 365 (1950)
23. Tang, M.S., Wang, C.Z., Chan, C.T., Ho, K.M.: Phys. Rev. B **53**, 979 (1996)
24. Wang, C.Z., Pan, B.C., Ho, K.M.: J. Phys. Condens. Matter **11**, 2043 (1999)
25. Yao, Y.X., Wang, C.Z., Zhang, G.P., Ji, M., Ho, K. M.: (to be published)
26. Son, Y.-W., Cohen, M.L., Louie, S.G.: Nature (London) **444**, 347 (2006)
27. Son, Y.-W., Cohen, M.L., Louie, S.G.: Phys. Rev. Lett. **97**, 216803 (2006)
28. Yan, Q.M., Huang, B., Yu, J., Zheng, F.W., Zang, J., Wu, J., Gu, B.L., Liu, F., Duan, W.H.: Nano Lett. **7**, 1469 (2007)
29. Han, M.Y., Ozyilmaz, B., Zhang, Y., Kim, P.: Phys. Rev. Lett. **98**, 206805 (2007)

Atomistic simulation studies of complex carbon and silicon systems using environment-dependent tight-binding potentials

Cai-Zhuang Wang · Gun-Do Lee · Ju Li · Sidney Yip · Kai-Ming Ho

Originally published in the journal Sci Model Simul, Volume 15, Nos 1–3, 97–121.
DOI: 10.1007/s10820-008-9109-x © Springer Science+Business Media B.V. 2008

1 Introduction

The use of tight-binding formalism to parametrize electronic structures of crystals and molecules has been a subject of continuous interests since the pioneer work of Slater and Koster [1] more than a half of a century ago. Tight-binding method has attracted more and more attention in the last 20 years due to the development of tight-binding potential models that can provide interatomic forces for molecular dynamics simulations of materials [2–7]. The simplicity of the tight-binding description of electronic structures makes the method very economical for large-scale electronic calculations and atomistic simulations [5,9]. However, studies of complex systems require that the tight-binding parameters be "transferable", [4] i.e., to be able to describe accurately the electronic structures and total energies of a material in different bonding environments. Although tight-binding molecular dynamics has been successfully applied to a number of interesting systems such as carbon fullerenes and carbon nanotubes, [10,11] the transferability of tight-binding potentials is still the major obstruction hindering the wide spread application of the method to more materials of current interest.

There are two major approximations made in a typical tight-binding representation of effective one-electron Hamiltonian matrix or band structure based on the Slater-Koster theory

C.-Z. Wang (✉) · K.-M. Ho
Ames Laboratory-U.S. DOE and Department of Physics and Astronomy, Iowa State University,
Ames, IA 50011, USA
e-mail: wangcz@ameslab.gov

G.-D. Lee
School of Materials Science and Engineering and Inter-university Semiconductor Research Center (ISRC),
Seoul National University, Seoul 151-742, Korea

J. Li
Department of Materials Science and Engineering, University of Pennsylvania, Philadelphia,
PA 19104, USA

S. Yip
Department of Nuclear Science and Engineering and Department of Materials Science and Engineering,
Massachusetts Institute of Technology, Cambridge, MA 02139, USA

[1]. One is the use of a fixed minimal basis set, i.e., the basis orbitals for a given atom type are not allowed to vary according to structures or bonding environments, and another is the two-center approximation which assumes that the crystal potential can be constructed as a sum of spherical potentials centered on atoms, and contribution from atom k's spherical potential on matrix element between two basis orbitals on atoms i and j can be neglected. Experiences from the tight-binding parametrizations have indicated that the transferability of tight-binding models are limited by both approximations [7, 12–14].

Several attempts to go beyond the above two approximations have been shown to improve the transferability of the tight-binding descriptions of electronic structures and total energies [12–14]. For example, Mehl and Papaconstantopoulos found that by incorporating a crystal-field like term into the tight-binding model and allowing the on-site atomic energies to vary according to the bonds surrounding of the atoms have significant improvement on the accuracy and transferability of the tight-binding models to describe metallic systems [12]. Wang et al. introduced an environment-dependent tight-binding model that allows the tight-binding parameters to vary not only with the interatomic distances but also according to the bonding environment around the interacting pair of atoms, and showed that the environment-dependent tight-binding model describes well not only the properties of the lower-coordinated covalent but also those of higher-coordinated metallic structures of carbon and silicon [13, 14].

In this article, we will review some progress on tight-binding modeling and simulations based on the environment-dependent tight-binding model. In particular, application of the EDTB potentials to the studies of complex systems such as vacancy diffusion and reconstruction in grapheme, dislocation climb and junction formation in carbon nanotubes, addimer diffusion on Si surfaces as well as grain boundary and dislocation core structures of in silicon will be discussed in more details.

2 Environment-dependent tight-binding potential model

2.1 General formalism of tight-binding potential model

The expression for the binding energy (or potential energy) of a system in a tight-binding molecular dynamics simulation is given by

$$E_{\text{binding}} = E_{\text{bs}} + E_{\text{rep}} - E_0 \tag{1}$$

The first term on the right hand side of Eq. (1) is the band structure energy which is equal to the sum of the one-electron eigenvalues ε_i of the occupied states given by a tight-binding Hamiltonian H_{TB} which will be specified later,

$$E_{\text{bs}} = \sum_i f_i \varepsilon_i \tag{2}$$

where f_i is the electron occupation (Fermi-Dirac) function and $\sum_i f_i = N_{\text{electron}}$.

The second term on the right hand side of Eq. (1) is a repulsive energy and can be expressed as a functional of sum of short-ranged pairwise interactions,

$$E_{\text{rep}} = \sum_i f \left(\sum_j \phi \left(r_{ij} \right) \right) \tag{3}$$

where ϕ_{ij} is a pairwise repulsive potential between atoms i and j, and f is a function which for example can be a 4th order polynomial [15] with argument $x = \sum_j \phi\left(r_{ij}\right)$,

$$f\left(x\right) = \sum_{n=0}^{4} c_n x^n \tag{4}$$

If $f\left(x\right) = x/2$, the repulsive energy is just a simple sum of pairwise potential $\phi\left(r_{ij}\right)$. In our environment-dependent tight-binding (EDTB) potential model that will be discuss in the following, we adopt the expression of Eq. (4) for the repulsive energy E_{rep}.

The term E_0 in Eq.(1) is a constant which represents the sum of the energies of the individual atoms. In our model, E_0 is absorbed into E_{bs} and E_{rep} respectively so that E_0 is set to be zero.

The tight-binding Hamiltonians H_{TB} for the electronic structure calculation is expressed as

$$H_{\mathrm{TB}} = \sum_i \sum_{\alpha=s,p} e_{i\alpha} a_{i\alpha}^+ a_{i\alpha} + \sum_{i,j} \sum_{\alpha,\beta=s,p} h_{i\alpha,j\beta} a_{i\alpha}^+ a_{j\beta} \tag{5}$$

where $e_{i\alpha}$ is the on-site energy of the α orbital on site i, $a_{i\alpha}^+$ and $a_{i\alpha}$ are the creation and annihilation operators, respectively. $h_{i\alpha,j\beta}$ is the hopping integral between α and β orbitals located at sites i and j, respectively. For a system described by only s and p orbitals, there are four types of hopping integrals $h_{ss\sigma}$, $h_{sp\sigma}$, $h_{pp\sigma}$, and $h_{pp\pi}$. In general, the Hamiltonian matrix elements between the orbitals on sites i and j should be dependent on the vector $\mathbf{R_j} - \mathbf{R_i}$. as well as the atomic configuration around these sites. However, under the two-center approximation made by Slater and Koster [1], the integrals are dependent only on the separation distance of the two atoms and can be parameterized by fitting to *ab initio* band structures. Once the hopping integrals are obtained, the TB Hamiltonian matrix can be constructed by linear combination of the hopping integrals using the direction cosines of the vector $\left(\mathbf{R_j} - \mathbf{R_i}\right)$ [1]

2.2 EDTB potential model formalism

In our EDTB potential model for carbon and silicon [13,14], the minimal basis set of sp^3 orbitals is taken to be orthogonal. The tight-binding Hamiltonian H_{TB} takes the form as in the Slater-Koster theory discussed above, but the effects of orthogonalization, three-center interactions and the variation of the local basis set with environment are taken into account empirically by renormalizing the interaction strength between atom pairs according to the surrounding atomic configurations. The TB hopping parameters and the repulsive interaction between atoms i and j depend on the environments of atoms i and j through two scaling functions. The first one is a screening function that is designed to weaken the interactions between two atoms when there are intervening atoms between them. Another is a bond-length scaling function which scales the interatomic distance (hence the interaction strength) between the two atoms according to their effective coordination numbers. Longer effective bond lengths are assumed for higher coordinated atoms.

Specifically, the hopping parameters and the pairwise repulsive potential as the function of atomic configurations are expressed as

$$h\left(r_{ij}\right) = \alpha_1 R_{ij}^{-\alpha_2} \exp\left(-\alpha_3 R_{ij}^{\alpha_4}\right)\left(1 - S_{ij}\right) \tag{6}$$

In Eq. (6), $h\left(r_{ij}\right)$ denotes the possible types of interatomic hopping parameters $h_{ss\sigma}, h_{sp\sigma}, h_{pp\sigma}, h_{pp\pi}$ and pairwise repulsive potential $\phi\left(r_{ij}\right)$ between atoms i and j. r_{ij} is the real distance and R_{ij} is a scaled distance between atoms i and j. S_{ij} is a screening function. The parameters $\alpha_1, \alpha_2, \alpha_3, \alpha_4$, and parameters for the bond-length scaling function R_{ij} and the screening function S_{ij} can be different for different hopping parameters and the pairwise repulsive potential. Note that expression Eq. (6) reduces to the traditional two-center form if we set $R_{ij} = r_{ij}$ and $S_{ij} = 0$.

The screening function S_{ij} is expressed as a hyperbolic tangent (tanh) function (i.e., $S_{ij} = \tanh\left(\xi_{ij}\right)$

$$S_{ij} = \frac{\exp\left(\xi_{ij}\right) - \exp\left(-\xi_{ij}\right)}{\exp\left(\xi_{ij}\right) + \exp\left(-\xi_{ij}\right)} \tag{7}$$

with argument ξ_{ij} given by

$$\xi_{ij} = \beta_1 \sum_l \exp\left[-\beta_2 \left(\frac{r_{il} + r_{jl}}{r_{ij}}\right)^{\beta_3}\right] \tag{8}$$

where β_1, β_2, and β_3 are adjustable parameters. Maximum screening effect occurs when atom l is situated close to the line connecting atoms i and j (i.e., $r_{il} + r_{jl}$ is a minimum). This approach allows us to distinguish between first and further neighbor interactions without explicit specification. This is well-suited for molecular dynamics simulations where it is difficult to define exactly which atoms are first-nearest neighbors and which are second-nearest neighbors.

The bond-length scaling function R_{ij} scales the interatomic distance between two atoms according to their effective coordination numbers. Longer effective bond lengths are assumed for higher coordinated atom pairs therefore interaction strength in larger-coordinated structures is reduced. The scaling between the real and effective interatomic distance is given by

$$R_{ij} = r_{ij}\left(1 + \delta_1\Delta + \delta_2\Delta^2 + \delta_3\Delta^3\right) \tag{9}$$

where $\Delta = \frac{1}{2}\left(\frac{n_i - n_0}{n_0} + \frac{n_j - n_0}{n_0}\right)$ is the fractional coordination number relative to the coordination number (n_0) of the diamond structure, averaged between atoms i and j. The coordination number can be modeled by a smooth function,

$$n_i = \sum_j \left(1 - S_{ij}\right) \tag{10}$$

with a proper choice of parameters for S_{ij} which has the form of the screening function described above (Eq. (7)). The parameters for the coordination number calculations in carbon and silicon will be given in next subsections, respectively.

Besides the hopping parameters, the diagonal matrix elements are also dependent on the bonding environments. The expression for the diagonal matrix elements is

$$e_{\lambda,i} = e_{\lambda,0} + \sum_j \Delta e_\lambda\left(r_{ij}\right) \tag{11}$$

where $\Delta e_\lambda\left(r_{ij}\right)$ takes the same expression as Eq. (6), λ denotes the two types of orbitals (s or p). $e_{s,0}$ and $e_{p,0}$ are the on site energies of a free atom.

Finally, the repulsive energy term is expressed in a functional of the sum of pairwise interactions as defined in Eq. (4) in the previous section.

To ensure that all interactions go to zero smoothly at the given cutoff distance r_{cut}, all the distance dependent parameters in the model are multiplied by an attenuation function of the form of $\cos^2 \theta$ with

$$\theta = \frac{\pi}{2} \frac{r - r_{match}}{r_{cut} - r_{match}} \tag{12}$$

when the distance is $r_{match} < r < r_{cut}$. This attenuation function can guarantee that the distance dependent parameters and their first derivatives are continuous at r_{match} and go to zero at r_{cut}.

3 EDTB potential for carbon and its applicationa

3.1 EDTB potential for carbon

Carbon is a strong covalently bonded material best described by the tight-binding scheme. In 1996, Tang, Wang, Chan, and Ho developed an environment-dependent tight-binding potential for carbon following the formalism described in the previous subsection [13]. The parameters of this potential are given in Tables 1 and 2. In addition to the parameters listed in the tables, the parameters for calculating the coordination number of carbon using Eq. (10) are $\beta_1 = 2.0$, $\beta_2 = 0.0478$, $\beta_3 = 7.16$. The cutoff distance for the interaction is $r_{cut} = 3.4\text{Å}$ and $r_{match} = 3.1\text{Å}$ (see Eq. (12). As shown in Fig. 1, the environment-dependent tight-binding potential model for carbon describes very well the binding energies not only for the covalent (diamond, graphite, and linear chain) structures, but also for the higher-coordinated metallic (bcc, fcc, and simple cubic) structures, as compared to the two-center tight-binding model developed by us earlier [15]. The EDTB potential is also more accurate for elastic constants and phonon frequencies of diamond and graphite structures as compare to the two-center tight-binding model (Tables 3, 4).

Table 1 The parameters of the EDTB model for carbon

	α_1	α_2	α_3	α_4	β_1	β_2	β_3	δ
$h_{ss\sigma}$	−8.9491	0.8910	0.1580	2.7008	2.0200	0.2274	4.7940	0.0310
$h_{sp\sigma}$	8.3183	0.6170	0.1654	2.4692	1.0300	0.2274	4.7940	0.0310
$h_{pp\sigma}$	11.7955	0.7620	0.1624	2.3509	1.0400	0.2274	4.7940	0.0310
$h_{pp\pi}$	−5.4860	1.2785	0.1383	3.4490	0.2000	8.500	4.3800	0.0310
ϕ	30.0000	3.4905	0.00423	6.1270	1.5035	0.205325	4.1625	0.002168
$\Delta e_s, \Delta e_p$	0.1995275	0.029681	0.19667	2.2423	0.055034	0.10143	3.09355	0.272375

The TB hopping integrals are in the unit of eV and the interatomic distances are in the unit of Å. ϕ is dimensionless

Table 2 The coefficients (in unit of eV) of the polynomial function $f(x)$ for the EDTB potential for carbon

c_0	c_1	c_2	c_3	c_4
12.201499972	0.583770664	$0.336418901 \times 10^{-3}$	$-0.5334093735 \times 10^{-4}$	$0.7650717197 \times 10^{-6}$

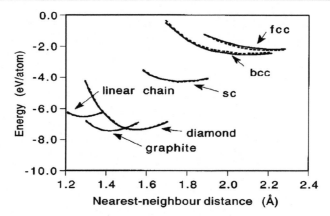

Fig. 1 Cohesive energies as a function of nearest neighbor distance for carbon in different crystalline structures calculated using the environment-dependent TB model are compared with the results from the first-principles DFT-GGA calculations. The solid curves are the TB results and the dashed curves are the GGA results (From Ref. [13])

		XWCH	EDTB	Experiment
Table 3 Elastic constants, phonon frequencies and Grünneisen parameters of diamond calculated from the XWCH-TB model [15] and the environment-dependent TB (EDTB) model [13] are compared with experimental results [16]	a(Å)	3.555	3.585	3.567
	B	4.56	4.19	4.42
	$c_{11}-c_{12}$	6.22	9.25	9.51
	c_{44}	4.75	5.55	5.76
	$v_{LTO}(\Gamma)$	37.80	41.61	39.90
	$v_{TA}(X)$	22.42	25.73	24.20
	$v_{TO}(X)$	33.75	32.60	32.0
	$v_{LA}(X)$	34.75	36.16	35.5
	$\gamma_{LTO}(\Gamma)$	1.03	0.93	0.96
	$\gamma_{TA}(X)$	−0.16	0.30	
Elastic constants are in units of 10^{12}dyn/cm^2 and the phonon frequencies are in terahertz	$\gamma_{TO}(X)$	1.10	1.50	
	$\gamma_{LA}(X)$	0.62	0.98	

Table 4 Elastic constants, phonon frequencies and Grüneisen parameters of graphite calculated from the XWCH-TB model [15] and the environment-dependent TB (EDTB) model [13] are compared with experimental results [17,18]

	XWCH	EDTB	Experiment
$c_{11}-c_{12}$	8.40	8.94	8.80
E_{2g2}	49.92	48.99	47.46
A_{2u}	29.19	26.07	26.04
$\gamma(E_{2g2})$	2.00	1.73	1.63
$\gamma(A_{2u})$	0.10	0.05	

Elastic constants are in units of 10^{12}dyn/cm^2 and the phonon frequencies are in terahertz

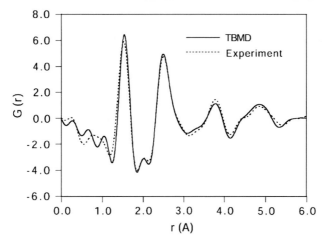

Fig. 2 Radial distribution functions G(r) of the tetrahedral amorphous carbon structure generated by tight-binding molecular dynamics using the environment-dependent TB potential (solid curve) are compared with the neutron scattering data of Ref. [22] (dotted curve). The theoretical result has been convoluted with the experimental resolution corresponding to the termination of the Fourier transform at the experimental maximum scattering vector $Q_{max} = 16 \, \text{Å}^{-1}$. (From Ref. [25])

Another example that demonstrates the better transferability of the EDTB model over the two-center model for complex simulations is the study of diamond-like amorphous carbon. Diamond-like (or tetrahedral) amorphous carbon consists of mostly sp^3 bonded carbon atom produced under highly compressive stress which promotes the formation of sp^3 bonds, in contrast to the formation of sp^2 graphite-like bonds under normal conditions [19–22]. Although the two-center XWCH carbon potential can produce the essential topology for the diamond-like amorphous carbon network [23], the comparison with experiment is not quite satisfactory as one can see from Fig. 2. There are also some discrepancies in ring statistics between the two-center potential generated and *ab initio* molecular dynamics generated diamond-like amorphous carbon model [24]. Specifically, a small fraction of 3 and 4-membered rings observed in the *ab initio* model is absent from the results of the two-center tight-binding model. These subtle deficiencies are corrected when the EDTB potential is used to generate diamond-like amorphous carbon [25,26]. The radial distribution function of the diamond-like a-c obtained from the EDTB potential is in much better agreement with experiment as one can see from Fig. 2.

3.2 TBMD simulation of vacancy diffusion and reconstruction in grapheme

Recently, the EDTB carbon potential by Tang et al. [13] has been further improved by Lee et al. by incorporating an angle dependence factor into the repulsive energy to describe correctly the diffusion of an adatom and a vacancy in carbon nanotubes and graphene [27–30]. The modified EDTB carbon potential has described successfully the reconstruction of vacancy defects in a graphene and carbon nanotubes [27–30].

Vacancy defects in graphene layers, which are usually generated by ion or electron irradiations of graphite or carbon nanotubes, have been an interesting subject of many studies, yet the dynamics and reconstruction of the defects in graphene layer are still not well understood [31–35]. Figure 3 illustrates the snapshots of the atomic processes of diffusion, coalescence,

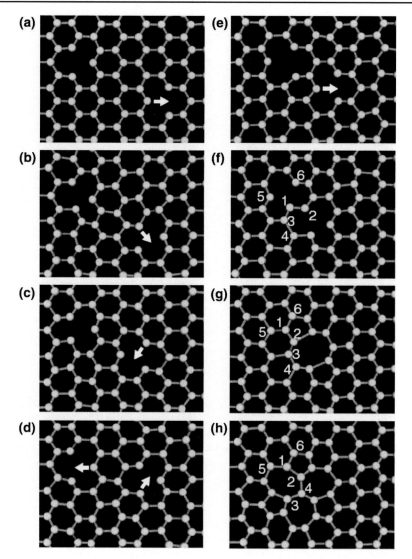

Fig. 3 Atomic processes from the TBMD simulations for vacancy diffusion in a graphene layer. (**a**) 0 K (at time $t = 0$ ps), (**b**) ~3000 K ($t = 2.7$ ps), (**c**) ~3000 K ($t = 3.0$ ps), (**d**) ~2900 K ($t = 3.3$ ps), (**e**) ~3000 K ($t = 5.0$ ps), (**f**) ~3100 K ($t = 6.0$ ps), (**g**) ~3100 K ($t = 6.5$ ps), and (**h**) ~3800 K ($t = 125$ ps). White arrows indicate the direction for the carbon atom to jump in the next step. The atoms involved in the diffusion process are labeled with the numbers. (From Ref. [27])

and reconstruction of vacancy defects in graphene layers during the simulation using molecular dynamics with the improved EDTB carbon potential. The TBMD simulations in Fig. 3 shows that two single vacancies diffuse and coalesce into a 5-8-5 double vacancy at the temperature of 3000 K, and it is further reconstructed into a new defect structure, the 555–777 defect, by the Stone-Wales type transformation at higher temperatures. The stability of the defect structures observed in the TBMD simulations is further examined by first-principles calculations which show that the graphene layer containing the 555–777 defect, as shown

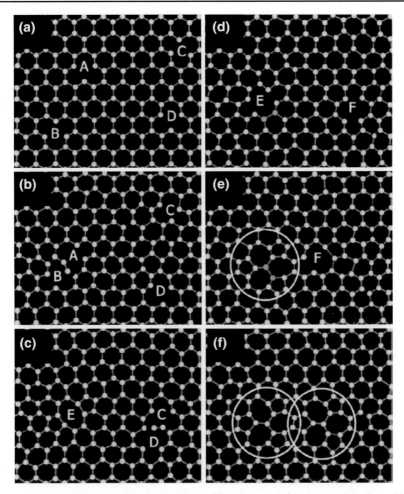

Fig. 4 Atomic processes from the TBMD simulations of four vacancy defects in a graphene layer. (a) 0 K (at time $t = 0$ ps); (b) ~3000 K ($t = 5.5$ ps); (c) ~3000 K ($t = 52.6$ ps); (d) ~3900 K ($t = 86.8$ ps); (e) ~3700 K ($t = 274.4$ ps); (f) ~3700 K ($t = 281.6$ ps)

in Fig. 3h, is most stable and its formation energy is lower than that of the 5-8-5 defect by 0.91 eV. The formation energy of the two separated single vacancies (Fig. 3a) is much higher than that of the 555–777 defect by 8.85 eV.

The simulations are also performed for four single vacancies in a graphene sheet. As shown in Fig. 4 the four single vacancies in the graphene layer first coalesce into two double vacancies, each consists of a pentagon-heptagon-pentagon (5-8-5) defective structure. While one of the 5-8-5 defects further reconstructs into a 555–777 defect, which is composed of three pentagonal rings and three heptagonal rings, the another 5-8-5 defect diffuses toward to the reconstructed 555–777 defect. During the 5-8-5 defect diffusion process, three interesting mechanisms, i.e., "dimer diffusion", "chain diffusion", and "single atom diffusion", are observed. Finally, the four single vacancies reconstruct into two adjacent 555–777 defects, forming a local haeckelite structure proposed by Terrones et al. [36].

3.3 TBMD simulation of junction formation in carbon nanotubes

The improved EDTB carbon potential has also been applied in tight-binding molecular dynamics simulation to study the junction formation through self-healing of vacancies, in single-walled carbon nanotubes (SWCNT) [29,30]. Figure 5 shows the atomic details of vacancy reconstruction in a (16,0) SWCNT with a six-vacancy hole by the TBMD simulation [29,30]. The TBMD simulation is performed starting from a relaxed six-vacancy hole geometry as shown in Fig. 5a. In the early stage of the simulation, the SWCNT is heated up to high temperature through a constant-temperature molecular dynamics simulation. It was found that rearrangement of carbon atoms around the vacancy hole starts to occur near 4500 K at the simulation time of 18 ps through the rotation of carbon dimers, i.e., Stone-Wales transformation. After 19 ps of the simulation time, three hexagons at the lower left corner of the vacancy hole (those containing the atoms 1-4 in Fig. 5b) are recombined into a pentagon-octagon-pentagon defect by successive Stone-Wales transformations of the dimers 1-2 and 3-4 as shown in Fig. 5b. After the simulation time of 20 ps, another two hexagons (containing the atoms 5-7) on the other side of the vacancy hole are also reconstructed into one pentagon and one heptagon by the Stone-Wales transformation of the dimer 5-6. In order to prevent the evaporation of carbon atoms, the system is then cooled down to 3,000 K for 4 ps and the vacancy hole is healed during this simulation period as shown in Fig. 5c. The structure immediately after the healing process consists of four pentagons and four heptagons with a two-fold rotation symmetry. The pentagon a and b and the heptagon c and d are related to the pentagon e and f and the heptagon g and h, respectively through the 2-fold axis which goes through the center of the carbon bond between atom 2 and 6. After the simulation time of 24 ps, the system is heated up again to 4,500 K for 7 ps and another structural reconstruction among the defects is observed. As shown in Fig. 5c and d, as the result of a Stone-Wales transformation of the dimer 1-3, the two heptagons (c and d in Fig. 5c) and one pentagon (b in Fig. 5c) on the left side of the 2-fold symmetry axis are transformed into three hexagons while one hexagonal ring containing the carbon atom 3 is transformed into a heptagonal ring. Finally a pentagon-heptagon pair defect, which has been observed in the experiment after the irradiation [37], is emerged through the reconstruction process. Since the dimer 5-7 is equivalent to the dimer 1-3 due to the 2-fold symmetry at the stage of Fig. 5c, the dimer 5-7 is expected to undergo a similar Stone-Wales transformation. Indeed, after 41 ps of simulation time, the Stone-Wales transformation happens to the carbon dimer 5-7. Consequently another pentagon-heptagon pair defect is formed at the right side of the 2-fold axis in the same way as the formation of the previous pentagon-heptagon pair on the left side of the 2-fold axis. The structure with two pentagon-heptagon pairs in Fig. 5e is very stable energetically and can sustain its shape without any changes for more than 20 ps in the simulation even at a temperature ∼4,500 K. At the final stage of the simulation, the system is gradually cooled down to 0 K in 12.5 ps and the structure with two pentagon-heptagon pair defects is found to maintain without any additional reconstruction as shown in Fig. 5f.

Because each pentagon-heptagon is a topological dislocation defect of the grapheme sheet, Fig. 5 in fact shows an elementary dislocation climb process. Unlike dislocation glide [38,39] which conserves the number of atoms while relaxing stress, dislocation climb is a non-conservative process that requires removal or addition of atoms by radiation knock-out, vacancy diffusion, evaporation, etc. Fig. 5 is akin to the collapse of a vacancy disk and the formation of Frank edge dislocation loop in metals [40]. We expect that once the edge dislocation dipole (two pentagon-heptagon of opposite polarity) is formed, as shown in Fig. 5, further mass removal and chirality change can occur in a steady fashion by a pentagon-heptagon defect moving up the tube axial direction. Dislocation climb is also stress-coupled,

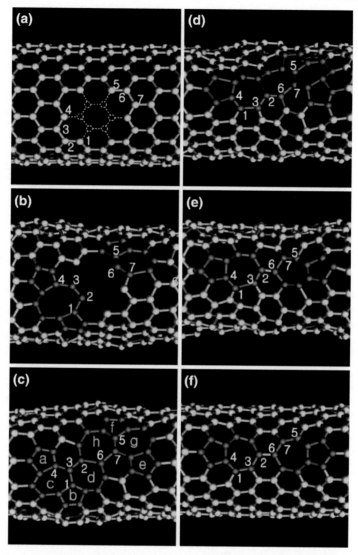

Fig. 5 Atomic processes from the TBMD simulations of a (16, 0) SWCNT with six vacancies. (**a**) 0 K (at time t = 0 ps); (**b**) ~4,500 K (t ≈ 20.2 ps); (**c**) ~3,100 K (t ≈ 23.2 ps); (**d**) ~4,400 K (t ≈ 32.3 ps); (**e**) ~4,700 K (t ≈ 41.5 ps); (**f**) ~90 K (t ≈ 53.9 ps). The carbon atoms on the rear side of the tube are concealed in figures in order to see the reconstruction of vacancies more clearly. Dotted circles in (A) indicate the positions of the six carbon vacancies in the perfect (16, 0) SWCNT. Yellow colors indicate carbon atoms and bonds in hexagonal rings. Blue colors indicate carbon atoms and bonds in non-hexagonal rings. See the text for small letters in (C) and numbers. (From Ref. [29])

which means tensile force on the nanotube can aid/impede climbing motion of the pentagon-heptagon and mass removal/addition, which may shift the semiconductor-metal junctions at high temperatures.

Figure 6 shows the front view of the initial and final structure from the TBMD simulation. The vacancy hole in the initial structure is healed up in the final structure and the radius of the tube in the middle section is reduced. The diameter and chirality in the center

Fig. 6 Front views of initial and final structure from TBMD simulation for (16, 0) SWCNT with six vacancies. The initial structure corresponds to Fig. 5a. The final structure corresponds to Fig. 1f. (From Ref. [29])

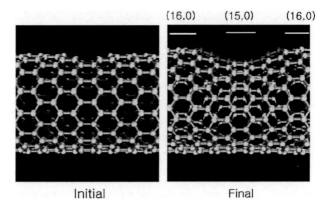

Initial   Final

part of the final structure is found to be (15, 0), which is one of the metallic SWCNTs. In order to understand the effects of the vacancy cluster size on the formation of junctions, they have also performed the TBMD simulation to study the junction formation dynamics of a (16, 0) SWCNT containing a hole of ten vacancies. The formation of two pentagon-heptagon pair defects is also observed, with the mechanism similar to that in the simulation of the (16, 0) SWCNT with six vacancies discussed earlier in this subsection. The most interesting difference between the simulation results of the ten and six vacancies is that the length of the (15, 0) tube section is longer with ten vacancies. These simulation results demonstrate that intramolecular semiconductor-metal junctions of SWCNTs can be produced by irradiation followed by a proper annealing which allow various vacancy defects generated by the irradiation to reconstruct into the pentagon-heptagon pairs at the junction. These simulations also suggest a mechanism for synthesis of carbon nanotube semiconductor-metal intramolecular junctions with specific locations and controlled sizes and show the possibility of application to nanoelectronic devices.

4 EDTB potential for silicon and its applications

4.1 EDTB potential for silicon

Although the diamond structure of Si also exhibits covalent sp^3 bonding configurations, the higher coordinated metastable structures of Si are metallic and with energies close to that of the diamond structure. Therefore, Si can be metallic under high pressures or at high temperatures. For example, the coordination of the liquid phase of Si is close to the coordination of the metallic structures (i.e., 6.5). These properties of Si pose a challenge for accurate tight-binding modeling of Si: it is difficult to describe the low-coordinated covalent structures and high-coordinated metallic structures with good accuracy using one set of tight-binding parameters. With the environment-dependent tight-binding formalism, Wang, Pan, and Ho show that this difficulty can be overcome [14]. The EDTB Si potential developed by them gives excellent fit to the energy vs interatomic distance of various silicon crystalline structures with different coordination as shown in Fig. 7. The EDTB Si potential also describes well the structure and energies of Si surfaces in addition to other bulk properties such as elastic constants and phonon frequencies [14]. These results can be seen from Tables 5 and 6. The parameters of the EDTB Si potential are listed in Tables 7 and 8. The parameters for calculating

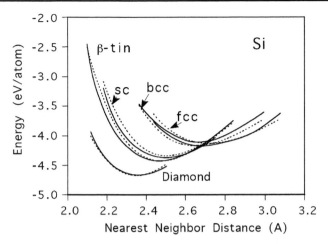

Fig. 7 cohesive energies as a function of nearest neighbor distance for silicon in different crystalline structures calculated using the environment-dependent TB model are compared with the results from the first-principles DFT-LDA calculations. The solid curves are the TB results and the dashed curves are the LDA results. (From Ref. [14])

the coordination number of Si using the Eq. (10) are $\beta_1 = 2.0$, $\beta_2 = 0.02895$, $\beta_3 = 7.96284$. The cutoff distances for the interaction are $r_{cut} = 5.2$ Å and $r_{match} = 4.8$ Å (see Eq. (12)).

A useful benchmark for Si interatomic potentials is a series of model structures for the $\Sigma = 13\{510\}$ symmetric tilt boundary structures in Si [41]. Eight different structures as indicated in the horizontal axis of Fig. 8 have been selected for the calculations. These structures were not included in the database for fitting the parameters. The structures are relaxed by steepest-decent method until the forces on each atom were less than 0.01 eV/Å. The energies obtained from the calculations using the EDTB Si potential are compared with the results from *ab initio* calculations, and from two-center Si tight-binding potentials [42], and classical potential calculations [43,44] as shown in Fig. 8. The energy differences for different structures predicted by the EDTB calculations agree very well with those from the *ab initio* calculations. The energies from the two-center tight-binding potentials and classical potentials do not give the correct results in comparison with the results from *ab initio* and environment tight-binding potential calculations even though the atoms in the structures are all four-fold coordinated.

4.2 TBMD simulation studies of addimer diffusion on Si(100) surface

The EDTB silicon potential has been used to investigate the diffusion pathways and energy barriers for Si addimer diffusion along the trough and from the trough to the top of dimer row on Si(100) surface [45,46]. Diffusion of Si addimers on the Si(100) surface have attracted numerous experimental and theoretical investigations [47–52] because it plays an essential role in the homoepitaxial growth of silicon films.

Clean Si(100) surfaces exhibit a c(4 × 2) reconstruction in which the surface Si atoms form a row of alternating buckled dimers along the [010] direction [53,54]. There are four principal addimer configurations [47,51] on the Si(100) as shown in Fig. 9. An addimer can sit on top of a dimer row (A and B) or in the trough between two rows (C and D), with its axis oriented either parallel (A and D) or perpendicular (B and C) to the dimer-row direction. All four

Table 5 The parameters obtained from the fitting for the EDTB model of Si [14]. The α_1 is in the unit of eV. Other parameters are dimensionless

	α_1	α_2	α_3	α_4	β_1	β_2	β_3	δ_1	δ_2	δ_3
$h_{ss\sigma}$	−5.9974	0.4612	0.1040	2.3000	4.4864	0.1213	6.0817	0.0891	0.0494	−0.0252
$h_{sp\sigma}$	3.4834	0.0082	0.1146	1.8042	2.4750	0.1213	6.0817	0.1735	0.0494	−0.0252
$h_{pp\sigma}$	11.1023	0.7984	0.1800	1.4500	1.1360	0.1213	6.0817	0.0609	0.0494	−0.0252
$h_{pp\pi}$	−3.6014	1.3400	0.0500	2.2220	0.1000	0.1213	6.0817	0.4671	0.0494	−0.0252
ϕ	126.640	5.3600	0.7641	0.4536	37.00	0.56995	19.30	0.082661	−0.023572	0.006036
$\Delta e_s, \Delta e_p$	0.2830	0.1601	0.050686	2.1293	7.3076	0.07967	7.1364	0.7338	−0.03953	−0.062172

Table 6 The coefficients of the polynomial function $f(x)$ for the EDTB potential of Si

	$c_0 (eV)$	c_1	$c_2 \left(eV^{-1}\right)$	$c_3 \left(eV^{-2}\right)$	$c_4 \left(eV^{-3}\right)$
$x \geq 0.7$	-0.739×10^{-6}	0.96411	0.68061	-0.20893	0.02183
$x < 0.7$	-1.8664	6.3841	-3.3888	0.0	0.0

Table 7 Elastic constants and phonon frequencies of silicon in the diamond structure calculated from the two-center TB model [42] and the environment-dependent TB (EDTB) model [14] are compared with experimental results [16]

	Two-center TB	EDTB	Experiment
a(Å)		5.450	5.430
B	0.876	0.90	0.978
$c_{11}-c_{12}$	0.939	0.993	1.012
c_{44}	0.890	0.716	0.796
$v_{LTO}(\Gamma)$	21.50	16.20	15.53
$v_{TA}(X)$	5.59	5.00	4.49
$v_{TO}(X)$	20.04	12.80	13.90
$v_{LA}(X)$	14.08	11.50	12.32

Elastic constants are in units of $10^{12} dyn/cm^2$ and the phonon frequencies are in terahertz

Table 8 Surface energies of the silicon (100) and (111) surfaces from the EDTB Si potential [14]

Structure	Surface energy	ΔE
Si(100)		
(1×1)-ideal	2.292	0.0
(2×1)	1.153	-1.139
$p(2 \times 2)$	1.143	-1.149
$c(4 \times 2)$	1.148	-1.144
Si(111)		
(1×1)-ideal	1.458	0.0
(1×1)-relaxed	1.435	-0.025
(1×1)-faulted	1.495	0.037
$\sqrt{3} \times \sqrt{3} - t_4$	1.213	-0.245
$\sqrt{3} \times \sqrt{3} - h_3$	1.346	-0.112
(2×1)-Haneman	1.188	-0.270
(2×1)-π-bonded chain	1.138	-0.320
(7×7)-DAS	1.099	-0.359

ΔE is the energy relative to that of the (1×1)-ideal surface. The energies are in the unit of $eV/(1 \times 1)$

configurations have been identified in scanning tunneling microscopy (STM) experiments [55]. Addimer configuration A is the lowest energy configuration. The relative energies of the four addimer configurations A, B, C, and D are 0.0, 0.02, 0.28, and 1.02 eV respectively from the tight-binding calculations as compare to 0.0, 0.03, 0.24, and 0.91 eV respectively from first principles calculations.

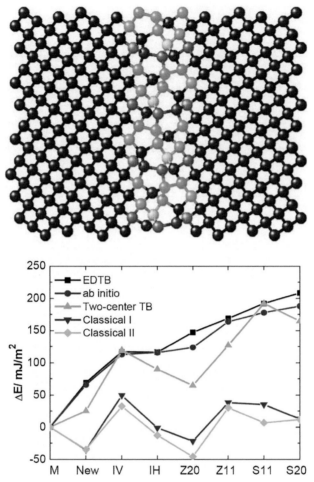

Fig. 8 Energies of the $\Sigma = 13\{510\}$ symmetric tilt boundary structures in Si. Eight different structures as indicated in the horizontal axis were selected for calculations. The energies are relative to that of the structure M which has been identified by experiment. The energies obtained from the calculations using the EDTB Si potential are compared with results from *ab initio* calculations, and from two-center Si tight-binding potentials [42], and classical potential calculations (classical I [43] and classical II [44]). The results of EDTB, *ab initio*, and classical I are taken from Ref. [41]

Experimental evidence and theoretical calculations [49,52] suggest that the diffusion of addimers has an anisotropic property: they prefer diffusion along the top of the dimer rows. However, using the atom tracking method [56], addimer diffusion along the troughs as well as crossing the trough to the next dimer row at a temperature of 450 K, in addition to the diffusion along the top of the dimer rows has also been observed [57]. The energy barrier for addimer to diffuse along the trough and to leave the trough to the top of the dimer row are estimated by STM experiment to be 1.21 ± 0.09 eV and 1.36 ± 0.06 eV respectively [57].

Because the unit cell used in such calculations contains a large number of atoms, a comprehensive search for the low energy barriers diffusion pathway is very expensive using *ab initio* methods. Here we have employed tight-binding molecular dynamics calculations to explore the possible diffusion pathways and select plausible candidate pathways for study by

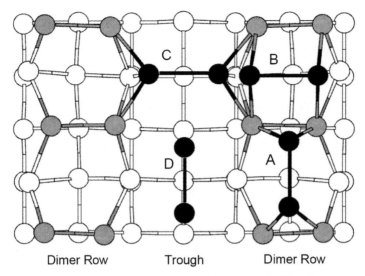

Dimer Row   Trough   Dimer Row

Fig. 9 Schematic drawing of the four principal dimer configurations on Si(100). Black circles represent the Si addimers, the gray circles represent the dimer atoms of the Si(100) substrate, and the open circles represent the subsurface atoms

more accurate *ab initio* calculations. The tight-binding studies reveal new pathways which have diffusion barriers in excellent agreement with the experimentally estimated values. The EDTB silicon tight-binding model reproduces excellently the experimental observation and the *ab initio* calculation results for addimer diffusion and opens up the possibility of studying surface dynamics on the Si(100) surface by using tight-binding molecular dynamics.

4.2.1 Diffusion between trough and the top of dimer row

Most of the previous calculations consider a straightforward pathway for addimer diffusion from trough to the top of dimer row by a direct translational motion of perpendicular addimer from C to B (path I). The energy as the function of addimer displacement along this pathway obtained by the tight-binding calculations is plotted in Fig. 10a (solid line) which shows that the energy barrier for diffusion of an addimer from C to B along this pathway is 1.72 eV, much larger than the experimental value of 1.36 eV.

The energy barrier for the diffusion of a parallel addimer from D to A has also been investigated. The energy as a function of addimer displacement for D to A along the straight pathway is plotted in Fig. 10b (solid line). The diffusion barrier from D to A is only 0.88 eV, which is much smaller than the experimental value of 1.36 eV. However, since the energy of the D configuration is 0.74 eV higher than that of the C configuration, the total energy barrier for diffusion from C to A via D (path II) is at least 1.62 eV which is also much higher than the experimental value.

Using tight-binding molecular dynamics as a search engine, Lee el al. discovered an unusual diffusion pathway for a Si addimer to diffuse between trough and the top of the dimmer row [45]. This pathway (path III) consists of rotation of the addimer along the diffusion pathway as shown in Fig. 11. The energy along this pathway is plotted in Fig. 10c (solid line). The tight-binding calculation gives an energy barrier of 1.37 eV for addimer diffusion from C to B, in excellent agreement with the experimental value of 1.36 eV [57].

Fig. 10 The total energy variations for (**a**) the direct translational diffusion of a perpendicular addimer (path I), (**b**) the direct translational diffusion of a parallel addimer (path II), and (**c**) the diffusion consisting of rotation of addimer (path III). In each figure, solid lines represent the calculations by our tight-binding model and dashed lines represent the LDA calculations. Energies are compared with respect to the energy of the dimer configuration A in Fig. 9. The abscissa is the position of the center of the addimer from the center-line of the trough to the center-line of the dimer row. Numbers over points in figure (**c**) indicate the geometries in Fig. 11

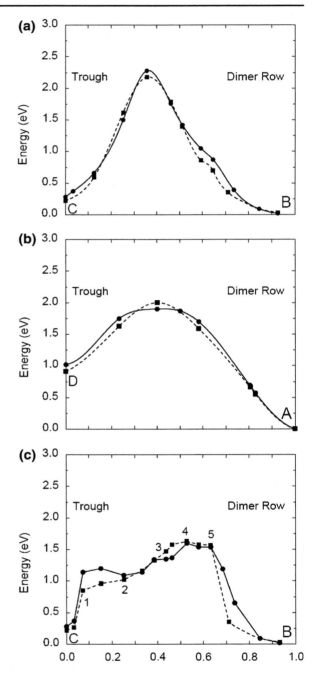

The diffusion containing rotation is more energetically favorable than the translational diffusion of the perpendicular addimer (path I) because a smaller number of broken bonds is involved along path III.

The above results from the tight-binding calculations are further confirmed by first-principles calculations as one can see the comparison plotted in Fig. 10.

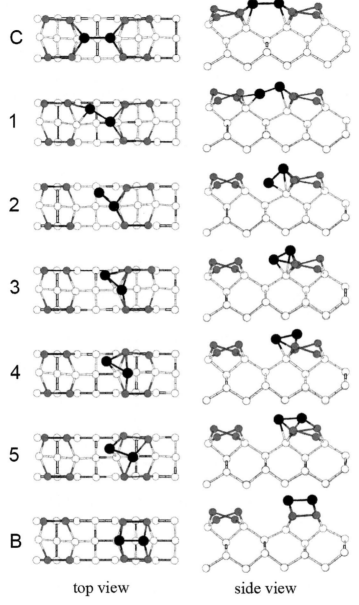

C

1

2

3

4

5

B

top view   side view

Fig. 11 Principal geometries by LDA calculations on the diffusion pathway III. Black circles represents the Si addimer, the gray circles represent the dimer atoms of the Si(100) substrate and the open circles represent the subsurface atoms. Numbering of each geometry corresponds to the number over points in Fig. 10c

4.2.2 Diffusion along the trough between the dimmer rows

Diffusion of an addimer can be viewed as a combination motion of the two individual adatoms as illustrated in top left corner of Fig. 12. The energy surface of the two silicon adatoms diffusion along the trough between the dimer rows are calculated using the environment-dependent silicon tight-binding potential. A contour plot of the resulting energy surface is shown in Fig. 12. The ζ and ξ axes of the contour indicate the displacements of each adatom along the trough from their initial positions in geometry C, the most stable position of the addimer in the trough.

The tight-binding calculations resolve three local minima M1, M2, and M3 on the left hand side of the symmetry line PQ and two saddle points, one on the symmetry line PQ (S1) and another below the line PQ (S2). The calculations also show that addimer diffusion along the trough without dissociation or rotation (along the line C-Q-C′) has very high

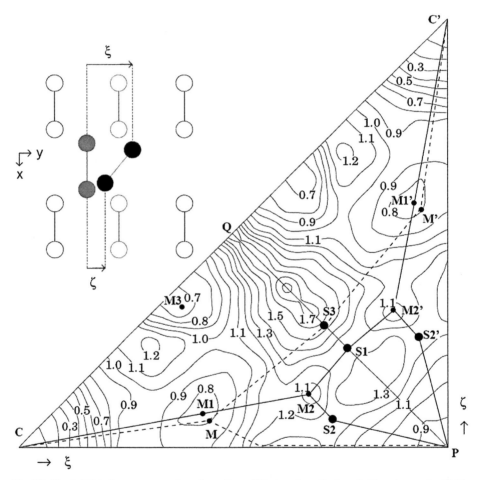

Fig. 12 The tight-binding energy contour for a dimer diffusing along the trough. Energies are in eV. The ζ and ξ axes are the displacements along the diffusion direction of the two adatoms, as illustrated in figure above (Black circles indicate the addimer, opaque circles indicate the substrate dimer atoms, and gray circles indicate the initial position of addimer). C is the initial position for the addimer, P indicates completely separated adatoms on neighboring sites, C′ indicates the addimer diffused to the neighboring site from the initial position. Line CC′ indicates the translational diffusion path without any rotation of the addimer. The line connecting Q to P indicates that the addimer on the line connecting surface dimers separates into two atoms on neighboring sites without any translation of the center of the addimer. The solid lines indicate the diffusion pathways by the tight-binding calculation. Small black spots and large black spots on the lines indicate local minima and saddle points, respectively. The dashed lines indicate the diffusion pathways by the LDA calculation. Note that the dashed lines are not related to the energy contour in this figure and are related to the position of adatoms

energy barrier of about 1.7 eV. From the resulting energy contour, one can identify two paths of diffusion which have similar low energy barriers. One path (1) follows the lines C → M1 → M2 → S2 → P → S2′ → M2′ → M1′ → C′. The other path (2) follows the lines C → M1 → M2 → S1 → M2′ → M1′ → C′. The local minima M1 and M2, which are surrounded by small energy barriers, are on the paths for addimer diffusion from C to C′. The energy barrier for path 1 is 1.26 eV. The addimer on path 1 dissociates into two monomers from M2 to P and reform at M2′. This pathway is similar to the dissociation pathway modeled by Goringe et al. [58]. The highest energy barrier along path 2 is 1.27 eV which is very similar to the energy barrier of path 1. However, the addimer along path 2 does not dissociate but instead rotates to minimize the energy barrier. Starting from the two best candidates for pathways predicted by the tight-binding calculations, more accurate first-principles calculations have been applied to further optimize the pathways and diffusion barriers. While the diffusion pathways after the optimization by first-principles calculations are slightly different from the tight-binding predictions, they are essentially similar and with almost identical diffusion barriers. It should be noted that without the comprehensive tight-binding calculations to search for the possible diffusion pathway, it would be very difficult for first-principles calculation to find out the correct diffusion paths and barriers for this complex system.

4.3 TBMD study of dislocation core structure in Si

The EDTB potential of Si also been applied to the study of dislocation in Si, a much more complex system. Crystalline structure of Si possess two types of (111) planes for inelastic shear, a widely separated shuffle-set (blue in Fig. 13 and a compact glide-set (red). In 2001, experiment using transmission electron microscopy (TEM) [59] show that at low temperature and high pressure, long undissociated screw dislocations appear, which cross-slip frequently. Suzuki et al. proposed [60,61] that these may be shuffle-set screw dislocations, centered at A

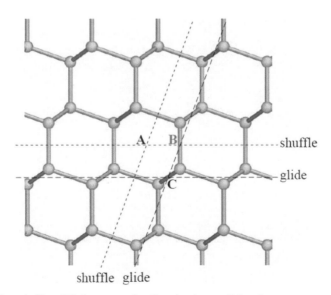

Fig. 13 Slip planes in Si, and likely centers of undissociated screw dislocations

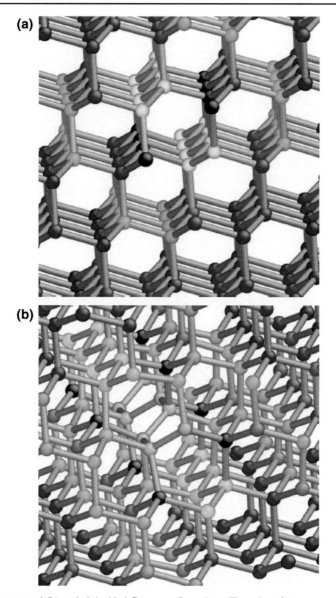

Fig. 14 (**a**) A core, and (**b**) period-doubled C core configurations. The color of atoms represents the local atomic shear strain

and B in Fig. 13 based on energy calculations using empirical classical potentials. Pizzagalli et al. then performed DFT calculations to show that the A core (Fig. 14a) has lower energy [62] than B, as well as a single-period glide-set full screw dislocation C.

Using tight-binding calculation with the EDTB silicon potential, we are able to investigate the core structure of this dislocation with larger number of atoms and with a calculation supercell that has more layers in the direction along the dislocation line [63]. We found that the C core has lower energy than A after period-doubling reconstruction in the direction along the dislocation line. This double-period C core structure is shown in Fig. 14b. Since

C can cross-slip between two glide-set planes, it satisfies all the experimental observations to date. We believe that the double-period C configuration, previously unstudied, may also play important transient roles in partial dislocation constriction and cross-slip at high temperature, and perhaps ductile-to-brittle transition [64]. The prediction from the tight-binding calculations are further confirm by first-principles calculations. We find that after the period-doubling reconstruction, the C core energy is lower than A by $0.16\,eV/\text{Å}$ in TB, and $0.14\,eV/\text{Å}$ in DFT. We also find that the double-period C is energetically favorable due to the electronic structure contribution. The single-period C core has a semi-metallic chain of dangling bonds which introduces electronic states near the Fermi level. This chain of dangling bonds is susceptible to Peierls distortion [65,66], leading to a period-doubling reconstruction along the chain (Fig. 14b) that opens up a wide band gap. Such an electronic mechanism is missed in the classical potential calculations.

5 Future perspective

Success of the EDTB modeling and simulations are not limited to the carbon and silicon systems as we discussed above. Its success has also been extended to the transition metal systems such as Mo [67–71], simple metal systems such as Al [72,73], and Pb [74], as well as two components systems such as Si-H [75].

In spite of these progresses, the development of environment-dependent tight-binding models so far still relies on empirical fitting to the band structure and total energies of some standard structures. The fitting procedure is quite laborious if we want to study a broad range of materials, especially in compound systems where different sets of interactions have to be determined simultaneously from a given set of electronic structures. Moreover, fundamental questions such as how and to what extent the approximations used in the Slater-Koster scheme influence the transferability of the tight-binding models are still not well understood from the empirical fitting approach. Information from first-principles calculations about these issues is highly desirable to guide the development of more accurate and transferable tight-binding models.

In general, overlap and one-electron Hamiltonian matrices from first-principles calculations cannot be used directly to infer the tight-binding parameters because fully converged first-principles calculations are done using a large basis set while tight-binding parameters are based on a minimal basis representation. Very recently, the authors and co-workers have developed a method for projecting a set of chemically deformed atomic minimal-basis-set orbitals from accurate first-principles wavefunctions [76–81]. These orbitals, referred to as "quasi-atomic minimal-basis-sets orbitals" (QUAMBOs), are highly localized on atoms and exhibit shapes close to orbitals of the isolated atom. Moreover, the QUAMBOs span exactly the same occupied subspace as the original first-principles calculation with a large basis set. Therefore, accurate tight-binding Hamiltonian and overlap matrix elements can be obtained directly from *ab initio* calculations through the construction of QUAMBOs. This new development enables us to examine the accuracy and transferability of the tight-binding models from a first-principles perspective.

Acknowledgment Ames Laboratory is operated for the U.S. Department of Energy by Iowa State University under Contract No. DE-AC02-07CH11358. This work was supported by the Director for Energy Research, Office of Basic Energy Sciences including a grant of computer time at the National Energy Research Supercomputing Center (NERSC) in Berkeley. Work of JL is supported by NSF CMMI-0728069, AFOSR, ONR N00014-05-1-0504, and Ohio Supercomputer Center.

References

1. Slater, J.C., Koster, G.F.: Phys. Rev. **94**, 1498 (1954)
2. Wang, C.Z., Chan, C.T., Ho, K.M.: Phys. Rev. B **39**, 8586 (1989)
3. Khan, F.S., Broughton, J.Q.: Phys. Rev. B **39**, 3688 (1989)
4. Goodwin, L., Skinner, A.J., Pettifor, D.G.: Europhys. Lett. **9**, 701 (1989)
5. Wang, C.Z., Chan, C.T., Ho, K.M.: Phys. Rev. B **42**, 11276 (1990)
6. Cohen, R.E., Mehl, M.J., Papaconstantopoulos, D.A.: Phys. Rev. B **50**, 14694 (1994)
7. Mercer, J.L. Jr., Chou, M.Y.: Phys. Rev. B **49**, 9366 (1993)
8. Mercer, J.L. Jr., Chou, M.Y.: Phys. Rev. B **49**, 8506 (1994)
9. Colombo, L.: In: Stauffer, D. (ed.) Annual Reviews of Computational Physics, vol. IV, p. 147. World Scientific, Singapore (1996)
10. Wang, C.Z., Chan, C.T., Ho, K.M.: Phys. Rev. Lett. **66**, 189 (1991)
11. Wang, C.Z., Chan, C.T., Ho, K.M.: Phys. Rev. B **45**, 12227 (1992)
12. Mehl, M.J., Papaconstantopoulos, D.A.: In: Fong, C.Y. (ed.) Topic in Computational Materials Science, pp. 169–213. World Scientific, Singapore (1997)
13. Tang, M.S., Wang, C.Z., Chan, C.T., Ho, K.M.: Phys. Rev. B **53**, 979 (1996)
14. Wang, C.Z., Pan, B.C., Ho, K.M.: J. Phys. Condens. Matter **11**, 2043 (1999)
15. Xu, C.H., Wang, C.Z., Chan, C.T., Ho, K.M.: J. Phys. Condens. Matter **4**, 6047 (1992)
16. Semiconductors: physics of Group IV elements and III–V compounds. In: Madelung, O., Schulz, M., Weiss, H. (eds.) Landolt-Börnstein New Series, vol. III/17a. Springer, Berlin (1982)
17. Semiconductors: intrinsic properties of Group IV elements and III–V, II–VI and I–VII Compounds. In: Madelung, O., Schulz, M. (eds.) Landolt-Börnstein New Series, vol. III/22a. Springer, Berlin (1987)
18. Dresselhaus, M.S., Dresselhaus, G.: In: Cardona, M., Guntherodt, G. (eds.) Light Scattering in Solids III, p. 8. Springer, Berlin (1982)
19. Robertson, J.: Adv. Phys. **35**, 317(1986)
20. Diamond and diamond-like films and coatings. In: Clausing R., et al. (eds.) NATO Advanced Study Institutes Series B, vol. 266, p. 331. Plenum, New York (1991)
21. McKenzie, D.R., Muller, D., Pailthorpe, B.A.: Phys. Rev. Lett. **67**, 773 (1991)
22. Gaskell, P.H., Saeed, A., Chieux, P., McKenzie, D.R.: Phys. Rev. Lett. **67**, 1286 (1991)
23. Wang, C.Z., Ho, K.M.: Phys. Rev. Lett. **71**, 1184 (1993)
24. Marks, N.A., McKenzie, D.R., Pailthorpe, B.A., Bernasconi, M., Parrinello, M.: Phys. Rev. Lett. **76**, 768 (1996)
25. Wang, C.Z., Ho K.M.: Structural trends in amorphous carbon. In: Siegal, M. P., et al. (eds.) MRS Symposium Proceedings, vol. 498 (1998)
26. Mathioudakis, C., Kopidakis, G., Kelires, P.C., Wang, C.Z., Ho, K.M.: Phys. Rev. B **70**, 125202 (2004)
27. Lee, Gun-Do, Wang, C.Z., Yoon, Euijoon, Hwang, Nong-Moon, Kim, Doh-Yeon, Ho, K.M.: Phys. Rev. Lett. **95**, 205501 (2005)
28. Lee, Gun-Do, Wang, C.Z., Yoon, Euijoon, Hwang, Nong-Moon, Ho, K.M.: Phys. Rev. B. **74**, 245411 (2006)
29. Lee, Gun-Do, Wang, C.Z., Yu, Jaejun, Yoon, Euijoon, Hwang, Nong-Moon, Ho, Kai-Ming: Phys. Rev. B **76**, 165413 (2007)
30. Lee, Gun-Do, Wang, C.Z., Yu, Jaejun, Yoon, Euijoon, Hwang, Nong-Moon, Ho, Kai-Ming: Appl. Phys. Lett. **92**, 043104 (2008)
31. Kaxiras, E., Pandey, K.C.: Phys. Rev. Lett. **61**, 2693 (1988)
32. Nordlund, K., Keinonen, J., Mattila, T.: Phys. Rev. Lett. **77**, 699 (1996)
33. Ewels, C.P., Telling, R.H., El-Barbary, A.A., Heggie, M.I., Briddon, P.R.: Phys. Rev. Lett. **91**, 025505 (2003)
34. Lu, A.J., Pan, B.C.: Phys. Rev. Lett. **92**, 105504 (2004)
35. Sammalkorpi, M., Krasheninnikov, A., Kuronen, A., Nordlund, K., Kaski, K.: Phys. Rev. B **70**, 245416 (2004)
36. Terrones, H., Terrones, M., Herna'ndez, E., Grobert, N., Charlier, J.-C., Ajayan, P.M.: Phys. Rev. Lett. **84**, 1716 (2000)
37. Hashimoto, A., Suenaga, K., Gloter, A., Urita, K., Iijima, S.: Nat. Lond. **430**, 870 (2004)
38. Yakobson, B.I.: Appl. Phys. Lett. **72**, 918 (1998)
39. Mori, H., Ogata, S., Li, S.J., Akita, S., Nakayama, Y.: Phys. Rev. B **76**, 165405 (2007)
40. Arakawa, K., Ono, K., Isshiki, M., Mimura, K., Uchikoshi, M., Mori, H.: Science **318**, 956 (2007)
41. Morris, J.R., Lu, Z.Y., Ring, D.M., Xiang, J.B., Ho, K.M., Wang, C.Z., Fu, C.L.: Phys. Rev. B **58**, 11241 (1998)
42. Kwon, I., Biswas, R., Wang, C.Z., Ho, K.M., Soukoulis, C.M.: Phys Rev B **49**, 7242 (1994)

43. Tersoff, J.: Phys. Rev. B **38**, 9902 (1988)
44. Lenosky, T.J., Sadigh, B., Alonso, E., Bulatov, V.V., Diaz de la Rubia, T., Kim, J., Voter, A.F., Kress, J.D.: Model. Simul. Mater. Sci. Eng. **8**, 825 (2000)
45. Lee, Gun-Do, Wang, C.Z., Lu, Z.Y., Ho, K.M.: Phys. Rev. Lett. **81**, 5872 (1998)
46. Lee, Gun-Do, Wang, C.Z., Lu, Z.Y., Ho, K.M.: Surf. Sci. **426**, L427 (1999)
47. Zhang, Z., Wu, F., Zandvliet, H.J.W., Poelsema, B., Metiu, H., Lagally, M.G.: Phys. Rev. Lett. **74**, 3644 (1995)
48. Swartzentruber, B.S., Smith, A.P., Jönsson, H.: Phys. Rev. Lett. **77**, 2518 (1996)
49. Dijkkamp, D., van Loenen, E.J., Elswijk, H.B.: In: Proceedings of the 3rd NEC Symposium on Fundamental Approach to New Material Phases, Springer Series on Material Science. Springer, Berlin (1992)
50. Bedrossian, P.J.: Phys. Rev. Lett. **74**, 3648 (1995)
51. Brocks, G., Kelly, P.J.: Phys. Rev. Lett. **76**, 2362 (1996)
52. Yamasaki, T., Uda, T., Terakura, K.: Phys. Rev. Lett. **76**, 2949 (1996)
53. Chadi, D.J.: Phys. Rev. Lett. **43**, 43 (1979)
54. Wolkow, R.A.: Phys. Rev. Lett. **68**, 2636 (1992)
55. Zhang, Z., Wu, F., Lagally, M.G.: Surf. Rev. Lett. **3**, 1449 (1996)
56. Swartzentruber, B.S.: Phys. Rev. Lett. **76**, 459 (1996)
57. Borovsky, B., Krueger, M., Ganz, E.: Phys. Rev. Lett. **78**, 4229 (1997)
58. Goringe, C.M., Bowler, D.R.: Phys. Rev. B **56**, R7073 (1997)
59. Rabier, J., Cordier, P., Demenet, J.L., Garem, H.: Mater. Sci. Eng. A **309**, 74 (2001)
60. Suzuki, T., Yasutomi, T., Tokuoka, T., Yonenaga, I.: Phys. Status Solidi A **171**, 47 (1999)
61. Koizumi, H., Kamimura, Y., Suzuki, T.: Philos. Mag. A **80**, 609 (2000)
62. Pizzagalli, L., Beauchamp, P.: Philos. Mag. Lett. **84**, 729 (2004)
63. Wang, C.Z., Li, Ju, Ho, K.M., Yip, S.: Appl. Phys. Lett. **89**, 051910 (2006)
64. Pirouz, P., Demenet, J.L., Hong, M.H.: Philos. Mag. A **81**, 1207 (2001)
65. Marder, M.P.: Condensed Matter Physics. 2nd edn. Wiley, New York (2000)
66. Lin, X., Li, J., Först, C.J., Yip, S.: Proc. Natl. Acad. Sci. USA **103**, 8943 (2006)
67. Haas, H., Wang, C.Z., Fahnle, M., Elsasser, C., Ho, K.M.: Phys. Rev. B **57**, 1461 (1998)
68. Haas, H., Wang, C.Z., Fahnle, M., Elsasser, C., Ho, K.M.: In: Turchi, P.E.A., et al. (eds.) MRS Symposium Proceedings, vol. 491, p. 327 (1998)
69. Haas, H., Wang, C.Z., Fahnle, M., Elsasser, C., Ho, K.M.: J. Phys. Condens. Matter **11**, 5455 (1999)
70. Haas, H., Wang, C.Z., Ho, K.M., Fahnle, M., Elsasser, C.: Surf. Sci. **457**, L397 (2000)
71. Li, Ju, Wang, C.Z., Chang, J.-P., Cai, W., Bulatov, V., Ho, K.M., Yip, S.: Phys. Rev. B **70**, 104113 (2004)
72. Chuang, Feng-chuan, Wang, C.Z., Ho, K.M.: Phys. Rev. B **73**, 125431 (2006)
73. Zhang, Wei, Lu, Wen-Cai, Sun, Jiao, Zhao, Li-Zhen, Wang, C.Z., Ho, K.M.: Chem. Phys. Lett. **455**, 232 (2008)
74. Lu, W.C., Wang, C.Z., Ho, K.M.: (to be published)
75. Tang, Mingsheng, Wang, C.Z., Lu, W.C., Ho, K.M.: Phys. Rev. B **74**, 195413 (2006)
76. Lu, W.C., Wang, C.Z., Schmidt, M.W., Bytautas, L., Ho, K.M., Ruedenberg, K.: J. Chem. Phys. **120**, 2629 (2004)
77. Lu, W.C., Wang, C.Z., Schmidt, M.W., Bytautas, L., Ho, K.M., Ruedenberg, K.: J. Chem. Phys. **120**, 2638 (2004)
78. Lu, W.C., Wang, C.Z., Chan, Z.L., Ruedenberg, K., Ho, K.M.: Phys. Rev. B (Rapid Commun.) **70**, 041101 (2004)
79. Lu, W.C., Wang, C.Z., Ruedenberg, K., Ho, K.M.: Phys. Rev. B **72**, 205123 (2005)
80. Chan, T.-L., Yao, Y.X., Wang, C.Z., Lu, W.C., Li, J., Qian, X.F., Yip, S., Ho, K.M.: Phys. Rev. B **76**, 205119 (2007)
81. Qian, X.-F., Li, J., Wang, C.-Z., Qi, L., Chan, T.-L., Yao, Y.-X., Ho, K.-M., Yip, S.: Phys. Rev. B (to be published)

First-principles modeling of lattice defects: advancing our insight into the structure-properties relationship of ice

Maurice de Koning

Originally published in the journal Sci Model Simul, Volume 15, Nos 1–3, 123–141.
DOI: 10.1007/s10820-008-9110-4 © Springer Science+Business Media B.V. 2008

Abstract   We discuss a number of examples that demonstrate the value of computational modeling as a complementary approach in the physics and chemistry of ice I_h, where real-life experiments often do not give direct access to the desired information or whose interpretation typically requires uncontrollable assumptions. Specifically, we discuss two cases in which, guided by experimental insight, density-functional-theory-based first-principles methods are applied to study the properties of lattice defects and their relationship to ice I_hs macroscopic behavior. First, we address a question involving molecular point defects, examining the energetics of formation of the molecular vacancy and a number of different molecular interstitial configurations. The results indicate that, as suggested by earlier experiments, a configuration involving bonding to the surrounding hydrogen-bond network is the preferred interstitial structure in ice I_h. The second example involves the application of modeling to elucidate on the microscopic origin of the experimental observation that a specific type of ice defect is effectively immobile while others are not. Inspired by previous suggestions that this defect type may be held trapped at other defect sites and our finding that the bound configuration is the preferred interstitial configuration in ice I_h, we use first-principles modeling to examine the binding energetics of the specific ice defect to the molecular vacancy and interstitial. The results suggest a preferential binding of the immobile defect to the molecular interstitial, possibly explaining its experimentally observed inactivity.

Keywords   Ice · Density-functional theory · Point defects

1 Introduction

Ice is arguably one of the most abundant, most studied and most fascinating crystalline solid materials on Earth. This fascination is shared by scientists from very different research areas.

M. de Koning (✉)
Instituto de Física Gleb Wataghin, Universidade Estadual de Campinas, Unicamp,
Caixa Postal 6165, 13083-970, Campinas, São Paulo, Brazil
e-mail: dekoning@ifi.unicamp.br

Geophysicists and glaciologists are interested in the dynamics of extended ice masses on the surface of Earth [1]. Engineers are concerned with the specific measures that need to be taken to handle the construction of structures in the presence of ice. Meteorologists and atmospheric scientists are involved in studying the influence of ice on the climatological conditions on our planet and its possible role in global warming. Condensed-matter physicists are, among other issues, attempting to understand the electrical, optical, mechanical and thermodynamic properties of bulk ice, the peculiar characteristics of its surface as well as the complexity of the solid-state part of water's phase diagram.

While waters phase diagram features at least 15 crystalline varieties, the by far most abundant ice form on Earth is hexagonal proton-disordered ice, also known as ice I_h [1]. In addition to being the water phase of the home-made ice cubes that we use to chill our drinks, it is the fundamental constituent of natural ice that appears in the form of glaciers, sea ice, and snow flakes, among others. In this light it is not surprising that the bulk of the scientific research efforts have focused primarily on this particular crystalline form of water.

From a crystallographic point of view, ice I_h is a molecular crystal characterized by the Wurtzite lattice structure [2], which consists of two interpenetrating hexagonal lattices. Each water molecule, whose oxygen atoms occupy the lattice sites, is linked to its four tetrahedrally positioned neighbors by hydrogen bonds [1]. A representative picture of the structure can be appreciated in Fig. 1, which depicts a view along a direction close to the hexagonal c-axis. An essential feature of the ice I_h structure is that there is no long-range order in the orientation of the water molecules on their lattice sites. For this reason, ice I_h is often referred to as a proton-disordered ice phase. The randomness is subject only to the two constraints formulated in terms of the Bernal-Fowler ice rules [1,3], which dictate that, (i), each molecule accepts/donates two protons from/to two nearest-neighbor molecules and, (ii), there be precisely one proton between each nearest-neighbor pair of oxygen atoms. The first, and nowadays rather obvious, rule implies that ice consists of water molecules, whereas the second imposes *local* restrictions on the possible orientations of nearest-neighbor molecules. To facilitate visualization [1], Panel a) of Fig. 2 depicts a square two-dimensional representation of a typical hydrogen-bond topology of a defect-free ice I_h structure that is consistent with these two prescriptions. Each molecule is hydrogen-bonded to four nearest neighbors by donating and receiving two protons, respectively.

In reality, however, defect-free ice crystals do not exist. As is the case for any crystalline solid, real ice I_h crystals are permeated by a variety of crystal defect species. In addition to the kind typically found in any crystalline solid, such as vacancies and interstitials, the properties of ice I_h are affected by defect species that are specific to its particular crystal structure, reflecting its molecular nature and representing violations of the two ice rules. The species of *ionic* defects represents violations of the first ice rule. The fundamental characteristics of these defects can be appreciated in in Panel b) of the two-dimensional square representations of Fig. 2. The ionic H_3O^+/OH^- defect pair, shown by the blue molecules in Panel b), is created by transferring a proton along a hydrogen bond and creating two faulted *molecules*. Successive proton transfer events may then further separate both ions, eventually creating an independent H_3O^+/OH^- defect pair. The second type of protonic defect pair, referred to as the Bjerrum D/L defect pair, originates from a violation of the second ice rule and is formed by the rotation of a molecule about one of its two molecular O-H axes. The resulting defect pair is characterized by two faulted *hydrogen bonds*, one featuring the presence of two protons, referred to as the D defect and depicted schematically as the bond between the blue and green molecules in Fig. 2b, and one without any protons, which is called an L defect and is represented schematically as the bond between the red and green molecules in Fig. 2c.

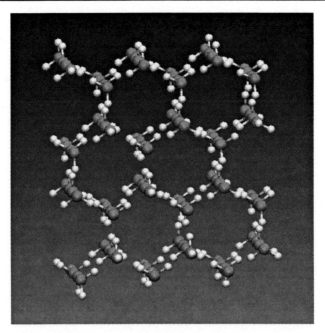

Fig. 1 Typical molecular arrangement of the ice I$_h$ Wurtzite structure. Shown view is along direction vicinal to the hexagonal c-axis

Similar to the case of the ionic defect pair, a series of successive molecular rotations may eventually create an independent pair of D and L defects.

The existence of the protonic defect species was proposed by Niels Bjerrum [4], inspired by the peculiar electrical properties of ice I$_h$. When a specimen of ice is subjected to a static electric field E, for instance, the material responds by electric polarization, mainly through the reorientation of molecular dipole moments in the crystal. This is a relatively slow process, in which the time evolution of the electrical polarization P is a so-called Debye relaxation process described by

$$\frac{dP}{dt} = \frac{1}{\tau_D}(P_s - P),$$ (1)

where P_s is the equilibrium polarization and τ_D is the Debye relaxation time. The phenomenology of the electrical polarization in the liquid phase of water is the essentially the same as that in ice. However, there is an enormous disparity in the time scales τ_D associated with molecular-dipole realignments in both phases. Whereas the typical Debye relaxation time in ice I$_h$ at a temperature of -10 °C is of the order of 10^{-5} s, its value in liquid water at $+10$ °C is approximately 6 orders of magnitude shorter, at 10^{-11} s. This tremendous timescale gap is due to the fact that, compared to the liquid state, it is much more difficult for the molecules to reorient themselves in the crystalline phase due to their bonding in the lattice and the ice rules. Indeed, as recognized by Bjerrum, in an ice crystal in which both ice rules are strictly obeyed, the molecular-dipole realignments would not be possible at all, given that the orientation of a given molecule is then strictly determined by its neighbors. This insight led Bjerrum to postulate the existence of protonic defects, whose thermally activated motion provides a mechanism for dipolar reorientations as can be inferred from Fig. 2.

Fig. 2 Square two-dimensional representation of the hydrogen-bond topology in ice I_h. **a** Defect-free ice. **b** Formation of ionic defect pair. **c** Formation of Bjerrum defect pair

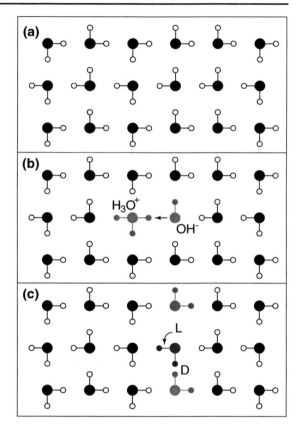

Subsequently, [5] developed a quantitative theory describing the electrical properties of ice in terms of the concentrations, formation energies, migration energies (mobilities) and the effective charges of the D and L Bjerrum defects and the H_3O^+/OH^- ionic defects. Since it provides a direct link between the microscopic structural properties of ice defects and the macroscopic electrical properties of ice I_h, Jaccards theory has been used extensively to interpret experimental observations in terms of the properties of the individual Bjerrum and ionic defects.

One of the main difficulties in this effort, however, is that the extraction of separate formation and migration energies requires measurements on *doped* ice samples, in which an *extrinsic* (i.e. athermal) concentration of protonic defects is introduced into the lattice [1]. Typical dopants are HCl, HF and NH_3, which, if incorporated substitutionally into the ice lattice, enable the release of different species of protonic defects into the crystal. An example is shown in Fig. 3, which depicts the substitutional incorporation and dissociation of an HCl molecule into an L-type Bjerrum defect and a hydronium ion. This doping is quite similar to the effect of doping a semiconductor with donor or acceptor atoms, leading to the release of electrons or holes and providing extrinsic carriers to participate in electrical conduction. Because of this similarity, ice I_h is sometimes referred to as a protonic semiconductor, given that the electrical activity in ice is mediated by protons rather than electrons. The interpretation of experimental results based on doped ice samples, however, calls for the introduction of a number of uncontrolled assumptions. One of them, for instance, is that in case of doping with the acids HCl or HF, the extrinsic L Bjerrum defect is assumed to be fully dissociated

Fig. 3 Substitutional incorporation and dissociation of an HCl molecule into an L-type Bjerrum defect and a hydronium ion

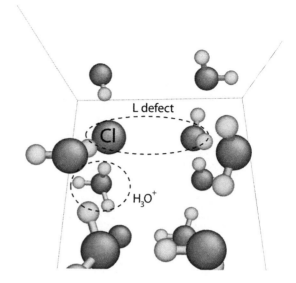

from the chlorine or fluorine sites, behaving as a free L defect. As a consequence of these assumptions, the values of the formation and migration energies of protonic defects extracted from experiments on doped ice samples are subject to large uncertainties. Moreover, different experiments have led to results that are mutually inconsistent.

Such issues are not restricted to protonic defects and their role in ice's electrical properties. Indeed, even the most basic crystalline defects in ice, namely the molecular vacancy and self interstitial, remain a subject of debate. [1] First, positron annihilation experiments [6,7] indicated that the vacancy should be the prevalent molecular point-defect species. Later on, a series of X-ray topographical studies of dislocation loops [8–12] provided convincing evidence for the opposite, indicating that for $T \gtrsim -50\,^{\circ}\mathrm{C}$ the self-interstitial is dominant. The structure of this self interstitial, however, remains unknown. In addition to a surprisingly high formation entropy of $\sim 4.9\,k_B$, the X-ray studies inferred a formation energy below the sublimation energy of ice I_h, which led to the suggestion [8] that its structure might involve bonding to the surrounding hydrogen-bond network. This idea, however, contrasted with the established consensus [1] that the relevant structures involve cavity-centered sites in the ice I_h lattice, [1], for which such bonding is not expected [1,13].

In view of the difficulties in the experiments and the interpretation of their results, modeling and simulation techniques provide an extremely useful complementary approach in studying the physics and chemistry of ice. Guided by experimental insight, it allows the design and execution of controlled computational "experiments" that provide information not otherwise accessible. In the remainder of this article, we will demonstrate the capabilities of this approach, discussing two examples in which first-principles modeling methods are applied to gain insight into the characteristics of defects in ice I_h and their relationship to ice I_hs macroscopic behavior. First, we address a question involving molecular point defects in ice I_h. Specifically, we examine the energetics of formation of the molecular vacancy and a number of different molecular interstitial configurations. The results indicate that, as suggested by earlier experiments, a configuration involving bonding to the surrounding hydrogen-bond network is the preferred interstitial structure in ice I_h. Moreover, they allude to a possible cross-over in dominance between the molecular vacancy and interstitial, with the former being the most abundant thermal equilibrium point-defect species at lower temperatures and the latter taking over close to the melting point, which was also suggested experimentally. The second example involves the application of modeling to elucidate on the microscopic origin of the robust experimental observation [1] that D-type Bjerrum defects are effectively immobile and that only L defects play a role in the Bjerrum-defect mediated electrical response of ice I_h. Inspired by the suggestion that D-defects may be held trapped at other defect sites, we use first-principles modeling to examine the binding energetics of D and L-type Bjerrum defects to the molecular vacancy and interstitial. Our previous finding of the bound interstitial structure plays a significant role here, with the results displaying a preferential binding of D-type Bjerrum defects to this type of interstitial as compared to the L-type Bjerrum defect. No such preferential binding is found for Bjerrum defect/vacancy defect complexes. A calculation of the corresponding equilibrium concentrations then shows that such a preferential binding may lead to conditions under which the concentration of free L defects is significantly larger than that of free D-defects, possibly explaining the experimentally observed activity asymmetry of both Bjerrum defect species.

2 Molecular point defects

In crystalline solids the process of self-diffusion is typically mediated by the migration of point defects such as vacancies and interstitials. In ice I_h this is no different, although the involved defects are not "point-like", in view of the molecular character of ice. In case of the vacancy this does not present any serious difficulties and its structure is readily understood, representing a site on which an entire water molecule is missing, as is shown in Panel (a) of Fig. 4. The situation for the interstitial is different, however. Although we know that it concerns a water molecule in excess of those positioned on regular crystal sites, it is not immediately clear where the additional molecule should be placed. Panel (b) of Fig. 4 depicts a view along a direction vicinal to the hexagonal c-axis of the ice I_h structure, with one excess water molecule. In this particular example, the interstitial molecule has been placed in the center of a 12-molecule cage, although other positions are possible. We will return to this point below.

From the experimental point of view, it has not yet been possible to obtain direct information as to the preferred structure of the self interstitial in ice I_h. In this context, computational modeling provides a useful complementary approach that allows one to consider a number of different interstitial configurations, optimizing their geometries and evaluating the respective formation energetics.

As a first step towards this goal we need to select the methodology used to model the ice I_h system. This choice involves essentially two possibilities. On the one hand we may choose a semi-empirical approach, in which the interactions between the water molecules is described in terms of interatomic potentials. There is a large variety of such models today, featuring different functional forms and fitting databases [14]. A second course of action is to embark on a first-principles electronic structure approach, in which the inter and intramolecular interactions are treated quantum-mechanically. In the present case we opt for the latter, in view of the prospect that the presence of defects may lead to intramolecular distortions, modifying the equilibrium values of both the O-H distance as well as the H-O-H bond angles. In addition to affecting the total energy of a configuration, such distortions are also expected to modify the intramolecular vibrational frequencies and the associated zero-point contributions to the defect formation energetics.

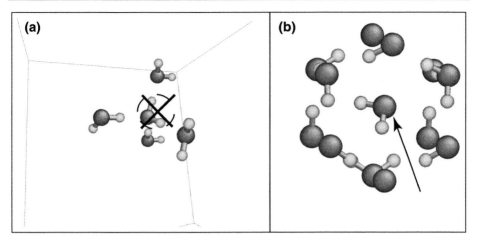

Fig. 4 Molecular point defects in ice I$_h$. **a** Molecular vacancy. **b** Molecular interstitial. Arrow indicates excess water molecule

For this purpose, we employ density-functional theory (DFT) [15,16] as implemented in the VASP package [17,18] to provide a description of the energetics of ice I$_h$. As our particular choice for the exchange-correlation functional we utilize the Perdew-Wang 91 generalized-gradient approximation and select the projector-augmented-wave [19] approach. We restrict Brillouin-zone sampling to the Γ-point and adopt a plane-wave cut-off of 700 eV. As our order model for ice I$_h$ we employ the 96-molecule periodic supercell that was created by Hayward and Reimers [20], keeping in mind that the two ice rules should be satisfied at each site and that there should be no long-range order in the molecular orientations.

This particular modeling approach has shown to provide a satisfactory description of different aspects of ice I$_h$. In addition to giving a reasonable sublimation energy of 0.69 eV (compared to the experimental estimate of 0.611 eV) [21,22], it yields a good estimate of the effective charge carried by the Bjerrum defects, one of the key parameters in Jaccards electrical theory of ice. To determine this quantity, we measured the excess energy (energy difference between defected and defect-free cells) of a D-L Bjerrum defect pair as a function of the distance between them[22]. For this purpose we created sequences of cells containing a D-L defect pair at different distances by sequences of molecular rotations of the type shown in Panel b) of Fig. 2. When the separation between the D and L defects is sufficiently large, the interaction between them is expected to become of the simple Coulombic type, with point charges $+q_{DL}$ (on the D defect) and $-q_{DL}$ (on the L defect) at a distance r, interacting in a dielectric continuum:

$$E_{\text{excess}}(r) = E_{DL} - \frac{q_{DL}^2}{4\pi\epsilon r}. \tag{2}$$

Here, E_{DL} is the excess energy in the limit $r \to \infty$ and ϵ is an appropriate dielectric constant. Fig. 5 shows results of the excess energy as a function of the inverse distance r^{-1} for two different D-L separation sequences. According to Eq. (2), this plot should be a straight line, of which the slope allows the determination of the magnitude of the effective charge. Indeed, both linear regressions provide good fits to the data, giving, respectively, $q_{DL} = (0.34 \pm 0.07)\,e$ and $q_{DL} = (0.35 \pm 0.05)\,e$, with e the electron charge. This value is in excellent agreement with the experimental estimate of $q_{DL} = 0.38\,e$ [1] and establishes confidence in the chosen modeling methodology.

Fig. 5 Bjerrum defect pair
excess energy as a function of the
inverse distance r^{-1} for two
different D-L separation
sequences

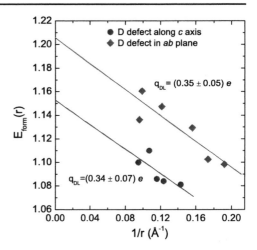

To address the question of the molecular point defects in ice, [23] we start by creating a number of different cells containing a single defect. For this purpose we always start from the defect-free 96-molecule reference cell, followed by the introduction of a single vacancy or interstitial. To capture the effect of proton disorder, which leads to variations in the local environment of each lattice site, we create various realizations of each defect type, placing it at different locations of the reference cell. For the case of the molecular vacancy this is straightforward, removing a single water molecule from different lattice sites of the defect-free cell. The situation involving the molecular interstitial, however, is somewhat more involved because of the *a priori* unknown location of the excess molecule. The most obvious point to start are the two cage center sites available in the ice I_h lattice structure. Placing the excess molecule in these positions, known as the uncapped trigonal (Tu) and capped trigonal (Tc) sites, respectively, [1] then leads to interstitial geometries of the kind shown in Fig. 6a and b, which were obtained by minimizing the forces on all atoms in the cell at fixed volume. The pictures clearly show that these two interstitial structures, although stable, do not involve any bonding to the surrounding hydrogen bonding network.

In the search for a possible interstitial configuration that might involve such bonding, we attempt a different initial geometry, in which we place the interstitial molecule midway between two water molecules that are hydrogen-bonded to each other in the defect-free crystal. This is shown in Fig. 7a, in which the interstitial molecule is indicated by the arrow. The two molecules originally hydrogen-bonded in the defect-free crystal have been displaced upward and downward respectively. After optimizing this geometry by minimizing the forces on all atoms in the cell, the structure relaxes into the configuration shown in Fig. 7b. It is clear that this interstitial structure, which is referred to as the bond-center (Bc) interstitial, now *does* involve bonding to the hydrogen-bond network. Specifically, the two molecules that were originally bound to each other in the defect-free crystal have now formed two hydrogen bonds with the interstitial molecule. In this manner, these two molecules still participate in four hydrogen bonds while the interstitial molecule takes part in two. In this respect, the Bc interstitial is fundamentally different from the Tc and Tu structures, in which the interstitial molecule does not participate in any hydrogen bonds.

In order to determine which of these is the dominant thermal equilibrium point defect in ice I_h, we need to analyze the formation energetics of the molecular vacancy and the three considered interstitial configurations. For this purpose, we first compute the formation

(a) **(b)**

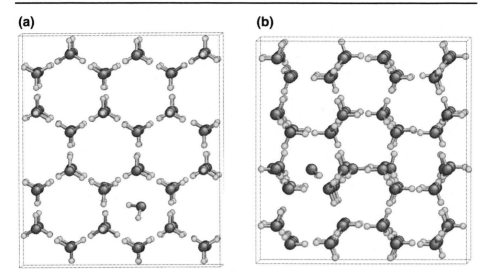

Fig. 6 Relaxed molecular interstitial structures. **a** Trigonal uncapped (Tu) interstitial. **b** Trigonal capped (Tc) interstial

(a) **(b)**

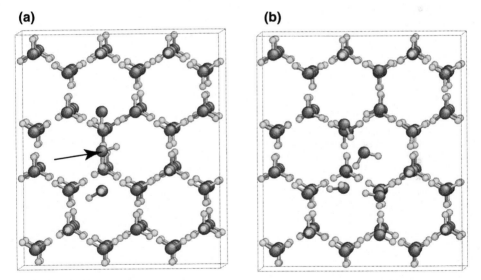

Fig. 7 Structures of the bond-center (Bc) interstitial in ice I_h. **a** Configuration before relaxation. Excess molecule is indicated by arrow. **b** Relaxed Bc interstitial structure

energies of the respective defect structures. For the molecular vacancy and interstitial, these are defined as

$$E_f^{\text{vac}} = E^{\text{vac}} - \frac{N-1}{N} E_0 \tag{3}$$

and

$$E_f^{\text{int}} = E^{\text{int}} - \frac{N+1}{N} E_0, \tag{4}$$

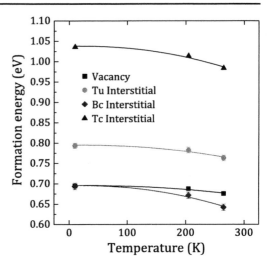

Fig. 8 Formation energies of molecular point defects in ice I$_h$ as a function of temperature. Temperature-dependence includes only implicit effects associated with thermal expansion

where E^{vac} and E^{int} are the total energies of the computational cells containing a vacancy and an interstitial, respectively. N is the total number of molecules in the defect-free ice cell, which has a total energy E_0. To take into account effects of thermal expansion with temperature, we compute the formation energies as a function of volume, using the experimental lattice parameters [1] at $T = 10$ K, $T = 205$ K and $T = 265$ K. To assess the effect of proton disorder, we average the formation energies over different realizations of each defect type, sampling different locations in the cell. The results are shown in Fig. 8. It is clear that the formation energy of the Tc interstitial is by far the highest among the four point-defect species and, therefore, is not expected to play a significant role. The formation energy of the Tu interstitial is substantially lower than that of the Tc interstitial but still larger than that of the Bc interstitial and molecular vacancy. Comparing the latter two, we find that the Bc interstitial is particularly sensitive to thermal expansion, showing a reduction of $\sim 7\%$ upon a linear lattice expansion of only $\sim 0.5\%$. This effect is much less pronounced for the vacancy, which shows a decrease of only $\sim 2\%$. If we consider the optimized defect structures we find that the strengthened bonding of the Bc interstitial is associated mainly with the relief of compressed hydrogen bonds upon thermal expansion. Since almost all the hydrogen bonds in the vicinity of the Bc interstitial are compressed, this effect is more pronounced than for the vacancy, for which only a third of the affected hydrogen bonds are compressed.

Since the creation of these defects involves stretched/compressed and/or broken hydrogen bonds, it is important to include zero-point vibrational contributions in the formation energetics. To do this, we compute the vibrational frequencies of the molecules in the vicinity of the respective defects before and after the introduction, where we consider both intra and intermolecular modes. The intramolecular stretching frequencies tend to increase when hydrogen bonds are broken because of the associated decrease of the intramolecular O-H distance. Given the elevated frequencies of these modes this leads to a significant zero-point contribution. After having computed the vibrational modes in the presence and absence of the defects, not only can we estimate the zero-point contributions, but we can also compute a full defect formation free energy within the harmonic approximation. To this end, we apply the local harmonic approximation [24], in which the defect formation free energy is given by

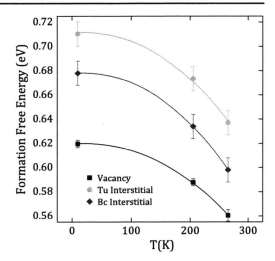

Fig. 9 Local harmonic formation free energies of molecular point defects in ice I$_h$ as a function of temperature

$$F_f = E_f + \frac{1}{2}\sum_i h(v_i^{\text{def}} - v_i^0)$$

$$+ k_B T \sum_i \ln\left[\frac{1 - \exp(-hv_i^{\text{def}}/k_B T)}{1 - \exp(-hv_i^0/k_B T)}\right], \tag{5}$$

where v_i^{def} and v_i^0 represent the frequencies in the presence and absence of the defect, respectively. The results for the molecular vacancy and the Tu and Bc interstitials is shown in Fig. 9. A comparison between the two interstitials shows that, consistent with the results of Fig. 8, the formation free energy of the Bc interstitial is lower than that of the Tu structure across the entire temperature interval. Comparing the vacancy and Bc interstitial, the downward shift of the vacancy curve with respect to the Bc interstitial in relation to Fig. 8 is due to the mentioned zero-point contributions, which lower the vacancy formation energy by ~ 0.08 eV compared to a reduction of only ~ 0.02 eV for the Bc interstitial. This means that the vibrations around the Bc interstitial are more "bulk-like" than in the vicinity of the vacancy, essentially because the Bc configuration features only one molecule that is not fully hydrogen bonded, compared to the four in case of the vacancy. As the temperature increases, however, the formation free-energy difference between the defects decreases due to the elevated formation entropy of Bc interstitial. At $T = 265$ K, for example, the formation entropy of the Bc interstitial reaches a value of $\sim 7 k_B$ compared to $\sim 5 k_B$ for the vacancy. This elevated formation entropy value [8] is mostly due to the appreciable reduction of the formation energy contribution upon thermal expansion as shown in Fig. 8.

With these results we are now in a position to evaluate the relative importance of the different point-defect species. For this purpose, we determine their thermal equilibrium concentrations c (per crystal lattice site) as a function of temperature, which is given by [1]

$$c = z N \exp(-F_f/k_B T). \tag{6}$$

Here N is the number of available defect sites per lattice site and, for the interstitial defects, z represents the number of possible orientations of the interstitial molecule on a given site. For the Tu and Bc interstitials we have $N_{\text{Tu}} = 1/2$ and $N_{\text{Bc}} = 2$, respectively, which follows from the number of cages and the number of hydrogen bonds per lattice site [1]. For the Tu

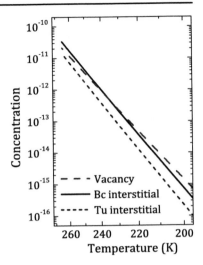

Fig. 10 Thermal equilibrium concentrations of the molecular point defects in ice I$_h$

interstitial we have observed that the orientation of an interstitial molecule on a given Tu site is always such that both of its O-H bonds tend to closely align with two oxygen atoms of the surrounding cage. Counting the number of different ways in which this can be realized, taking into account the fact that the H-O-H angle of the interstitial molecule is approximately 104°, we deduce $z_{Tu} \approx 40$. From different realizations of the Bc interstitial on a given site we estimate $z_{Bc} = 4$. For instance, considering Fig. 7b, a different stable orientation can be obtained by rotating the interstitial molecule such that its dangling hydrogen bond points along the c-axis, out of the paper. Two more stable orientations with similar energies were then created by a mirror symmetry operation with respect the plane containing the line connecting molecules 1 and 2 and the c axis. For the molecular vacancy we have $N_V = z_V = 1$.

Assuming these values for the site multiplicities, Fig. 10 shows the corresponding Arrhenius plot for the concentrations of the molecular point defects. It is found that the concentration of Tu interstitials is significantly lower than that of Bc interstitials across the entire temperature interval. These modeling results thus suggests that a structure different from the established Tc and Tu interstitials is the preferred interstitial configuration in ice I$_h$. Indeed, the structural properties of the Bc interstitial provided by our model are coherent with the experimental suggestion of a bound self interstitial [8]. Furthermore, the modeling results indicate that there is a cross-over in the dominant point-defect species. While the vacancy dominates for temperatures below $T \sim 200$ K, the Bc interstitial takes over for temperatures above $T \gtrsim 230$ K. This prediction is also consistent with the crossover scenario suggested in Ref. [8], in which the interstitial is assumed to dominate for temperatures above −50 °C whereas the vacancy becomes the principal thermal equilibrium point defect at lower temperatures.

3 Bjerrum defect/molecular point defect interactions

The discussion in the previous section represents an example of the capabilities of molecular modeling in situations where no direct experimental information is available. In addition, its specific findings, namely that a configuration involving bonding to the surrounding hydrogen-bond network is the preferred structure of the molecular interstitial in ice I$_h$, motivated another modeling study related to the activity of Bjerrum defects. As we mentioned in the Introduction, there is robust experimental evidence that, of the two Bjerrum defect species, only the L-type is active [1]. The main idea that has been put forward to explain this effect is that D-defects, somehow, are trapped at other defects. In first place, there would only be a need for such an explanation if the intrinsic mobility of D and L defects were to be effectively the same. And second, even if their mobilities were to be the same, the entrapment by other defects might only explain the observations if the interactions with these defects were to be significantly different for D and L defects. Given that experiments have so far been unable to address these questions in a direct manner, computational modeling is the only available tool that allows a direct investigation of this issue.

As a first step, we investigate whether the idea of entrapment at other defect sites is at all necessary, considering the intrinsic mobilities of L and D-type defects. For this purpose we determine the migration barrier for L and D-defect motion within the same DFT modeling approach of the previous section. Figure 11 shows two sequences of configurations depicting, respectively, the mechanisms of D and L-defect migration. Full details can be found in Refs. [22] and [25]. Panels (a–c) demonstrate a typical realization of a D-defect migration event. In Panel (a), the D defect is located on the indicated water molecule. The event occurs by the rotation of this molecule, provoking the rotation of one of its neighbors. Panel (b) shows the corresponding saddle-point configuration, while Panel (c) shows the final structure, in which the D-defect has moved to the adjacent molecule. Panels (d) and (e) show a typical L-defect migration event. The L-defect motion occurs through the rotation of a single molecule, with the saddle point configuration shown in Panel e). Considering a number of different realizations for both migration processes, the adopted DFT model gives migration barrier values in the ranges 0.06–0.13 eV [25] and 0.10-0.14 eV [22] for D-defect and L-defect motion, respectively. This result indicates that the intrinsic mobility of D-defects is certainly not expected to be any smaller than that of L-defects. This strongly suggests that the observed asymmetry in the activity of D and L-type defects is not caused by any intrinsic mobility factors.

Having verified the possibility of intrinsic mobility differences, we now turn to the idea of Bjerrum defect entrapment at other defect sites. For this purpose, the two most natural candidates are the molecular vacancy and interstitial. In particular, the finding of the bound interstitial configuration in the previous section opens the way for the possibility of Bjerrum-defect/interstitial interactions, given that the bound interstitial forms hydrogen bonds with the surrounding ice lattice. Such interactions are not expected, on the other hand, for the Tu and Tc structures, whose configurations involve isolated molecules that remain at relatively large distances from the hydrogen-bond network. To examine the strength of the molecular vacancy and interstitial as trapping centers to Bjerrum defects we compute the binding energies of four molecular point defect/Bjerrum defect defect complexes: (1) The D defect with a vacancy (DV), (2) the L defect with a vacancy (LV), (3) the D defect with a bond-center interstitial (DB) and (4) the L defect with a bond-center interstitial [26]. The topological features of these structures are shown in the two-dimensional square ice representation of Fig. 12. The binding energies are defined as the difference between the formation energy of a cell containing the defect complex and the sum of the formation energies of an isolated Bjerrum defect pair and of the molecular point defect in question. In case of the defect complex formed by a D defect and an interstitial, for instance, it is given by

$$\Delta E_{\mathrm{b}}^{\mathrm{DB}} = E_{\mathrm{f}}^{\mathrm{DB}} - E_{\mathrm{f}}^{\mathrm{DL}} - E_{\mathrm{f}}^{\mathrm{B}},$$

where $E_{\mathrm{f}}^{\mathrm{DB}}$ is the formation energy of cell containing a DB complex and an isolated L defect and $E_{\mathrm{f}}^{\mathrm{DL}}$ and $E_{\mathrm{f}}^{\mathrm{B}}$ are the formation energies of a Bjerrum defect pair and a bond-center

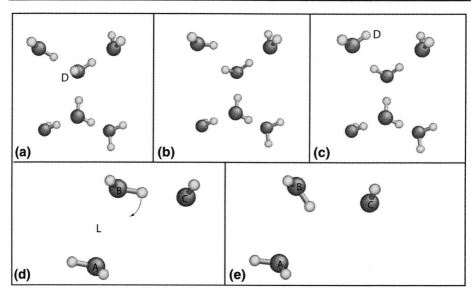

Fig. 11 Free Bjerrum defect migration mechanisms. Sequence shown in panels **a–c** shows typical D-defect migration event. Panels **d** and **e** show mechanism for L-defect motion

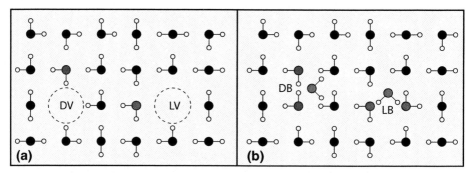

Fig. 12 Topological features of Bjerrum defect/molecular point defect defect complexes in two-dimensional square ice representation. **a** Defect complexes involving the molecular vacancy. **b** Defect complexes involving the Bc interstitial

interstitial, respectively. In the cells containing the isolated defects, both Bjerrum defects as well as the interstitial are created such that they are located at the same lattice position as they occupy in the cells containing the respective defect complexes. In order to probe the effect of the proton disorder, several realizations of each defect complex are considered.

Figure 13 shows typical atomistic configurations of the two defect-complex types involving the molecular interstitial. Figure 13a depicts the defect agglomerate formed by the D defect and the bond-center interstitial. The interstitial molecule receives two hydrogen bonds from the two molecules hosting the D-defect. Part (b) shows the defect complex involving the L defect and the bond-center interstitial. Similarly to the case of the DB complex, the interstitial is bonded to the two molecules hosting the L defect. The difference, however, is that in this case both hydrogen bonds are donated by the interstitial molecule.

Table 1 provides a summary of the energetics of the molecular point defect/Bjerrum defect agglomerates. It contains the binding energies of the four defect complexes as a function of

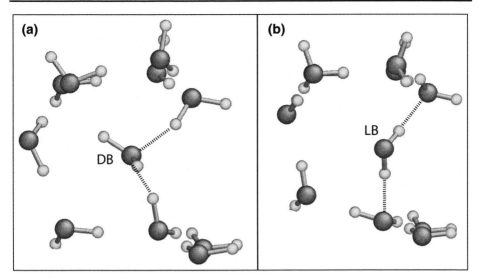

Fig. 13 Relaxed structures of DB and LB complexes. **a** DB complex **b** LB complex

Table 1 Average binding energies of the LV, DV, LB and DB defect complexes as a function of temperature

T	LV(3)	DV(3)	LB(4)	DB(5)
10	−0.41 ± 0.01	−0.38 ± 0.02	−0.20 ± 0.02	−0.36 ± 0.07
205	−0.41 ± 0.01	−0.38 ± 0.02	−0.19 ± 0.02	−0.35 ± 0.06
265	−0.40 ± 0.01	−0.36 ± 0.02	−0.17 ± 0.03	−0.33 ± 0.06

Values and error bars correspond to the mean value and the standard deviation in the mean obtained from a number of different realizations. Number of realizations is indicated by number between parentheses. Energies are given in eV and temperatures in K

temperature. Here, as in the case of the Fig. 8, the temperature dependence involves only the influence of thermal expansion, not including explicit temperature effects. These effects are seen to be small for all defect complexes, with variations of the order of only a few hundredths of an eV across the considered temperature interval. The reported values and corresponding error bars are determined as the average value and standard deviation of the mean as obtained from a set of distinct realizations of each defect complex. The numbers of realizations used for each defect complex are marked in parentheses. First, the results clearly show that both molecular point defects serve as trapping centers for Bjerrum defects, with all binding energies being negative. Secondly, the vacancy is a stronger trapping center than the molecular interstitial, with all binding energies of the complexes involving the vacancy being larger, in absolute value, than those with the interstitial. Third, the results indicate that both D and L Bjerrum traps are essentially equally strongly trapped by the vacancy, with both binding energies being practically equal, within the considered error bars. This suggests that the vacancy is not responsible for the observed inactivity of D-type Bjerrum defects. The situation is different, however, for the defect complexes based on the interstitial. The data in Table 1 imply that the interstitial represents a stronger trapping center for the D defect than it is for the L defect, with the binding energy of the former being larger than that of the latter beyond the error bars.

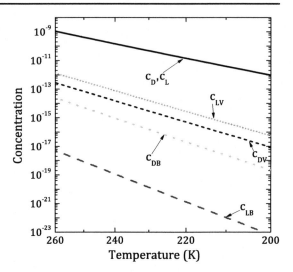

Fig. 14 Thermal equilibrium concentrations of free Bjerrum defects and their complexes based on DFT energetics

The data in Table 1, together with the point-defect [23] and Bjerrum defect [22] (M. de Koning et al. (unpublished)) formation free energies enable a calculation of the thermal equilibrium concentrations of free Bjerrum and molecular point defects, as well as those of the various defect complexes. Similar to the standard calculation [2] of the thermal equilibrium concentration of vacancies in a crystalline solid, this is done by minimizing the free energy of an ice I_h crystal with respect to the numbers of the different defect species. The details of this calculation can be found in Ref. [26].

Figure 14 shows the thermal equilibrium concentrations of free Bjerrum defects, as well as those of the four considered defect complexes, as obtained from the formation and binding energetics provided by the used DFT model. Despite the fact that the D defects are more strongly trapped at the bond-center interstitial compared to L defects, the corresponding concentrations of free D and L defects remain essentially the same. Given that the migration barriers of both free Bjerrum defects are essentially equal, this result, at least in quantitative terms, does not explain the inactivity of D defects in ice I_h. At this point, however, it is important to emphasize the limitations of the utilized modeling methodology. Our DFT approach is known to give discrepancies of the order of 0.1 eV for formation energies in water-based systems. [21,27] Although errors of such magnitude are acceptable in many systems, in the case of ice I_h the results tend to be highly sensitive to such variations, mostly due to the small values of the involved formation/binding energies as well as the low temperatures. To gain an appreciation of these variations, we recomputed the equilibrium concentrations of the free Bjerrum defects as well as the Bjerrum defect/point defect complexes using (1) the lower ends of the confidence interval in Table 1 for the binding energies and, (2) the lower limit of the experimental estimate of 0.35 eV for the interstitial formation energy. The results are shown in Fig. 15. Even though the absolute values of the energetics parameters have changed only by amounts of the order of 0.1 eV, the equilibrium concentration results are significantly different. Specifically, in this situation, the concentration of free L defects is between one and two orders of magnitude larger than that of free D-defects, which would be consistent with the experimental observation of D-defect inactivity.

These results show that a *quantitative* prediction regarding thermal equilibrium defect concentrations in ice I_h is very challenging. On the other hand, from a *qualitative* standpoint, the DFT results do suggest that the molecular interstitial, in contrast to the case of the

Fig. 15 Thermal equilibrium concentrations of free Bjerrum defects and the associated Bc-interstial complexes based on a combination of DFT and experimental energetics (see text)

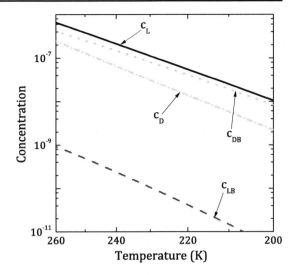

molecular vacancy, displays an energetic preference for the formation of complexes involving the D-type defect. This element, combined with the experimentally observed dominance of the interstitial as the predominant molecular point defect species in ice I_h, may then lead to conditions in which the concentration of free D defects becomes considerably smaller than that of free L defects. Such a scenario could possibly be involved in the experimentally observed inactivity of D-type Bjerrum defects.

4 Summary

In this paper we have demonstrated two examples of situations in which computational modeling provides a useful tool in studying the physics and chemistry of ice. In particular, these examples illustrate how computational modeling provides a means for the execution of controlled computational "experiments" that give access to information that cannot be obtained otherwise. In the first example, we address the question involving molecular point defects in ice I_h, which has been a longstanding issue from the experimental point of view. Specifically, we examine the energetics of formation of the molecular vacancy and a number of different molecular interstitial configurations using a DFT modeling framework. The results indicate that, as was suggested experimentally, [8] a configuration involving bonding to the surrounding hydrogen-bond network is the preferred interstitial structure in ice I_h. Moreover, the DFT results suggest a cross-over in dominance between the molecular vacancy and interstitial, which is consistent with experimental suggestions.

The second example involves a DFT modeling effort to elucidate on the microscopic origin of the robust experimental observation [1] that D-type Bjerrum defects are effectively immobile and that only L defects play a role in the Bjerrum-defect mediated electrical response of ice I_h. Inspired by the earlier suggestion that D-defects may be entrapped at other defect sites, we use DFT modeling to investigate the binding energetics of D and L-type Bjerrum defects to the molecular vacancy and interstitial. The previous finding of the bound interstitial structure plays a significant role here, with the results displaying a preferential binding of D-type Bjerrum defects to this type of interstitial. A calculation of the corresponding equilibrium concentrations then shows that such a preferential binding may lead to conditions under which the concentration of free L defects is significantly larger than that of free D-defects, possibly explaining the experimentally observed activity asymmetry of both Bjerrum defect species.

The two discussed examples illustrate the capabilities of DFT-based computational modeling involving defects in bulk ice I_h. But certainly, the approach is not limited to this area. Perhaps an even more important application involves one of the least-understood aspects of the physics and chemistry of ice: its interfaces. Aside from the structure of the free ice surface, there is significant interest, for instance, in the adsorption of gas molecules on the ice interface and interactions of the ice surface with other substances. In many situations experimental results alone are insufficient to arrive at a satisfactory understanding of the phenomena at hand. Under such circumstances, modeling efforts of the kind described above can offer valuable complementary insight.

Acknowledgements The author gratefully acknowledges financial support from the Brazilian agencies FAPESP, CNPq and CAPES.

References

1. Petrenko, V.F., Whitworth, R.W.: The Physics of Ice. Oxford University Press, Oxford (1999)
2. Ashcroft, N.W., Mermin, N.D.: Solid State Physics. Thomson Learning, USA (1976)
3. Pauling, L.: The structure and entropy of ice and other crystals with some randomness of atomic arrangement. J. Am. Chem Soc **57**, 2690 (1935)
4. Bjerrum, N.: Structure and property of ice. Science **115**, 385–390 (1952)
5. Jaccard, C.: Étude théorique et expérimentale des propriétés de la glace. Helv. Phys. Acta **32**, 89 (1959)
6. Mogensen, O.E., Eldrup, M.: Vacancies in pure ice studied by positron annihilation techniques. J. Glaciology **21**, 85 (1978)
7. Eldrup, M., Mogensen, O.E., Bilgram, J.H.: Vacancies in HF-doped and in irradiated ice by positron annihilation techniques. J. Glaciology **21**, 101–113 (1978)
8. Hondoh, T., Itoh, T., Higashi, A.: Behavior of point defects in ice crystals revealed by X-ray topography. In: Takamura, J., Doyama, M., Kiritani, M. (eds.) Point Defects and Defect Interactions in Metals, p. 599. University of Tokyo Press, Tokyo (1982)
9. Goto, K., Hondoh, T., Higashi, A.: Experimental determinations of the concentration and mobility of interstitials in pure ice crystals. In: Takamura, J., Doyama, M., Kiritani, M. (eds.) Point Defects and Defect Interactions in Metals, p. 174. University of Tokyo Press, Tokyo (1982)
10. Goto, K., Hondoh, T., Higashi, A.: Determination of diffusion coefficients of self-interstitials in ice with a new method of observing climb of dislocations by X-ray topography. Jpn. J. Appl. Phys **25**, 351–357 (1986)
11. Hondoh, T., Azuma, K., Higashi, A.: Self-interstitials in ice. J. Phys. Paris **48**, C1 (1987)
12. Oguro, M., Hondohin, T.: Interactions between dislocations and point defects in ice crystals. In: Higashi, A. (ed.) Lattice Defects in Ice Crystals, p. 49. Hokkaido University Press, Sapporo (1988)
13. Fletcher, N.H.: The Chemical Physics of Ice. Cambridge University Press, London (1970)
14. Guillot, B.: A reappraisal of what we have learnt during three decades of computer simulations on water. J. Mol. Liq **101**(1–3), 219 (2002)
15. Martin, R.M.: Electronic Structure: Basic Theory and Practical Methods. Cambridge University Press, Cambridge (2004)
16. Kohanoff, J.: Electronic Structure Calculations for Solids and Molecules. Cambridge University Press, Cambridge (2006)
17. Kresse, G., Hafner, J.: Ab initio molecular dynamics for liquid metals. Phys. Rev. B **47**, 558–561 (1993)
18. Kresse, G., FurthMüller, J.: Efficiency of ab-initio total energy calculations for metals and semiconductors using a plane-wave basis set. Comp. Mat. Sc **6**, 15–50 (1996)
19. Kresse, G., Joubert, J.: From ultrasoft pseudopotentials to the projector augmented-wave method. Phys. Rev. B **59**, 1758–1775 (1999)
20. Hayward, J.A., Reimers, J.R.: Unit cells for the simulation of hexagonal ice. J. Chem. Phys **106**, 1518 (1997)

21. Feibelman, P.J.: Partial dissociation of water on Ru(0001). Science **295**, 99–102 (2002)
22. de Koning, M., Antonelli, A., da Silva, A.J.R., Fazzio, A.: Orientational defects in ice I_h: an interpretation of electrical conductivity measurements. Phys. Rev. Lett **96**, 075501 (2006a)
23. de Koning, M., Antonelli, A., da Silva, A.J.R., Fazzio, A.: Structure and energetics of molecular point defects in ice I_h. Phys. Rev. Lett **97**, 155501 (2006b)
24. LeSar, R., Najafabadi, R., Srolovitz, D.L.: Finite-temperature defect properties from free-energy minimization. Phys. Rev. Lett **63**, 624–627 (1989)
25. de Koning, M., Antonelli, A.: On the trapping of Bjerrum defects in ice I: The case of the molecular vacancy. J. Phys. Chem. B **111**, 12537 (2007)
26. de Koning, M., Antonelli, A.: Modeling equilibrium concentrations of Bjerrum and molecular point defects and their complexes in ice Ih. J. Chem. Phys **128**, 164502 (2008)
27. Hamann, D.R.: H_2O hydrogen bonding in density-functional theory. Phys. Rev. B **55**, R10157 (1997)

Direct comparison between experiments and computations at the atomic length scale: a case study of graphene

Jeffrey W. Kysar

Originally published in the journal Sci Model Simul, Volume 15, Nos 1–3, 143–157.
DOI: 10.1007/s10820-008-9105-1 © Springer Science+Business Media B.V. 2008

Abstract   This paper discusses a set of recent experimental results in which the mechanical properties of monolayer graphene molecules were determined. The results included the second-order elastic modulus which determines the linear elastic behavior and an estimate of the third-order elastic modulus which determines the non-linear elastic behavior. In addition, the distribution of the breaking force strongly suggested the graphene to be free of defects, so the measured breaking strength of the films represented the intrinsic breaking strength of the underlying carbon covalent bonds. The results of recent simulation efforts to predict the mechanical properties of graphene are discussed in light of the experiments. Finally, this paper contains a discussion of some of the extra challenges associated with experimental validation of multi-scale models.

Keywords   Graphene · Intrinsic strength · Elastic properties · Ab initio methods · Multi-scale simulations

1 Introduction

Computer-based modeling and simulation have revolutionized our ability to quantitatively investigate all scientific and engineering phenomena from weather prediction, to study of earthquakes, to the properties of both biological and non-biological materials. The common thread that runs through all such simulations is that an actual physical system must first be idealized as a model, after which the fundamental governing equations are solved numerically subject to boundary conditions and initial conditions. Approximations and assumptions are inevitably made in every step of the simulation process and, ultimately, the validity of the simulations must be established by comparing the results to experiments.

The material properties we can predict with modeling and simulation can be categorized in traditional ways, such as electrical, optical, magnetic, mechanical, etc. A broader

J. W. Kysar (✉)
Department of Mechanical Engineering, Columbia University, New York, NY, USA
e-mail: jk2079@columbia.edu

characterization, though, is that material and system properties can be sorted into: (i) those systems for which the atoms retain their nearest neighbors and, (ii) those systems for which the atoms change nearest neighbors.

An impressive list of properties can be accurately predicted from the first category which include [1]: surface energies, thermal conductivity, electrical conductivity, magnetic properties, heat capacity, thermal expansion, phonon spectra, crystal structure and interatomic spacing, surface structure, interface structure, phase transitions, equations of state, melting temperatures, and elastic moduli. Such properties are calculated from first principles based predominantly upon solution of Schrödinger's equation, with atomic positions at their equilibrium positions or displaced slightly to induce a small mechanical strain.

The second category includes systems such as irreversible deformation of materials, turbulent fluids, earthquakes, weather prediction, etc., for which accurate predictions remain elusive. One reason is the difficulty of precisely defining the initial conditions of such systems because of rearrangement of nearest atomic neighbors that have occurred in the past. In addition, the behavior of such systems depends upon competitions between various mechanisms that contribute to relative motion. One such example, for fluids, is the competition between the intermolecular attractions through atomic potentials in the fluid which promotes laminar flow and the inertia of the flow which promotes turbulent flow. Another example, for the irreversible deformation of materials, is the competition between, say, dislocation-mediated deformation and twin-mediated deformation, which is determined by the details of the interatomic potentials in conjunction with the local stress evolution. In both cases, the details of the forces between atoms at relative distances far from equilibrium—even to the breaking point—play a decisive role in determining the transition in the behavior. In addition different mechanisms can occur simultaneously and with different characteristic time and length scales, each of which must be accounted for explicitly in order to accurately model observed behavior. Thus, these models are called multi-scale simulations.

This article is concerned predominantly with discussion of a recent set of experiments in which the elastic properties and intrinsic strength of free-standing monolayer graphene were obtained [1]. The results lend direct insight into the behavior of atomic bonds stretched to their breaking points and have implications in terms of validation of atomic-scale simulations.

In what follows, Sect. 2 discusses briefly the traditional paradigm for performing multi-scale simulations which attempt to predict the mechanical behavior of plastically deforming metals. Section 3 gives an overview of requirements of mechanical experiments at small length scales. Section 4 describes the recently performed experiments of monolayer graphene. Section 5 gives an overview of the analysis of those experiments. Recent simulations are compared to the experiments and suggestions for more detailed models and simulations of the graphene experiments are made in Sect. 6. Finally, conclusions are drawn in Sect. 7.

2 Overview of multi-scale simulations in ductile metals

There has long been a concerted effort to obtain a physics-based description of the mechanical constitutive properties of ductile materials [2]. Such knowledge is critical for prediction of the conditions under which a material deforms plastically, develops fatigue cracks, spalls, fractures via void growth and coalescence, or fails by some other mechanism. Of particular interest is the behavior of materials under extreme conditions: at high pressures, high strain-rates, high strains, high temperatures, and under conditions of high strain gradient. It is recognized that such material behavior is governed by physical phenomena which act over several different length and time scales [3–24].

In order to develop physics-based multi-scale models it is necessary to identify and model the dominant deformation mechanism at each of the pertinent length and time scales. The models at smaller length scales and/or shorter time scales are then used to "inform" the models at larger scales in a hierarchical manner from the atomic length scale to the continuum length scale. Typically the physical phenomena at the smaller scales are represented by the evolution of internal variables in models at the larger scales. The task of developing multiscale constitutive models is complicated by the fact that experiments must be performed at all relevant length and time scales in order to provide feedback during the modeling process as well as to validate the models.

At the atomic length scale, elastic properties are the main concern. At very high pressures, first-principles calculations can predict the pressure-volume response in terms of an equation of state (EOS) as well as the elastic shear modulus [25–29]. Other research efforts to predict elastic modulus include adding the effects of alloying elements, dislocations, atomic vacancies, and grain boundaries in the simulations [3]. Other properties of interest which must be calculated at the atomic length scale include properties of dislocations such as energetics, mobility and dislocation-dislocation interactions [30–33]. The properties of atomic scale phenomena are then used to develop dislocation dynamics models which treat large groups of individual dislocations that are assumed to exist within a continuum elastic background [34–36]. Of interest to the present paper are recent simulations to predict the elastic response of monolayer graphene films at very large strains, even to the breaking point [37].

Information from atomic scale simulations of the dynamics of individual dislocations is employed to elucidate appropriate values for internal variables such as dislocation mobility which are necessary for mean-field continuum models at the micrometer length scale to predict the current strength of plastic slip systems [13–15, 38–41]. Such continuum models demonstrate that slip system hardening behavior can be predicted under certain conditions over various temperature ranges and strain rates. However these models can not yet predict the behavior beyond relatively small strains [3]. It will be necessary to carry out experiments and dislocation dynamics models to higher strains in order to parameterize the continuum models for higher levels of strain.

3 Mechanical experiments at small length scales

As stated previously, multi-scale simulations must be validated at all pertinent time and length scales. It goes without saying that the most that can be claimed of unvalidated models is that the results demonstrate possible deformation mechanisms and behavior, rather than actual behavior. It has proven very challenging to perform mechanical experiments at the smallest length scales, which has inhibited opportunities for feedback and validation of small scale models. This section addresses some of the conceptual difficulties in performing small length scale mechanical experiments.

Experiments should measure physics-based, intrinsic variables rather than derived quantities or phenomenological quantities. For example, mechanical experiments that strive to characterize the continuum properties of, say, metals, quantify the yield stress and the hardening behavior beyond the yield point. However useful these quantities are in terms of describing the evolution of deformation from an engineering perspective, they are still phenomenological [42]. Instrumented indentation tests have been invaluable as a means of characterizing deformation at the micrometer length scale and below, but hardness too is a phenomenological quantity.

Physics-based theories of the mechanical behavior of materials must account for the tensorial nature of the various quantities. As such, every effort should be made to measure tensor components, rather than averaged quantities. Also, every attempt should be made to measure spatially and temporally resolved quantities rather than averaged quantities. A recent example of this is the ability to measure directly components of the dislocation density tensor (also called the incompatibility tensor) associated with imposed strain gradients in plastically deforming metal crystals [43]. A counter example is the method of instrumented indentation which reports a force versus displacement behavior of the indenter tip as it plunges into the material of interest to investigate the elastic-plastic response of a bulk material. The force is, in essence, an average of the three-dimensional second-rank stress tensor field and the displacement is a similarly averaged quantity derived from the strain field. There is typically not enough richness in the data to be able to provide feedback as to the evolution of individual variables used in small scale simulations, nor to distinguish between good and bad constitutive models. Recent developments of micro-column compression [44] have been a significant advance, but it should be kept in mind that the output variables are still averaged quantities.

The goal of all experiments is to quantify physical phenomena of interest to serve as a benchmark for the development of theory. Ultimately any mechanics theory will be expressed in terms of a boundary value problem which incorporates kinetics, kinematics and constitutive relationships. For the *kinetics* and *kinematics*, it is necessary to determine the shape, volume and mass of the specimen, its displacement and traction boundary conditions, as well as the initial positions, velocities and the like. For the *constitutive relationship*, it is necessary to determine the initial state of the material with as much precision as possible. At minimum, the main elemental constituents of the material, the crystallographic orientation and the temperature of the environment must be known in order to determine elastic properties. If deformation is to be modeled beyond the elastic regime, much more information is required, such as temperature, magnetic, electrical fields and the reactive elements, etc. Most important, though, is that the initial defect structure of the material must be determined as precisely as possible, including dislocations, grain boundaries, second-phase particles, previous irradiation, etc. Spatially-resolved information of this type is extremely difficult to obtain. As a consequence, most constitutive models incorporate this information only in an average sense.

Mechanical experiments at the micrometer length scale and below are very difficult to perform. Traditionally, mechanical specimens are first fabricated and then mounted in the loading device. The gripping issues inherent in loading small scale specimens often introduce uncertainties in the displacement boundary conditions. For indentation methods, the shape of the indenter and the radius of curvature of its tip are important—the latter of which evolves with extended usage—so the traction boundary conditions often are not well-defined. In order to circumvent these difficulties, novel test methods such as the micro-compression methods [44] have been developed in which the specimen is carved out of a bulk specimen using Focused Ion Beam (FIB) in an electron microscope and the resulting micrometer length scale column is loaded in compression with a flat-tipped indenter. Even this method has difficulties, because the cylinders as fabricated are often slightly conical so the stress distribution is not constant within the specimen. In addition, plastic slip often initiates preferentially at the top and bottom corners of the cylinders due to stress concentration. Also, the slip boundary conditions at the bottom of the cylinders where they meet the underlying bulk specimen are not well-defined because plastic slip is able to propagate from the cylinder into the bulk. Nevertheless, these new methods represent significant advances in experimental capability.

Thus, the boundary conditions and initial conditions on the kinetics and kinematics can be very difficult to either prescribe or describe in small scale mechanical experiments. Likewise the description of the initial state of the material for the constitutive relationship can be very challenging. To begin, a precise characterization of the specimen is necessary. At small length scales, it may not be sufficient to simply state the volume of the material. Rather it may be necessary to specify the mass, or equivalently, the number of atoms in the system. Graphene films are an example of this, as will be discussed further in Sect. 4. Likewise, at sufficiently small length scales, the behavior of the specimen is extremely sensitive to the details of the initial defects. Therefore, two otherwise identically specimens can exhibit dramatically different behavior which depends exclusively on the initial defect structure. To make matters worse, the very steps involved in the fabrication of small scale specimens can introduce defects so the essence of the material can change during the fabrication process. Therefore, the defect structure of the as-fabricated specimen must be determined and ultimately modeled. Finally, since the constitutive behavior of small scale specimens is so highly dependent on the initial defect structure—which is statistically distributed—the observed behavior of different specimens often has a wide range. Therefore in materials for which the defect distribution is unknown, often the best that can be measured from a set of mechanical specimens is the range of possible behaviors which must be quantified statistically.

4 Mechanical experiments on monolayer graphene

Graphene is a molecule of carbon atoms arranged in a close-packed hexagonal planar structure; as such it is a two-dimensional material. It is the basic structural element of carbon nanotubes and other fullerenes and can, in principle, be arbitrarily large. Graphene is also the basic structural element for graphite. The covalent carbon bonds in graphene endow it with both high strength and high stiffness. A set of experiments [1] on free-standing graphene to determine experimentally its strength and stiffness will be described in this section.

A silicon substrate with a 300 nm SiO_2 epilayer was patterned using nanoimprint lithography and reactive ion etching to create a 5 mm by 5 mm array of wells (either 1 μm or 1.5 μm in diameter) with a depth of 500 nm. Graphite flakes from a kish[1] graphite [45] source were mechanically exfoliated [46] and placed on the substrate over the array of wells using transparent adhesive tape, as shown schematically in Fig. 1. In essence, the adhesive side of the tape was placed on the graphite source and pulled off, which cleaved graphite flakes from the surface. The tape was then removed from the graphite source and the adhesive surface was placed on the previously prepared substrate. After removing the tape from the substrate, many flakes of graphite remained. The randomly shaped graphite flakes had an average size of several tens of micrometers and were randomly distributed over the wells. The thicknesses of the flakes varied, but a few flakes were graphene monolayers, bilayer, trilayers, etc.[2] A preliminary evaluation of the thicknesses was made using optical microscopy to identify the thinnest flakes. A monolayer of graphene sufficiently changes the optical path length of light to cause a slight change of intensity due to interference effects. In fact, with practice, it is possible to distinguish between monolayers, bilayers, trilayers and higher integer number

---

[1] Kish graphite is a product of the steel smelting process. The slag covering molten steel for some smelting processes is known as kish, and graphite is one its constituents. Kish graphite is renowned for its tendency to flake, perhaps indicating a large grain size.

[2] Strictly speaking, a graphene molecule is an individual atomic layer of carbon atoms. Nevertheless, the term graphene will be employed herein to designate a graphite flake that contains an integer number of atomic layers.

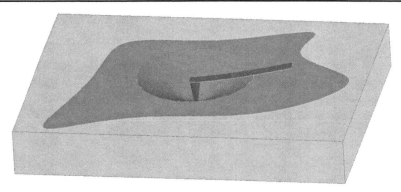

Fig. 1 Schematic of experimental set-up showing graphene flake extending over circular well on substrate. The cantilevered atomic force microscope tip is shown indenting the graphene film

of layers based upon the contrast under the microscope. Raman spectroscopy was then used on candidate flakes of various thicknesses to confirm the integer number of atomic layers. There is yet no method to use mechanical exfoliation to control systematically the position and pre-tension of any individual graphene flake.

The graphene flakes adhere to the SiO_2 epilayer on the substrate via van der Waals interactions. Each flake is large enough to cover many wells in the substrate. Thus, each flake results in many graphene films suspended over individual wells. The profile of the suspended graphene films were characterized by Atomic Force Microscopy (AFM) in a non-contact mode. The results showed that the graphene was "sucked" down into the well a few nanometers, presumably due to van der Waals interactions between the periphery of the well and the suspended film, which induces a pre-tension in the suspended film. It is reasonable to assume that the pre-tensions of the suspended films from a given flake would be approximately the same. Therefore, each specimen to be tested can be thought of as being a tensioned "drum head" that is one atomic layer thick, with a diameter of either 1 μm or 1.5 μm.

The mechanical loading was performed via nanoindentation with an AFM (XE-100, Park Systems). Two micro-cantilevers with diamond[3] tips were employed, with tip radii of 16.5 nm and 27.5 nm, as measured by Transmission Electron Microscopy (TEM). Just prior to the loading, the suspended graphene films were scanned with the AFM tip, after which the tip was immediately moved to within 50 nm of the center of the film and loading commenced. The loading was performed under position control at the rate of 1.4 μm/s and the reaction force on the tip was measured. Two to three load-unload cycles were made to multiple depths (from 20 nm to 100 nm) for each of the 23 graphene films from two different flakes. In that way, a total of 67 force-displacement responses were measured. A representation of the deformed graphene near the indenter tip is shown in Fig. 2. The force-displacement responses were independent of the rate of indentation and showed negligible hysteresis, which demonstrated that the graphene did not slip around the periphery of the well.[4] Immediately after measuring the elastic response, the graphene films were loaded to failure and the breaking force on the AFM tip was recorded.

---

[3] Diamond tips were necessary because traditional silicon tips broke at forces lower than the breaking load of the monolayer graphene films.

[4] The force-displacement response of some films did exhibit hysteresis upon unloading. AFM topography measurements of such films after a hysteretic response indicated that sliding of the graphene had occurred around the periphery of the wells. Data from these specimens were not considered in the analysis, but they do offer a glimpse of how the van der Waals interactions can be measured quantitatively.

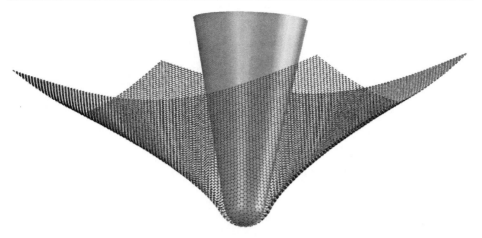

Fig. 2 Schematic representation of highest strained region of graphene under the indenter tip

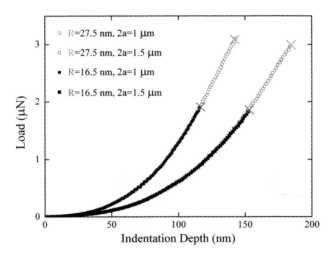

Fig. 3 Representative force-displacement data showing that the force-displacement response is independent of indenter tip radius, and the breaking force (shown by the × symbols) is independent of the film radius

As shown in Fig. 3, the force-displacement responses were a strong function of the diameter of the well, as expected, but were essentially independent of the diamond tip radius. This suggests that the system can be modeled as a circular film with a point load at the center. The measured force was linear with displacement at very small deflections and cubic at large deflections, as would be expected for a linear-elastic material [47,48].

The breaking force depended strongly on the radius of the AFM tip, but was independent of the diameter of the graphene film. This suggests that the very high stresses in the material under the indenter tip depend mainly upon the details of the geometry and size of the indenter tip. The average breaking load for the 16.5 nm radius tip was 1.8 μN and for the 27.5 nm tip was 2.9 μN and the deflections at failure were greater than 100 nm.

It is interesting to return attention briefly to the discussion in Sect. 3 about experimental requirements for mechanical experiments at small length scales. One issue is to ensure a correct description of the boundary conditions and initial conditions. The graphite flakes are placed randomly on the substrate and are maintained in position by van der Waals interactions. As a consequence, one can assume zero displacements around the periphery of the film during mechanical loading—assuming the van der Waals forces are sufficiently strong—which was borne out in the experiments. Thus the initial position of the entire film is known and the initial velocities are zero. The displacement boundary conditions at the load point are mostly determined by the radius of curvature of the diamond tip and the displacement rate of loading. There is one caveat however: the nature of the van der Waals interactions between the diamond tip and the film are not obvious. Experimentally there is a "snap-in" event as the graphene film is attracted up to and into contact with the descending diamond tip; the measured force upon attachment of the graphene to the diamond tip is between 5 and 10 nN. This value can be used to garner information about the van der Waals attraction, but it does not give information about the sliding forces, which require lateral force measurements.

Another important issue is to minimize damage induced into the specimen during fabrication. The graphene was obtained by exfoliation during which individual atomically thin graphene layers were cleaved from the graphite source. Since the substrate was prepared prior to the deposition of graphene, the graphene films were never exposed to highly reactive chemicals. Therefore, it is unlikely that any new defects were introduced into the specimen during fabrication. One important uncertainty, though, is that the number of atoms in the specimen is not known (although it is of the order of 10^8 atoms) due to the presence of the pre-tension from the adherence of the graphene film to the interior surface of the well. In principle, the number of atoms could be calculated from a measurement of the relative atomic spacing in the suspended graphene; glancing angle x-ray methods offer a potential means of measuring the lattice spacing. Contamination of the graphene must also be considered, especially given the fact that every atom in the graphene specimen is on a surface. It is apparent that the adhesive tape used to transfer the graphite flakes from the source to the substrate initially must have cleaved the flakes from the source. Although the details are unclear, it is likely that the flakes cleave again when the adhesive tape is removed from the substrate. Thus, it is plausible that both sides of the monolayer graphene films are newly-cleaved surfaces, which reduces the probability of surface contamination.

It is also necessary to describe the initial defect structure of the graphene films which acts as the initial conditions for the constitutive properties. The experiments showed conclusively that graphene is a brittle material in the sense that final failure occurs suddenly and catastrophically. Thus, the breaking strength of graphene is expected to be very sensitive to the presence and the details of pre-existing defects. The lowest stresses in the system are at the periphery of the well, so any imperfections near the edges are not expected to dominate the behavior of the system. Detailed continuum-based simulations of the stress state in the indented graphene film indicated that the region of highest stress is limited to only about 1% of the area of the graphene film in the immediate neighborhood of the indenter tip. Therefore, only defects that exist within this small region, which amounts to only a few thousand square nanometers, are expected to influence the breaking strength.

The sensitivity of the breaking force to the presence of defects was investigated using Weibull statistics. The statistical distribution of the breaking force for each tip radius was very narrow and the Weibull modulus [1,49] strongly suggested the graphene films to be defect-free, at least in the most highly stressed region of the films. Several observations support this conclusion. First, atomic scale resolution experiments [50] with Scanning Tunneling Microscopy (STM) on a graphene monolayer film—from the same kish graphite source used in these experiments—showed there to be no defects over an area of many hundreds of square nanometers. Second, the concentration of atomic vacancies in a graphite crystal is expected to be exceedingly small, based on the formation energy of up to 7.4 eV to introduce one atomic vacancy in graphite [51,52]. Thus, graphene films exfoliated from the crystal should have a negligible concentration of atomic vacancies based upon elementary thermodynamic considerations. Third, the slip plane of dislocations that exist in graphite is typically the basal plane [53,54]. Thus the dislocations tend to exist between sheets of graphene in the graphite crystal so the dislocations disappear upon exfoliation. Other potential defects include grain boundaries or twin boundaries in the basal plane, but these are more likely to be at the boundaries of an exfoliated flake than at the middle. Finally, interstitial and substitutional atoms may very well exist, however they would likely not cause as significant a degradation of strength as one of the other types of defects. Therefore it is not surprising that such a small region of graphene could be defect free. The stress experienced by the graphene under the indenter tip at the point of failure is expected to be the intrinsic strength, which is the maximum stress that can be supported by the graphene prior to failure.

5 Analysis of experiments

As stated in the previous section, the force-displacement response is insensitive to the radius of the indenter tip. In addition, *ab initio* calculations of the bending stiffness of graphene films relative to the in-plane stiffness strongly suggest that to a very good approximation, the bending contribution to the elastic strain energy at the radii of curvatures of the indenter tips can be neglected [55]. Hence, the system can be analyzed by treating it as an initially flat circular membrane indented by a center point load. It is known that at large displacements, the force scales with the cube of displacement in the limit of zero pre-tension in the film [48], but the response is more complicated for a non-zero pre-tension [47]. Nevertheless, a closed form solution of the force-displacement response has not yet been obtained. Therefore a detailed finite element analysis of the graphene film was performed by treating the graphene as a membrane subject to a pre-tension to investigate the force-displacement response. An approximate expression for force that has both linear and a cubic term in displacement was then determined from the simulations that deviated from the numerical results significantly less than the experimental uncertainty. This expression was used as a tool to extract values of pre-tension and stiffness of the graphene based upon least-squares curve fit of the data with the approximate force versus displacement response. The average Young's modulus from all 67 measurements is 340 N·m^{-1}, where the stiffness is reported as force per length of the graphene.

Another result from the experiments is that the breaking force of the films depends predominantly upon the radius of curvature of the indenter tip and is independent of the film diameter. Therefore a mechanics analysis of the stresses in the graphene film in the contact region was performed using finite elements in the limit of zero friction between the indenter and the graphene film. The maximum stress attained in the film agreed well with an analytical solution [56] when a linear stress-strain response is employed. The result showed, however, that the strains in the graphene near the indenter reached values of well over 0.2, which is far beyond the linear-elastic regime.

Therefore, a non-linear elastic constitutive behavior [57–60] was adopted which is predicated on expanding the elastic strain energy density, Φ, of a material in a Taylor series in terms of powers of strain that has the first two terms

$$\Phi = \frac{1}{2}C_{ijkl}\eta_{ij}\eta_{kl} + \frac{1}{6}D_{ijklmn}\eta_{ij}\eta_{kl}\eta_{mn} \tag{1}$$

where η_{ij} is the Lagrangian elastic strain, C_{ijkl} are the second-order elastic moduli and D_{ijklmn} are the third-order elastic moduli; the summation convention is observed for repeated subscripts. The symmetric second Piola-Kirchhoff stress tensor, Σ_{ij}, is then determined as

$$\Sigma_{ij} = \frac{\partial \Phi}{\partial \eta_{ij}} = C_{ijkl}\eta_{kl} + \frac{1}{2}D_{ijklmn}\eta_{kl}\eta_{mn} \tag{2}$$

which is work conjugate to the Lagrangian strain. The third-order elastic moduli are effectively negative so there is a decreased tangent modulus at high tensile strains and an enhanced tangent modulus at high compressive strains. The third-order elastic moduli, D_{ijklmn}, is a sixth-rank tensor quantity and its independent tensor components depend upon both the symmetry of the crystal lattice [58] of interest as wall as the symmetry in indices due to thermodynamic considerations.

Casting this relationship in a uniaxial context, the stress, σ, and strain, ε, response can be written as $\sigma = E\varepsilon + D\varepsilon^2$, where E is Young's modulus and D is the effective uniaxial third-order elastic constant. It is evident that for a negative value of D the maximum stress that material can achieve in uniaxial tension, here called the intrinsic strength, is $\sigma_{int} = -E^2/4D$, and at a corresponding strain of $\varepsilon_{int} = -E/2D$. Since $D < 0$, both quantities are non-negative.

It is tempting to assign a thickness to the graphene monolayer. A number of researchers have discussed an appropriate effective thickness by treating a graphene sheet as a continuum shell to simulate carbon nanotubes. In a small-strain continuum context a shell with thickness, h, has bending stiffness $h^3 E/12(1 - \nu^2)$, where E and ν are the in-plane Young's modulus and Poisson's ratio, respectively. In a monolayer graphene film, the bending stiffness and the in-plane stiffness are related, but do not necessary follow the same relationship as the continuum formulation. As a consequence, effective values of E and h were found by other researchers (e.g. [61,62] and references therein) in the spirit of a multi-scale model in order to allow reasonable continuum-based simulations of carbon nanotubes. However it bears emphasis that the effective thickness is not an intrinsic property of a graphene sheet, and that the appropriate way to present quantities with units of stress is on a force per length basis rather than a force per area basis. Therefore, quantities that are obtained from the experiments are the two-dimensional Young's modulus, E^{2D}, the two-dimensional third-order stiffness, D^{2D}, and the, two-dimensional intrinsic strength, σ_{int}^{2D}, each with units of force per length. They can be thought of as the product of effective thickness, h, and the respective effective three-dimensional quantities with units of force per area.

The value of $E^{2D} = 340 \text{ N} \cdot \text{m}^{-1}$ was determined from the force-displacement response of the graphene films. The value of D^{2D} was determined by modeling the deformation in the film under the indenter tip while assuming the non-linear stress-strain constitutive relationship using the finite element method. The strain in the graphene directly under the tip increased monotonically with depth of indentation. Eventually the highest strains exceeded ε_{int}, after which the stress in the most highly strained region began to decrease. This induced an instability in the deformation and the finite element simulation was unable to converge to an equilibrium solution. The force on the indenter in the simulation at the point of instability was deemed to be the breaking force of the graphene film. The value of $D^{2D} = -690 \text{ N} \cdot \text{m}^{-1}$ led to a predicted breaking force very similar to the experimentally observed breaking force for both indenter tip radii. This corresponds to an intrinsic strength of $\sigma_{int}^{2D} = 42 \text{ N} \cdot \text{m}^{-1}$.

It is important to note that the predicted force-displacement response of the graphene film with the non-linear elastic constitutive description was virtually indistinguishable from simulations for which a linear constitutive description was employed. This is due to the fact that only about 1% of the film nearest the indenter tip had sufficiently large strains to experience a non-linear response, and the additional displacement accrued in this region as the tangent modulus decreased was very small compared to the overall displacement.

6 Suggestions for further simulations

The goal of the analysis of the experiments presented above was to determine the elastic properties and the intrinsic strength based upon elementary continuum concepts. As such, the analysis purposely avoided any reference to discrete methods of analysis such as molecular dynamics based upon empirical potentials or to *ab initio* methods. However, the main utility of the experiments is to serve as a benchmark for the discrete methods, as will be discussed in this section.

Ab initio methods based upon quantum mechanics have been used for a plethora of different problems. The underlying theory uses Schrödinger's equation so errors in the results creep in due to approximations used in the calculations. Therefore, comparison of the predictions to experimental results should be considered as a validation of the solution methods inherent in the simulation, rather than the validity of the underlying governing equation given the accuracies that can be currently obtained experimentally in mechanical experiments. Thus, development of methods that are both accurate and robust for relatively simple materials such as graphene will give more confidence when the same methods are applied to more complex materials [63].

There are a number of different *ab initio* methods [63], including post-Hartree-Fock (HF) quantum chemistry, quantum Monte Carlo (QMC) simulations, and various types of Density Functional Theory (DFT). Two recent papers have been published on the mechanical properties of graphene at large strains. One [37] employed Density Functional Perturbation Theory (DFPT) to study the phonon modes associated with graphene as it is stretched uniaxially for two different crystallographic directions, from which they predicted the stress-strain response and the Poisson's ratio in the so-called zigzag and armchair directions. The predicted behavior deviates significantly from linear behavior beyond a strain of about 0.05, after which the tangent modulus decreases. Furthermore, the predicted elastic behavior in the two directions at strains less then 0.15 is virtually indistinguishable. The tangent modulus of the predicted response is zero at the highest strain that the graphene can support, which is the intrinsic strength. The uniaxial prediction in the armchair direction suggests the intrinsic strength of the graphene to be $40.4 \text{ N} \cdot \text{m}^{-1}$ at a strain of 0.266, as compared to an intrinsic strength in the zigzag direction to be $36.7 \text{ N} \cdot \text{m}^{-1}$ at a strain of 0.194. Another *ab initio* calculation [64] probed the sensitivity of the breaking strength of graphene to the presence of atomic vacancies and small tears (or cracks). Quantum mechanical DFT calculations were carried out in the region surrounding the defects in the graphene, while molecular dynamics using an empirical potential were used in regions further away, while a continuum mechanical description was used in regions even further away. This multi-scale simulation predicted the intrinsic strength to be $38.4 \text{ N} \cdot \text{m}^{-1}$, and the breaking strength dropped off dramatically upon the introduction of defects. The intrinsic strength predicted from these simulations compares well with the experimental value of $42 \pm 3 \text{ N} \cdot \text{m}^{-1}$ at a strain of about 0.25. Hence it appears that *ab initio* methods are able to predict accurately the experimentally observed behavior, at least for idealized load states.

There are many different Molecular Dynamics (MD) potentials that can be used to model mechanical behavior (see the review in [65]). The interatomic potentials are empirical and typically contain a large number of parameters that are fit to experiment as well as from *ab initio* calculations. Furthermore, potentials specialized for studies of mechanical properties are typically specialized for small strains. As such, MD potentials are at best an approximation of the actual physical behavior. Estimates [66] of the error introduced by the empirical nature of the potentials are as large as 20%. Nevertheless, the allure of MD methods is that the computational expense is much lower than that of *ab initio* methods. Therefore, future work to devise MD potentials that closely approximate the behavior from *ab initio* methods, especially at very high strains where the elastic behavior is clearly non-linear, would constitute a significant advance.

In addition to discrete methods of simulation, a proper continuum description of the non-linear elastic behavior in Eq. 2 requires that the components of both the second-order elastic moduli, C_{ijkl}, as well as the third-order elastic moduli, D_{ijklmn}, be determined. The second-order elastic moduli tensor is isotropic due to the symmetries inherent in the unstrained graphene lattice. The third-order elastic modulus, however, is anisotropic because the symmetries are broken at high strains. Therefore, one direction of future research is to determine the values of all non-zero components of the third-order elastic modulus tensor for graphene. This would allow detailed simulations of various devices made of monolayer graphene, such as mechanical resonators, as long as the radius of curvature of the deformed graphene is sufficient small that the bending stiffness can be neglected.

7 Conclusions

A set of experiments in which the mechanical properties of monolayer graphene are measured was discussed in detail. The pertinent properties were the Young's modulus, the intrinsic strength, and an estimate of the third-order elastic modulus for uniaxial tension. In addition, a number of different directions of future research were delineated.

One was to determine the values of all components of the sixth-rank, third-order elastic moduli tensor of graphene. This is important for two reasons. First, the stress-strain response of graphene is known to be non-linear beyond about a strain of 0.05. Second, the mechanical response of graphene at small strains is isotropic, but is anisotropic at larger strains. The non-linear, anisotropic behavior at large strains is embedded within the framework of the third-order elastic modulus.

Two *ab initio* simulations of the mechanical properties of graphene were reviewed. The intrinsic strength predictions of both corresponded to the experimental value within experimental uncertainty. However, *ab initio* methods are very expensive computationally. Thus, it would be beneficial to develop molecular dynamics interatomic potentials that are valid at large strains, which would dramatically reduce computational time.

Simulations of graphene sheets that contain tears and other defects rely on a multi-scale methodology that uses *ab initio* methods in the region nearest the defect, a continuum mechanics description far away from the defect, and a molecular dynamics formulation at intermediate distances. Interatomic potentials that are valid for large strains would allow a reduction in the size of the *ab initio* region. Likewise, a continuum description valid at large strains would allow a reduction in the size of the molecular dynamics region. This would result in less expensive computations. In addition, validation of the interatomic potentials will bring more accuracy to simulations at the smallest length scales, which will cascade out through the larger length scales.

A number of additional experiments are currently under way. One set of experiments is to measure the mechanical properties of other two-dimensional molecules that can be cleaved from macroscopic materials, including: BN, MoS_2, $NbSe_2$, $Bi_2Sr_2CaCu_2O_x$, among others. The methods to exfoliate these films and deposit them on substrates are described in [46]. It is important to note that these new molecules consist of at least two different types of atoms, which will allow validation of more complex models suitable for these materials.

Other experiments underway include a determination of both the maximum van der Waals force exerted between graphene and a substrate as well as the indenter tip. In addition to van der Waals forces, experiments are underway to measure the *energy* associated with the van der Waals interactions. Another set of experiments to measure the friction properties is being performed.

It should be emphasized that the experiments and simulations reviewed herein were for the in-plane mechanical properties. Experiments to probe the bending behavior should also be performed.

Acknowledgements The author is grateful for conversations with C. Lee, J. Hone and X. Wei, as well as support from the National Science Foundation through grants DMR-0650555 and CMMI-0500239, the Air Force Office of Scientific Research through grant FA9550-06-1-0214 and a Department of Energy grant administered through Lawrence Livermore National Laboratory for a PECASE award.

References

1. Lee, C., Wei, X.D., Kysar, J.W., Hone, J.: Measurement of the elastic properties and intrinsic strength of monolayer graphene. Science **321**(5887), 385–388 (2008)
2. Pablo, J.J., Curtin, W.A.: Multiscale modeling in advanced materials research: challenges, novel methods, and emerging applications. MRS Bull. **32**, 905–909 (2007)
3. Becker, R.: Developments and trends in continuum plasticity. J. Comput-Aided Mater. Des. **9**(2), 145–163 (2002)
4. Chandler, E., Moriarty, J., Rubia, T.D.de la , Couch, R.: LLNL's dynamics of metals program: multi-scale modeling of plasticity and dynamic failure. Abstr. Pap. Am. Chem. Soc **222**, U13–U13 (2001)
5. Buehler, M.J., Hartmaier, A., Gao, H.: Hierarchical multi-scale modelling of plasticity of submicron thin metal films. Model. Simul. Mater. Sci. Eng. **12**(4), S391–S413 (2004)
6. Clayton, J.D., McDowell, D.L.: Homogenized finite elastoplasticity and damage: theory and computations. Mech. Mater. **36**(9), 799–824 (2004)
7. Hao, S., Liu, W.K., Moran, B., Vernerey, F., Olson, G.B.: Multi-scale constitutive model and computational framework for the design of ultra-high strength, high toughness steels. Comput. Methods Appl. Mech. Eng. **193**(17–20), 1865–1908 (2004)
8. Khan, S.M.A., Zbib, H.M., Hughes, D.A.: Modeling planar dislocation boundaries using multi-scale dislocation dynamics plasticity. Int. J. Plast. **20**(6), 1059–1092 (2004)
9. Belak, J.: Multi-scale applications to high strain-rate dynamic fracture. J. Comput. Aided Mater. Des. **9**(2), 165–172 (2002)
10. Curtin, W.A., Miller, R.E.: coupling in computational materials science. Model. Simul. Mater. Sci. Eng. **11**(3), R33–R68 (2003)
11. Zbib, H.M., Rubia, T.D. de la : A multiscale model of plasticity. Int. J. Plast. **18**(9), 1133–1163 (2002)
12. Hartley, C.S.: Multi-scale modeling of dislocation processes. Mater. Sci. Eng. A Struct. Mater. Prop. Microstruct. Process. **319**, 133–138 (2001)
13. Stainier, L., Cuitino, A.M., Ortiz, M.: A micromechanical model of hardening, rate sensitivity and thermal softening in bcc single crystals. J. Mech. Phys. Solids **50**(7), 1511–1545 (2002)
14. Stainier, L., Cuitino, A.M., Ortiz, M.: Multiscale modelling of hardening in BCC crystal plasticity. J. Phys. Iv **105**, 157–164 (2003)
15. Cuitino, A.M., Stainier, L., Wang, G.F., Strachan, A., Cagin, T., Goddard, W.A., Ortiz, M.: A multiscale approach for modeling crystalline solids. J. Comput. Aided Mater. Des. **8**(2–3), 127–149 (2002)
16. Cuitino, A.M., Ortiz, M.: Computational modeling of single-crystals. Model. Simul. Mater. Sci. Eng. **1**(3), 225–263 (1993)
17. Horstemeyer, M.F., Baskes, M.I., Prantil, V.C., Philliber, J., Vonderheide, S.: A multiscale analysis of fixed-end simple shear using molecular dynamics, crystal plasticity, and a macroscopic internal state variable theory. Model. Simul. Mater. Sci. Eng. **11**(3), 265–286 (2003)
18. Baskes, M.I.: The status role of modeling and simulation in materials science and engineering. Curr. Opin. Solid State Mater. Sci. **4**(3), 273–277 (1999)

19. Horstemeyer, M.F., Baskes, M.I.: Atomistic finite deformation simulations: a discussion on length scale effects in relation to mechanical stresses. J. Eng. Mater. Technol. Trans. ASME **121**(2), 114–119 (1999)
20. Campbell, G.H., Foiles, S.M., Huang, H.C., Hughes, D.A., King, W.E., Lassila, D.H., Nikkel, D.J., Rubia, T.D. de la , Shu, J.Y., Smyshlyaev, V.P.: Multi-scale modeling of polycrystal plasticity: a workshop report. Mater. Sci. Eng. A Struct. Mater. Prop. Microstruct. Process. **251**(1–2), 1–22 (1998)
21. Hansen, N., Hughes, D.A.: Analysis of large dislocation populations in deformed metals. Phys. Status Solidi A Appl. Res. **149**(1), 155–172 (1995)
22. Horstemeyer, M.F., Baskes, M.I., Godfrey, V., Hughes, D.A.: A large deformation atomistic study examining crystal orientation effects on the stress–strain relationship. Int. J. Plast. **18**(2), 203–229 (2002)
23. Godfrey, A., Hughes, D.A.: Physical parameters linking deformation microstructures over a wide range of length scale. Scr. Mater. **51**(8), 831–836 (2004)
24. Rubia, T.D. de la , Bulatov, V.V.: Materials research by means of multiscale computer simulation. Mater. Res. Soc. Bull. **26**(3), 169–175 (2001)
25. Soderlind, P., Moriarty, J.A.: First-principles theory of Ta up to 10 Mbar pressure: structural and mechanical properties. Phys. Rev. B **57**(17), 10340–10350 (1998)
26. Ogata, S., Li, J., Hirosaki, N., Shibutani, Y., Yip, S.: Ideal shear strain of metals and ceramics. Phys. Rev. B. **70**(10):104104 (2004)
27. Ogata, S., Li, J., Yip, S.: Ideal pure shear strength of aluminum and copper. Science **298**(5594), 807–811 (2002)
28. Shibutani, Y., Krasko, G.L., Sob, M., Yip, S.: Atomic-level description of material strength of alpha-Fe. Mater. Sci. Res. Int. **5**(4), 225–233 (1999)
29. Widom, M., Moriarty, J.A.: First-principles interatomic potentials for transition-metal aluminides. II. Application to Al-Co and Al-Ni phase diagrams. Phys. Rev. B **58**(14), 8967–8979 (1998)
30. Moriarty, J.A., Belak, J.F., Rudd, R.E., Soderlind, P., Streitz, F.H., Yang, L.H.: Quantum-based atomistic simulation of materials properties in transition metals. J. Phys. Condens. Matter **14**(11), 2825–2857 (2002)
31. Moriarty, J.A., Vitek, V., Bulatov, V.V., Yip, S.: Atomistic simulations of dislocations and defects. J. Comput. Aided Mater. Des. **9**(2), 99–132 (2002)
32. Yang, L.H., Soderlind, P., Moriarty, J.A.: Atomistic simulation of pressure-dependent screw dislocation properties in bcc tantalum. Mater. Sci. Eng. A Struct. Mater. Prop. Microstruct. Process. **309**, 102–107 (2001)
33. Schiotz, J., Jacobsen, K.W.: A maximum in the strength of nanocrystalline copper. Science **301**(5638), 1357–1359 (2003)
34. Bulatov, V.V.: Current developments and trends in dislocation dynamics. J. Comput. Aided Mater. Des. **9**(2), 133–144 (2002)
35. Hiratani, M., Bulatov, V.V.: Solid-solution hardening by point-like obstacles of different kinds. Philos. Mag. Lett. **84**(7), 461–470 (2004)
36. Cai, W., Bulatov, V.V.: Mobility laws in dislocation dynamics simulations. Mater. Sci. Eng. A Struct. Mater. Prop. Microstruct. Process. **387–389**, 277–281 (2004)
37. Liu, F., Ming, P., Li, J.: Ab initio calculation of ideal strength and phonon instability of graphene under tension. Phys. Rev. B **76**, 064120 (2007)
38. Arsenlis, A., Wirth, B.D., Rhee, M.: Dislocation density-based constitutive model for the mechanical behaviour of irradiated Cu. Philos. Mag. **84**(34), 3617–3635 (2004)
39. Arsenlis, A., Parks, D.M., Becker, R., Bulatov, V.V.: On the evolution of crystallographic dislocation density in non-homogeneously deforming crystals. J. Mech. Phys. Solids **52**(6), 1213–1246 (2004)
40. Arsenlis, A., Tang, M.J.: Simulations on the growth of dislocation density during Stage 0 deformation in BCC metals. Model. Simul. Mater. Sci. Eng. **11**(2), 251–264 (2003)
41. Arsenlis, A., Parks, D.M.: Modeling the evolution of crystallographic dislocation density in crystal plasticity. J. Mech. Phys. Solids **50**(9), 1979–2009 (2002)
42. Kysar, J.W.: Energy dissipation mechanisms in ductile fracture. J. Mech. Phys. Solids **51**(5), 795–824 (2003)
43. Larson, B.C., El-Azab, A., Yang, W.G., Tischler, J.Z., Liu, W.J., Ice, G.E.: Experimental characterization of the mesoscale dislocation density tensor. Philos. Mag. **87**(8–9), 1327–1347 (2007)
44. Uchic, M.D., Dimiduk, D.M., Florando, J.N., Nix, W.D.: Sample dimensions influence strength and crystal plasticity. Science **305**(5686), 986–989 (2004)
45. Nicks, L.J., Nehl, F.H., Chambers, M.F.: Recovering flake graphite from steelmaking kish. J. Mater. **47**(6), 48–51 (1995)
46. Novoselov, K.S., Jiang, D., Schedin, F., Booth, T.J., Khotkevich, V.V., Morozov, S.V., Geim, A.K.: Two-dimensional atomic crystals. Proc. Natl. Acad. Sci. USA **102**(30), 10451–10453 (2005)
47. Komaragiri, U., Begley, M.R.: The mechanical response of freestanding circular elastic films under point and pressure loads. J. Appl. Mech. Trans. ASME **72**(2), 203–212 (2005)

48. Schwerin, E.: Über Spannungen und Formänderungen kreisringförmiger Membranen. Z. Tech. Phys. **10**(12), 651–659 (1929)
49. Barber, A.H., Andrews, R., Schadler, L.S., Wagner, H.D.: On the tensile strength distribution of multi-walled carbon nanotubes. Appl. Phys. Lett. **87**, 203106 (2005)
50. Stolyarova, E., Rim, K.T., Ryu, S.M., Maultzsch, J., Kim, P., Brus, L.E., Heinz, T.F., Hybertsen, M.S., Flynn, G.W.: High-resolution scanning tunneling microscopy imaging of mesoscopic graphene sheets on an insulating surface. Proc. Natl. Acad. Sci. USA **104**(22), 9209–9212 (2007)
51. Coulson, C.A., Santos, E, Senent, S., Leal, M., Herraez, M.A.: Formation energy of vacancies in graphite crystals. Proc. R. Soc. Lond. A Math. Phys. Sci. **274**, 461–479 (1963)
52. El-Barbary, A.A., Telling, R.H., Ewels, C.P., Heggie, M.I. and Briddon,P.R.: Structure and energetics of the vacancy in graphite. Phys. Rev. B. **68**(14) (2003). Article Number 144107
53. Grenall, A.: Direct observation of dislocations in graphite. Nature **182**(4633), 448–450 (1958)
54. Williamson, G.K.: Electron microscope studies of dislocation structures in graphite. Proc. R. Soc. Lond. A Math. Phys. Sci. **257**(1291), 457–& (1960)
55. Yakobson, B.I., Brabec, C.J., Bernholc, J.: Nanomechanics of carbon tubes: instabilities beyond linear response. Phys. Rev. Lett. **76**(14), 2511–2514 (1996)
56. Bhatia, N.M., Nachbar, W.: Finite indentation of an elastic membrane by a spherical indenter. Int. J. Nonlinear Mech. **3**(3), 307–324 (1968)
57. Brugger, K.: Thermodynamic definition of highter order elastic coefficients. Phys. Rev. A Gen. Phys. **133**(6A), A1611–A1612 (1964)
58. Brugger, K.: Pure modes for elastic waves in crystals. Journal of Applied Physics **36**(3), 759–768 (1965)
59. Brugger, K.: Determination of 3rd-order elastic coefficients in crystals. J. Appl. Phys. **36**(3), 768–773 (1965)
60. Lubarda, V.A.: Apparent elastic constants of cubic crystals and their pressure derivatives. Int. J. Non-Linear Mech. **34**(1), 5–11 (1999)
61. Huang, Y., Wu, J., Hwang, K.C.: Thickness of graphene and single-wall carbon nanotubes. Phys. Rev. B **74**, 245413 (2006)
62. Pantano, A., Parks, D.M., Boyce, M.C.: Mechanics of deformation of single- and multi-wall carbon nanotubes. J. Mech. Phys. Solids **52**, 789–821 (2004)
63. Carter, E.A.: Challenges in modeling materials properties without experimental input. Science **321**(5890), 800–803 (2008)
64. Khare, R., Mielke, S.L., Paci, J.T., Zhang, S., Ballarini, R., Schatz, G., Belytschko, T.: Coupled quantum mechanical/molecular mechanical modeling of the fracture of defective carbon nanotubes and graphene sheets. Phys. Rev. B **75**, 075412 (2007)
65. Stone, A.J.: Intermolecular potentials. Science **321**, 787–789 (2008)
66. Porter, L.J., Li, J., Yip, S.: Atomistic modeling of finite-temperature properties of b-SiC. I. Lattice vibration, heat capacity and thermal expansion. J. Nucl. Mater. **246**, 53–59 (1997)

Shocked materials at the intersection of experiment and simulation

H. E. Lorenzana · J. F. Belak · K. S. Bradley · E. M. Bringa · K. S. Budil ·
J. U. Cazamias · B. El-Dasher · J. A. Hawreliak · J. Hessler · K. Kadau ·
D. H. Kalantar · J. M. McNaney · D. Milathianaki · K. Rosolankova ·
D. C. Swift · M. Taravillo · T. W. Van Buuren · J. S. Wark · T. Diaz de la Rubia

Originally published in the journal Sci Model Simul, Volume 15, Nos 1–3, 159–186.
DOI: 10.1007/s10820-008-9107-z © US government 2008

Abstract Understanding the dynamic lattice response of solids under the extreme conditions of pressure, temperature and strain rate is a scientific quest that spans nearly a century.
Critical to developing this understanding is the ability to probe and model the spatial and
temporal evolution of the material microstructure and properties at the scale of the relevant
physical phenomena—nanometers to micrometers and picoseconds to nanoseconds. While
experimental investigations over this range of spatial and temporal scales were unimaginable
just a decade ago, new technologies and facilities currently under development and on the
horizon have brought these goals within reach for the first time. The equivalent advancements in simulation capabilities now mean that we can conduct simulations and experiments
at overlapping temporal and spatial scales. In this article, we describe some of our studies
which exploit existing and new generation ultrabright, ultrafast x-ray sources and large scale
molecular dynamics simulations to investigate the real-time physical phenomena that control
the dynamic response of shocked materials.

H. E. Lorenzana (✉) · J. F. Belak · K. S. Bradley · E. M. Bringa · K. S. Budil · B. El-Dasher ·
J. A. Hawreliak · D. H. Kalantar · J. M. McNaney · D. Milathianaki · D. C. Swift ·
T. W. Van Buuren · T. D. de la Rubia
Lawrence Livermore National Laboratory, 7000 East Avenue, Livermore, CA, USA
e-mail: HLorenzana@llnl.gov

J. U. Cazamias
University of Alabama at Birmingham, Birmingham, AL, USA

J. Hessler
Argonne National Laboratory, Argonne, IL, USA

K. Kadau
Los Alamos National Laboratory, Los Alamos, NM, USA

K. Rosolankova · J. S. Wark
University of Oxford, Oxford, UK

M. Taravillo
Universidad Complutense de Madrid, Madrid, Spain

Keywords   Shock · Phase transformations · Damage · Shock diagnostic tools · Molecular
dynamics materials simulation · Ultrafast phenomena

1 Introduction

Materials dynamics, particularly the behavior of solids under extreme dynamic compression,
is a topic of broad scientific and technological interest [16,43]. It is well-established that the
bulk material response is strongly dependent on the processing history and is affected by physical processes that encompass a wide range of temporal and spatial scales [1]. In this work,
we consider the lattice-level response of a solid to shock compression. It is widely accepted
that the material morphology and timescales of atomistic phenomena have a profound impact
on bulk properties, such as plasticity, phase transformations, and damage [50]. Yet, despite
the acknowledged importance of these ultrafast microscopic processes, few studies have cast
light on their nature or provided details of how they govern material response during the
passage of the shock.

Consider a notional crystalline solid, as illustrated in Fig. 1. In the case of a planar shock
denoted here, the initial response is a compression of the crystal along lattice planes whose
unit normals are partially aligned with the direction of shock propagation. This uniaxial
response can remain elastic; that is, once the disturbance is removed, the lattice will relax
back to its original configuration. However, under high-stress conditions and given sufficient
time, the lattice will undergo an irreversible response. The local nucleation and kinetics
of defects and phases leads to plasticity, melting, resolidification, or solid-solid structural
transformations. These atomistic changes can have dramatic and important consequences
to most macroscopically observable behavior such as the material's thermodynamic state
(pressure, temperature, density), strength, and failure. Little or no data exist concerning the
nature of defect generation and mobility, phase nucleation and growth, and void formation
under these highly dynamic stress conditions.

Historically, two approaches have been employed to address this lack of data: sample
recovery [60] and in situ bulk property measurements. The first approach emphasizes microstructural analysis but does not allow direct probing of the samples under dynamic loading.
The examination of shock recovered specimens provides end-state information that can be
useful in *inferring* dynamic behavior [44,59]. In the second approach, real-time measurements of the bulk response are recorded with fast diagnostics such as surface velocimetry
(VISAR) and internal stress gauges [18]. While such approaches have proven themselves to
be valuable and are largely responsible for our current understanding, these continuum level
methodologies are limited in the insight that they can provide about lattice level processes
under dynamic conditions. As an illustration of this point, a phase transformation event and
its kinetics are inferred from the change in slope of the VISAR wave profile in Fig. 1, effectively measuring a change in the compressibility of the material but providing no insight as
to the atomic level details.

It is clear that a fundamental understanding requires the direct probing and study of
relevant, transient physical processes at their characteristic lattice length scales. We are at
a unique crossroad in materials science where theory, experiments, and simulations have
progressed to the point that such investigations at overlapping temporal and spatial scales
are now possible. This manuscript describes some of our efforts enabled by new advances
in the technology of ultrabright, pulsed x-ray sources and massively parallel computation.
We describe two classes of experimental approaches: laser-based, which offer seamless
timing precision and proven shock loading capabilities, and accelerator-based, which possess

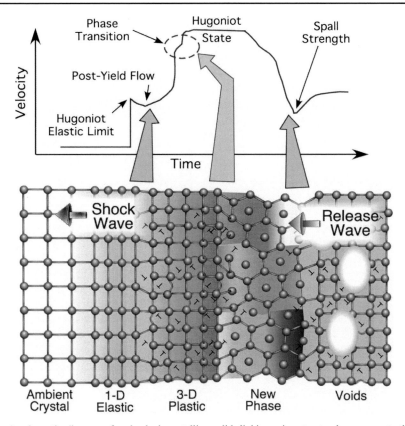

Fig. 1 A schematic diagram of a shocked crystalline solid, linking microstructural processes to the bulk behavior measured through conventional velocimetry. Material properties are typically inferred from the change in velocity at the surface of a target sample (top panel). The rearrangement of the atoms as the lattice relaxes plastically or undergoes a phase-transition alters the response of the material causing a change in surface velocity (bottom panel). While a change in slope indicates a change in material response, it does not provide any information on the microscopic lattice details

unparalleled x-ray beam quality and brightness. These approaches are complementary in their capacity to probe from the nano-to-macroscale in the spatial and picosecond-to-nanosecond in the temporal regimes. We discuss our recent experiments and simulations, casting light on elastic-to-plastic relaxation and solid-solid phase transformations during the shock. We also introduce future areas of investigation of melt and damage phenomena under dynamic compression conditions.

2 Approaches to in situ studies of atomic processes under dynamic compression

It is now routine to subject macroscopic samples (~ 0.1 mm to 1 cm) to energy densities exceeding 10^{12} ergs/cm^3 (~ 100 GPa pressures) using high-energy laser and multiple-stage gas guns. Experiments are currently being designed to investigate solids at pressures significantly exceeding 100 GPa at ultrafast timescales using next generation high energy density (HED) facilities, such as the National Ignition Facility (NIF at Lawrence Livermore National Laboratory) and the Linac Coherent Light Source (LCLS at Stanford University) [13,51]. In parallel, computational molecular dynamics (MD) methodologies have now progressed to the point where the simulation of $\sim \mu$m size samples out to ~ 100 ps is now possible using state-of-the-art computers [8]. Because of the extreme challenges in conducting experiments under HED conditions, insight into the atomistic response of the solid under shock conditions has been limited. We now briefly outline both the experimental and computational approaches used in this work to perform the real-time and in situ investigations of the dynamic material processes.

2.1 X-ray techniques—Diffraction and scattering

X-ray diffraction and scattering provide a non-perturbative probe into the lattice level response of the shocked solid [14,33]. Diffraction in solids relies on the coherent scattering of x-rays from a periodic array of atoms to produce spectra that can be correlated with the atomic structure and described by Bragg's law. X-ray scattering, on the other hand, is sensitive to the larger-scale density variations associated with void nucleation and growth. In principle, temporal resolution of shock loading can be obtained through two basic approaches: (a) gating or streaking the detector and/or (b) pulsing the x-ray probe. The first approach could be implemented to as fast as 50ps but with a narrow field of view and has been demonstrated to ~ 50 ns resolutions in powder gun experiments [31,52] with 2 to 4 ns resolution when streaked [53]. In contrast, current state-of-the-art, pulsed x-ray studies are sub-nanosecond with the ability to use wide-angle detectors [34,63]. Femtosecond resolution is on the horizon with the development of new accelerator-based sources.

2.1.1 Laser-based systems for x-ray diffraction

Current high-energy laser facilities have been shown to be excellent venues to generate both high pressure loading [6,12,55] and implement x-ray-based diagnostics [45]. Shock pressures between 10 and 1000 GPa are easily accessible through direct ablative drive. Temporal variation of the drive can be tailored by either laser pulse shaping or target design, while spatial uniformity of the drive is ensured through the use of phase plates. Recently, quasi-isentropic drives capable of producing shockless loading have been demonstrated at pressures between 10 and 200 GPa [19,40]. This shockless loading is expected to allow access to strain rates from as low as 10^6sec^{-1} up to shock regimes $> 10^{10} \text{sec}^{-1}$. Laser sources can also thermally excite a wide range of x-ray source characteristics from inner core atomic transitions [47]. Source fluence can exceed 10^{15} photons/pulse of isotropically illuminating x-rays, orders of magnitude larger than the $\sim 10^9$ photons required for single-pulse x-ray diffraction, with energies in the range of 500 eV to 22 keV [46].

Multi-beam laser facilities provide exquisite control over timing between drive and diagnostic functions. A typical experimental geometry, as illustrated in Fig. 2, offers a largely unobstructed experimental view for large-area x-ray detectors [35,61,63]. The sample is driven by a set of coordinated laser pulses that launch a shock into the sample at pressure $P \sim I_L^{2/3}$, where I_L is the laser intensity [39]. A separate set of beams are used to generate a thermal plasma via laser heating that serves as a point source of diagnostic x-rays, termed a backlighter. To provide the time evolution measurements, these x-rays are delayed relative to the beams that drive the shock. The backlighting beams are of ~ 1 ns duration and are short enough to effectively freeze some of the lattice dynamics of interest in a single image. The lattice compression shifts the angle at which the diffracted signal is observed, as shown

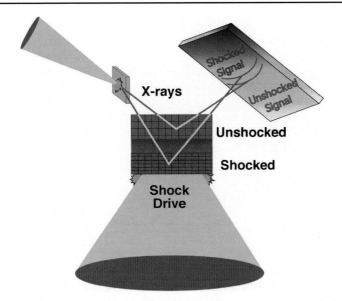

Fig. 2 A schematic diagram of the laser-based in situ x-ray diffraction technique for shock-loaded single-crystal solids. A pulsed beam is used to generate ablative pressure on a surface of a sample. As the shock propagates, another beam irradiates a metal foil whose pulsed K-shell x-rays are used to probe the crystal. By setting the relative timing of the pump-probe experiment, it is possible to record both shocked and unshocked signal. Due to the geometry of the instrument, the diffraction from the lattice planes generate arcs whose curvature and position give insight into plane spacings and orientations

in Fig. 2. This separation of signal in angle and space makes it possible to perform a measurement in which both the shocked and unshocked states of the crystal can be diagnosed simultaneously. The detector for these measurements covers a large field of view, allowing for the collection of diffraction from multiple lattice planes.

2.1.2 Accelerator-based light source for x-ray scattering

Synchrotrons have traditionally offered significant advantages for x-ray experiments including brightness, monochromaticity, coherence, collimation, and stability. Synchrotron radiation is pulsed, since the electrons in the storage ring are bunched and can be used to generate ultrashort, ultrabright pulses of x-rays for time-resolved measurements. To optimize the timing and brightness for pump-probe experiments, the storage ring fill pattern can be adjusted to run in a hybrid mode which consists of a single isolated electron bunch separated in time from a train of bunches. Such fill patterns can change the timing between electron bunches to the order of microseconds and allow the use of gated cameras to time-resolve signals. We note that the time resolution is currently ~ 100 picosecond (sub-nanosecond) but soon, with the development of fourth generation synchrotrons, is expected to be ~ 100 femtoseconds [37].

2.1.3 Computation-simulation

MD simulations solve the equations of motion for a set of atoms mediated by interatomic potentials [25]. Such interatomic potentials are generated by fitting to a data bank of electronic structure calculations for the energies and atomic forces of relevant crystal structures. For our purposes, samples are prismatic with free surfaces parallel and periodic boundary conditions perpendicular to the shock direction. The first few surface layers on one of the free surfaces are taken to act as a piston and constrained to move at a constant or ramped velocity. At each time step, the stress and strain state, local phase and dislocation densities can be extracted.

Post-processing these simulations allows the calculation of the expected x-ray diffraction and scattering signals [56,57]. To simulate these observables, we perform a Fourier transform on the calculated atomic positions. Since the positions of the atoms are given by arbitrary and not discretized x, y, and z coordinates, special techniques are required to take advantage of fast Fourier transform methods, which are not discussed here. We calculate the Fourier transform using

$$I\left(\vec{k}\right) = \left| \sum_{n=0}^{N-1} e^{-i(\vec{k}\cdot\vec{r}_n)} \right|^2$$

where r_n is the position of the nth atom and N is the total number of atoms in a given test volume or crystal. The solution to this equation for a lattice of atoms is a periodic distribution of intense scattering peaks in k-space. The location of these resonances is given by $\vec{k} = h\vec{b}_1 + k\vec{b}_2 + l\vec{b}_3$ where $\vec{b}_n (n = 1, 2, 3)$ are the reciprocal lattice vectors that define the crystallography and (h, k, l) are the standard Miller indices that describe the diffraction plane.

Modeling simulated x-ray scattering in this fashion has three distinct and important advantages. First, it provides a method for analyzing the vast amounts of output from an MD simulation. The location and intensity of the spots can be used to determine atomic structure, and more subtle features, like broadening and shifts, can give information about dislocation density through techniques applied to standard high resolution x-ray diffraction experiments. Second, it allows for the optimization of future experiments by identifying specific physical phenomena and relevant experimental conditions to target for investigation. Last, but potentially most important, it allows a direct comparison between experiment and simulation. Good agreement provides higher confidence in the simulations, resulting in an enhanced understanding of the measured lattice kinetics and generating improved potentials.

3 Materials response to shock loading

3.1 Inelastic response to shock loading (1D to 3D transition)

The initial response of a crystalline material to shock loading is uniaxial, elastic compression (1D), leading to a strong shear stress on the lattice. If the shock wave propagating through the lattice exceeds the Hugoniot elastic limit (HEL), this shear stress leads to nucleation of dislocations, allowing the lattice to plastically relax to a more quasi-hydrostatic compression (3D) on some characteristic timescale. The rate of relaxation to hydrostatic conditions is controlled by the nucleation rates and mobility of dislocations, which in turn depend upon microstructural parameters such as grain boundaries, barriers, etc.

Experimentally, this relaxation of the lattice behind a shock has been studied with time-resolved diffraction for FCC metals. As an example, such kinetics have been studied in single crystal copper shocked to 18Gpa [41]. The degree of lattice relaxation was determined through diffraction recorded from (200) planes both parallel and perpendicular to the shock loading direction. These samples were found to have approached a fully 3D state in less than 400 ps after the passage of the shock front. The observations in copper are broadly supported by

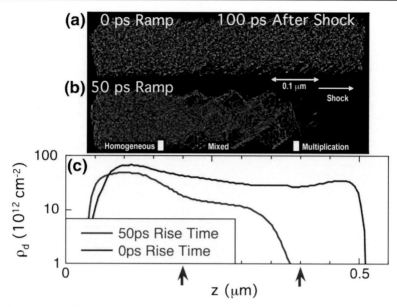

Fig. 3 Contrast in the dislocation behaviors between the 0 ps and 50 ps ramp compressions in single-crystal copper. **a** Shows a real space plot of the dislocation of atoms for a 0 ps rise time compression, 100 ps after the start of the shock. Dislocation activity appears to be fairly uniform and continuous throughout. **b** A similar plot for the 50 ps rise time case with two distinguishable regions of dislocation activity, a multiplication and homogenous nucleation regimes with a transition region of mixed mechanisms. **c** Dislocation density for both simulations. In both cases, the copper crystal included pre-existing dislocation sources located at the arrows. Marked differences in ρ_d occur in the vicinity of the initial dislocation sources. Peak particle speed is 0.75 km/s. Reprinted with permission from [9]. Copyright 2006, Nature Publishing Group

results from our MD simulations [9], which are discussed next and show relaxation occurring over a period of less than 100 ps.

With the advent of large-scale MD simulations, we can now directly compare results from simulations to time-resolved in-situ x-ray diffraction experiments. The simulations used a $\sim$ 1 μm long single crystal copper comprised of up to 352 million atoms as the starting material with defect sources in the form of prismatic dislocation loops. Two large scale MD simulations were conducted, both with a peak piston velocity of $U_p = 0.75$ km s^{-1} (peak pressures of approximately 35 GPa), but with different loading profiles. One simulation ramped the piston to the peak velocity over 50 ps while the other used an instantaneous "shock" loading. The ramped velocity was aimed at forming a link to experiments, which have been conducted with rise times ranging from several picoseconds to several nanoseconds. A snapshot of both simulations at 100 ps shows a large dislocation network in Fig. 3. The 0 ps rise time shows only homogeneous nucleation (Fig. 3a), while the 50 ps rise time exhibits three regions of dislocation activity (Fig. 3b).

The total dislocation density (ρ_d) versus depth is plotted for 100 ps in Fig. 3c. The zero rise-time case saturates at a steady-state value of 3×10^{13} cm^{-2}. This value is very large compared with the density of pre-existing sources or the dislocation density typically found in shock recovered samples [44]. Interestingly, we observe that the final state and micro-structure in the 0 ps rise time simulations appeared not to be sensitive to the initial defect concentration. In contrast, for the ramped case, the pre-existing defects led to significant dislocation multiplication and partial stress relaxation throughout the simulated sample.

Fig. 4 Plastic response of the
0 ps and 50 ps ramp compressions
in single-crystal copper. **a** Shear
stress relaxation under loading
conditions of 0 and 50 ps rise
time. **b** The corresponding lattice
compression from simulated
x-ray diffraction for lattice planes
whose normals are parallel and
orthogonal to the shock direction.
The shock front is located at
$z = 0$. The black vertical line is
at the limit of previous
simulations while the horizontal
shows the hydrostatic limit of
relaxation. The 'jitter' of the
curve corresponds to the
magnitude of the error in our
simulations and the dependence
on local structure. Reprinted with
permission from [9]. Copyright
2006, Nature Publishing Group

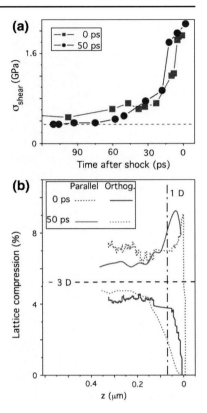

A fully 3D relaxed state is ideally defined as having zero shear stress and hydrostatic compression. Fig. 4 shows that the shear stress behind the shock front markedly decreases over approximately 60 ps for both types of loadings with more than half of shear stress relieved within ∼10 ps after the shock passage. The shear stress evolves to an asymptotic value of 0.43 GPa for the zero rise-time simulation and 0.34 GPa for the ramped loading simulation and is comparable to that inferred from the in situ diffraction experiments on copper [44]. Using the MD simulation, we calculated the expected x-ray diffraction pattern. We analyzed such patterns with tools used to quantify lattice changes in experimental data. These results are summarized in Fig. 5, which indicate significant lateral relaxation only at times much greater than 10 ps. Due to the significant stress relaxation that has occurred, the simulated x-ray diffraction shows nearly uniform 3D compression, i.e. 5% in each direction. These large scale simulations are the first to capture the scope of this lattice relaxation, as previous work was limited to timescales of ∼10 ps, too short to follow fully the evolution.

On the basis of our simulations, we can anticipate two interesting implications for future experiments. The first is that relaxation phenomena in copper can be observed only at the picosecond timescale, with little or no relaxation expected until after ∼10 ps after shock passage. This suggests that we would need to experimentally probe with picosecond resolution, currently an experimental challenge. The second is that the influence of pre-existing defects on the final microstructure should be negligible for strong shocks above the limit for homogeneous nucleation of dislocations. This latter observation poses a testable prediction that can be addressed by appropriate selection of drive conditions and engineering of the initial microstructure of samples.

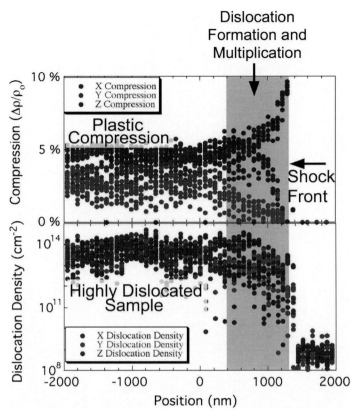

Fig. 5 Using simulated x-ray diffraction, we estimate the stacking fault density and lattice compression with the same tools used to analyze experimental data. The agreement of these estimates with those determined from the MD simulations using other approaches serves to validate our tools. These results show that the loading close to the shock front is initially elastic, followed by a rapid formation of dislocations that eventually relieve the shear stress. Fluctuations are primarily due to statistical variations throughout the simulated sample. Directions X and Y are orthogonal and Z is parallel to the shock direction

While copper is an example of an FCC material that shows ultrafast plastic relaxation, iron is a BCC material that exhibits notably longer timescale kinetics for plasticity and appears to be sensitive to the starting microstructure. Iron MD simulations performed by Kadau et al. [32] generated a shock by using the momentum mirror method [25] and relied on periodic boundary conditions in the lateral directions (x and y). The duration of each simulation was 10 ps, the time taken for the shock to transit the 80 nm simulated single crystal. This work predicted the $\alpha-\varepsilon$ transition under shock compression at a pressure of just over 15 GPa, reasonably consistent with gas gun experiments [5].

Using these MD simulations, we calculated the expected x-ray diffraction pattern for single crystal iron. As evident from Fig. 6, there was no lattice relaxation observed below the transition on the timescale of the simulations. The generation and motion of dislocations associated with plastic flow would have resulted in both a reduction in the mean lattice parameter perpendicular to the shock propagation direction and an expansion of the reciprocal lattice, which were not observed.

For highly pure, melt-grown single crystals, the experimental results are consistent with the simulation, namely that no lattice relaxation is observed before the onset of the phase

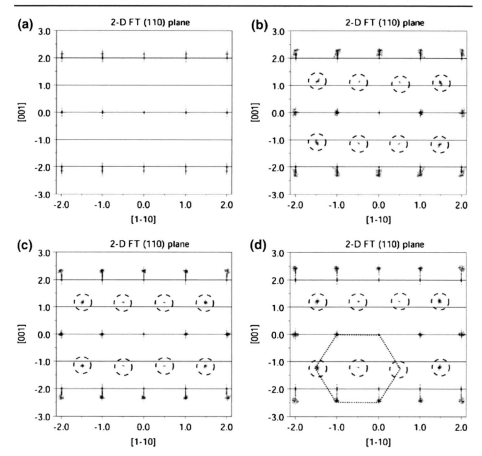

Fig. 6 Simulated x-ray diffraction patterns from iron MD simulations performed in Kadau et al. 2002. The pressure corresponds to 15 GPa, 19.6 GPa, 28.8 GPa and 52.9 GPa for **a**, **b**, **c** and **d**, respectively. The discrete points correspond to different lattice planes. Uniaxial compression causes expansion along the [001] axis, while lattice relaxation results in expansion along the [110] axis. For these simulations, we do not see any change in lateral compression. At the $\alpha–\varepsilon$ transition, we see a doubling of the lattice, indicated by the appearance of the circled points. The k space units are $2/a_o$, where a_o is the lattice constant of the BCC unit cell. Reprinted with permission from [22]. Copyright 2006, American Physical Society

transition (data shown later in Fig. 12). For data at 12 GPa just below the transition, the diffraction lines associated with the (002), (112), and (1$\bar{1}$2) indicate a compression along the shock direction of 5.2% based on the change in diffraction of the (002) plane. The lateral deformation, in contrast, can be estimated at 0.0% ± 0.6% based on the (112) and (1$\bar{1}$2) planes. A similar analysis of data recorded at 13 GPa shows compression in the shock direction of 5.8% ± 0.3% with the lateral compression estimated to be 0.0% ± 0.6% based on the same diffraction planes [22]. Within the uncertainty of the measurement, the experiments demonstrate only uniaxial, elastic compression before the phase transition, which agrees well with the MD simulations. We note two key points and differences: first that the experimental observations are made on a timescale that are approximately a nanosecond or two orders of magnitude longer than that of the simulation, and second, that the measured pressure far exceeds the accepted HEL.

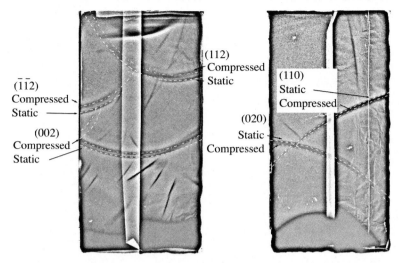

Fig. 7 X-ray diffraction data from a vapor-deposited, single-crystal foil of iron shock-compressed to 11 GPa. In contrast to melt-grown single crystals, these vapor-deposited samples exhibited lattice relaxation after the shock that was faster than the nanosecond timescale of the measurements These experiments show that the initial defected state of the sample can strongly affect the lattice kinetics. Reprinted with permission from [22]. Copyright 2006, American Physical Society

This nanosecond elastic response of iron all the way up to the transition pressure of 13 GPa contrasts with the plastic behavior inferred from continuum wave profiles of shocked iron at the microsecond timescales [3,38]. In continuum experiments, a three-wave structure is often observed—the first being associated with the HEL, the second with an elastic-plastic transition, and the final with the phase transition itself. While further work is needed to understand the elastic behavior in iron vis-à-vis the plastic response at different timescales, we can outline two important conclusions from the existing work that begin to address this complicated scientific area.

First, we note that the laser experiments are at a time scale intermediate between the picosecond MD simulations and microsecond shock-wave experiments. Because we see no plastic response in high-quality single crystals, it is evident that our laser-based results are serving as an important window into the nucleation and mobility of defects under shock, as both these processes play key roles in sample relaxation. We conclude that the combined kinetics of these phenomena appear to be longer than ~1 ns.

Second, our results also indicate that the elastic/plastic response in iron appears to be significantly affected by the initial microstructure of the sample itself. Most previous work on iron has focused on polycrystalline samples, while the diffraction experiments described here utilized single crystals. En route to systematically identifying the effect of microstructure, we have studied two kinds of single-crystal iron samples in this work, thick samples (~400 of μm) grown from a melt process and thin samples (~ 10μm) vapor deposited onto a single crystal substrate. Perhaps surprisingly, the diffraction patterns from the vapor deposited samples shown in Fig. 7 exhibited lattice relaxation at a timescale faster than the ~1 ns probe. At 11 GPa, the vapor deposited samples show a compression of $3.8\% \pm 0.2\%$ of the (002) plane along the shock direction a $2.8\% \pm 0.6\%$ compression of (112), ($\bar{1}\bar{1}2$), (110) and (020) planes in the lateral direction, exhibiting a degree of plastic deformation [22]. We speculate that the difference between the elastic response of the melt grown and plastic behavior of the vapor deposited crystals is due to the lower crystalline quality of the latter samples.

Results such as outlined here are only now beginning to provide some insight into the atomistic processes that control the evolution of the shocked solid. It is clear that substantial work remains in order to understand the complicated and interrelated effects of the initial state of the crystals, the temporal and spatial drive characteristics, strain rate, temperature, sample size, and other parameters on the dynamic plastic response of the shocked solid.

3.2 Phase transformations

Few properties of the ordered solid are as fundamental and important as its crystallographic structure. The ramifications of the crystal phase are multifold, fundamentally affecting thermodynamic behavior, electronic structure, shock response, strength, and many other properties. Under shock compression, solids generally exhibit a rich gamut of solid-to-solid phase transformations, some of which are metastable and, due to the uniaxial and ultrafast nature of the compression, may differ from those under longer timescale and more hydrostatic conditions. Carbon, for example, demonstrates a wide range of structural states along the Hugoniot, including cubic, hexagonal, nanocrystalline, and amorphous diamond [10,64]. Solid-to-melt transformations can also occur under sufficiently strong shocks.

Shock-induced phase transitions represent a fruitful scientific research area where material kinetics can play a key role. One of the most well known and widely studied of these shock-induced transformations is the $\alpha-\varepsilon$ transition in iron, where the BCC α transforms to hexagonal close-packed (HCP) ε structure. This transition was initially discovered in shock compression work by Bancroft [3]. These researchers associated the splitting of velocity waves with the elastic-plastic response and the subsequent transformation of iron to a different phase. Analysis of the wave profiles implied that the onset of the phase transition occurred at a pressure of 13 GPa. At that time, the crystallographic structure of this identified phase in iron was unknown. Subsequent static high-pressure x-ray diffraction measurements of iron showed a transformation at about 13 GPa from the BCC to the HCP structure [42], which has been assumed to be the same transition as in shock experiments due to the coincidence in pressure. Recently, iron was the first metal studied with dynamic x-ray diffraction techniques to establish the high pressure phase in situ [22,36]. Some of this work is outlined here.

3.2.1 Phase transition pathways

We now discuss two candidate transformation pathways from the high pressure BCC to HCP phase in iron. These phases have certain transformation relations that were first pointed out by Burgers [11]. Specifically, the $(0002)_{HCP}$ basal planes in ε phase HCP structure correspond to what were $(110)_{BCC}$ planes in the low pressure BCC phase. Similarly we find the $[2\overline{1}10]_{HCP}$ axis is aligned with the original $[001]_{BCC}$ axis, $[0002]_{HCP}$ with $[110]_{BCC}$ and $[2\overline{1}10]_{HCP}$ with $[002]_{BCC}$. The exact alignment of these orientations relative to the initial crystal depends on the detailed pathways of the transition. Figs. 8 and 9 give a description of the two pathway mechanisms being compared in this discussion [62]. One mechanism consists of a compression and shuffle that yields alignment between the low pressure BCC and high pressure HCP planes. Described in detail in Fig. 8, this mechanisms involves a compression along the [001] direction accompanied by an expansion in [110] to generate a hexagon, followed by a shuffle to cause the period doubling for the HCP structure. The other possible mechanism involves a shear and shuffle that introduces a rotation between the corresponding lattice planes, described in Fig. 9 as a shear and slip mechanism that generates a hexagon with a rotation relative to the corresponding BCC planes.

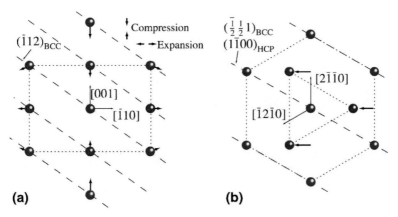

Fig. 8 Schematic diagram showing the transformation from BCC to HCP following the compression transition mechanism. The $(110)_{BCC}$ and $(0002)_{HCP}$ planes are in the plane of the paper. The blue and red circles denote atoms in the page and above the page, respectively. Arrows denote the direction of motion of the atoms. **a** Defines the coordinate system in terms of the BCC lattice and has arrows showing the expansion/compression and dashed lines indicating the location of the $(\bar{1}12)_{BCC}$ planes. **b** Defines the coordinate system for the HCP with arrows showing the shuffle of atoms with the $(1\bar{1}00)_{HCP}$ planes labeled which are equivalent to $(\frac{\bar{1}}{2}\frac{1}{2}1)_{BCC}$. Reprinted with permission from [22]. Copyright 2006, American Physical Society

3.2.2 Phase transition under uniaxial shock compression along $[001]_{BCC}$ direction

In the case of uniaxial shock compression along a particular crystallographic direction, the atomic rearrangements required to reach a HCP structure may be influenced by the anisotropies induced by uniaxial compression. In the case of the compressive mechanism illustrated in Fig. 8, there is no expected change in the transformation pathway as the compression can continue along the uniaxial direction to peak compression without requiring any change in the lateral direction. In contrast, there is marked change to the slip/shear mechanism in Fig. 9, as the initial uniaxial compression will change the needed rotation to get from the BCC to HCP state during the initial shear. The amount of rotation is shown in Fig. 9b. Static diamond anvil cell experiments using extended x-ray absorption fine structure measurements have indicated a 5° rotation is needed [62]. Due to the cubic symmetry of the BCC lattice and the cylindrical symmetry imposed by the uniaxial compression of the shock, both transition mechanisms have degenerate crystallographic directions for the end state. There are four degenerate pathways leading to two distinguishable states for the compressive mechanism, whereas the slip shear mechanism has 4 degenerate states that are all distinguishable.

3.2.3 Calculated observables for the α–ε phase transition

We now discuss predictions from the simulations that can be directly compared to experimental observations, specifically crystal rotations and volume changes.

Postprocessing of iron MD simulations performed by Kadau et al. [32], as shown in Figs. 6 and 10, yields calculated reciprocal lattice space corresponding to the uniaxially compressed BCC lattice. We observe that the BCC lattice reaches a maximum uniaxial compression of $7.0 \pm 0.5\%$ for all of the simulations. In contrast, the $(002)_{BCC}$ HCP peak varies in our simulations from 11.5 to $17.4 \pm 0.5\%$ compression upon transition to the HCP phase. These compressions, deduced from the shift of the spots in reciprocal space, agree with the direct

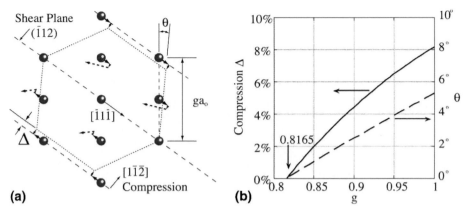

Fig. 9 Transformation from BCC to HCP following the slip/shear mechanism. **a** Schematic diagram showing the formation of hexagons in the $(110)_{BCC}$ plane. The blue and red circles denote atoms in the page and above the page, respectively. The $(\bar{1}1\bar{1})_{BCC}$ shear plane, the $[1\bar{1}2]_{BCC}$ direction, and the $[1\bar{1}\bar{2}]_{BCC}$ direction are labeled. The solid arrows denote the direction of the atomic motion due to the shear/compression. The dashed arrows represent the effective overall movement of the atoms when the crystal structure undergoes period doubling. **b** Rotation of the HCP unit cell, θ, and required compression along $[1\bar{1}\bar{2}]_{BCC}$ to form the hexagon in the initial step as functions of the initial elastic compression, g, where g = 1 for hydrostatic conditions. Δ, g, and θ are defined in panel **a**. Reprinted with permission from [22]. Copyright 2006, American Physical Society

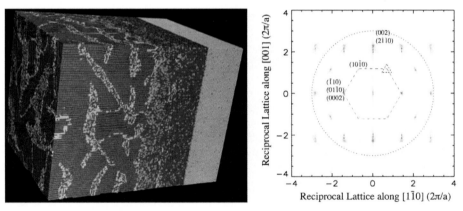

Fig. 10 This figure schematically shows the method used in calculating the expected diffraction. The atomic positions from a select portion or entire MD simulation are Fourier transformed into reciprocal lattice space. The high intensity peaks represent diffraction planes in the crystal structure. The units are based on the inverse of the original cubic BCC cell. Three points are labeled with the plane indices. One point contains three labels to show that there is a BCC component and 2 degenerate HCP components. A dashed hexagon shows that the reciprocal lattice points are approaching a hexagonal symmetry. The dotted circle gives the limits that can be experimentally probed using 0.185 nm K-shell radiation from iron. MD simulation image reprinted with permission from [32]. Copyright 2002, AAAS. Fourier transform reprinted with permission from [23]. Copyright 2006, American Institute of Physics

density measurements from the simulations obtained trivially by counting the number of atoms per unit volume.

The simulations predict a compression wave of ∼7% propagating through the BCC crystal before the transformation to an HCP-like structure. For the compressive mechanism, no rotation of the crystal lattice would be expected even with uniaxial compression. For the

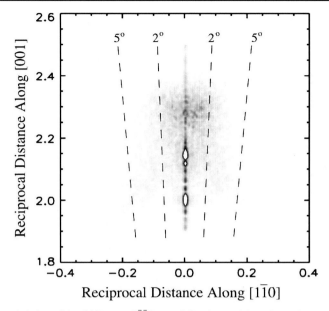

Fig. 11 An expanded view of the $(002)_{BCC}/(2\bar{1}\bar{1}0)_{HCP}$ diffraction peak in reciprocal space. The broadening of the spot associated with the HCP phase is not consistent with a rotation between the HCP and BCC lattice orientation required by the slip/shear mechanism. Reprinted with permission from [22]. Copyright 2006, American Physical Society

slip/shear mechanism, in contrast, we would expect the rotation of the lattice to decrease from 5% to 3.5%, in order to form perfect hexagons following uniaxial compression. A rotation in real space is manifested in reciprocal space as a rotation in the pattern of peaks. In Fig. 11, we show an expanded view of the simulated $(002)_{BCC}/(2\bar{1}\bar{1}0)_{HCP}$ spot in reciprocal space. Also shown on the plot are dashed lines oriented $2°$ and $5°$ relative to the $[001]_{BCC}$ axis. With two possible directions that the rotation can occur, we would expect to see the HCP feature composed of two spots, each slightly shifted off the axis. However, the spot is a single diffuse spot (of order $1°$ FWHM) centered along the $[001]_{BCC}$ axis. We conclude from the simulations that the α–ε transition is predicted to occur by the compressive mechanism.

3.2.4 In situ, Real-time diffraction measurements during the shock

The experiments described here were performed using the OMEGA [4], Janus, and Vulcan [15] lasers. Melt-grown single-crystal iron samples with a purity of 99.94% were polished to a thickness of $\sim 250\mu m$ with the surfaces aligned along the [001] direction. These samples were shock loaded by direct laser irradiation at 10^{10} to $10^{12} W/cm^2$ using 2 to 6 ns constant intensity laser pulses with a laser focal spot size of 2 to 3 mm in diameter and probed with 1.85 angstrom iron K-shell x-rays. Due to absorption in these thick samples, x-rays were diffracted from the shocked-side of the iron crystal in reflection geometry—which we refer to as the Bragg geometry. Experiments were also conducted using 10μm thick single-crystal samples of iron. For these thinner samples it was also possible to perform diffraction in transmission geometry, which we refer to as Laue geometry. Due to the divergence of the x-rays and the large angle which the crystal subtends to the x-ray source, the x-rays are simultaneously collected from many different lattice planes in the crystal. Two wide angle

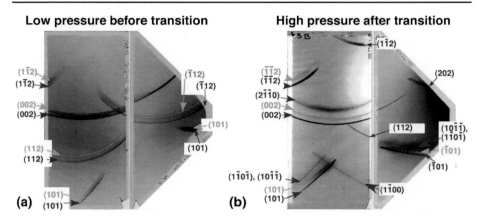

Fig. 12 In situ diffraction data from melt-grown, single-crystal iron shocked along the [100] direction. In a single pulse measurement, we obtain signal from many compressed and static planes, which are labeled with their corresponding Miller indices. Some features are identified with more than one label due to degeneracy of planes. The static BCC and HCP lattice lines are denoted by the blue, green, and red lines, respectively. **a** 5 GPa below the transition. The lattice in these samples shows no plastic relaxation below the transition. **b** 28 GPa above the transition. Reprinted with permission from [36]. Copyright 2005, American Physical Society

multiple film packs covering a total of nearly 2π steradians recorded the diffracted signal in both the Bragg and Laue geometries.

A typical diffraction pattern below the α–ε transition is shown in Fig. 12a from both the shocked and unshocked BCC lattice at a pressure of 5.4 GPa. In contrast, Fig. 12b shows data for a pressure of 26 GPa above the α–ε transition. In this latter case, we note that the $(002)_{BCC}$ plane splits into three components—the first corresponding to diffraction from unshocked BCC, the second to shock-compressed BCC, and the third to a further degree of compression which we have attributed to the HCP phase, as discussed next.

3.2.5 The transformation mechanism

The experimental data analysis supports the conclusion that the high pressure ε phase is HCP and that it is polycrystalline with two preferentially ordered variants highly aligned relative to the original crystal. Out of the list of possible transformation mechanisms from the BCC to HCP structure that are consistent with static experiments [62], only the two candidate pathways discussed previously are reasonably consistent with the number of observed variants and rotation of the lattice. We now contrast these two pathways.

Due to the degeneracy of the shift along the $(110)_{BCC}$ family of planes, there are 4 variants of the rotation of the $[2\bar{1}\bar{1}0]_{HCP}$ plane. The experimental Laue diffraction has shown that the transformed HCP crystal is polycrystalline suggesting that, if the slip/shear mechanism is responsible for the transition, then all four rotations would have occurred. We compare the data with an overlay of the 4 rotations assuming a single compression. As discussed earlier, a rotation of 3.5° would still be required by the slip/shear mechanism to form a hexagonal structure when the BCC crystal is uniaxially compressed by $\sim 7\%$. Figure 13 shows line outs from the data with overlays of calculated lineshape, assuming all four degenerate lattice rotations exist. The arrows denote the center position of the diffraction lines. The simulated line outs are broadened and scaled to obtain a best fit with the data. It is apparent that a rotation larger then 2° would not be consistent with our data. This small rotation appears to rule out

Fig. 13 Lineouts from experimental data showing the compression of the $(002)_{BCC}$ lattice plane fitted by various degrees of rotation between the HCP and BCC lattices. We observe that the width of the fitted line exceeds that of the experimental lineout beyond $2°$. The vertical arrows denote the center position of each diffraction line. Reprinted with permission from [22]. Copyright 2006, American Physical Society

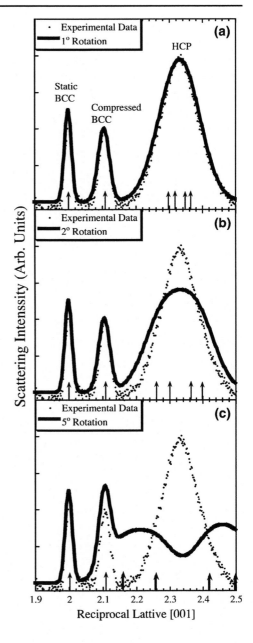

the slip/shear mechanism, even taking into account an imperfect hexagonal structure upon transformation. Thus we conclude that the experimentally measured pathway is consistent with only the compressive mechanism of those listed by Wang et al. and as predicted by the MD simulations.

These results represent one of the first detailed studies of the atomistic dynamics of a shock-induced transformation. We have been able to clearly identify the high-pressure crystallographic structures as well as the probable transformation pathways for atomistic motion.

For the first time, we have demonstrated the existence of variants in the high-pressure phase, vital information predicted in MD simulations but that had not been previously observed in recovery experiments. In future work, we will strive to study transformation timescales through combined simulation and experiments.

4 Future work

4.1 Dynamic melt: simulation and experiment

The kinetics of dynamically driven phase transitions, specifically melt, has been an important topic of experimental and theoretical investigation for several decades. Few empirical models have attempted to explain the processes occurring at the atomic level in materials shock driven to a Hugoniot state beyond the solid-liquid phase boundary [30,58]. Kinetics models based on a simple picture of nucleation and growth, such as the Johnson-Mehl-Avrami-Kolmogorov (JMAK) [2] model, have been employed to describe a variety of materials, but determination of the rate with which the melt phase grows is based on assumptions regarding the homogeneous or heterogeneous nature of the nucleation rate. As such, there could be a significant difference between materials of polycrystalline or single crystal structure. In the former, the grain boundaries may play the primary role in the kinetics of the melt transition, whereas in the latter, dislocations and crystal orientation may be the determining factor of the transition kinetics. Lastly, the loading history of the sample under investigation, namely the shock pressure duration and amplitude, is critical for the transition kinetics and final microstructure.

Combining results of dynamic compression experiments with large scale MD simulations is beginning to provide knowledge on the non-equilibrium processes occurring during ps-ns time-scales leading to melt. An example of simulated shock-induced melt in single crystal Cu is presented here (Fig. 14), where the behavior of the [100] oriented crystal shocked to 300 GPa and $T = 6000$ K is shown [7]. The material exhibits melt within nanometers of the shock front, corresponding to a time-scale of a few ps. A plot of the calculated scattering function S(q) depicted in Fig. 15 from which information about the interatomic spacing in the liquid Cu can be extracted from the series of maxima.

Coupling MD results to experimental observables requires a dynamic technique capable of revealing transient structural information with ps-ns temporal resolution. Only recently have dynamic x-ray diffraction techniques managed to demonstrate such capability. Specifically, we have developed a novel cylindrical pinhole powder camera (CPPC) suitable for laser-based x-ray backlighting and shock wave generation that has successfully captured signal from polycrystalline and amorphous materials [24].

The camera consists of a cylindrical detector arrangement with a point x-ray source of He-like α or $K - \alpha$ radiation produced at the entrance of the camera along the cylinder axis as shown in Fig. 16. Samples are interrogated by the quasi-collimated x-ray beam obtained by a suitable pinhole arrangement. Diffracted x-rays reach the detector, consisting of image plates positioned around the cylinder walls, in both transmission and reflection. Information is recorded in the azimuthal angle direction ϕ and the scattering angle 2θ, thus making the instrument capable of a variety of material property measurements under shock conditions including texture as well as phase transitions. Proof-of-principle experiments demonstrating the camera's ability in measuring amorphous signals, such as those expected from shock-melted samples, have been successfully carried out. Fig. 17 shows x-ray diffraction from the surro-

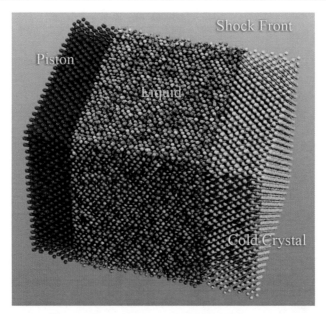

Fig. 14 Snapshot of MD simulation of Cu shocked to 300 GPa along the [100] direction. The blue region corresponds to the piston region constrained to move at a fixed speed of 3 km/sec. The yellow is the pristine unshocked copper atoms, and the orange (color scaled by the centro-symmetry parameter) shows the region where the periodic lattice has melted into a liquid

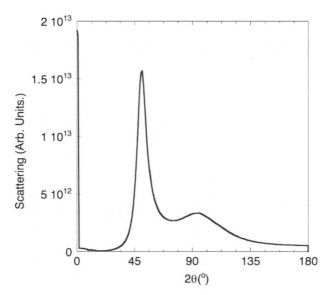

Fig. 15 Simulated x-ray scattering from the molten region of the MD simulation, showing broad features associated with amorphous systems. These features give insight into the correlations in the high pressure fluid

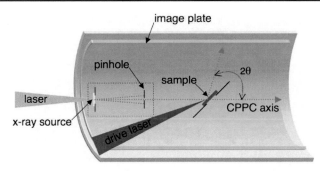

Fig. 16 Cross sectional view of the cylindrical pinhole powder camera designed for dynamic x-ray diffraction studies of polycrystalline and amorphous materials [24]. The cylindrical design can record diffraction rings from angles as low as $2\theta = 30°$ to as high as $150°$. The design uses a laser produced plasma as an x-ray source and an ablative laser drive to shock compress the sample

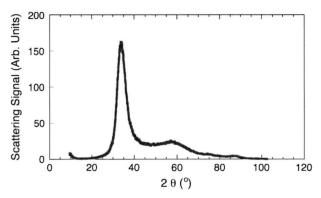

Fig. 17 Nanosecond single-shot amorphous diffraction from uncompressed metallic glass. The signal-to-noise of the instrument suggests it should be possible to experimentally measure the amorphous features of a fluid system under dynamic compression. Reprinted with permission from [24]. Copyright 2007, American Institute of Physics

gate amorphous material Metglas [$Ni_{55}Fe_{27}Co_{15}Si_{1.6}Al_{1.4}$], where peak positions and widths corresponding to the He-like Fe x-ray source were observed.

With feasibility demonstrated, we plan to implement this diffraction geometry under dynamic conditions to record the melt signature from a variety of materials. Of great interest will be the connection of our MD simulation results in shock-melted Cu with corresponding experiments, although the required shock pressure of >200 GPa may present a challenge. Additionally, bounding information on the melt kinetics may be provided by a simple pump-probe experiment where the relative timing of the x-ray source and shock wave is varied, perhaps to as fast as ~fs at new generation facilities such as LCLS. Such future investigations should offer a glimpse into the dynamics and kinetics of ultrafast melt phenomena via a multi-faceted theoretical and experimental effort coupling MD and potentially hydrocode simulations to dynamic diffraction techniques such as the CPPC described above.

4.2 Damage: in situ void nucleation and growth

The non-equilibrium nucleation of phase transformations and submicron porosity behind the shock wave result in a microscopically heterogeneous material. When release waves interact,

Fig. 18 Micrograph of voids in shock-recovered aluminum. During passage of the release wave, these voids grow and may coalesce, resulting in failure

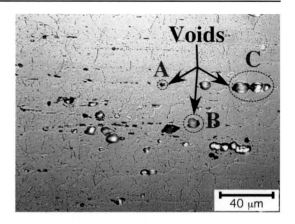

the metal may be driven into tension leading to the nucleation and growth of voids from these heterogeneities. Small angle x-ray scattering (SAXS), which arises due to scattering from electron density fluctuations, is particularly sensitive to the abrupt change in electron density accompanying voids and thus is a powerful tool to follow the evolution of this process during shock compression. We discuss here our effort aimed at conducting real-time, in situ void studies during the shock release process.

As mentioned earlier, shock wave loading is accompanied by considerable defect production during both the compression and unloading portions of the process. Defects and other microstructure (e.g. grain boundaries, second phase regions) may act as nucleation sites for the formation of voids. These voids can grow and coalesce under tension, resulting in material failure as exemplified in the micrograph of a shock recovered sample shown in Fig. 18. The time to failure can be fast from ∼sub-ns to ∼ μs, depending upon the drive conditions in the experiment. The entirety of this process is termed spall.

Most experimental observations of the spall process have been generated using two methods: velocimetry and recovery. These types of measurements, while valuable, are inherently limited. Velocimetry measurements are continuum measurements that cannot identify or provide direct insight into the underlying processes. Similarly, shock recovered samples [21] have proven essential in identifying microscopic flaws and residual void distributions, but they are limited by the difficulty of or uncertainty in arresting the damage evolution during high strain-rate dynamic loading. Temporally resolved in situ studies offer the opportunity to potentially answer key questions in the nucleation, early void growth, and final void linkage regimes.

The SAXS technique is well-established and has been applied to problems ranging from biology to materials science. In its simplest form, the SAXS scattered intensity is the square of the Fourier Transform of the scattering length density [20, 54]. In the dilute approximation, scattering is only a function of the form factor, or size and shape of the particles or voids.[1]

---

[1] In this approximation, and for a single population of scatterers, the scattering can be written in the following form [28], [29]: $I(q) = |\Delta\rho|^2 \int_0^\infty |F(q,r)|^2 V^2(r) N P(r) \, dr$ where $q = \frac{4\pi}{\lambda} \sin(\theta)$, λ is the wavelength of scattered radiation, θ is the scattering half, r is the size of the scattering particle, $\Delta\rho$ is the scattering contrast (related to the difference in electron density) between the minority and majority phases, $F(q,r)$ is the scattering form factor, $V(r)$ is the volume of the particle, N is the total number of particles, and $P(r)$ is the probability of having a minority phase particle of size r.

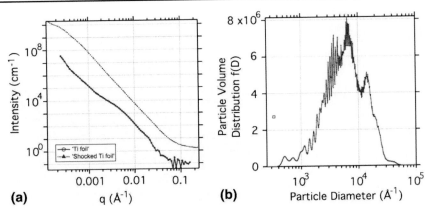

Fig. 19 Static SAXS measurements of recovered titanium and analysis of data. **a** SAXS data of recovered laser shocked and unshocked Ti samples. **b** Void distribution assuming spherical voids calculated using maximum entropy algorithm

Although a given electron density distribution gives a unique scattering intensity profile, the inversion is not unique and requires model assumptions about microstructure.

To make our initial assessment, we performed SAXS experiments using incipiently spalled Ti samples recovered from laser shock experiments. The resulting static small-angle scattering data[2] is shown in Fig. 19 [26,27]. The maximum entropy method, implemented in the "Irena" package for SAXS data analysis, was used to determine scatterer size distributions [28]. For this study, the voids were modeled using a form factor for spheres, and a maximum entropy algorithm was employed to calculate size distributions of the voids [28]. This iterative, linear inversion method, which does not require an *a priori* guess as to the shape of the distribution, converges to a unique distribution of sizes for a given form factor by maximizing the configurational entropy subject to fitting the measured scattering profile [28,48,49].

We believe that time dependent SAXS is an ideal technique to quantify the nucleation and growth of sub-micron voids under shock compression. To make our initial assessment, we performed SAXS experiments at the Advanced Photon Source (APS), whose single bunch mode capability allows the isolation of signal from only one bunch for time dependent x-ray experiments. To further increase the x-ray fluence, the APS beamline was operated with a $\sim$ 4% energy spread around the characteristic energy, which increases the fluence by a factor of 10 yet does significantly degrade our desired angular resolution.

To demonstrate the ability to perform dynamic SAXS measurements, we used the unique time-resolved SAXS detector built by Dr. Jan Hessler at APS [17]. Small angle scattering signals were acquired through the use of a specialized solid state CCD that has ring shaped pixels that assume circumferential symmetry of the scattering signals but markedly increase the collected signal. The detector can record a single 120 ps x-ray pulse with a temporal resolution of 300 ns, which is sufficient to gate against subsequent x-ray pulses. We show data in Fig. 20 from a single pulse measurement of a Ti foil to demonstrate that time resolved SAXS measurements are feasible.

---

[2] The Ultra small angle x-ray scattering (USAXS) data were acquired using the beam line 32-ID at the Advanced Photon Source, Argonne National Laboratory, Argonne, Illinois, U.S.A. The endstation consists of a Bonse-Hart camera, which can measure scattering vectors (q) from about 0.0001 to 10Å^{-1}. The monochromator was positioned at about 11 keV (1.13 Å). Data were processed using the codes developed for use on this USAXS instrument, and included absolute scattering intensity calibration and slit desmearing.

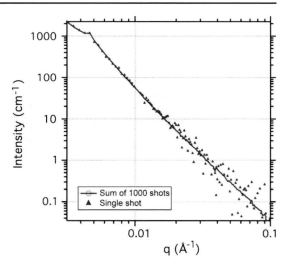

Fig. 20 Comparison of SAXS spectra collected on an unshocked Ti foil in a single 100 ps shot versus a sum of 1000 shots

In Fig. 21, we show an MD simulation of spall fracture and its corresponding calculated SAXS signal in dynamically compressed single crystal copper loaded to ∼90 GPa in 60 ps and then released. During the loading process, stacking faults and dislocations are formed to relieve the flow stress. MD simulations can identify the primary mechanisms for nucleation, growth and coalescence of voids, as well as determine Burgers vectors and habit planes of the observed dislocations. Such results can be parametrized and transitioned to higher-level statistical models such as dislocation dynamics or level-set coalescence calculations. Simulating the x-ray scattering using a Fourier transform of the MD simulation output shows the resolution required to measure the voids at this early stage of material failure.

Our combined experiment and simulation studies strongly indicate that single shot SAXS has the ability to provide relevant and valuable void formation data in dynamically loaded metals. With the goal of developing this in situ approach, we have demonstrated several key intermediate steps. We have shown, through shock-recovered measurements, that SAXS can provide valuable information about size distribution of microvoids. We have established that time resolved measurements at a third generation x-ray source, such as the APS, can potentially enable such measurements under shock conditions. These static data are showing direct evidence of structure in shock recovered samples at length scales less that 1μm. In parallel, we have simulated the growth of voids during spallation by using large scale MD simulations and then calculating the experimental observables. Being able to quantify the temporal evolution of these voids under shock conditions by using both experiment and simulation would be exceedingly valuable in ultimately understanding the processes controlling void phenomenology.

5 Conclusion

This ongoing work couples together large-scale simulations that model the atomistic level response of materials to state-of-the-art experiments in order to address the multi-scale problem of building predictive models of material response under highly dynamic environments. Applying these capabilities to the shock process, we have investigated and obtained interesting insight into the 1D to 3D transition and the generation and motion of dislocations.

Fig. 21 A plot of the simulated
SAXS from MD simulations with
voids being generated in single
crystal copper released after
being dynamically loaded to
~90 GPa. The scattering is shown
13 ps and 21 ps after the pressure
has been relieved. At top are real
space images of the MD
simulation showing the void
morphology. The atom color is
scaled by central symmetry
parameter a to accentuate the
defects and voids

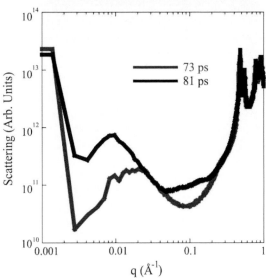

While laser-based experiment and simulation agree very well for iron, the discrepant results
of short timescale laser-based experiments and long timescale gun experiments have raised
intriguing questions as to the processes that affect the kinetics of the plastic response. Using
in situ x-ray diffraction, we have studied the shock-induced high pressure phase of iron and
found both the pathway and structure to be in good agreement with MD simulations. In
addition, we are developing techniques that can be used both to diagnose material melting

under dynamic loading and that can be used to address polycrystalline solids where grain boundaries and the interaction of differently oriented crystallites introduce fascinating and complicated phenomena. And finally, we are investigating void formation and material failure during the dynamic tensile stress conditions often generated under shock release. These efforts are poised to provide a more complete picture of the shock response of crystalline materials than has been previously possible.

Acknowledgements This work was funded by the Laboratory Directed Research and Development Program at LLNL under project tracking code 06-SI-004 and under the auspices of the U.S. Department of Energy by Lawrence Livermore National Laboratory under Contract DE-AC52-07NA27344. Work at the Advanced Photon Source was supported by the U.S. Department of Energy, Office of Science, Office of Basic Energy Sciences, under Contract No. DE-AC02-06CH11357. The authors would like to thank the staff at the Vulcan Laser Facility at the Rutherford Appleton Laboratory, the Janus laser at Lawrence Livermore National Laboratory, and the University of Rochester Laboratory for Laser Energetics under the NLUF grants program. MT acknowledges the Secretaría de Estado de Educación y Universidades of the Government of Spain for a postdoctoral fellowship and the MEC for partial support through grant MAT2006-13548-C02-01.

References

1. Asay, J.R.: The use of shock-structure methods for evaluating high-pressure material properties. Int. J. Impact Eng. **20**(1–5), 27–61 (1997)
2. Avrami, M.: Kinetics of phase change I—general theory. J. Chem. Phys. **7**(12), 1103–1112 (1939)
3. Bancroft, D., Peterson, E.L., Minshall, S.: Polymorphism of iron at high pressure. J. Appl. Phys. **27**(3), 291–298 (1956)
4. Boehly, T.R., Craxton, R.S., Hinterman, T.H., Kelly, J.H., Kessler, T.J., Kumpan, S.A., Letzring, S.A., McCrory, R.L., Morse, S.F.B., Seka, W., Skupsky, S., Soures, J.M., Verdon, C.P.: The upgrade to the OMEGA laser system. Proceedings of the tenth topical conference on high temperature plasma diagnostics, vol. 66, p. 508. AIP, Rochester, New York (1995)
5. Boettger, J.C, Wallace, D.C.: Metastability and dynamics of the shock-induced phase transition in iron. Phys. Rev. B **55**(5), 2840–2849 (1997)
6. Boustie, M., De Resseguier, T., Hallouin, M., Migault, A., Romain, J.P., Zagouri, D.: Some applications of laser-induced shocks on the dynamic behavior of materials. Laser Part. Beams (Print) **14**(2), 225–235 (1996)
7. Bringa, E.M., Cazamias, J.U., Erhart, P., Stolken, J., Tanushev, N., Wirth, B.D., Rudd, R.E., Caturla, M.J.: Atomistic shock Hugoniot simulation of single-crystal copper. J. Appl. Phys. **96**(7), 3793–3799 (2004)
8. Bringa, E.M., Caro, A., Wang, Y.M., Victoria, M., McNaney, J.M., Remington, B.A., Smith, R.F., Torralba, B.R., Van Swygenhoven, H.: Ultrahigh strength in nanocrystalline materials under shock loading. Science **309**, 1838–1841 (2005)
9. Bringa, E.M., Rosolankova, K., Rudd, R.E., Remington, B.A., Wark, J.S., Duchaineau, M., Kalantar, D.H., Hawreliak, J., Belak, J.: Shock deformation of face-centred cubic metals on subnanosecond timescales. Nat. Mater. **5**, 805–809 (2006)
10. Bundy, F.P., Bassett, W.A., Weathers, M.S., Hemley, R.J., Mao, H.K., Goncharov, H.K.: The pressure-temperature phase and transformation diagram for carbon; updated through 1994. Carbon **34**(2), 141–153 (1996)
11. Burgers, W.G.: On the process of transition of the cubic-body-centered modification into the hexagonal-close-packed modification of zirconium. Physica **1**(7–12), 561 (1934)
12. Colvin, J.D., Ault, E.R., King, W.E., Zimmerman, I.H.: Computational model for a low-temperature laser-plasma driver for shock-processing of metals and comparison to experimental data. Phys. Plasmas **10**, 2940 (2003)
13. Cornacchia, M., Arthur, J., Bane, K., Bolton, P., Carr, R., Decker, F.J., Emma, P., Galayda, J., Hastings, J., Hodgson, K., Huang, Z., Lindau, I., Nuhn, H.D., Peterson, J.M., Pellegrini, C., Reiche, S., Schlarb, H., Stöhr, J., Stupakov, G., Walz, D., Winick, H.: Future possibilities of the Linac coherent light source. J. Synchrotron. Rad. **11**(Part 3), 227–238 (2004)
14. d'Almeida, T., Gupta, Y.M.: Real-time x-ray diffraction measurements of the phase transition in KCl shocked along [100]. Phys. Rev. Lett. **85**, 330–333 (2000)

15. Danson, C.N., Barzanti, L.J., Chang, Z., Damerell, A.E., Edwards, C.B., Hancock, S., Hutchinson, M.H.R., Key, M.H., Luan, S., Mahadeo, R.R.: High contrast multi-terawatt pulse generation using chirped pulse amplification on the VULCAN laser facility. Opt. Commun. **103**(5–6), 392–397 (1993)

16. Davidson, R.C., Arnett, D., Dahlburg, J., Dimotakis, P., Dubin, D., Gabrielse, G., Hammer, D., Katsouleas, T., Kruer, W., Lovelace, R., Meyerhofer, D., Remington, B., Rosner, R., Sessler, A., Sprangle, P., Todd, A., Wurtele, J.: Frontiers in high energy density physics: the X-games of contemporary science. The National Academies Press. Washington D.C. (2003)

17. De Lurgio, P.M., Hessler, J.P., Weizeorick, J.T., Kreps, A.S., Molitsky, M.J., Naday, I., Drake, G.R., Jennings, G.: A new detector for time-resolved small-angle X-ray scattering studies. Nucl. Sci. Symp. **2**, 1215–1222 (2005)

18. Duff, R.E., Minshall, F.S.: Investigation of a shock-induced transition in bismuth. Phys. Rev. **108**(5), 1207–1212 (1957)

19. Edwards, J., Lorenz, K.T., Remington, B.A., Pollaine, S., Colvin, J., Braun, D., Lasinski, B.F., Reisman, D., McNaney, J.M., Greenough, J.A., Wallace, R., Louis, H., Kalantar, D.: Laser-driven plasma loader for shockless compression and acceleration of samples in the solid state. Phys. Rev. Lett. **92**(7), 075002 (2004)

20. Glatter, O., Kratky, O. (eds.): Small Angle X-ray Scattering. Academic Press Inc, London (1982)

21. Gray, G.T. III.: Influence of shock-wave deformation on the structure/property behavior of materials. In: Asay, J.R., Shahinpoor, M. (eds.) High Pressure Shock Compression of Solids, pp. 187–215. Springer-Verlag, New York (1983)

22. Hawreliak, J., Colvin, J.D., Eggert, J.H., Kalantar, D.H., Lorenzana, H.E., Stolken, J.S., Davies, H.M., Germann, T.C., Holian, B.L., Kadau, K., Lomdahl, P.S., Higginbotham, A., Rosolankova, K., Sheppard, J., Wark, J.S.: Analysis of the x-ray diffraction signal for the alpha-epsilon transition in shock-compressed iron: Simulation and experiment. Phys. Rev. B (Condensed Matter and Materials Physics) **74**(18), 184107 (2006a)

23. Hawreliak, J., Rosolankova, K., Belak, J., Collins, G., Colvin, J., Davies, H., Eggert, J., Germann, T., Holian, B., Kalantar, D., Kadau, K., Lomdahl, P., Lorenzana, H.E., Sheppard, J., Stolken, J., Wark, J.: Shock induced alpha-epsilon phase change in iron analysis of MD simulations and experiment. AIP Conf. Proc. **845**, 220–223 (2006b)

24. Hawreliak, J., Lorenzana, H.E., Remington, B.A., Lukezic, S., Wark, J.S.: Nanosecond x-ray diffraction from polycrystalline and amorphous materials in a pinhole camera geometry suitable for laser shock compression experiments. Rev. Sci. Instrum. **78**(8), 083908 (2007)

25. Holian, B.L., Lomdahl, P.S.: Plasticity induced by shock waves in nonequilibrium molecular-dynamics simulations. Science **280**(5372), 2085–2088 (1998)

26. Ilavsky, J., Allen, A.J., Long, G.G., Jemian, P.R.: Effective pinhole-collimated ultrasmall-angle x-ray scattering instrument for measuring anisotropic microstructures. Rev. Sci. Instrum. **73**, 1660 (2002)

27. Ilavsky, J., Jemian, P., Allen, A.J., Long, G.G.: Versatile USAXS (Bonse-Hart) facility for advanced materials research. AIP Conf. Proc. **705**, 510 (2004)

28. Ilavsky, J., Jemian, P.R.: Irena and Indra SAXS data analysis macros, including maximum entropy (2005)

29. Jemian, P.R.: Characterization of steels by anomalous small-angle X-ray scattering. PhD Thesis, Department of Materials Science and Engineering, Northwestern University, IL, USA (1990)

30. Jin, Z.H., Gumbsch, P., Lu, K., Ma, E.: Melting mechanisms at the limit of superheating. Phys. Rev. Lett. **87**(5), 055703-1 (2001)

31. Johnson, Q., Mitchell, A.C.: First X-ray diffraction evidence for a phase transition during shock-wave compression. Phys. Rev. Lett. **29**(20), 1369 (1972)

32. Kadau, K., Germann, T.C., Lomdahl, P.S., Holian, B.L.: Microscopic view of structural phase transitions induced by shock waves. Science **296**(5573), 1681–1684 (2002)

33. Kalantar, D.H., Chandler, E.A., Colvin, J.D., Lee, R., Remington, B.A., Weber, S.V., Wiley, L.G., Hauer, A., Wark, J.S., Loveridge, A., Failor, B.H., Meyers, M.A., Ravichandran, G.: Transient x-ray diffraction used to diagnose shock compressed Si crystals on the Nova Laser. Rev. Sci. Instrum. **70**, 629–632 (1999)

34. Kalantar, D.H., Remington, B.A., Colvin, J.D., Mikaelian, K.O., Weber, S.V., Wiley, L.G., Wark, J.S., Loveridge, A., Allen, A.M., Hauer, A.A.: Solid-state experiments at high pressure and strain rate. Phys. Plasmas **7**, 1999 (2000)

35. Kalantar, D.H., Bringa, E., Caturla, M., Colvin, J., Lorenz, K.T., Kumar, M., Stolken, J., Allen, A.M., Rosolankova, K., Wark, J.S., Meyers, M.A., Schneider, M., Boehly, T.R.: Multiple film plane diagnostic for shocked lattice measurements (invited). Rev. Sci. Instrum. **74**, 1929–1934 (2003)

36. Kalantar, D.H., Belak, J.F., Collins, G.W., Colvin, J.D., Davies, H.M., Eggert, J.H., Germann, T.C., Hawreliak, J., Holian, B.L., Kadau, K., Lomdahl, P.S., Lorenzana, H.E., Meyers, M.A., Rosolankova, K., Schneider, M.S., Sheppard, J., Stolken, J.S., Wark, J.S.: Direct observation of the alpha-epsilon transition in shock-compressed iron via nanosecond X-ray diffraction. Phys. Rev. Lett. **95**(7), 075502 (2005)

37. Kim, K.J., Mills, D.M.: Workshop for the Generation and Use of Short X-ray Pulses at APS. Advanced Photon Source, Argonne National Laboratory (2005)

38. Kozlov, E.A., Telichko, I.V., Gorbachev, D.M., Pankratov, D.G., Dobromyslov, A.V., Taluts, N.I.: The metastability and incompleteness of the alpha-epsilon phase transformation in unalloyed iron loading pulses: Specific features under the effect of threshold of the deformation behavior and structure of armco iron. Phys. Met. Metallogr. **99**(3), 300–313 (2005)

39. Lindl, J.: Development of the indirect-drive approach to inertial confinement fusion and the target physics basis for ignition and gain. Phys. Plasmas **2**(11), 3933–4024 (1995)

40. Lorenz, K.T., Edwards, M.J., Glendinning, S.G., Jankowski, A.F., McNaney, J., Pollaine, S.M., Remington, B.A.: Accessing ultrahigh-pressure, quasi-isentropic states of matter. Phys. Plasmas **12**(5), 056309 (2005)

41. Loveridge-Smith, A., Allen, A., Belak, J., Boehly, T., Hauer, A., Holian, B., Kalantar, D., Kyrala, G., Lee, R.W., Lomdahl, P., Meyers, M.A., Paisley, D., Pollaine, S., Remington, B., Swift, D.C., Weber, S., Wark, J.S.: Anomalous elastic response of silicon to uniaxial shock compression on nanosecond time scales. Phys. Rev. Lett. **86**(11), 2349–2352 (2001)

42. Mao, H., Bassett, W.A., Takahashi, T.: Effect of pressure on crystal structure and lattice parameters of iron up to 300 kbar. J. Appl. Phys. **38**(1), 272–276 (1967)

43. Meyers, M.A.: Dynamic Behavior of Materials. Wiley, New York (1994)

44. Meyers, M.A., Gregori, F., Kad, B.K., Schneider, M.S., Kalantar, D.H., Remington, B.A., Ravichandran, G., Boehly, T., Wark, J.S.: Laser-induced shock compression of monocrystalline copper: characterization and analysis. Acta Mater. **51**(5), 1211–1228 (2003)

45. Ochi, Y., Golovkin, I., Mancini, R., Uschmann, I., Sunahara, A., Nishimura, H., Fujita, K., Louis, S., Nakai, M., Shiraga, H.: Temporal evolution of temperature and density profiles of a laser compressed core (invited). Rev. Sci. Instrum. **74**, 1683 (2003)

46. Park, H.S., Maddox, B.R., Giraldez, E., Hatchett, S.P., Hudson, L.T., Izumi, N., Key, M.H., Le Pape, S., MacKinnon, A.J., MacPhee, A.G., Patel, P.K., Phillips, T.W., Remington, B.A., Seeley, J.F., Tommasini, R., Town, R., Workman, J., Brambrink, E.: High-resolution 17–75 KeV backlighters for high energy density experiments. Phys. Plasmas **15**, 072705 (2008)

47. Phillion, D.W., Hailey, C.J.: Brightness and duration of X-ray-line sources irradiated with intense 0.53µm laser-light at 60 and 120 Ps pulse width. Phys. Rev. A **34**(6), 4886–4896 (1986)

48. Potton, J.A., Daniell, G.J., Rainford, B.D.: A new method for the determination of particle size distributions from small-angle neutron scattering measurements. J. Appl. Cryst. **21**, 891–897 (1988)

49. Potton, J.A., Daniell, G.J., Rainford, B.D.: Particle size distributions from SANS data using the maximum entropy method. J. Appl. Cryst. **21**, 663–668 (1988)

50. Preston, D.L., Tonks, D.L., Wallace, D.C.: Model of plastic deformation for extreme loading conditions. J. Appl. Phys. **93**, 211 (2002)

51. Remington, B.A., Cavallo, R.M., Edwards, M.J., Ho, D.D.M., Lasinski, B.F., Lorenz, K.T., Lorenzana, H.E., McNaney, J.M., Pollaine, S.M., Smith, R.F.: Accessing high pressure states relevant to core conditions in the giant planets. Astrophys. Space Sci. **298**(1–2), 235–240 (2005)

52. Rigg, P.A., Gupta, Y.M.: Real-time x-ray diffraction to examine elastic–plastic deformation in shocked lithium fluoride crystals. Appl. Phys. Lett. **73**(12), 1655–1657 (1998)

53. Rigg, P.A., Gupta, Y.M.: Time-resolved x-ray diffraction measurements and analysis to investigate shocked lithium fluoride crystals. J. Appl. Phys. **93**(6), 3291–3298 (2003)

54. Roe, R.J.: Methods of X-ray and Neutron Scattering in Polymer Science. Oxford University Press, Oxford (2000)

55. Romain, J.P., Cottet, F., Hallouin, M., Fabbro, R., Faral, B., Pepin, H.: Laser shock experiments at pressures above 100 Mbar. Physica B + C **139**, 595–598 (1986)

56. Rosolankova, K.: Picosecond x-ray diffraction from shock-compressed metals: experiments and computational analysis of molecular dynamics simulations. Dphil Thesis, Physics Department, University of Oxford, Oxford, UK (2005)

57. Rosolankova, K., Wark, J.S., Bringa, E.M., Hawreliak, J.: Measuring stacking fault densities in shock-compressed FCC crystals using in situ x-ray diffraction. J. Phys. Condens. Matter **18**(29), 6749–6757 (2006)

58. Ross, M.: Generalized Lindemann melting law. Phys. Rev. **184**(1), 233–242 (1969)

59. Schneider, M.S., Kad, B.K., Gregori, F., Kalantar, D., Remington, B.A., Meyers, M.A.: Laser-induced shock compression of copper: orientation and pressure decay effects. Metall. Mater. Trans. A Phys. Metall. Mater. Sci. **35A**(9), 2633–2646 (2004)

60. Smith, C.S.: Metallographic studies of metals after explosive shock. Trans. Metal. Soc. AIME **212**, 574–589 (1958)

61. Swift, D.C., Ackland, G.J., Hauer, A., Kyrala, G.A.: First-principles equations of state for simulations of shock waves in silicon. Phys. Rev. B 6421(21):art. no.-214107 (2001)
62. Wang, F.M., Ingalls, R.: Iron bcc-hcp transition: local structure from X-ray absorption fine structure. Phys. Rev. B **57**(1), 5647–5654 (1998)
63. Wark, J.S., Whitlock, R.R., Hauer, A.A., Swain, J.E., Solone, P.J.: Subnanosecond x-ray diffraction from laser-shocked crystals. Phys. Rev. B **40**(8), 5705–5714 (1989)
64. Yamada, K., Tanabe, Y., Sawaoka, A.B.: Allotropes of carbon shock synthesized at pressures up to 15 GPa. Philos. Mag. A **80**(8), 1811–1828 (2000)

Calculations of free energy barriers for local mechanisms of hydrogen diffusion in alanates

Michele Monteferrante · **Sara Bonella** ·
Simone Meloni · **Eric Vanden-Eijnden** ·
Giovanni Ciccotti

Originally published in the journal Sci Model Simul, Volume 15, Nos 1–3, 187–206.
DOI: 10.1007/s10820-008-9097-x © Springer Science+Business Media B.V. 2008

Abstract   Brute force histogram calculation and a recently developed method to efficiently reconstruct the free energy profile of complex systems (the *single-sweep method*) are combined with ab initio molecular dynamics to study possible local mechanisms for the diffusion of hydrogen in sodium alanates. These compounds may help to understand key properties of solid state hydrogen storage materials. In this work, the identity of a mobile species observed in experiments characterizing the first dissociation reaction of sodium alanates is investigated. The activation barrier of two suggested processes for hydrogen diffusion in Na_3AlH_6 is evaluated and, by comparing our results with available experimental information, we are able to discriminate among them and to show that one is compatible with the observed signal while the other is not.

Keywords   Hydrogen storage · Sodium Alanates · Activated processes · Free energy reconstruction · TAMD · Single-sweep method · Radial basis reconstruction · String method

1 Introduction

Safe and efficient storage is one of the major challenges for using hydrogen as a sustainable energy carrier in technological applications with potentially zero greenhouse gas emissions.

M. Monteferrante
Dipartimento di Fisica, Università La Sapienza, P.le A. Moro 2, 00185 Rome, Italy

S. Bonella (✉) · G. Ciccotti
Dipartimento di Fisica and CNISM Unità Roma 1, Università La Sapienza, P.le A. Moro 2, 00185 Rome, Italy
e-mail: sara.bonella@roma1.infn.it

S. Bonella · G. Ciccotti
Centro Interdisciplinare Linceo "B. Segre", Accademia dei Lincei, Via della Lungara 10, 00165 Rome, Italy

S. Meloni
CASPUR, SuperComputing Consortium, c/o Università La Sapienza, P.le A. Moro 5, 00185 Rome, Italy

E. Vanden-Eijnden
Courant Institute of Mathematical Sciences, New York University, New York, NY 10012, USA

In the past few years, solid state materials have emerged as plausible candidates to solve this problem. In particular, experiments indicate that alanates, a class of aluminum and light-metals hydrides, could provide a viable and convenient mean to store and release hydrogen. Sodium alanates are recognized as the most interesting among these compounds. They dissociate via three chemical reactions:

$$3NaAlH_4 \Leftrightarrow Na_3AlH_6 + 2Al + 3H_2 \qquad (1)$$

$$2Na_3AlH_6 \Leftrightarrow 6NaH + 2Al + 3H_2$$

$$2NaH \Leftrightarrow 2Na + H_2$$

The first reaction occurs at $T = 350\,K$ and releases a quantity of molecular hydrogen equal to 3.7 percent of the weight of the reactant ($3.7\,wt\%$); the second happens at $T = 423\,K$ and with $1.9\,wt\%$ hydrogen release. These reactions occur close enough to ambient conditions to be considered interesting for technological applications, while the third takes place at a much higher temperature ($T = 689\,K$, with an additional $1.9\,wt\%$ production of hydrogen) and it is not. Recently, it was observed [1] that the addition of a catalyst such as Ti makes the first two reactions reversible and sustainable over several hydrogen charge and discharge cycles thus opening the possibility of reuse of the material. The presence of the catalyst also reduces the decomposition temperatures and enhances the kinetics of dehydrogenation and rehydrogenation of the compound. In spite of several experimental and theoretical studies [2–6], however, little is known about the mechanism of the reaction and the nature of the intermediate products of the transformation from tetra to hexa-hydride, both in the presence of the catalyst and for pure samples. In doped compounds, the role of Ti, or even its location (interstitial or substitutional in the lattice, or on the surface), is essentially unknown along with the characteristics of the aluminum-titanium alloys formed during repeated hydrogen cycling of the material. There is also a debate on the role of hydrogen and/or sodium vacancies in facilitating the diffusion of atomic or molecular species within the crystal [7].

In this paper we shall address a controversy that has recently emerged with regards to the first dissociation reaction of sodium alanates. Quasi-elastic neutron scattering [8] has confirmed the existence, first revealed by anelastic spectroscopy experiments [9,10], of a mobile species that appears both in doped and pure samples as soon as the reaction produces Na_3AlH_6. It is known that the species diffuses after overcoming an activation barrier of about $0.12\,eV$ but its identity has not yet been established beyond doubt. Neutron scattering measurements, combined with some *ab initio* calculations, suggest that the signal corresponds to sodium atoms, or rather to sodium vacancy, migration within the lattice. The same calculations reject two possible motions involving hydrogen, a local diffusion in a defective Al site (AlH_{6-x} with $x = 1$) and the diffusion of an H from a AlH_6 unit to the defect, due to high activation energies estimated from a Nudged Elastic Band (NEB) simulation [11–13] (0.41 and $0.75–1.0\,eV$ respectively). The anelastic spectroscopy result, on the other hand, is interpreted precisely as the local diffusion process mentioned above. In the following, we shall reconsider the possibility that the mobile species is related to the hydrogen atoms present in the system by employing an accurate method to calculate free energy barriers.

Two numerical challenges must be faced in this calculation. First of all, hydrogen diffusion involves breaking and forming of chemical bonds that cannot be captured reliably by simple phenomenological potentials. Consequently, it is necessary to describe the process at the quantum level at least as far as the electronic structure is concerned. A satisfactory representation of the system requires considering a relatively large number of atoms (on the order of one hundred) and the best compromise among accuracy and efficiency in *ab initio* molecular dynamics simulations of alanates is obtained by describing the interactions using density functional theory (DFT). The numerical cost associated to DFT calculations, however, limits the timescales accessible with a reasonable investment of CPU time. The second, more significant, challenge stems from the fact that, in the conditions of interest, the event under investigation is activated. In general, activated processes can be modeled by introducing a set of collective variables that characterize the initial and final state of the system as local minima in the free energy, and correspond to transitions between the basins around the minima over a free energy barrier higher than the available thermal energy. As the time necessary to escape from the metastable states increases exponentially with the height of the barrier, a brute force simulation of the event is doomed to fail and special techniques must be deployed to explore the relevant regions of the free energy profile.

The contents of this paper can be summarized as follows. Section 2 starts with a description of the model of the system that we employ. The model is based on one main hypothesis, namely that the process can be studied in a Na_3AlH_6 crystal where an H vacancy is created. This hypothesis, a drastic simplification with respect to the experimental set up, is compatible with the available information. In fact, in experiments, the signal assigned to the mobile species is observed as soon as the hexa-coordinated Al is formed and persists until completion of the dissociation when Na_3AlH_6 is essentially the only compound left in the sample. This points to this compound as the crucial environment for diffusion. The interactions are described at the DFT level and appropriate collective variables are introduced to characterize the two suggested mechanisms related to hydrogen diffusion. These are presented in subsections 2.2 and 2.3 for the local diffusion in the defective Al site and for the diffusion from the AlH_6 unit to the defect (henceforth, we shall refer to the latter as "non-local" diffusion). The behavior of the collective variables is monitored in exploratory *ab initio* molecular dynamics simulations of the system at $T = 380$ K. This temperature was chosen because it is slightly above that of the first dissociation reaction and in the experimentally interesting range. From the experiment it is known that, at $T = 380$ K, the height of the barrier is about four times the available thermal energy so the expectation is that transitions among the values of the collective variables characterizing the different minima in the corresponding free energy will be rare events in the simulation. As shown in Sect. 4, the results of the exploratory runs are quite different in the two cases considered, allowing for a direct exploration of all relevant states in the free energy landscape in the case of the local process, while indicating the existence of a significant activation barrier in the case of the non-local diffusion. This is an indication that the local vacancy diffusion does not correspond to the experimental signal. To assess the possibility that the non-local hydrogen motion is responsible for the signal, an accurate estimate of the activation barrier associated to the process is needed for comparison with the experimental result. To obtain it, we reconstruct the free energy landscape via the single-sweep method recently introduced by Maragliano and Vanden-Eijnden [14] and described in the Method section. After reconstruction, the value of the activation barrier for the non-local diffusion is evaluated as discussed in Sect. 4. Once the barrier height is available, it is confronted with the experimental result and conclusions on the possible identity of the mobile species are drawn.

2 Model and computations

As mentioned in the Introduction, we assume that the diffusion process can be modeled as happening in a Na_3AlH_6 crystal where a hydrogen vacancy is created. This set up is the same as that used in previous calculations on the problem and it has two main advantages: It allows for a well defined (and numerically affordable) atomistic model of the system—see Sect. 2.1—and it is the most suitable for comparison with the NEB results.

The local and non-local diffusion processes that we wish to study correspond to changes in the arrangement of the hydrogen atoms in the system. Since these changes can involve more than one H, and in general require cooperative motion of different degrees of freedom, it is more convenient to construct collective variables to discriminate among them rather than to follow the motion of individual atoms. Once the collective variables are defined, the steps we perform to identify the mobile species are: (1) to compute the free energy as a function of the collective variables that characterize the local and non-local diffusion; (2) to identify the minima corresponding to the initial and final state of the process; (3) to determine the saddle point between the two. The activation barrier can then be estimated from the difference among the values of the free energy at the minima and at the saddle point. The similarity or difference among the computed values and the experiment will provide us with an indication on the nature of the mobile species.

In the space of the collective variables, the free energy of the system is defined as

$$F(z) = -\beta^{-1}\ln Z^{-1} \int dx \ e^{-\beta V(x)} \prod_{j=1}^{p} \delta(\theta_j(x) - z_j) \qquad (2)$$

where $\theta_j(x)$ are the p collective variables, functions of the coordinates of the N Cartesian coordinates of the particles of the system $x = \{x_1, \ldots, x_N\}$, and $z = \{z_j\}$ is a specific set of values for the collective variables. The definition above amounts to counting the number of configurations of the system corresponding to a given value of z. This is equivalent to assigning the probability

$$P(z) = e^{-\beta F(z)} \qquad (3)$$

to the macroscopic state $\theta(x) = z$ of the system. In principle, this probability density function can be computed directly by running a long molecular dynamics trajectory for the system and monitoring the values of $\theta_j(x(t))$ generated by any dynamics that preserves the canonical probability. For example, one could use

$$m\ddot{x} = -\nabla V(x) + \text{thermostat at } \beta \qquad (4)$$

to evolve the system. In the equation above m is the mass of the particles, and the thermostat can be any standard thermostat such as Langevin, Nosé-Hoover, isokinetic, etc. The probability density can be obtained from the trajectory via the frequency estimator by binning the collective variables and constructing a histogram of their values along the trajectory. Once $P(z)$ is known, inversion of Eq. (3) immediately gives the free energy. There are however two problems in implementing this method. The first stems from the fact that the number of bins that must be included in the calculation scales exponentially with the number of collective variables. The second problem, perhaps even more serious, arises when activated processes are of interest. The procedure outlined above, in fact, will produce a global, unbiased, estimate of $F(z)$ only if the trajectory visits all relevant regions of the space of the collective variables a number of times sufficient to ensure convergency of the frequency estimator. When activated processes are important, this condition will never be met since the system will be trapped in metastable minima of the free energy from which it cannot escape in a time compatible with current trajectory lengths. In fact, it can be estimated from the Arrhenius formula that the time scale necessary to go over a barrier of height ΔF_b is roughly proportional to $e^{\beta \Delta F_b}$. To accurately sample the values of the collective variables, the trajectory must be much longer than this characteristic time and this will not be possible if the barrier is large compared to the typical thermal energy available to the system. This is indeed the situation that we expect to encounter in studying the diffusion process in sodium alanates since, from the experiment, it is known that the height of the activation barrier is about four times the available thermal energy of the system. In Sect. 3 we summarize the method adopted to overcome this problem. In the remainder of this section, we describe the model built to mimic and study the process. In 2.1 we specify the interactions and the simulations set up, while the collective variables chosen to describe the local and non-local diffusion are introduced in Sects. 2.2 and 2.3. As always, the choice of the collective variable is a delicate point in the procedure since no rigorous way to identify the most appropriate set of functions exists. The only available approach is to select them on the basis of physical, or chemical, intuition and verify *a posteriori* [15,16], that they provide an accurate description of the activated process.

2.1 The simulation set up and its validation

To determine the initial conditions in our simulations, we started by constructing a $(2 \times 2 \times 1)$ supercell for pure Na_3AlH_6 crystal. The supercell contained a total of 80 atoms. The characteristics of this crystal are well known, both experimentally [17] and from simulations [2,6]. Na_3AlH_6 has monoclinic structure of $P2_{1/n}$ symmetry (space group number 14) with two chemical units per primitive cell. The symmetry, lattice parameters, and the subset of Wyckoff coordinates for the atoms in Table 1, combined with Wyckoff tables available on the web [18], allow to generate the initial positions for the atoms in the supercell. In the pure crystal, each aluminum is surrounded by six hydrogen atoms with which it forms a slightly distorted octahedron. The nature of the bond in the octahedron is not clear, some calculations showing a covalent character [3], other finding a strong ionic component [2]. The ionic bonding between the Na^+ and the negatively charged AlH_6 complex, on the other hand, is well established.

Table 1 Cell parameters (upper panel) and Wyckoff positions of the atoms necessary to build the pure Na_3AlH_6 cell (lower panel), $\alpha = \gamma = 90$ for monoclinic crystals

Cell parameters	a(Å)	b(Å)	c(Å)	β
Experimental	5.39	5.514	7.725	89.86
Calculated	5.27	5.46	7.60	89.99
This work	5.43	5.55	7.65	89.59

	Wyckoff positions
Al(2a)	0, 0, 0
Na(2b)	0, 0, 0.5
Na(4e)	−0.010, 0.454, 0.255
H(4e)	0.102, 0.051, 0.218
H(4e)	0.226, 0.329, 0.544
H(4e)	0.162, 0.269, 0.934

The experimental cell parameters are as given in ref. [17], while the calculated ones are those obtained in ref. [8]

As for the calculation of the electronic structure, required by *ab initio* MD, the orbitals were expanded in plane waves, with a spherical cut-off of $E_c = 1088\,\text{eV}$ (80 Rydberg). Only orbitals corresponding to the Γ point [19] have been used. We have used the BLYP [20,21] form of the generalized gradient approximation for the density functional theory. We adopted Troullier-Martins pseudopotentials [22], with nine electrons in the valence state of sodium, three electrons in the valence state of aluminum and one in that of hydrogen. Inclusion of the semi-core states of sodium was necessary to reproduce the configuration and the cell parameters of the crystal. We validated this set up by performing atomic positions and cell optimization in a simulation for pure Na_3AlH_6.

Having verified that the optimized lattice parameters and interatomic distances obtained for the pure crystal compared very well with available experimental values and numerical results (see Table 1), we used the same set up to construct a Na_3AlH_6 crystal with one hydrogen defect. In addition to removing one H atom, we adjusted the lattice parameters by scaling them by a common factor so that the pressure in short dynamical runs approached zero. The adjustment was performed to mimic more closely the experimental conditions. The corresponding simulation cell, containing 24 Na, 7 AlH_6 and 1 AlH_5 groups, is shown in Fig. 1. This cell was used as the initial condition in a set of *ab initio* MD runs in which nuclear constant temperature dynamics (T = 380 K) was obtained with a Nosé-Hoover chain [23] with four thermostats whose inertia, in frequency units, had characteristic frequency $\omega = 10^3\,\text{cm}^{-1}$. The collective variables, instead, were evolved according to Langevin dynamics at a temperature $\bar{T} = 6500$ K with (fictitious) mass equal to $\bar{m} = 250[t^2(\text{fs})]$, and friction $\bar{\gamma} = 4.2 \times 10^{-3}\,\text{fs}^{-1}$. The integration scheme for this calculation, accurate to order $dt^{1/2}$, can be found in ref. [24]. The timestep was set to dt = 0.1 fs for all degrees of freedom in all molecular dynamics runs. The asymptotic value of the force constant in Eq. (14) (see Sect. 3.2) is $k = 5.44\,\text{eV}$. All calculations were performed with CPMD [25].

2.2 Collective variables for the local H-vacancy diffusion

From inorganic chemistry we know that, in vacuum, there are two possible equilibrium structures for compounds such as AlH_5. The first is a pyramid with square base, the second a triangular bi-pyramid, side (A) and (B) respectively in Fig. 2. The two structures can mutate into one another via a relatively minor rearrangement in which the base of the bi-pyramide, defined by hydrogens 1,2,3 in side B of the figure, changes so that the angle H_1-Al-H_3 goes from 120 to 180 degrees, while H_2 does not move. The square base of the new structure is identified by hydrogens 1,3,4,5. Both structures were observed in preliminary *ab initio* MD runs for the system and can be used to determine two classes of possible configurations for the defective aluminum group. Within each class, different orientations of the AlH_5 group with respect to the other AlH_6 molecules in the crystal, or to a fixed reference frame, are possible. The local hydrogen diffusion process amounts to changes of the geometry of the AlH_5 group leading to transitions among the configurations within each class and across the two sets.

To discriminate among these configurations, we introduced the following three dimensional vector whose components will be used as the collective variable

$$\vec{W} = \frac{\sum_{i=1}^{5} w_i \vec{r}_{iAl_v}}{\sum_{i=1}^{5} w_i} \tag{5}$$

where $\vec{r}_{iAl_v} = \vec{r}_i - \vec{r}_{Al_v}$ ($\vec{r}_{Al_v}$ is the position of the defective aluminum, and $\vec{r}_i$ are the coordinates of the defective group's hydrogen atoms), while w_i is a constant weight. Let us

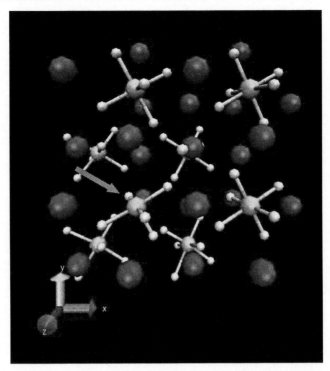

Fig. 1 Simulation ($2 \times 2 \times 1$) supercell modeling Na_3AlH_6 in the presence of a hydrogen vacancy. The 79 atoms in the cell are represented as spheres color coded as follows: Al green, Na blue, and H white. Bonds in the molecular AlH_6 groups are shown as white and green sticks. The defective aluminum group is indicated by the red arrow

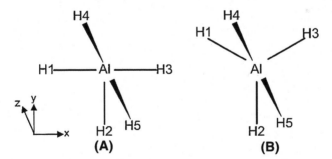

Fig. 2 Schematic representation of the possible equilibrium structures of AlH_5 in vacuum. Side (**A**) shows the square base pyramid configuration, side (**B**) the triangular bi-pyramid. Out of plane bonds are represented as thick lines, bonds in plane are simple lines

consider first the case $w_i = 1$ for all i. In this case, $|\vec{W}| = 0$ for configurations in class B due to the symmetric disposition of the hydrogen atoms around the aluminum. On the other hand, a finite value of the modulus of the vector indicates that the configuration belongs to class A. In Figs. 3 and 4, some of the features of the collective variable for configurations belonging to the two different classes are shown. The symmetry of the arrangements of the hydrogens is reflected in the very small value of the length of $\vec{W}$, as shown in the two snapshots in Fig. 3.

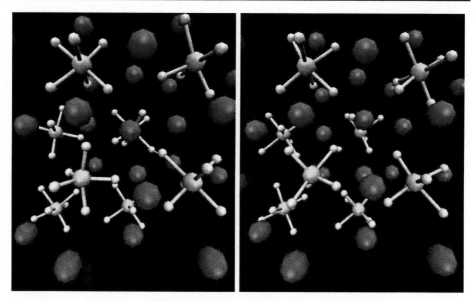

Fig. 3 Snapshots of typical configurations of the system with the defective aluminum group approaching the by-piramidal structure. The collective variable $\vec{W}$ is represented as a red stick. The vector is barely visible due to the small value of the modulus in the bi-pyramidal arrangement. For graphical purposes, in the figure the vector has been scaled by multiplying its components by a factor of five

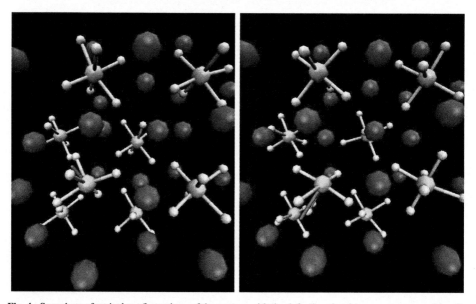

Fig. 4 Snapshots of typical configurations of the system with the defective aluminum group approaching the square base pyramid structure. The collective variable $\vec{W}$ is represented as a red stick. For graphical purposes, in the figure the vector has been scaled by multiplying its components by a factor of five

In the figure, the origin of the vector coincides with the position of Al_v, and $\vec{W}$ is represented as a red stick protruding from the aluminum. Of course, the anisotropy of the environment and the thermal fluctuations in the system will prevent the group from reaching the exactly symmetric configuration of Fig. 2, however the difference in the length of the vector for configurations in class A or B is apparent by comparing Figs. 3 and 4. In the latter, snapshots of the defective group in configurations belonging to class A are shown. Typical values of $\rho = |\vec{W}|$ measured in this case range between 0.05 Å and 0.26 Å. The direction of the vector carries information both on the orientation of the defective group and on the deformation of the square base of the pyramid. In the ideal case, the vector would be aligned with the bond connecting the aluminum with the azimuthal hydrogen, while distortions of the geometry of the base modify both its orientation and magnitude, as shown in the two examples in Fig. 4. If all weights in Eq. (5) are equal, no information can be obtained about the different orientations of the by-pyramidal structure defining class B. As mentioned above, in this case the vector has length zero so its direction cannot be defined. Rigid rotations of the symmetric structure in the space of configurations can be discriminated by our collective variable if a set of different values for the weights is introduced. For example, it can be shown by direct inspection that there is a one to one correspondence among orientations of the symmetric group and values of the angles in the polar representation of the vector $\vec{W}$ if the choice $w_i = i$ with $i = 1, \ldots, 5$ is made. However, analysis of the preliminary MD runs showed that fully symmetric configurations were never maintained for a time sufficient to observe rotations. For that reason, in the following we shall limit the definition of the collective variable to the case of equal, unit, weights.

2.3 Collective variables for the non-local H-vacancy diffusion

Non-local hydrogen diffusion, i.e. the transfer of one hydrogen from an hexa to a penta coordinated aluminum, can be described as a change in the coordination number of the donor and acceptor Al with respect to the hydrogens.

A smoothed expression of the coordination of Al_α as a function of the hydrogen's positions is [26]

$$C_\alpha = \sum_{i=1}^{N_H} \frac{1}{1 + e^{\lambda(r_{i\alpha} - r_0)}} \qquad (6)$$

with $r_{i\alpha} = |\vec{r}_i - \vec{r}_\alpha|$ (the index i runs from 1 to the number of hydrogens N_H, $\vec{r}_i$ is the position of hydrogen i, and $\vec{r}_\alpha$ is the position of aluminum α). λ and r_0 define the characteristics of the sigmoid: r_0 identifies the inflection point of the function, while λ determines the steepness of the decay. The specific choice of these parameters for our calculations is discussed in the Results section.

We used the coordination numbers of all Al atoms in the simulation cell as collective variables. The presence of all coordinations avoids an a priori choice of the donor and acceptor AlH_n groups and any assumption on the identity of the H atoms that participate in the diffusion process.

3 Methods

As mentioned in the previous section, the reconstruction by direct histogramming methods of the free energy of a system in which activated processes are relevant must tackle two challenges. First, the number of bins that need to be considered scales exponentially with the number of collective variables and, second, due to the presence of energy barriers, the dynamical trajectory is unable to explore all relevant states. In this paper we shall use the single-sweep method, introduced in [14], that offers a way out from both difficulties. The single-sweep method is based on two steps. The first employs a temperature accelerated evolution of an extended system that includes the collective variables as dynamical variables (TAMD) [27] to overcome the barriers and efficiently explore the free energy landscape. The second chooses points along the TAMD trajectory of the collective variables as centers of an interpolation grid. The free energy is represented by a set of radial basis functions centered on the interpolation grid. The coefficients of the representation can be obtained by least square fitting. The two steps of the single-sweep method were described in detail in [14]. For the reader's convenience we summarize next the main features of the approach.

3.1 Temperature accelerated molecular dynamics

Temperature accelerated molecular dynamics begins by defining an extended version of the physical system in which new variables z are coupled to the physical variables via the collective variables $\theta(x)$ and the potential energy $V(x)$ is extended to include a quadratic term that couples $\theta(x)$ and z. The new total potential is written as

$$U_k(x, z) = V(x) + \frac{k}{2} \sum_{j=1}^{p} (\theta_j(x) - z_j)^2 \tag{7}$$

It is further assumed that the z variables, to which a fictitious mass $\bar{m}$ can be assigned, evolve at a fictitious temperature $\bar{T}$ (and with a friction coefficient $\bar{\gamma}$ if one is using Langevin dynamics) different from that of the physical system. The evolution of the extended system is governed by the following set of coupled equations (compare (4))

$$m\ddot{x} = -\nabla_x V(x) - k \sum_{j=1}^{p} (\theta_j(x) - z_j) \nabla_x \theta_j(x) + \text{thermostat at } \beta$$

$$\bar{m}\ddot{z} = k(\theta(x) - z) - \bar{\gamma}\dot{z} + \sqrt{2\bar{\beta}^{-1}\bar{\gamma}}\eta^z \tag{8}$$

where $\bar{\beta} = 1/k_B\bar{T}$, k is a constant, and η^z is the white noise associated with the Langevin evolution of the new variables. The extended system so defined has the remarkable property that if we increase the value of $\bar{m}$ and $\bar{\gamma}$ (with the condition that $\bar{m} = O(\bar{\gamma}^2)$) so as to induce adiabatic separation of the motion of the physical and fictitious systems, the variables z evolve, on the slower time scale of their motion, according to the effective equation

$$\bar{m}\ddot{z} = -\nabla_z F_k(z) - \bar{\gamma}\dot{z} + \sqrt{2\bar{\beta}^{-1}\bar{\gamma}}\eta^z \tag{9}$$

where (compare (2))

$$F_k(z) = -\beta^{-1} \ln Z_k^{-1} \int dx e^{-\beta U_k(x,z)} \tag{10}$$

with $Z_k = \int dx dz e^{-\beta U_k(x,z)}$. In the limit of large k, the quantity above is the free energy of the *physical* system at inverse temperature β. The time evolution of the fictitious variables then explores the relevant regions of the free energy landscape at the physical temperature. Moreover, the evolution equation holds for any value of $\bar{T}$, and this quantity can be increased to a point when the thermal energy of the fictitious variables is high enough to overcome the physical system's free energy barriers and sample all metastable and transition regions.

Equation (9), could in principle be used to sample directly the free energy landscape thanks to the accelerated dynamics of the fictitious variables. This is in fact why the extended system was originally introduced. Although not affected by the metastability problem, this approach would still require to construct $F(z)$ via histograms and, as such, it suffers from the exponential growth with the dimensionality of the collective variable space of the number of bins required. In the single-sweep strategy, this difficulty is circumvented by substituting the histogram with the best fitting procedure for the free energy, using points sampled along the z trajectory as described in the next subsection.

3.2 Radial basis representation of the free energy

In this section, following [14], we describe how to use TAMD within the single-sweep method to reconstruct the free energy. Let us introduce the radial basis set in the space of functions of, real, variable z given by Gaussian functions (a convenient choice, but not the only possible)

$$\phi_\sigma(|z - z_j|) = e^{-\frac{|z-z_j|^2}{2\sigma^2}} \tag{11}$$

where $\{z_j\}$ is a set of "centers" in the z-space and $\sigma > 0$ is an adjustable parameter to be discussed in a moment. In single-sweep these centers are chosen among the points along a TAMD trajectory, $z(t)$, according to a distance criterion: a new center is dropped when the distance of $z(t)$ from all previous centers exceeds a given threshold d. In this basis, the fitting free energy can be written as

$$\tilde{F}(z) = \sum_{j=1}^{J} a_j \phi_\sigma(|z - z_j|) + C \tag{12}$$

where C is an additive constant that does not affect the properties of the system and J is the number of centers (equal to the number of basis functions). The coefficients a_j and σ are adjustable parameters which can be determined by minimizing the objective function

$$E(a, \sigma) = \sum_{j=1}^{J} \left| \sum_{j'=1}^{J} a_{j'} \nabla_z \phi_\sigma(|z_j - z_{j'}|) + f_j \right|^2 \tag{13}$$

This equation requires only computing the gradient of the free energy, indicated as f_j, at the points z_j. This gradient is called the mean force and it can be calculated locally as the conditional expectation

$$f_j = -\lim_{k \to \infty} \nabla_{z_j} F_k(z) = -\lim_{k \to \infty} \lim_{T \to \infty} \frac{k}{T} \int_0^T (z_j - \theta_j(x(t))) dt \tag{14}$$

Here we used the ergodic hypothesis to substitute the conditional average expressing the mean force

$$f_j = -\lim_{k \to \infty} \frac{1}{Z'_k} \int dx \; k(z_j - \theta_j(x)) e^{-\beta U_k(x,z)} \tag{15}$$

where $Z'_k = \int dx \, e^{-\beta U_k(x,z)}$, with a time average along the trajectory of the coordinates of the system as determined by the first line of Eq. (8) with fixed $z = \{z_j\}$.

The minimization of the objective function is performed in two steps. First, keeping the value of σ fixed, the function is minimized with respect to the coefficients of the Gaussians by solving the linear algebraic system

$$\sum_{j'=1}^{J} B_{j,j'}(\sigma) a_{j'} = c_j(\sigma) \tag{16}$$

where

$$B_{j,j'}(\sigma) = \sum_{j''=1}^{J} \nabla_z \phi_\sigma(|z_j - z_{j''}|) \cdot \nabla_z \phi_\sigma(|z_{j''} - z_{j'}|)$$

$$c_j(\sigma) = \sum_{j'=1}^{J} \nabla_z \phi_\sigma(|z_j - z_{j'}|) \cdot f_{j'} \tag{17}$$

In the following, the solutions of this system are called $\tilde{a}_j$. Once the coefficients of the Gaussians at a given variance are computed, the optimal value of σ is determined by finding $\min_\sigma E(\tilde{a}_j(\sigma), \sigma)$. The minimum is obtained by performing a one-dimensional scan on the values of the objective functions as a function of the variance and computing the residual value of $E(\tilde{a}(\sigma), \sigma)$ for increasing values of σ starting with the value of d.

The radial basis representation has several advantages, the most important being that it seems to converge fast with the number of centers compared to regular grids. This suggests that the method can be applied to reconstruct high dimensional free energy landscapes with reasonable numerical cost. Furthermore, the centers $\{z_j\}$ do not have to be located on a regular grid so that data collected anywhere on the landscape can be used. Note also, that unlike the histogram approach described previously, the z-trajectory does not have to visit several times the same region of the free energy. Here, the only requirement is that regions are visited once so that a center can be dropped and used in the basis set.

4 Results

As a preliminary step in the study of the diffusion processes, we performed an exploratory *ab initio* 40 ps molecular dynamics run for the physical system at a temperature T = 380 K in which we monitored the behavior of all collective variables described in Sect. 2. The analysis of the time evolution of the collective variables showed very different results for the local and non-local diffusion. In the case of the local diffusion, the full space was spanned during the run and no appreciable barrier was detected. Since this excluded local hydrogen diffusion as the origin of the experimental signal we were investigating, we limited ourselves to a relatively rough estimate of the free energy profile obtained by building histograms based on the *ab initio* molecular dynamics trajectory. Results are shown in the next subsection. The coordination numbers of the Al atoms, on the other hand, did not change during the exploratory run pointing to the fact that the non-local diffusion is indeed an activated process and that the system remained trapped in a metastable state for the duration of the run. We therefore applied the single-sweep strategy outlined in Sect. 3 to calculate the free energy profile. Results for this case are presented in subsection 4.2.

4.1 Local hydrogen diffusion

The free energy profile of the system was roughly estimated by evaluating the probability

$$P(\vec{W}) = e^{-\beta F(\vec{W})} \tag{18}$$

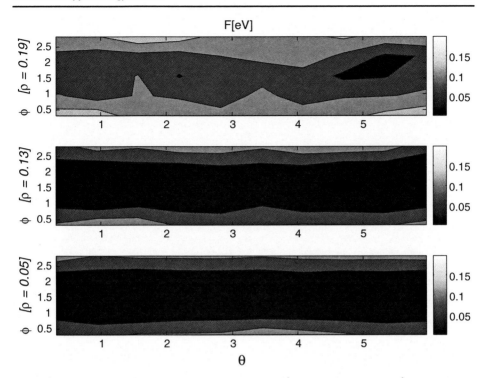

Fig. 5 Cuts of the contour plots of the free energy for $\rho = 0.19$ Å (upper figure), $\rho = 0.13$ Å (middle figure), and $\rho = 0.05$ Å (bottom figure) as a function of the angular components, θ and ϕ, of $\vec{W}$. The free energy is quite flat for all values of ρ

where we introduced a polar representation of the vector $\vec{W} = (\rho, \theta, \phi)$ and discretized the values of $\rho \in (0, 0.3)$, $\theta \in (0, 2\pi)$ and $\phi \in (0, \pi)$, on a $(10 \times 10 \times 5)$ grid. The *ab initio* molecular dynamics gave us the sample used to estimate the probability density above, and 40 ps were sufficient to give converged results. The estimated free energy, see Fig. 5, shows an essentially flat profile with features that depend slightly on the value of the length of $\vec{W}$. Since several non-local diffusion events occur in the exploratory trajectory, the barriers relevant for this process, if any, cannot be much higher than $k_B T = 0.03$ eV, about four times smaller than the experimental value (0.126 eV). Therefore, according to our simulations, the local hydrogen diffusion is not the process observed in the experiments.

The results of this run disagree with the estimate, $\Delta E = 0.3$ eV, of the activation energy obtained by Voss et al. [8]. In our opinion, the difference arises because the process they investigate to get the local rearrangements of the bond structure of the defective Al group in their work is not the most energetically favorable. As shown in the bottom panel of Fig. 6, they choose, to go from one square base pyramid configuration to another, a local diffusion mechanism that occurs by flipping a single Al-H bond, while leaving the positions of all other hydrogens essentially unchanged. According to our calculations, the process proceeds instead via a collective rearrangement of the bonds passing through the bi-pyramidal configuration, see upper panel of Fig. 6, for which the free energy barrier is much lower.

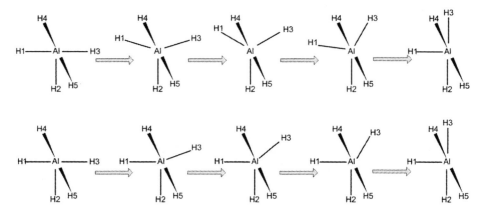

Fig. 6 Schematic representation of the local diffusion mechanism suggested in this work, upper panel, compared with the one hypothesized by Voss et al. [8], bottom panel. The leftmost figure is the initial configuration (square pyramid with base identified by hydrogens 1,3,4,5) while the final configuration (square base pyramid with base identified by hydrogens 2,3,4,5) is at the far right. Our simulations show that the rearrangement occurs by first changing the angle H_1-Al-H_3 to 120 degrees, then visiting the bi-pyramidal configuration—middle state in the sequence specified by the arrows—and finally rotating further hydrogens 1 and 3. In the final configuration, the angles H_1-Al-H_2 and H_1-Al-H_3 are equal to 90 degrees, while H_3-Al-H_2 is 180 degrees. These rearrangements involve collective motion of the hydrogens. In the NEB path only H_3 changes position (see Fig. 8 in ref. [8])

4.2 Non-local hydrogen diffusion

4.2.1 The TAMD trajectory

In order to explore the free energy surface and to find a convenient set of centers in preparation of the radial basis reconstruction, a 100 ps TAMD trajectory associated with the coordination numbers C_α was run. The fictitious mass and friction chosen (see Sect. 3) guaranteed the adiabatic separation between the evolution of the physical system and that of the collective variables. This was verified by computing the Fourier transform of the velocity autocorrelation function: the average frequency of the atomic motion was about two order of magnitude greater than that of the collective variables. We also have $\beta k \gg 1$ so that the force on the collective variables approximates the gradient of the physical free energy. Though these conditions are not strictly necessary since the main purpose of the trajectory is to provide a set of centers, they ensure that only relevant regions of the free energy landscape are visited and give meaning to a qualitative analysis of the dynamics.

Some comments on the collective variables are in order. The coordination number (see Eq. (6)) is a smoothed step function, characterized by two flat regions before and after a sudden drop that depends on r_0 and λ. This poses two problems. First, as described in Eq. (8), one of the force terms in the evolution equation of the coordinates of the system depends on the gradient of the collective variables with respect to the physical positions. If the parameter λ is too large, this force term becomes stiff and this complicates the numerical integration of the dynamics. On the other hand, a small value of λ results in a very slow decay of the function that blurs the meaning of the collective variable since, in this case, the coordination of an atom cannot be uniquely defined. Second, to change the collective variables the force must drive the system from one stable state to the other. In the present case, however, the force is non-zero only in a region of space determined by λ and r_0. A combination of values capable of driving the necessary changes in the

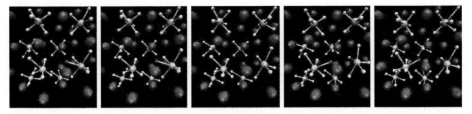

Fig. 7 Typical non-local diffusion event in the TAMD trajectory. The transferring hydrogen is tagged as the red sphere. The sequence, to be read from left to right, follows the event from the initial configuration in which the hydrogen is bound to the donor aluminum group to the accepting defective aluminum group

configuration of the system may not exist in general, although in our case it was found by setting $\lambda = 10\,\text{Å}^{-1}$ and $r_0 = 2\,\text{Å}$. This choice places the inflection point of the step function around the equilibrium Al-H distance and the force decays over a range of about 1 Å around this distance. Consequently, the boundaries of the coordination spheres of neighboring Al overlap, so that the force itself or the thermal fluctuations in the system can always drive the hydrogen exchange. Furthermore, the function is smooth enough for numerical integration.

With this choice of the parameters, the accelerated dynamics produced a considerable number of hydrogen hops between the aluminum atoms and, during the overall TAMD run, seven out of eight alumina became penta-coordinated at least once. Figure 7 shows a typical event. Reading the cartoon from left to right, the initial configuration of the defective Al group is a roughly a square pyramid. The donor AlH_6 group has one hydrogen, tagged as the red sphere in the figure, pointing in the direction of the base of the pyramid. It is this hydrogen that will hop between the two Al groups. As the hop proceeds there is very little rearrangement of the surrounding sodium atoms or of the bond structure of the two groups. The putative transition state, shown in the middle panel of the figure, is in fact quite symmetric with the transferring H midway between the groups. As the diffusion event proceeds, the hydrogen is captured by the acceptor that then rearranges very slightly to assume the hexa-coordinated octahedron geometry while the donor is now in a square base pyramid conformation. The analysis of the hydrogen transfer events in the TAMD trajectory shows three main features: (1) the hopping involves only two Al atoms per event with no apparent cooperative effects among different alumina; (2) there are no appreciable differences in the transfer dynamics based on the identity of the pair of aluminum atoms participating in the event; (3) most of the reactive events occur between a specific pair of aluminum atoms, involved in 18 hops over a total of 44. The average number of hops among other active pairs is 3. Thus we are led to focus on the pair of alumina among which the largest number of hydrogen hops occur since they can be considered as representative of all hops. Therefore, we reduced the number of collective variables from the eight coordination numbers of all Al in the cell to the two coordination numbers of the pair that participated in most hops. In the following, these alumina are called Al_1 and Al_2. Given the computational cost of the estimates of the gradient of the free energy in *ab inito* molecular dynamics, this reduction is important to make the analysis affordable. We now proceed to apply single-sweep to reconstruct a two dimensional free energy surface.

4.2.2 Radial basis reconstruction of the free energy

Due to an error in our implementation of the distance criterion described in the Method section, we produced a biased set of centers along the TAMD trajectory that suffered from clustering of the points. Since we verified that our centers were nonetheless placed in mean-ingful regions of the free energy landscape, we decided not to repeat the expensive calculation of the mean force on a set of new, equally spaced, points but rather to improve the characteris-tics of the available set. First, we added to the 43 TAMD centers 36 new points placed by hand, then we extracted subsets of points that respected the distance criterion for given choices of d. Three subsets where obtained this way using d = 0.1 (which selected 55 centers), d = 0.15 (37 centers), d = 0.2 (25 centers). The free energy landscape was then reconstructed with the three different subsets following the single-sweep procedure described in Sect. 3.2. The gradient of the free energy at the centers was computed using Eq. (14) with restrained *ab initio* molecular dynamics runs. The equilibrium averages were obtained with runs of $T \simeq 2.2$ ps, that ensured an error on the measure, as estimated by the variance associated to the average, of about 10^{-2} eV. With this information, the linear system (16) was solved and the variance of the Gaussians in the basis was then optimized.

The free energy profile reconstructed with 55 centers is shown in Fig. 8. To test the accu-racy of this calculation, the free energy surface reconstructed with increasing number of centers, and the results for the free energy barriers, were compared. In Fig. 9 we show the absolute value of the difference in the reconstructed free energies with 25 and 37 centers, and with 37 and 55 centers, respectively. As the profiles are defined within a constant, the figure was obtained by shifting the reconstructed surfaces so that the values at the minimum corresponding to an hexa-coordinated Al_2 coincided. As it can be seen, the convergence with number of centers is quite fast, and the difference among the free energy calculated with 37 and 55 centers is less than 0.02 eV in the regions involved in the non-local diffusion, giving an accuracy sufficient for comparison with the experimental data. As a further test of con-vergence, we evaluated the relative residual, defined in terms of the objective function (13) and the mean forces (14) as

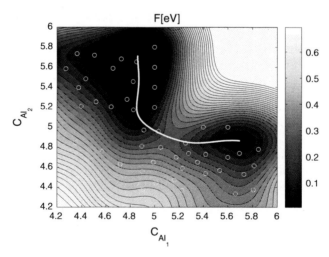

Fig. 8 Contour plot of the free energy reconstructed with 55 centers as a function of the coordination numbers of alumina Al_1 and Al_2. The white circles superimposed to the plot are the positions of the centers used in the single-sweep reconstruction. The white curve is the converged steepest descent path computed using the string method

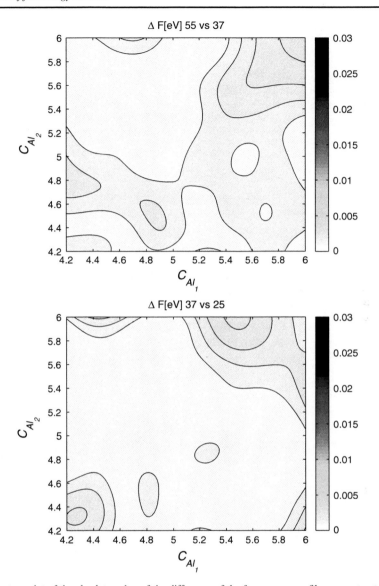

Fig. 9 Contour plot of the absolute value of the difference of the free energy profiles reconstructed via the single-sweep method with increasing number of centers. The free energy scale is the same as in Fig. 8 for easy comparison, but the colormap has been inverted (here, white is low and dark is high whereas it is the opposite in Fig. 8) for esthetical purposes. In the bottom panel, the difference among the profiles obtained with 25 and 37 centers is shown. The upper panel presents the difference among the reconstructions with 37 and 55 centers

$$\text{res} = \sqrt{\frac{E(\tilde{a}, \tilde{\sigma})}{\sum_{j=1}^{J} |f_j|^2}} \qquad (19)$$

where, as before, $\tilde{a}_j$ and $\tilde{\sigma}$ denote the optimized paramters. With $J = 55$ centers, the residual was res $= 0.2$.

4.2.3 Calculation of the activation barrier

As described in the Model section, we shall estimate the activation barrier by measuring the difference between the value of the free energy at the minima and at the saddle point. To determine the position of these points we computed the steepest descent path connecting the two metastable basins. By definition, along this path we have

$$\left(\nabla_z \tilde{F}(z)\right)^{\perp} = 0 \tag{20}$$

where $\left(\nabla_z \tilde{F}(z)\right)^{\perp}$ is the projection of the gradient in the plane orthogonal to the path. To compute this path we used the string method [28,29] with 200 discretization points. The maximum of the free energy along the path identifies the minimum free energy barrier to the hydrogen non-local diffusion process. In Fig. 8 the contour lines of the free energy reconstructed with 55 centers are shown. Overlaid to them are white circles indicating the positions of the centers used in the reconstruction of the profile. The white curve shows the converged steepest descent path. The activation free energy barrier for the non-local diffusion event was then calculated as

$$\Delta F_i = \tilde{F}(z_b) - \tilde{F}(z_i) \tag{21}$$

($i = 1, 2$). The results are shown in Table 2 for different number of centers in the radial basis set. The values for the energy barrier in the Table are in the range of that reported in the experiments. This indicates that, according to our calculation, the observed process can be identified with hydrogen transfer from hexa to penta coordinated aluminum. Here too, we disagree with the conclusions of Voss et al.[8], who have calculated an energy barrier ranging from 0.75 eV to 1 eV for different NEB paths. Also in this case we presume that the difference is due to the bias imposed by the initial guess for the NEB trajectory, resulting in too high an activation barrier.

We conclude with some information on the computational cost of the *ab initio* calculations associated to the reconstruction of the free energy. One picosecond of dynamics requires about 100 hours of CPU time on a Power5 IBM cluster running at 1.9 GHz. The overall time needed for the different calculations presented in this work amounted to about 30.000 hours. While significant, this cost is still affordable because the single-sweep method is trivially parallelizable.

Table 2 Non-local diffusion results as a function of the deposition threshold, d

d	Centers	$C^*_{Al_1}$	$C^*_{Al_2}$	SP	$\Delta F_1[eV]$	$\Delta F_2[eV]$
0.2	25	(5.68,4.84)	(4.85,5.68)	(5.17,4.94)	0.14	0.26
0.15	37	(5.72,4.86)	(4.86,5.73)	(5.11,4.93)	0.13	0.24
0.1	55	(5.69,4.86)	(4.86,5.71)	(5.14,4.89)	0.13	0.27

The second column contains the number of centers. Third, fourth and fifth column are the coordinates of the two minima, $C^*_{Al_i}$, and of the saddle point, SP. The ΔF_i are the height of the barriers as measured from the first and second minima respectively. The experimental value is $\Delta F = 0.126 eV$

5 Conclusions

In this paper, two local mechanisms for hydrogen diffusion in a Na_3AlH_6 crystal with one H vacancy were investigated with the intent to identify a mobile species that appears in experiments on these materials and whose identity is controversial. The two mechanisms are the local rearrangement of H positions around a defective Al group and transfer of one hydrogen from an hexa-coordinated aluminum to the defective group. In our calculations, they were modeled by introducing collective variables whose behavior was monitored in a set of *ab initio* molecular dynamics simulations. The calculated free energy landscapes showed characteristics that make it possible to discriminate among the two processes. The local rearrangement of the hydrogen's positions in a defective Al group, turned out not to be activated at the experimental temperature. In this case standard histogram methods based on the *ab initio* trajectory were sufficient to reconstruct, roughly, the free energy profile. The second process, on the other hand, was activated and the single-sweep method was used for reconstruction. The free energy barriers to diffusion for both processes were evaluated and compared with available experimental and theoretical information. The barrier corresponding to the non-local diffusion mechanism is quite close to the one observed in experiments so our calculations indicate that this may indeed be the origin of the signal. Before a definite answer can be given, however, a third hypothesis involving the diffusion of a sodium vacancy in the Na_3AlH_6 crystal must be investigated. The corresponding calculations, and refinements of the ones presented in this paper, will be the subject of future work.

Acknowledgements The authors are grateful to R. Cantelli, O. Palumbo and A. Paoloni for pointing out the experimental problem at the basis of this work and for discussions related to the model employed here, and to L. Maragliano for useful discussions about the single-sweep method and TAMD. We also thank R. Vuilleumier for contributing to the definition of the collective variables and of the pseudopotentials, and for suggestions on the most efficient implementation of the TAMD procedure in CPMD. Funding from the Ministero dell'Ambiente e della Tutela del Territorio e del Mare is gratefully acknowledged along with the computational resources provided by CASPUR. E. V.-E. acknowledges support from NSF grants DMS02-09959 and DMS02-39625, and ONR grant N00014-04-1-0565. Finally, the authors wish to acknowledge C. Dellago and the inspiring hospitality of the Erwin Schrödinger Institut in Vienna during the program "Metastability and rare events in complex systems".

References

1. Bogdanovi, B., Schwickardi, M.: Ti-doped alkali metal aluminium hydrides as potential novel reversible hydrogen storage materials. J. Alloys Compd. **1**, 253–254 (1997)
2. Peles, A., Alford, J.A., Ma, Z., Yang, L., Chou, M.Y.: First-principles study of $NaAlH_4$ and Na_3AlH_6 complex hydrides. Phys. Rev. B **70**, 165105–165112 (2004)
3. Li, S., Jena, P., Ahuja, R.: Effect of Ti and metal vacancies on the electronic structure, stability, and dehydrogenation of Na_3AlH_6: supercell band-structure formalism and gradient-corrected density-functional theory. Phys. Rev. B **73**, 214107–214114 (2006)
4. Kiran, B., Kandalam, A.K., Jena, P.: Hydrogen storage and the 18-electron rule. J. Chem. Phys. **124**, 224703–224706 (2006)
5. Kiran, B., Jena, P., Li, X., Grubisic, A., Stokes, S.T., Gantefor, G.F., Bowen, K.H., Burgert, R., Schnockel, H.: Magic rule for Al_nH_m magic clusters. Phys. Rev. Lett. **98**, 256802–256804 (2007)
6. Ke, X., Tanaka, I.: Decomposition reactions for $NaAlH_4$ and Na_3AlH_6, and NaH: first-principles study. Phys. Rev. B **71**, 024117–024133 (2005)
7. Araujo, C.M., Li, S., Ahuja, R., Jena, P.: Vacancy-mediated hydrogen desorption in $NaAlH_4$. Phys. Rev. B **72**, 165101–165107 (2005)
8. Voss, J., Shi, Q., Jacobsen, H.S., Zamponi, M., Lefmann, K., Vegge, T.: Hydrogen dynamics in Na_3AlH_6: a combined density functional theory and quasielastic neutron scattering study. J. Phys. Chem. B **111**, 3886–3892 (2007)

9. Palumbo, O., Cantelli, R., Paolone, A., Jensen, C.M., Srinivasan, S.S.: Point defect dynamics and evolution of chemical reactions in alanates by anelastic spectroscopy. J. Phys. Chem. B **109**, 1168–1173 (2005)

10. Palumbo, O., Paolone, A., Cantelli, R., Jensen, C.M., Sulic, M.: Fast H-vacancy dynamics during alanate decomposition by anelastic spectroscopy. Proposition of a model for Ti-enhanced hydrogen transport. J. Phys. Chem. B **110**, 9105–9111 (2006)

11. Mills, G., Jonsson, H.: Quantum and thermal effects in H_2 dissociative adsorption—evaluation of free-energy barriers in multidimensional quantum-systems. Phys. Rev. Lett. **72**, 1124–1127 (1994)

12. Jonsson, H., Mills, G., Jacobsen, K.W.: Nudged elastic band method for finding minimum energy paths of transitions. In: Berne, B.J., Ciccotti, G., Coker, D.F. (eds.) Classical and Quantum Dynamics in the Condensed Phase, pp. 385–404. World Scientific, Singapore (1998)

13. Henkelman, G., Uberuaga, B.P., Jonsson, H.: A climbing image nudged elastic band method for finding saddle points and minimum energy paths. J. Chem. Phys. **113**, 9901–9904 (2000)

14. Maragliano, L., Vanden-Eijnden, E.: Single-sweep methods for free energy calculations. J. Chem. Phys. **128**, 184110 (2008)

15. Maragliano, L., Fischer, A., Vanden-Eijnden, E., Ciccotti, G.: String method in collective variables: minimum free energy paths and isocommittor surfaces. J. Chem. Phys. **125**, 024106–024115 (2006)

16. Vanden Eijnden, E.: Transition path theory. In: Ferrario, M., Ciccotti, G., Binder, K. (eds.) Computer Simulation in Condensed Matter: From Materials to Chemical Biology, pp. 349–391. Springer, Berlin (2006)

17. Ronnebro, E., Noreus, D., Kadir, K., Reiser, A., Bogdanovic, B.: Investigation of the perovskite related structures of $NaMgH_3$, NaMgF, and Na_3AlH_6. J. Alloys Compd. **299**, 101–106 (2000)

18. http://www.cryst.ehu.es/cryst/getwp.html

19. Vuilleumier, R.: Density functional theory based *ab initio* MD using the Car-Parrinello approach. In: Ferrario, M., Ciccotti, G., Binder, K. (eds.) Computer Simulation in Condensed Matter: From Materials to Chemical Biology, pp. 223–285. Springer, Berlin (2006)

20. Becke, A.D.: Density-functional exchange-energy approximation with correct asymptotic behavior. Phys. Rev. A **38**, 3098–3100 (1988)

21. Lee, C., Yang, W., Parr, R.G.: Development of the Colle-Salvetti correlation-energy formula into a functional of the electron density. Phys. Rev. B **37**, 785–789 (1988)

22. Troullier, N., Martins, J.L.: Efficient pseudopotentials for plane-wave calculations. Phys. Rev. B **43**, 1993–2006 (1991)

23. Martyna, G.J., Klein, M.L., Tuckerman, M.: Nose-Hoover chains: the canonical ensemble via continuous dynamics. J. Chem. Phys. **97**, 2635–2643 (1992)

24. Vanden-Eijnden, E., Ciccotti, G.: Second-order integrator for Langevin equations with holonomic constraints. Chem. Phys. Lett. **492**, 310–316 (2006)

25. CPMD, Copyright IBM Corp. 1990–2001, Copyright MPI für Festkorperforschung Stuttgart 1997–2004

26. Sprik, M.: Computation of the pK of liquid water using coordination constraints. Chem. Phys. **258**, 139–150 (2000)

27. Maragliano, L., Vanden-Eijnden, E.: A temperature accelerated method for sampling free energy and determining reaction pathways in rare events simulations. Chem. Phys. Lett. **426**, 168–175 (2006)

28. Weinan, E., Ren, W., Vanden-Eijnden, E.: String method for the study of rare events. Phys. Rev. B **66**, 052301–052305 (2002)

29. Weinan, E., Ren, W., Vanden-Eijnden, E.: Simplified and improved string method for computing the minimum energy paths in barrier-crossing event. J. Chem. Phys. **126**, 164103–164108 (2007)

Concurrent design of hierarchical materials and structures

D. L. McDowell · G. B. Olson

Originally published in the journal Sci Model Simul, Volume 15, Nos 1–3, 207–240.
DOI: 10.1007/s10820-008-9100-6 © Springer Science+Business Media B.V. 2008

Abstract   Significant achievements have been demonstrated in computational materials design and its broadening application in concurrent engineering. Best practices are assessed and opportunities for improvement identified, with implications for modeling and simulation in science and engineering. Successful examples of integration in undergraduate education await broader dissemination.

Keywords   Materials design · Multiscale modeling · Robust design · Concurrent engineering

1 Introduction

Viewing a material as a hierarchical structure having design degrees of freedom associated with composition and microstructure morphology opens new vistas for improving products. As traditionally taught and practiced in engineering, product design involves materials selection as part of the process of satisfying required performance specifications [1]. For example, an aluminum alloy might be selected instead of a medium strength steel in a given application by virtue of high specific stiffness or strength, along with secondary considerations such as corrosion resistance. Titanium alloys might be selected in golf club head inserts instead of fiber reinforced composites due to higher impact resistance and lower cost of fabrication. Material choices are typically listed in catalogs by material suppliers, with various properties for these nominal material forms available in databases and research or trade literature.

   In this conventional paradigm, the role of the materials engineer is largely that of independent materials development. This involves process-structure experiments and modeling to understand structure and properties of candidate microstructures, followed by time-consuming

D. L. McDowell (✉)
Georgia Institute of Technology, Atlanta, GA, USA
e-mail: david.mcdowell@me.gatech.edu

G. B. Olson
Northwestern University and QuesTek Innovations LLC, Evanston, IL, USA

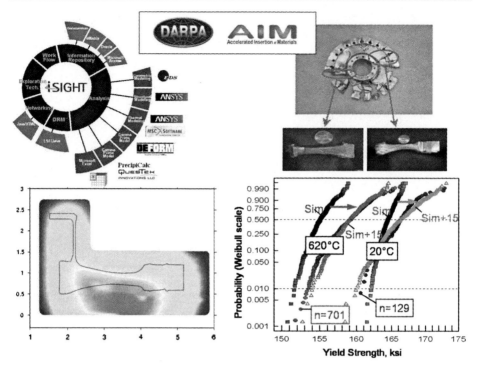

Fig. 1 Elements of Ni-base superalloy concurrent process, microstructure and gas turbine engine design with objective of increasing burst speed and decreasing disk weight in the AIM program [2,3]

experimental studies to quantify stability of structure and verify or certify relations of structure to properties. This sequence of stages to develop and certify a new material have often been too long (20 years) for a new material or alloy system to be conceived as part of the systems design process. However, with emerging computational modeling and simulation tools on the one hand, and increasingly high resolution and rapid characterization instruments and methods on the other, the goal of accelerating the insertion of new or improved materials into next generation transportation vehicles and propulsion systems is beginning to be realized. The DARPA Accelerated Insertion of Materials (AIM) program [2,3] from 2000–2003 offered insight into how computational materials science and engineering can be harnessed in the future to assist in developing and certifying materials in a shorter timeframe to more closely match the duration of the systems design cycle. AIM was a bold initiative that assembled materials developers, original equipment manufacturers (OEMs), and government and academic researchers in a collaborative, distributed effort to build designer knowledge bases comprised of the various elements of systems design such as databases, digital realizations of microstructure, modeling and simulation tools that addressed various level of materials hierarchy and interplay with products, experiments, materials characterization, statistical approaches to uncertainty, metamodeling, and information protocols for managing workflow and communications.

The AIM program was focused on metallic systems (Ni-base superalloys for gas turbine engine disks) and composite airframe materials. As shown in Fig. 1, iSIGHT information management software (upper left) was employed in the metals program to integrate various codes and databases used to predict precipitate strengthened $\gamma - \gamma'$ Ni-base superalloy microstructures and estimate the yield strength based on relatively simple dislocation-based models for strengthening [3]. The disk microstructure was coupled to thermomechanical process history using predictive codes that draw from fundamental thermodynamic calculations of phase and interface properties. The model prediction of the cumulative distribution of yield strength as a function of temperature and microstructure, coupled with computer-aided design (CAD) tools and optimization routines, enabled accelerated design of a process route, microstructure, and disk geometry with a significant increase of the burst speed in a spin test (upper right), while simultaneously reducing overall disk weight. The disk design demonstration also allowed experimental validation of predicted spatial variation (lower left) in disk forging microstructure and properties. Final simulation of process variation over six stages of manufacturing successfully predicted measured property distributions (lower right) for room temperature and 620°C strength with efficient fusion of minimal (n = 15) datasets. All of this was done concurrently over a three year period, indicating the feasibility and payoff of concurrent design of process route, material microstructure, and component geometry.

The DOE-sponsored USCAR program from 1995–2000 provided an earlier indication of the feasibility of obtaining substantial improvements by coupling structure-property relations and component level design [4–7]. In this program, an increase of cast automotive vehicle component fatigue strength was achieved with reduction of component weight based on these ideas of concurrent design of microstructure and associated structure-property relations in a suspension "A" arm. Computational micromechanics was employed to characterize, to first order, cyclic plasticity and fatigue processes associated with casting inclusions over a wide range of length scales, from several microns to the order of one millimeter. These simulations involved FE calculations on actual and idealized microstructures using concepts of volume averaging to address scale effects and damage nonlocality, and were aimed at a very different goal from typical fatigue analyses, namely understanding the sensitivity of various stages of fatigue crack formation and early growth at hierarchical levels of microstructure. Thresholds for forming small cracks, for small crack growth out of the influence of micronotches, and microstructurally small crack propagation in eutectic regions and dendrite cells were all considered in simulations of microstructure attributes at different length scales. The critical issue of dependence of component fatigue strength on casting porosity and eutectic structure was addressed by employing numerical micromechanical simulations using FE methods for a hierarchy of five inclusion types [7] for cast Al-Si alloy A356-T6, spanning the range of length scales relative to the secondary dendrite arm spacing or dendrite cell size (DCS). Resulting predictions of high cycle and low cycle fatigue resistance for a range of initial inclusion types are shown in Fig. 2; the predicted range of fatigue lives for microstructures with extremal inclusions established by metallographic characterization conform to measured variation of the experimental fatigue lives. In the LCF regime, multisite fatigue damage and crack impingement/coalescence is taken into account. This microstructure-sensitive multistage fatigue model [7] formed the basis for estimating fatigue resistance of cast components. Similar casting porosity-sensitive strength models were developed to couple with process models for casting porosity levels to design component geometry and processing for A356-T6 to reduce weight and gain strength. As in AIM, results were validated using full scale demonstrations.

One major branch of the genesis of the idea of using computational modeling and simulation tools in concert with experimental characterization of materials to design a material with targeted property sets traces back to the mid-1980s with the inception of the Steel Research Group at Northwestern University [8]. Figure 3 summarizes a comparison between the experiment-intensive traditional process of empirical materials development and the new analysis-intensive process of "materials by design." Where the input of scientific knowledge

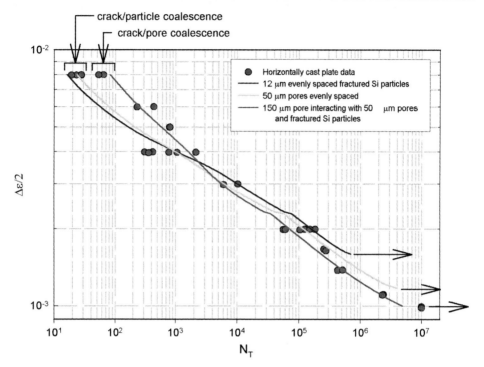

Fig. 2 Completely reversed, uniaxial applied strain-total life behavior as a function of inclusion type and size for cast A356-T6 Al, including coalescence effects in the LCF regime [7]. Arrows denote non-propagating crack limits (fatigue limits) for small cracks formed at inclusions

is only qualitative in the first approach, the new approach builds on the numerical implementation of quantitative scientific knowledge to minimize costly experimentation through validation of model predictions by iterative evaluation of a limited number of prototypes. Acknowledging the intrinsic hierarchical character of material process-structure-property-performance relations, and the reciprocal nature of the top-down design problem framed by mapping the required performance requirements onto properties, then to structure, and finally into process route and composition, Olson and colleagues constructed system diagrams of the type shown in Fig. 4. This is a foundational step in systems-based materials design to achieve target property sets. It is an exercise that is entirely material-specific and application/property-specific. It sets the stage for addressing true design of material systems as an information management exercise, instantiated by specific types of simulations and validation experiments that convey the necessary information to support decision-making at various points within the process.

For example, Fig. 4 expresses that tempering dominantly affects dispersed strengthening phases, and must be designed to minimize larger carbides that would compromise ductility (and thereby toughness). Toughness is sensitive to a variety of microstructure related aspects, including lath martensite structure, incoherent carbides, grain size, resistance to nucleation of microvoids, and amount and stability of precipitated austenite, as well as characteristics of transformation strain. Some of these are geometric (i.e., stereological) attributes of microstructure, while others involve dynamic evolutionary responses such as fracture or phase transformation. Hydrogen resistance is directly related to grain boundary chemistry, with

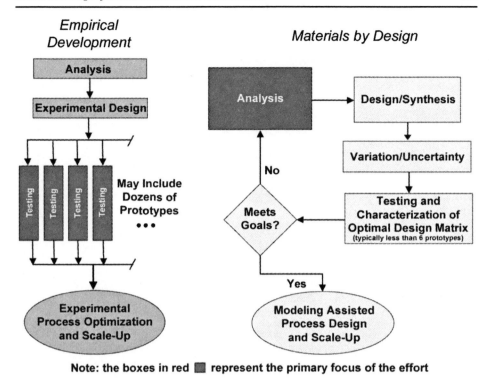

Note: the boxes in red ▉ represent the primary focus of the effort

Fig. 3 Comparison of experiment-intensive traditional empirical development (left) and analysis-intensive "materials by design" approach (right)

impurities controlled by refining and deoxidation processes. Pursuit of this design problem requires a combination of expert estimates based on prior experience, experiments, and models and simulations at virtually every mapping (branch connections) shown in Fig. 4. Note also that multiple elements populate each level of the hierarchy (six processing steps, five dominant levels of microstructure, and three target properties to affect). Since the entire system is coupled by virtue of cause-and-effect interactions, modification of each element propagates changes through the system, with the dominant interactions shown in the diagram. This complicates design optimization, as it is generally not possible to optimize the entire system by focusing on optimization of a given element within a given layer of the hierarchy. In certain cases, sub-systems can be optimized in decoupled manner from the overall system, greatly accelerating design exploration. Identification of the degree of coupling of sub-systems is indeed an important aspect of design of hierarchical material systems.

In tension with this need for searching for acceptable design solutions to the global systems problem, there are limitations on the timeframe involved for each of the processing steps. Moreover, if models do not exist to relate process path to microstructure, or various relations of microstructure to properties, then costly and time consuming experiments are necessary. In the worst case, iterations of the system shown in Fig. 4 are not practical, and perhaps only a part of the system flow diagram can be attempted. This is in fact the historical reason why traditional material development (Fig. 3 left) has been highly empirical, and dramatic compromise tradeoffs of competing property objectives such as strength and toughness have been widely accepted and taught in textbooks as a matter of fact, giving metals the reputation

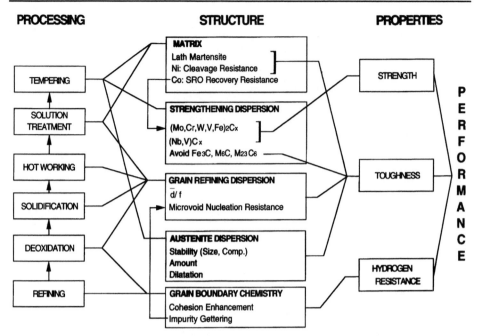

Fig. 4 Process-structure-property-performance hierarchy for design of high strength steels for multiple objectives of strength, toughness and hydrogen resistance. From the SRG at Northwestern University [8,9]

as a mature and heavily constrained/limited class of materials. This does not have to be the case; by increasing the fraction of decision points or connections in Fig. 4 with support from modeling and simulation tools, even in the preliminary design exploration stage, multiple iterations of the entire systems framework can be achieved in modern simulation-assisted materials design. We will discuss the example of high toughness steel design later in this regard. Over time, it has already been observed that increasing capabilities and availability of accurate models has enabled many iterations in efficient multi-objective design of material systems for required performance objectives.

Figure 5 serves as the general foundation for the specific example of materials design shown in Fig. 4. As discussed by Olson [9], it emphasizes the design hierarchy and clearly distinguishes the exercise of top-down, goals/means, inductive systems engineering from bottom-up, cause and effect, deductive, sequential linkages. The bottom-up approach has been the historical model for empirical materials development, with limited top-down feedback to guide the process. Materials design rests on the twin pillars of process-structure and structure-property relations. The process of relating properties to performance is effectively a selection-compromise exercise. For example, the Ashby materials selection charts [1] enable identification of existing material systems and properties that meet required performance indices for specified application, which is always an initial step in framing any materials design problem. This is conventionally done by searching databases for properties or characteristics of responses that best suit a set of specified performance indices [1], often using combinatorial search methods [10]. At this point, we note that identification of scales of material hierarchy are orthogonal to the linear design information flow shown in Fig. 5. Figure 6 presents an example of the hierarchy of computational models that support the predictive design in the system flowchart of Fig. 4. Acronyms of modeling methods and associated software

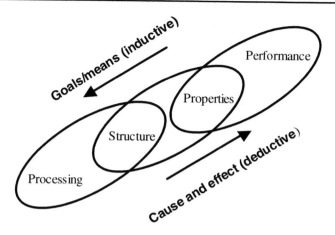

Fig. 5 Olson's linear concept of 'Materials by Design' [9]

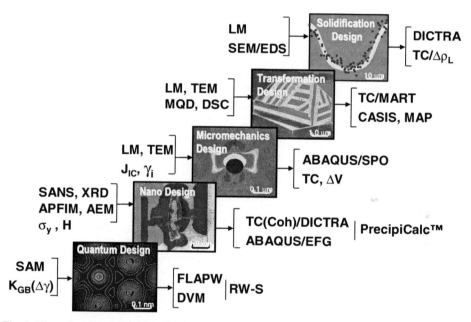

Fig. 6 Hierarchy of model length scales for system of Figure 4, from bottom-up angstroms (interfaces and lattice faults), nanoscale (coherent precipitates), sub-micron (grain refining dispersoids), microns (phase and grain boundaries), and tens or hundreds of microns (dendrites, inclusions, grains)

tools appropriate to each scale are shown to the right in each case, while corresponding characterization tools are shown at left (see [9] for details). Clearly, the notion of combinatorial design whereby an atomic (e.g., crystal) structure would be searched to meet mechanical property requirements at the scale of hundreds of microns is not particularly useful because of the influence of the various intermediate scales that affect macroscopic properties. On the other hand, if for this same hierarchical system it is established that environmental effects on ductility are limited mainly by the structure of grain boundaries and corresponding fracture susceptibility to segregation, effects of hydrogen, etc., then that aspect of the design problem

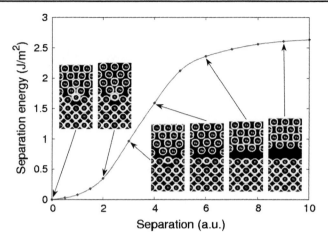

Fig. 7 Interfacial separation energy as a function of separation distance between interfacial layers in the semicoherent Fe/TiC system, with application to quantum-engineered steels using the FLAPW [15] plane wave DFT code

draws attention to detailed simulations of quantum and/or atomistic simulations of these interfaces. As represented by the charge density contour plot for a simulated grain boundary at the lowest level of Fig. 6, application of the FLAPW density functional total-energy quantum mechanical code to precise calculation of the relative energies of grain boundaries and their corresponding fracture surfaces has successfully predicted the relative embrittlement potencies of boundary segregants [11], including environment-induced hydrogen [12], and has enabled mechanistic prediction of desirable alloying elements for enhanced grain boundary cohesion [13]. Integration of these predictions with those of the full set of models in Fig. 6 in the design of the full system of Fig. 4 has yielded a new generation of "quantum steels" that have eliminated intergranular embrittlement [9,14].

Combinatorial searches typically focus on data mining and visualization, and providing convenient and powerful interfaces for the designer to support materials selection. In problems involving design of interfaces, nanostructures for catalysis, actuation or sensing, or molecular structures that serve as effective virus blockers, for example, the function to be delivered is delivered at the nanostructure scale. A further example of such a problem is the design of fracture-resistant interfaces at the nanoscale, as demonstrated in Fig. 7 by the variation of work to separate Fe-TiC interfaces. Application of these methods to the adhesion of interphase boundaries has aided both the selection of optimal phases for submicron grain refining dispersions with enhanced resistance to microvoid nucleation during ductile fracture and the prediction of force-distance laws for input into quantitative fracture simulations as represented in the middle level of Fig. 6.

For typical cases in which structure at multiple length scales affects properties, as represented in the examples in Figs. 4–6, it is essential to pursue a systems approach that targets application of models at various length scales to yield useful information to support design decisions. Indeed, recent federal initiatives emphasize the interdisciplinary collaboration of materials modeling and simulation, high performance computing, networking, and information sciences to accelerate the creation of new materials, computing structure and properties using a bottom-up approach. For example, the NSF vision for *Materials Cyber-Models for Engineering* is a computational materials physics and chemistry perspective [16] on using quantum and molecular modeling tools to explore potentially new materials and compounds, making the link to properties. The NSF *Blue Ribbon Panel on Simulation Based Engineering Science* [17] issued broad recommendations regarding the need to more fully integrate modeling and simulation within the curriculum of engineering to tackle a wide range of interdisciplinary and multiscale/multiphysics problems.

We advocate an approach that embeds material processing/supply, manufacturing, computational materials science, experimental characterization and systems engineering and design, similar to the conceptualization of *Integrated Computational Materials Engineering* (ICME) being pursued by a NAE National Materials Advisory Board study group [18]. ICME is an approach to concurrent design of products and the materials which comprise them. This is achieved by linking materials models at multiple length and time scales to address problems relevant to specific products and applications. ICME hearkens back to Olson's hierarchical scheme of Figs. 4–5 [9], with the understanding that top-down strategies are essential to supporting goal/means design of materials to meet specific performance requirements. This was defined over a decade ago for the academic and research communities at a 1998 NSF-sponsored workshop [19] entitled "New Directions in Materials Design Science and Engineering (MDS&E)". That workshop report concluded that a change of culture is necessary in U.S. universities and industries to cultivate and develop the concepts of simulation-based design of materials to support integrated design of material and products. It also forecasted that the 21st century global economy would usher in a revolution of the materials supply/development industry and realization of true virtual manufacturing capabilities (not just geometric modeling but also realistic material behavior). It was recommended to establish a national roadmap addressing (i) databases for enabling materials design, (ii) developing principles of systems design and the prospects for hierarchical materials systems, and (iii) identifying opportunities and deficiencies in science-based modeling, simulation and characterization "tools" to support concurrent design of materials and products.

2 What is materials design?

The term may have different meaning to different people and audiences. Our use of the term *materials design* (or *Materials by Design^{TM}*) *implies the top-down driven, simulation-supported, decision-based, concurrent design of material hierarchy and product or product family with a ranged set of performance requirements.* In this sense, our definition is more closely aligned, in general, with the aforementioned comprehensive notion of ICME [18] than perhaps more narrowly defined bottom-up cyberdiscovery, datamining, or simulation-based engineering science based on multiscale modeling. In our view, materials design is not just:

- materials selection (although often taught as such)
- computational materials science
- materials informatics, data mining or combinatorics
- multiscale modeling
- an intuitive exercise
- experiential learning applied to new materials and products
- performed by a single group or division
- materials development performed in isolation from product development
- artificial intelligence aiming at eliminating human intervention

The last point is quite important. Modeling and simulation tools provide support for design decisions but does not replace informed human decision-making in concurrent design of materials and products.

2.1 Hierarchy of scales in concurrent design of materials and products

Multiscale modeling contributes in a substantial way to the pursuit of materials design in many practical cases where different levels of hierarchy of material structure contribute to targeted ranges of material properties and responses. However, materials design is effectively a multilevel, multiobjective pareto optimization problem in which ranged sets of solutions are sought that satisfy ranged sets of performance requirements [20–33]. It does not rely on the premise of explicit linkage of multiple length scales via numerical or analytical means. In fact, it is often preferred to introduce rather more elementary, proven model concepts at each scale than abide the uncertainty of complex, coupled multiscale models for which parameter identification and validation are difficult [20]. Identification of sub-systems in the material hierarchy with weak coupling to responses of other sub-systems is an important step [29], as these sub-systems can be analyzed and optimized independently. Recent efforts within the Systems Realization Laboratory at Georgia Tech [21–23] have cast such design problems in terms of robust multilevel decision-based design. Multiscale modeling typically refers to a means of linking models with different degrees of freedom either within overlapping spatial and temporal domains or in adjacent domains.

Models at multiple scales can be executed concurrently or sequentially, with the former necessitating full coupling among scales and the latter only a one way coupling from the bottom-up. Examples of bottom-up modeling include multiresolution or overlapping domain decomposition approaches for passing from discrete to continuous models such as a dynamically equivalent continuum [34,35] informed by atomistics, coarse-grained molecular dynamics [36–38], and domain decomposition methods [39–41] that exchange lower frequency dynamic response between atomistic and coarse-grained MD or continuum domains.

Self-consistent schemes constitute another approach for multiscale homogenization. Bottom-up methods such as methods of Eshelby-Kröner type [42–44] are mainly focused on embedding effects of fine scale microstructure on higher length scale response at the scale of a representative volume element. For evolution of microstructure, the principle of virtual work has been generalized to incorporate rearrangement of microstructure as part of the kinematic structure [45–47]. Some authors have introduced concurrent multiscale schemes based either on finite element analyses that pass boundary conditions among meshes at various scales with different resolution and constitutive equations [48], or deformation gradient averaging approaches with higher order conjugate (e.g., couple) microstresses or micropolar formulations [49–51].

Statistical continuum theories have been framed at different scales, including dislocation field mechanics [52–54], internal state variable models [55,56], nonlocal reaction-diffusion models [57,58], and transition state theory models that employ kinetic Monte Carlo methods [59,60].

Often, these models involve statistical description of evolving microstructure at finer scales, based on "handshaking" methods for informing continuum models from high resolution models or experiments. These handshaking methods can range from intuitive formulation of constitutive equations to estimates of model parameters in coarse grain models based on high resolution simulations (e.g., atomistics or discrete dislocation theory) to development of metamodels or response surface models that reflect material behavior over some parametric range of microstructure and responses. An example of a combined bottom-up homogenization and handshaking among scales is found in work of McDowell and colleagues [61–63] on multiscale models for cyclic behavior of Ni-base superalloys. In these models, dislocation density evolution equations are formulated at the scale of either precipitates or homogenized

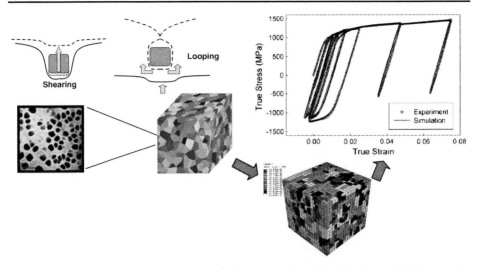

Fig. 8 Hierarchical multiscale models for $\gamma - \gamma'$ Ni-base superalloys [62,63], with handshaking from fine scale mechanisms to inform grain level responses, which are then subjected to random periodic boundary conditions to achieve description of macroscopic response at the scale of structural components

grains, which are then calibrated with experimental elastic stiffness and stress-strain data on single crystals and polycrystals. Figure 8 shows how such polycrystalline models are then used for purposes of both simulating stress-strain behavior for complex loading histories as well as distribution of slip among grains in a polycrystal to establish potency for nucleating and growing small fatigue cracks.

Another good example of a hierarchical multiscale model that involves a combination of handshaking between constitutive models at different scales, some of which are calibrated to experiments and others of purely predictive character, is the 'multiscale fracture simulator' developed by Hao and colleagues [64]. Figure 9 shows a progression of length scales considered in a multiscale modeling strategy for designing fracture resistant materials. In this way, material structure at various levels of hierarchy can be tailored to contribute to enhanced resistance to shear localization at the higher length scales. This kind of one-way hierarchical approach can be quite useful for purposes of materials design, whereas concurrent methods offer utility in modeling the interplay of microstructure rearrangement and structural response in applications. In contrast to the usual premise of materials selection based on properties, one nuance of Fig. 9 is that the resulting "properties" at the macroscale are more complex than captured by single parameter descriptions such as fracture toughness, strength, ductility, etc.

It is important to point out that there is considerable uncertainty in any kind of multiscale modeling scheme, including selection of specific scales of hierarchy, approximations made in separating length and time scales in models used, model uncertainty at various scales, approximations made in various scale transition methods, and lack of complete characterization of initial conditions and process history effects. Moreover, material microstructure typically has random character, leading to stochastic process-structure and structure-property relations. Accordingly, stochastic models and methods for scale transitions (statistical mechanics) are often necessary to indicate expected ranges of structure and properties in spite of a dominant focus on deterministic methods. These sources of uncertainty give rise to the need to consider sensitivity of properties and responses of interest to variation of microstructure at various scales.

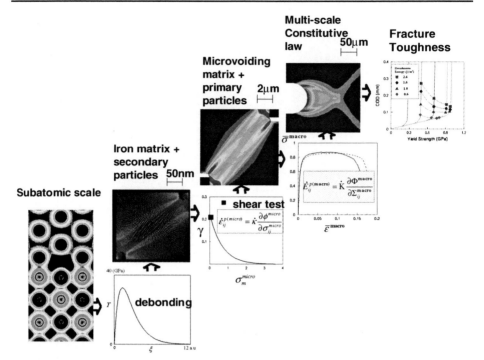

Fig. 9 Multiscale fracture simulation of an alloy system, with first principles computation to quantify separation energy of interfaces, which informs continuum separation laws for Fe matrix and secondary particles at scales of hundreds of nm that are incorporated into porosity-dependent constitutive laws at scales of primary particles (mm), and at the scale of structural stress raisers such as notches [64]

Olson's hierarchy in Figs. 4–5 should not be confused with multiscale modeling strategies. It has much more comprehensive nature, embedding multiscale modeling as part of the suite of modeling and simulation tools that provide decision-support in design. For example, structure-property relations can involve the full gamut of length and time scales, as can process-structure relations. In other words, the levels in Olson's linear design strategy of Fig. 5 do not map uniquely to levels of material hierarchy. Even concurrent multiscale models which attempt to simultaneously execute models at different levels of resolution or fidelity do not serve the full purpose of top-down materials design. Materials design and multiscale modeling are not equivalent pursuits. This distinction is important because it means that notions of cyberdiscovery of new or improved materials must emphasize not only modeling and simulation tools but also systems design strategies for using simulations to support decision-based concurrent design of materials and products.

Figure 10 conceptualizes how already established methods of design-for-manufacture of parts, sub-assemblies, assemblies and overall systems may be extended to address the multiple length and time scales of material structure and responses that govern process-property-performance relations. The objective of tailoring the material to specific applications (to the left of the vertical bar in Fig. 10) is distinct from traditional materials selection. The basic challenges revolve around the fact that hierarchical modeling of materials is still in its infancy, and systems-based design methods have not been widely applied to the region left of the vertical bar in Fig. 10. From a reductionist, bottom-up perspective, many would regard the hierarchy of scales from quantum to continuum on the left in Fig. 10 as a multiscale modeling

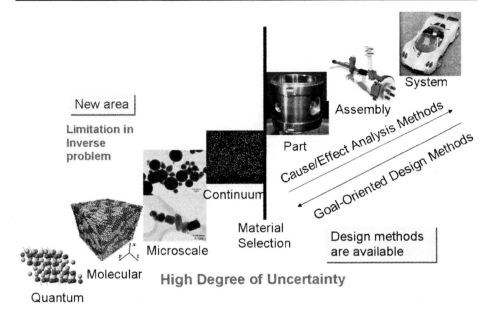

Fig. 10 Hierarchy of levels from atomic scale to system level in concurrent materials and product design. Existing systems design methods focus on levels to the right of the vertical bar, addressing mainly the materials selection problem, only one component in multilevel materials design

problem. The materials design challenge is to develop methods that employ bottom-up modeling and simulation, calibrated and validated by characterization and measurement to the extent possible, facilitated by top-down, requirements-driven exploration of the hierarchy of material length scales shown in Fig. 10. Moreover, aforementioned sources of uncertainty require more sophistication than offered by naïve optimization of limited objectives at either individual levels of hierarchy or at the systems level. Principles of multiobjective design optimization that recognize the need for ranged sets of performance requirements and ranged sets of potentially acceptable solutions are essential. It is a challenging multilevel robust design problem.

Figure 11 provides a path for materials design, whereby the structure of Fig. 5 is decomposed as a set of multilevel mappings (*Process-Structure (PS) relations, Structure-Property (SP) relations, and Property-Performance (PP) relations*) [65]. These mappings (represented by arrows) can consist of models, characterization and experiments, or some combination. Lateral movement at each level of hierarchy is associated with reducing model degrees of freedom, for example through multiscale material modeling shown in Fig. 9. It is also noted that the shaded red area at the upper right in Fig. 11 represents the materials selection problem, which occurs at just one or two levels of hierarchy; it involves selection based on tabulated data from models or experiments, and may be approached using informatics, e.g., data mining, combinatorics, and so forth [1,10,16]. In Fig. 11, each arrow can also be accompanied by a design decision.

2.2 Goals of materials design

Numerous applications can benefit from strategies of concurrent design of materials and products. It is important to be realistic about goals for leveraging modeling and simulation into

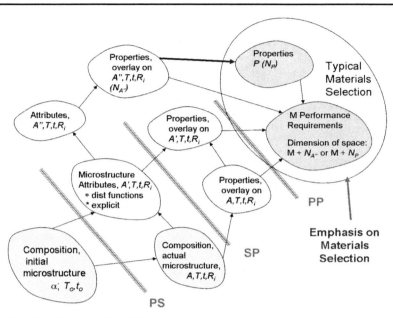

Fig. 11 Hierarchy of mappings in multi-level materials design

design decision-making. Traditional empirical design of materials has made use of extensive experience in material processing methods for a limited range of material compositions. Property goals have been achieved by adjusting compositions and process route in a rather incremental fashion. As this has been the dominant method for many years, it has imparted a sort of artificial notion that alloy development is a mature technology, and that there is not much to be gained by devoting further attention to property enhancement. In other words, the limitations of the material development cycle, combined with a focus on materials selection for "characteristic" properties of classes of materials, have led to conventional wisdom regarding limits on properties of multicomponent alloy systems.

The prospect of systems-based concurrent design of materials and products provides impetus for breaking through these limitations and preconceived notions. We will describe examples of such breakthroughs in a later section. In addition to enhanced performance, there are other important issues that can be addressed by simulation-assisted materials design. These relate to accelerating the materials discovery and development process and consideration of applications in which materials must meet multifunctional performance requirements, among others.

- To what degree can empiricism be replaced by information from simulations? How can we track the fraction of design decisions informed by modeling and simulation? If it is presently only 10%, can it be increased to 15%? 30%? Significant time and cost savings could result.
- To what extent can phenomena in different physical domains (mechanical, thermal, chemical, etc.) be considered simultaneously rather than sequentially? How does this affect constraints on design problems?

By definition, a multifunctional material is one for which performance dictates multiple property or response requirements. Often these properties conflict in terms of microstructure requirements, for example the classical trade-off of strength and ducility. By setting property targets in multiobjective design rather than constraint allowables on minimum properties, systems design offers a means for pushing boundaries. This is also the case for multifunctional materials with property requirements in different physical domains, for example conductivity, oxidation resistance, tensile strength, elastic stiffness, and fatigue resistance in gas turbine engine disk or blade materials. Multiple property goals cannot be met by optimizing individual models at different levels of hierarchy in Fig. 9, for example, but only by considering the entire hierarchy of scales. Material systems have high complexity; optimizing relational subsets within complex, nonlinear, coupled systems does not assure optimization of the overall system. Hence, a systems approach is essential.

3 Some aspects of systems approaches for materials design

3.1 Role of thermodynamics

The essential elements of modeling and simulation invariably fall into one of three categories: (i) thermodynamics, (ii) kinetics, and (iii) kinematics. Since feasible (realizable) structures of materials are established by either absolute or (more commonly) constrained minimization of thermodynamic free energy, *we regard thermodynamics as the fundamental building block for simulation-supported materials design.* Thermodynamics provides information on stable and metastable phases, characterization of structures and energies of interfaces, and driving forces (transition states) for rearrangement of structure due to thermally activated processes. As such, it facilitates preliminary design exploration for candidate solutions to coupled material and product design. First principles calculations are indispensible in this regard, and support exploration of multicomponent systems for which little if any empirical understanding has been established. More important than the calculation of constrained equilibria, thermodynamics defines the forces driving the kinetics of systems far from equilibrium.

Kinetics plays an important role as a further step in screening candidate solutions in preliminary design exploration. For example, stability of phases and interfaces at finite temperature in operating environments requires assessment prior to expensive investment in extensive computation and/or experimental characterization of candidate solutions. Moreover, upper bounds can sometimes be estimated on potential properties using thermodynamics, but kinetics dictates feasibility of the transition state pathways required for structure-property relations. Kinetics is often a stumbling block from a modeling perspective as methods for predicting mobilities are not fully developed. Currently, mobility databases compiled from empirical diffusivity data have proved to be quite accurate in the prediction of diffusion-based kinetics in metals.

Kinematics relates to the relative contributions of different attributes (defects, phases) of microstructure in contributing to overall rearrangement during deformation and failure. Kinematics can be approached both at the unit process level (individual defect or boundary segments) or from the many-body level; the latter is necessary at higher stages of hierarchy shown in Fig. 9, for example.

3.2 Challenges for top-down, inductive design

As previously mentioned, although design is necessarily a top-down exercise, material process-structure and structure-property relations are intrinsically bottom-up in character. In fact, elements of thermodynamics, kinetics and kinematics are built up from the unit process level. There are limited examples for which it is possible to comprehensively invert structure-property relations; the groups of Adams [66–68] and Kalidindi [69–71] have tackled the problem of finding crystallographic textures that deliver certain requirements on macroscopic anisotropic elastic stiffness of structures. To first order, this depends only on the orientation distribution function of polycrystals. Adams et al. [66–68] have introduced the notion of property closures, prominent in composites material structure-property relations, which bound the set of feasible properties that can be achieved by available microstructures. Zabaras and co-workers have developed a reduced order polycrystal plasticity model for such purposes [72], as well as approaches for dealing with estimation of PDFs for properties from microstructure ensemble calculations [73]. The assessment of the necessary process path to achieve target textures is another matter, requiring bottom-up simulation [74], in general. Lack of invertibility of process-structure and structure-property relations in modeling and simulation is typical, engendered by:

- Nonlinear, nonequilibrium path dependent behavior, limiting parametric study and imparting dependence upon initial conditions.
- Dynamic to thermodynamic model transitions in multiscale modeling, with non-uniqueness associated with reduction of model degrees of freedom.
- Wide range of suboptimal solutions that can be pursued.
- Approximations made in digital microstructure representation of material structure.
- Dependence of certain properties such as ducility, fracture toughness, fatigue strength, etc. on extreme value distributions of microstructure.
- Microstructure metastability and long term evolution.
- Uncertainty of microstructure, models, and model parameters.
- Lack and variability of experimental data.

3.3 Uncertainty in materials design

Uncertainty dominates the process of simulation-supported materials design. There are various sources of uncertainty, including [24]:

- Parameterizable (errors induced by processing, operating conditions, etc.) and unparameterizable (e.g., random microstructure) natural variability
- Incomplete knowledge of model parameters due to insufficient or inaccurate data
- Uncertain structure of a model due to insufficient knowledge (approximations and simplifications) about a system.
- Propagation of natural and model uncertainty through a chain of models.

Ultimately, design is a decision-making process, whether we are designing materials, systems, or both in concurrent fashion. As in manufacturing process design, the notion of *robust* design [27] appears to be central to any reasonable approach. Designs must be robust against variation of initial microstructure, variation of usage factors and history, variation of design goals, and various forms of uncertainty listed above, including the models, tools, and methods used to design. This includes issues such as the distribution of the design effort, level of expertise and knowledge of modelers and designers, and other human factors such as degree of interaction and information-sharing in the design process. There are important practical implications, namely that robust solutions do not necessarily involve large numbers of iterations, are not focused on excessive optimization searches at individual levels, and involve the human being as an interpreter of value of information. This means that ranged sets of solutions, rather than point solutions, are of practical interest. It also means that system performance requirements should be specified as ranges rather than single values. Moreover, systems performance should be specified rather than property requirements; in other words,

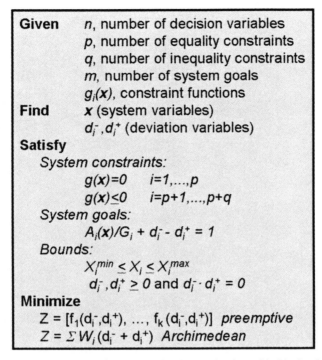

Given	n, number of decision variables
	p, number of equality constraints
	q, number of inequality constraints
	m, number of system goals
	$g_i(x)$, constraint functions
Find	x (system variables)
	d_i^-, d_i^+ (deviation variables)

Satisfy

System constraints:

$$g(x)=0 \quad i=1,...,p$$
$$g(x) \leq 0 \quad i=p+1,...,p+q$$

System goals:

$$A_i(x)/G_i + d_i^- - d_i^+ = 1$$

Bounds:

$$X_i^{min} \leq X_i \leq X_i^{max}$$
$$d_i^-, d_i^+ \geq 0 \text{ and } d_i^- \cdot d_i^+ = 0$$

Minimize

$$Z = [f_1(d_i^-,d_i^+), ..., f_k(d_i^-,d_i^+)] \ preemptive$$
$$Z = \Sigma W_i(d_i^- + d_i^+) \ Archimedean$$

Fig. 12 Compromise Decision Support Problem (cDSP) formulation for multi-objective design, with deviations from multiple goals minimized within constraints [25]

ranged sets of multiple properties can usually satisfy ranged sets of performance requirements, expanding the potential range of acceptable materials solutions.

A practical approach is to quantify the uncertainty to the extent possible and then seek robust solutions that are less sensitive to variation of microstructure and various other sources of uncertainty. To this end, the compromise Decision Support Problem (cDSP) protocol [25] has been introduced, shown in Fig. 12, as the primary decision support tool. It is based on goal programming rather than standard linear programming, and is a result of negotiation between multiple designers and analysts regarding assignment of goals, constraints and bounds. In the cDSP, multiple design objectives are set as targets, with deviations from these goals minimized subject to user preferences to select from among a family of solutions, subject to a set of constraints (cf. [25,26]).

For multiple design objectives, robustness establishes preference among candidate solutions [25–27]; we seek solutions with less sensitivity to variation of noise and control parameters. In addition, we seek designs that are robust against variability associated with process route and initial microstructure, forcing functions, cost factors, design goals, etc. Collaborative efforts at Georgia Tech have yielded new methods to deal with uncertainty due to microstructure variability and models [20,28] as well as chained sequences of models in a multi-level (multiscale) context [29].

There are several categories of robust design that deal with different types of uncertainty. Type I robust design, originally proposed by Taguchi [27], focuses on achieving insensitivity of performance with respect to noise factors—parameters that designers cannot control in a system. Type II robust design relates to insensitivity of a design to variability or uncertainty

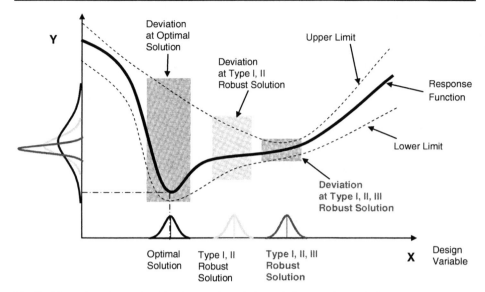

Fig. 13 Illustration of Types I and II robust design solutions relative to optimal solution based on extremum of objective function. Type III robust design minimizes deviation (dashed line) of the objective function from the flat region associated with model and microstructure uncertainty [20, 28]

associated with design variables—parameters that a designer can control in a system. A method for Types I and II robust design has been proposed, namely the Robust Concept Exploration Method [26]. These types of robust design have recently been extended to include Type III [20], which considers sensitivity to uncertainty embedded within a model (i.e., model parameter/structure uncertainty). Figure 13 clarifies the application of Types I-III robust design, showing that while application of traditional Types I-II robust design methods seek solutions that are insensitive to variations in control or noise parameters, Type III robust design additionally seeks solutions that have minimum distance between upper and lower uncertainty bounds on the response function(s) of interest associated with material randomness and model structure/parameter uncertainty. These bounds are determined from the statistics obtained from application of models over a parametric range of feasible microstructures and process conditions relevant to the simulations necessary to support design decisions (cf. [20, 28]).

This notion of goal programming to identify candidate ranged solutions must be couched in the context of a top-down strategy in the design system of Fig. 5. An iterative approach is essential for bottom-up information flow (simulations, experiments), combined with top-down guidance from applications and associated performance requirements. To this end, there are opportunities for developing efficient strategies for design exploration. Choi et al. [20, 22] have developed an approach called the Inductive Design Exploration Method (IDEM), schematically shown in Fig. 14. IDEM has two major objectives: (i) to explore top-down, requirements-driven design, guiding bottom-up modeling and simulation, and (ii) to manage uncertainty propagation in model chains. As illustrated in Fig. 14, IDEM requires initial configuration of the design process. An example of initial configuration of the design system is shown at a high level in Fig. 4 for a steel design problem. Implementation of IDEM necessitates identifying the connections of inputs and outputs of models, simulations, experiments, and databases, and insertion of decision-compromise such that a complete graph

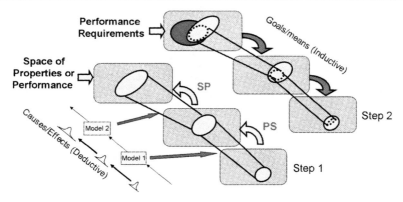

Fig. 14 Schematic of Steps 1 and 2 in IDEM. Step 1 involves bottom-up simulations or experiments, typically conducted in parallel fashion, to map composition into structure and then into properties, with regions in yellow showing the feasible ranged sets of points from these mappings. Step 2 involves top-down evaluation of points from the ranged set of specified performance requirements that overlap feasible regions established by bottom-up simulations in Step 1

of information flow is achieved. This configuration of information flow and decision points (the "design process") is reconfigurable and therefore constitutes an important element of the design itself. If certain models have greater certainty, they can be more heavily weighted. Quality experimental information can be factored in as desired. It is essentially an instantiation of the balanced decision-making process that has been employed in design for many years, biased towards quality information and insights. The difference is that it remains open to input regarding regimes of structure or behavior for which little prior empirical understanding is available, and can be reconfigured as necessary to adapt to new insight or opportunities. Step 1 in Fig. 14 is very important; it involves the pursuit of modeling and simulation to map potential process-structure and structure property relations over a sufficiently broad parametric range of compositions, initial structures, process-structure and structure-property assessments. It involves evaluation of discrete points at each level of hierarchy corresponding to the various mappings in Figs. 4 and 11, and is amenable to massive parallelization of simulations and database mining since each of these can be mapped independently without regard to a specific design scenario. In step 2, the results of the step 1 are inverted to inductively explore the feasible design spaces of properties, structure, and compositions, working backwards from ranged sets of performance requirements. After step 2, we obtain robust ranged solutions that include consideration of model uncertainty.

Applications of Types I-III robust design methods described above to design of extruded prismatic metals for multifunctional structural and thermal applications [30–32] and design of multiphase thermite metal-oxide mixtures for target reaction initiation probability under shock compression have been described elsewhere [20, 28], further summarized by McDowell [65]. We will discuss examples related to design of high strength, high toughness steels in Section 4.

3.4 Microstructure-mediated design

In practice, the duration and number of iterative cycles, mix of computation and experiments, and time frame of process-structure relations in materials development do not match those of structure-property relations. Moreover, product design is often conducted with systems level considerations by distinct groups separated by a "fence" (bold dashed line to the right in

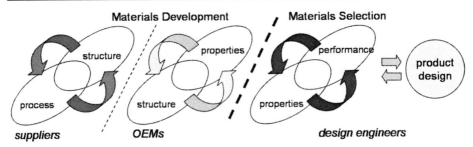

Fig. 15 Typical time scale and organizational separation of materials development and materials selection components of the materials design system of Fig. 5

Fig. 15), employing materials selection. As such, the property requirements may be specified but there is a lack of integration of the degrees of freedom of material structure into the overall systems design problem. In addition, the cooperative materials development process between materials suppliers and OEMs typically involves substantially different timeframes for materials processing and structure-property relations, often carried out in different organizations. In analogy to multiscale modeling, both time (cycle duration for processing and certification experiments) and length (organizational distribution) scales serve as limiting factors in coordinating the design process.

This is essentially just a somewhat more detailed decomposition of the issues addressed by the AIM program discussed in the Introduction. On the other hand, the approach suggested here is to effectively remove the bold dashed line to the right in Fig. 15, effectively integrating structure-property relations with materials selection by coupling computational materials simulation with the systems product design process. If this is realized, then the dashed line is moved to the left between the materials suppliers and the OEMs. The mediatory "language" for communicating design variables then becomes the material structure. The focus for accelerating the insertion of new and improved materials into products then is on balancing the timeframe for materials processing with the structure-properties-performance hierarchy; the latter is closely integrated, while the former must consider details of microstructure as the targets of processing rather than just properties. Within the ICME rubric, there is evidence that this is happening in OEMs [18]. As stated in the report of the 1998 MDS&E workshop [19], "The field of materials design is entrepreneurial in nature, similar to such areas as microelectronic devices or software. MDS&E may very well spawn a "cottage industry" specializing in tailoring materials for function, depending on how responsive large supplier industries can be to this demand. In fact, this is already underway."

4 Applications of materials design

The challenge is to extend these kinds of concurrent material and product systems design concepts to tailor microstructures that deliver required performance requirements in a wide range of problems, for example:

- Phase morphologies, precipitate/dispersoid distributions, texture, and grain boundary networks in alloy systems for multifunctional performance in terms of strength, ductility, fracture, fatigue, corrosion resistance, etc.
- Process path and in-service evolution of microstructure (e.g. plasticity, phase transformation, diffusion, etc.).

- Resistance to or preference for shear banding.
- Formable materials that employ transformation- or twinning-induced plasticity.
- Fine and coarse scale porosity control in castings.
- Surface treatment, heat treatment, and residual stresses in alloys with primary inclusions.

The foregoing systems engineering concepts have ushered in a first generation of designer "cyberalloys" [14,75,76]; these cyberalloys have now entered successful commercial applications, and a new enterprise of commercial materials design services has steadily grown over the past decade.

4.1 High strength and toughness steels

Building on the design methodology demonstrated at Northwestern University [9], QuesTek Innovations LLC (Evanston, IL) has integrated modeling of process-structure-property-performance relations in several major design programs for over a decade, with emphasis on proprietary high performance alloys suited to advanced gears and bearings and stainless steels for landing gear applications. Returning to the steel design example in Fig. 4, as explained by Olson [9], the objective was to develop martensitic steels with combinations of strength, toughness and stress corrosion resistance that would allow a major advance in the useable strength level of structural steels, beyond the levels that could be achieved by empirical alloy development over the same timeframe. Pursuant to a materials design approach to this problem, basic science modeling tools (Fig. 6) such as quantum mechanics and continuum micromechanics were used to facilitate the evaluation and analysis of microstructure 'subsystems' that relate to interface strength and the effects of strain-induced phase transformations. Certain subsystems control strength and others control toughness, for example. Diffusionless martensitic transformations occur at a length scale on the order of micrometers. To refine alloy carbide precipitate size at the nanometer scale, characterization tools such as x-ray diffraction, small-angle neutron scattering, atom-probe field-ion microscopy and analytical electron microscopy were combined with elastic energy calculations from continuum mechanics along with thermochemical software and related database to compute interfacial energies. Enhancing control of particle size facilitated development of efficient strengthening dispersions, leading to 50% increase in strength at a given alloy carbon content. Toughness subsystems of material architecture are dominated by yet another characteristic length scale; continuum mechanics analyses can be performed for ductile fracture associated with microvoid formation and growth at the interfaces on the order of 100 nanometers; these particles are introduced to decrease grain size in order to inhibit competing brittle fracture mechanism (cf. Fig. 9). The measured fracture energy and strain localization in shear are used to validate the results of the models. Finally, embrittlement resistance subsystems that govern environmental cracking are manifested at atomic scales of 0.1 nm through the effects of environmental hydrogen and the prior segregation of embrittling impurities, acting in concert to produce intergranular fracture. As described earlier, quantum mechanical calculations were employed to predict the segregation energy difference necessary to evaluate embrittlement potency. These quantum-based tools enabled designs in which grain boundaries are doped to attain desired electronic structures to enhance intrinsic cohesion and alter impurity interactions demonstrating significant improvements of environmental resistance.

Key concepts enabling early development of a design strategy for available computational tools were (a) subsystem "decoupling" as advocated in the general axiomatic design approach of Suh [77], and (b) establishing a "parametric" design approach where desired behaviors could be effectively mapped to predictable independent fundamental parameters. Panchal [21,29] discusses the desirability of decoupling sub-systems to the greatest extent possible,

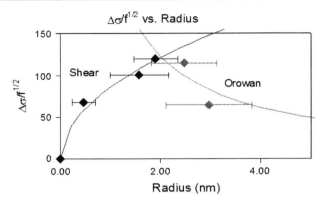

Fig. 16 Compromise of mechanisms of resistance to dislocation bypass via particle shearing and looping for coherent nm-scale precipitates [78]

as dictated by model utility in decision-making. An example of the decoupling is given by the experimentally calibrated analysis of the size dependence of coherent precipitation strengthening in Fig. 16 [78]. Here the identified optimum particle size, corresponding to the dislocation shear-bypass transition, gives a quantitative size goal for maximizing strengthening efficiency, evaluated independently of the requirements of other microstructure subsystems. Parametric control to refine particle size to this optimum is achieved by choosing to operate in a high supersaturation precipitation regime corresponding to a nucleation and coarsening behavior for which the trajectory of precipitation is well described by an evolving unstable equilibrium. This in turn enables a space-time separation in which the time constant of precipitation can be independently controlled by an extension of classical coarsening theory to multicomponent systems, while particle size is controlled through the thermodynamic driving force for coherent precipitation. Employing the science foundation described in [8,9], similar strategies were devised to exert independent parametric control of the subsystems of Fig. 4, following a mapping very similar to the schema of Fig. 11 [79]. The strongest nonlinear interactions between subsystems arise from multicomponent solution thermodynamics, which is well described by computational thermodynamics. Computational thermodynamics has thus served as the primary system integration tool for this efficient parametric approach to materials design. Using the same tools for a deterministic sensitivity analysis, the design output consists of a specified chemical composition and set of processing temperatures with allowable tolerances for each.

The first demonstration of this parametric design approach yielded a high performance stainless bearing steel for a very specific space shuttle application [80]. This was soon followed by a family of high performance carburizing steels [81,82], which has been successfully commercialized by QuesTek [75]. Exploratory design research has tested the generality of the design approach, demonstrating feasibility in case-hardenable polymers [83], hydrate cements [84], and oxidation resistant high temperature niobium-based alloys [85].

Following the iterative process of Fig. 3 (right), parametric design has typically achieved fairly ambitious property objectives within three iterations, employing as few as one experimental prototype per iteration. With steadily improving accuracy of design models and their supporting fundamental databases, an important milestone has been the demonstration of a successful design within a single iteration of prototyping. Described in detail elsewhere [86,87], a weldable high strength plate steel for naval blast protection applications was designed to maintain the impact toughness of the Navy's current HSLA 100 steel out to

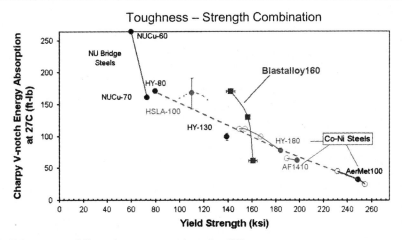

Fig. 17 Enhancement of Charpy impact energy absorption [87]

significantly higher strength levels. Adopting a tight carbon limit to promote high weldability, highly efficient strengthening was designed through a combination of copper precipitation and optimized alloy carbide precipitation. Superposition of active transformation toughening was achieved through designed precipitation of an optimal stability austenite dispersion nucleated on the copper dispersion using a two-step tempering treatment to balance particle size and composition. Process optimization and microstructural characterization of a single 66 kg slab cast prototype demonstrated the remarkable toughness-strength combination labeled "Blastalloy 160" shown in Fig. 17. High resolution microanalysis confirmed that the predicted strengthening and toughening dispersions were achieved, including the desired optimal austenite composition [87]. Continued development of variants of this steel by QuesTek has already demonstrated exceptional ballistic performance for fragment protection.

4.2 Integrating advances in 3D characterization and modeling tools

The success of computational materials design established the basis for the previously described DARPA-AIM initiative which broadened computational materials engineering to address acceleration of the full materials development and certification cycle. The central microstructural simulation engine of the AIM methodology is the PrecipiCalc code [3] developed under QuesTek-Northwestern collaboration, integrating precise calibration via high-resolution microanalysis. Employing the data fusion strategy for probabilistic modeling summarized in Fig. 1, the first demonstration of the AIM method in qualifying a new alloy is the just-completed specification of QuesTek's Ferrium S53 (AMS5922) corrosion-resistant steel for aircraft landing gear applications [76]. The project demonstrated both successful anticipation of process scaleup behavior, and employed data from 3 production-scale heats to fine tune processing in order to meet specified minimum design allowable properties, subsequently validated at the 10 production heat level.

An NMAB study [2] has documented the success of the DARPA-AIM program, highlighting the role of small technology startup companies in enabling this technology, and summarizing the commercial computational tools and supporting databases currently available. While the methods and tools of parametric materials design are now well established and undergoing wide application under QuesTek's commercial design services, the broadening

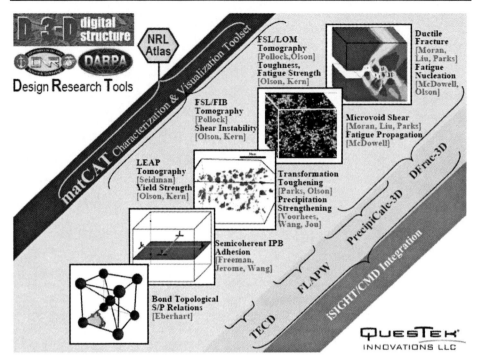

Fig. 18 Hierarchical material levels and tools for modeling (labels to right) and characterization (labels to left)

application of computational materials engineering in the materials-aware manufacturing context of both AIM accelerated qualification and ICME concurrent engineering practices drives the demand for even higher fidelity integrated simulation and characterization tools. A new level of science-based modeling accuracy is now being achieved under the ONR/DARPA "D3D" Digital Structure consortium. A suite of advanced 3D tomographic characterization tools are used to calibrate and validate a set of high fidelity explicit 3D microstructure simulation tools spanning the hierarchy of microstructure scales. Figure 18 provides an overview of the QuesTek-led university consortium component of the D3D program, supporting design of fatigue and fracture resistant high strength steels. This program is integrated with other aspects of D3D, including visualization systems, statistical analysis of distributed microstructure, integration of an archival 3D microstructure "atlas" at the Naval Research Laboratory, and ultimate iSIGHT-based integration of the full toolset in both computational materials design and AIM qualification. As examples, Fig. 19 shows how both multi-micron inclusions and submicron scale carbides that affect microvoid nucleation and growth can be characterized via microtomography for purposes of supporting multiscale strain localization and fracture models as shown in Fig. 9. Three-dimensional LEAP tomography (atom probe) shown at right in Fig. 19 renders quantitative information regarding the size and distribution of nanoscale dispersed carbides and precipitates, a key element in designing alloys for maximum strength. Figures 20 and 21 respectively show the 3D multiscale modeling strategies to account for effects of realistic distributions of submicron scale carbides on shear localization, and the potency of primary nonmetallic inclusions with respect to nucleation of cracks in high cycle fatigue, including effects of process history (carburization and shot peening) on shifting the critical location for nucleation to the subsurface.

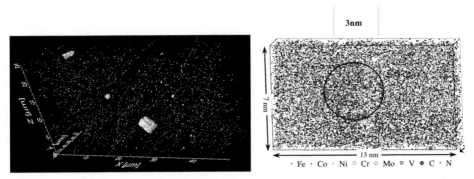

Fig. 19 Focused Ion Beam tomographic 3D reconstruction (left) of primary inclusion and submicron carbide distribution in modified 4330 steel, and (right) 3D Local Electrode Atom Probe tomography identifying a 3 nm strengthening carbide in the Ferrium S53 steel

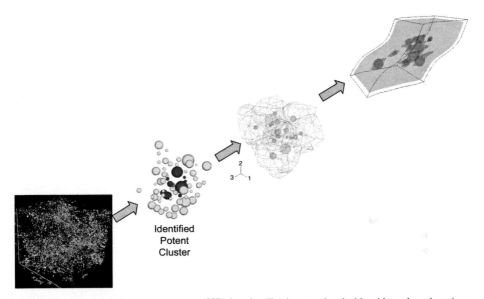

Fig. 20 Multiscale models for enhancement of 3D shear localization associated with voids nucleated at micron scale carbide particle dispersions in martensitic gear steels, offering substantial improvement compared to previous continuum models based on porous plasticity

5 Educational imperatives for materials design

Clearly, the concept of top-down systems-based robust materials design is an engineering exercise with several key characteristics:

- Strong role of materials simulation.
- Strong role of engineering systems design.
- Integration of multiple disciplines (materials science, applied mechanics, chemistry, physics, etc.).
- Drawing on science to support physically-based models, characterization of hierarchical materials, and bottom-up modeling.

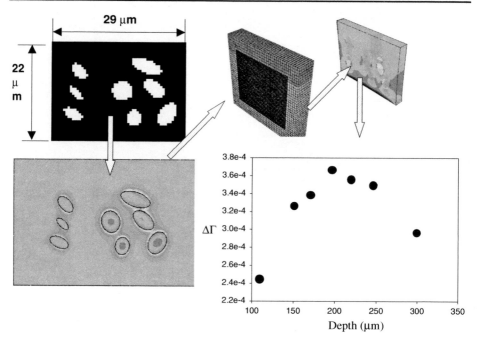

Fig. 21 Methodology for 3D modeling fatigue crack nucleation potency for clusters of nonmetallic inclusions in gear steels subjected to carburization and shot peening

An important issue relates to the need for integration of engineering design with the various other elements of simulation-assisted materials design. The report of the 1998 NSF MDS&E workshop [19] stated that "The systems integration that is necessary to conduct materials design must be recognized as part of the materials education enterprise. This has not generally been the case." This remains true to a large extent, in spite of fairly widespread materials design courses offered in undergraduate curricula in materials science in the United States. However, the necessary change of culture in US universities noted in that 1998 MDS&E workshop report is underway as more emphasis is being placed on related initiatives such as AIM and ICME. It is likely necessary for capstone courses in larger engineering disciplines such as mechanical and civil engineering to address elements of materials design in collaboration with MSE departments to inculcate many of the philosophies and insights expressed in this paper. Moreover, focused investigator grants for undergraduate and graduate program development to provide support for formulating systems-based materials science and engineering design curricula would likely accelerate this process. An attractive feature of this kind of systematic approach to materials design is that it can be applied to intriguing problem sets that excite undergraduate students and are relevant to real product needs, providing a creative, entrepreneurial product design environment based on modeling and simulation in addition to intuitive creativity.

Building on the parametric materials design approach developed by Northwestern's Steel Research Group, an upper level undergraduate materials design course has been taught at Northwestern since 1989 [88]. Given the challenge to undergraduates of the technical level of materials design, it has been found essential to implement a hierarchical coaching system [89] enabled by drawing projects from funded graduate research. As summarized by the array

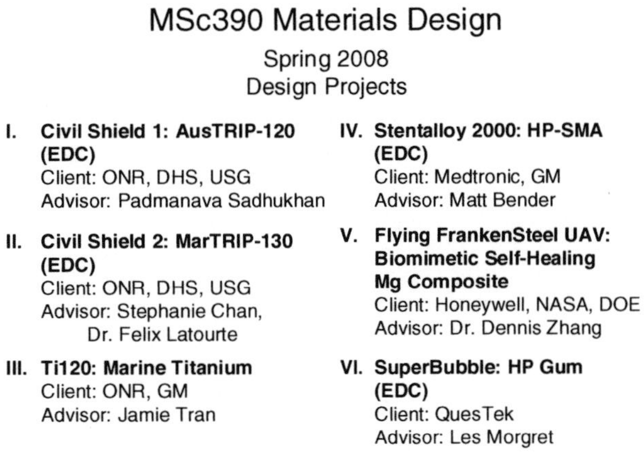

Fig. 22 Examples of design projects in undergraduate materials design course at Northwestern University

MSc390 Materials Design
Spring 2008
Design Projects

I. **Civil Shield 1: AusTRIP-120 (EDC)**
Client: ONR, DHS, USG
Advisor: Padmanava Sadhukhan

II. **Civil Shield 2: MarTRIP-130 (EDC)**
Client: ONR, DHS, USG
Advisor: Stephanie Chan,
Dr. Felix Latourte

III. **Ti120: Marine Titanium**
Client: ONR, GM
Advisor: Jamie Tran

IV. **Stentalloy 2000: HP-SMA (EDC)**
Client: Medtronic, GM
Advisor: Matt Bender

V. **Flying FrankenSteel UAV: Biomimetic Self-Healing Mg Composite**
Client: Honeywell, NASA, DOE
Advisor: Dr. Dennis Zhang

VI. **SuperBubble: HP Gum (EDC)**
Client: QuesTek
Advisor: Les Morgret

of projects listed in Fig. 22, each project is coached by a graduate student or post-doctoral researcher actively engaged in the multiyear iterative design of a material system.

Student teams are given access to the latest data and refined models to experience design integration in the latest iteration of a real design project, with the coaching support to operate at a high technical level. The technically ambitious projects listed span a range from next generation blast protection steels, high-performance low-cost titanium alloys, high-strength fatigue-resistant shape memory alloys, self-healing metallic composites, to high-performance bubble gum.

The design projects undertaken by materials majors are now being coordinated with engineering schoolwide interdisciplinary design project courses ranging from the freshman [90] to senior [91,92] levels. Building on the hierarchical coaching model, these undergraduate initiatives are enabling an integration of separately funded graduate research in different disciplines while allowing undergraduates to participate in the frontier of concurrent design of materials and structures. The "Civil Shield" projects listed in Fig. 22 integrate ONR-supported research in materials science and mechanical engineering on both materials and structures for blast protection. The undergraduate teams in multiple courses explore civilian applications of this integrated technology for anti-terrorism bomb mitigation [91]. Through the undergraduate-driven collaboration, blast simulations of novel panel structures have defined entirely new property objectives motivating new directions in materials design. Under industry gift support, demonstrated biomimetic self-healing behavior [93] motivated in part by protein transformation phenomena [94] is being integrated by mechanical engineering students in the design of self-repairing wingspars of unmanned aerial vehicles in the "Flying FrankenSteel UAV" project of Fig. 22, also defining new objectives for design of Mg-matrix self-healing composites. Having demonstrated a new generation of high strength shape memory alloys [78,95], undergraduates are now integrating projected property capabilities in the design of medical devices such as self-expanding endovascular stents.

6 Future prospects

Systems engineering approaches for concurrent design of hierarchical materials and structures are made feasible by the confluence of several fields:

- Computational materials science and ubiquitous computing.

- Advances in high resolution materials characterization and *in situ* measurements.
- Advances in micromechanics of materials.
- Information Technology (information theory, databases, digital interfaces, web protocols).
- Decision theory (game theory, utility theory, goal programming).

In addition to concurrent design of material and component/structure to meet specified performance requirements, there are other important capabilities that accrue to this technology, including but not limited to:

- Prioritizing models and computational methods in terms of measures of utility in supporting design decisions.
- Prioritizing mechanisms and materials science phenomena to be modeled for a given design problem.
- Conducting feasibility studies to establish probable return on investment of candidate new material systems.

In materials design problems, one often finds that models are either nonexistent or insufficiently developed to support decision-making. This includes both models for process-structure relations and 3D microstructure, as well as associated 3D models for structure-property relations. Of particular need is the coordination of model respositories for rapid availability to design search. A complicating factor that is rarely addressed is the quantification of uncertainty of model parameters and structure that is necessary in robust design of materials. Another very important consideration is that mechanistic models are often the limiting factor in applying decision-based design frameworks; however, guidance is required to decide how to best invest in model development that will maximize payoff or utility in the design process. Not all models are equally important in terms of their role in design, and this depends heavily on the design objectives and requirements.

On the other hand, one can readily identify gaps in multiscale modeling methods without regard to utility in design. One example is the gap in reliable, robust models between the level of atomistics and polycrystal plasticity. This gap is closing each year with advances in discrete dislocation plasticity, but progress in predictive methods for dislocation patterning at mesoscales has been slow, in part due to the lack of top-down calibration compared to polycrystal plasticity. On the other hand, from the perspective of decision support in materials design, much can be done using models at lower and higher scales of the hierarchy without a requirement to accurately predict these substructures. The relative need to bridge this gap is problem dependent.

Where are the opportunities for improvement in materials design? Several can be identified:

- Rapid methods for feasibility study and robust concept exploration – Early stage exploration of ranges of potential solutions to specific requirements, beyond experience or intuition. This requires assessment of value-of-information metrics (utility theory), and identification where models are needed, establishing model/database priorities.
- Microstructure-mediated design - Balancing process development iteration with structure-property iteration – managing assets and deciding on nature of interfaces between processing and structure-property relations (cf. Fig. 15); distinguishing design exploration from detail design.
- Parallel processing algorithms for robust concept exploration – Materials design is an ideal candidate for parallelization in the initial design exploration process (cf. IDEM Step 1 in Fig. 15). Such searching is normally mentioned in connection with data mining, but we believe the wider exploration of potential design space is a daunting task worthy of massively parallel computing.

The linkage between integrated material and product design and the information sciences is perhaps rather obvious in view of the foregoing discussion. There is ample room for creative contributions in this regard. The emerging field of materials informatics (JOM 60(3) 2008) embodies elements such as knowledge discovery extracted from databases via data mining in interdisciplinary areas such as statistics, materials databases, and results of material modeling to assist in discovery of new materials concepts. These ideas are particularly attractive for cases in which well-established theories and models do not exist, i.e., high uncertainty and little intuitive guidance. Scaling laws that arise from considering various relations between data may offer insight into physical relationships and dominant mechanisms across length and time scales, thereby providing support for metamodeling and simulation in lieu of high degree of freedom models. The development of networked cyberinfrastructure is an important aspect of realizing the potential of informatics, which purports to examine existing self-organized materials systems, even biological systems [94,96], arguing that the hierarchy shown in Fig. 4 is the materials science equivalent of a biological regulatory network. This is an interesting, potentially powerful assertion, and time will tell of its utility in pursuing systems-based robust design of materials. It is likely to be of utility mainly in the preliminary design exploration stage in which new or improved materials solutions are searched for feasibility. From a systems perspective, as in synthetic designed materials, understanding the structure of materials in nature requires a rather thorough understanding of the functions that are required. In biology, this can be a complex issue indeed.

Closure

Elements of systems approaches for designing material microstructures to meet multiple performance/property requirements of products have been outlined, distinguishing multilevel design of hierarchical materials from multiscale modeling. Robust design methods are preferred owing to the prevalence of uncertainty in process route, stochasticity of microstructure, and nonequilibrium, path dependent nature of inelastic deformation and associated constitutive models. Challenges for design of practical alloy systems and inelastic deformation and damage mechanisms are outlined, and successful examples of simplified parametric design are provided. Concurrent design of hierarchical materials and structures is facilitated by the confluence of engineering science and mechanics, materials science/physics, and systems engineering. Examples are presented. Continued improvement is a worthy pursuit of multi-physics modeling and simulation. Materials design exploration that requires intensive computation (e.g., bottom-up Step 1 in IDEM) is an excellent candidate for petascale computing.

The future of simulation-assisted materials design is promising, particularly with recent initiatives such as ICME that reinforce its value in industry. We envision that planning processes for materials development programs in the future will draw on this emerging multidiscipline. For materials design to realize its full potential, collaborative models must address intellectual property issues of data/model sharing or purchase. Perhaps one direction a bit further in the future is widespread availability of special purpose models or datasets which can be searched on the web and purchased for use in specific, targeted applications to complement use of general purpose analysis software, proprietary codes, and databases. Certainly, standards for verification of tool validity, as well as specifications of uncertainty would be elements of this distributed framework.

Acknowledgements The co-authors are grateful for funding that has supported their collaboration in several programs in recent years, including the DARPA AIM program (Dr. L. Christodoulou) and the ONR/DARPA D3D tools consortia (Dr. J. Christodoulou).

DLM especially wishes to thank his many Georgia Tech colleagues (Systems Realization Laboratory faculty F. Mistree and J.K. Allen, and former graduate students in materials design C. Seepersad, H.-J. Choi and J. Panchal) in collaborating to develop systems design concepts such as Type III robust design and IDEM outlined here. Early stages of this work were sponsored by DSO of DARPA (N00014-99-1-1016) and ONR (N0014-99-1-0852) on Synthetic Multifunctional Materials, monitored by Dr. L. Christodoulou of DARPA and Dr. S. Fishman of ONR. Support of an AFOSR Multi-University Research Initiative (1606U81) on Design of Multifunctional Energetic Structural Materials (Dr. C.S. Hartley, J. Tiley and B. Connor) is also acknowledged. Many elements described in this paper related to robust design are being further pursued with support of the Center for Computational Materials Design (CCMD), a NSF I/UCRC jointly founded by Penn State and Georgia Tech (DLM Co-Director), http://www.ccmd.psu.edu/.

GBO is especially grateful for long term research support from ONR. Combined with the career of his mentor, the late Morris Cohen of MIT, recent design achievements are the fruits of a continuous line of ONR-supported research beginning in 1948. Since the inception of the Steel Research Group materials design research effort in 1985, significant research support has also been provided by NSF, ARO, AFOSR, DARPA, NASA, and DOE with industry matching funds.

References

1. Ashby, M.F.: Materials Selection in Mechanical Design, 2nd edn. Butterworth-Heinemann, Oxford (1999)
2. Apelian, D.: National research council report. In: Accelerating Technology Transition. National Academies Press, Washington (2004)
3. Jou, H.-J., Voorhees, P., Olson, G.B.: Computer simulations for the prediction of microstructure/property variation in aeroturbine disks. In: Green, K.A., Pollock, T.M., Harada, H., Howson, T.E., Reed, R.C., Schirra, J.J., Walston, S. (eds.) Superalloys 2004, pp. 877–886 (2004)
4. Gall, K., Horstemeyer, M.F., McDowell, D.L., Fan, J.: Finite element analysis of the stress distributions near damaged Si particle clusters in cast Al-Si alloys. Mech. Mater. **32**(5), 277–301 (2000)
5. Gall, K., Horstemeyer, M.F., Degner, B.W., McDowell, D.L., Fan, J.: On the driving force for fatigue crack formation from inclusions and voids in a cast A356 aluminum alloy. Int. J. Fract. **108**, 207–233 (2001)
6. Fan, J., McDowell, D.L., Horstemeyer, M.F., Gall, K.: Cyclic plasticity at pores and inclusions in cast Al-Si alloys. Eng. Fract. Mech. **70**(10), 1281–1302 (2003)
7. McDowell, D.L., Gall, K., Horstemeyer, M.F., Fan, J.: Microstructure-based fatigue modeling of cast A356-T6 alloy. Eng. Fract. Mech. **70**, 49–80 (2003)
8. Olson, G.B.: Brains of steel: mind melding with materials. Int. J. Eng. Educ. **17**(4–5), 468–471 (2001)
9. Olson, G.B.: Computational design of hierarchically structured materials. Science **277**(5330), 1237–1242 (1997)
10. Shu, C., Rajagopalan, A., Ki, X., Rajan, K.: Combinatorial materials design through database science. In: Materials Research Society Symposium—Proceedings, vol. 804, Combinatorial and Artificial Intelligence Methods in Materials Science II, pp. 333–341 (2003)
11. Wu, R., Freeman, A.J., Olson, G.B.: First principles determination of the effects of phosphorous and boron on iron grain-boundary cohesion. Science **266**, 376–380 (1994)
12. Zhong, L., Freeman, A.J., Wuand, R., Olson, G.B.: Charge transfer mechanism of hydrogen-induced intergranular embrittlement of iron. Phys. Rev. B **21**, 938–941 (2000)
13. Geng, W.T., Freeman, A.J., Olson, G.B.: Influence of alloying additions on grain boundary cohesion of transition metals: first-principles determination and its phenomenological extension. Phys. Rev. B **63**, 165415 (2001)
14. Olson, G.B.: Designing a new material world. Science **288**, 993–998 (2000)
15. Lee, J.-H., Shishidou, T., Zhao, Y.-J., Freeman, A.J., Olson, G.B.: Strong interface adhesion in Fe/TiC. Philos. Mag. **85**, 3683–3697 (2005)
16. Billinge, S.J.E., Rajan, K., Sinnot, S.B.: From Cyberinfrastructure to Cyberdiscovery in Materials Science: Enhancing Outcomes in Materials Research, Education and Outreach. Report from NSF-sponsored workshop held in Arlington, Virginia, August 3–5. http://www.mcc.uiuc.edu/nsf/ciw_2006/ (2006)
17. Oden, J.T., Belytschko, T., Fish, J., Hughes, T.J.R., Johnson, C., Keyes, D., Laub, A., Petzold, L., Srolovitz, D., Yip, S.: Simulation-Based Engineering Science: Revolutionizing Engineering Science Through Sim-

ulation. Report of NSF Blue Ribbon Panel on Simulation-Based Engineering Science, May. http://www.nsf.gov/pubs/reports/sbes_final_report.pdf (2006)

18. Pollock, T.M., Allison, J.: Committee on Integrated Computational Materials Engineering: Developing a Roadmap for a Grand Challenge in Materials. National Materials Advisory Board, National Academy of Engineering. http://www7.nationalacademies.org/nmab/CICME_home_page.html (2007)

19. McDowell, D.L., Story, T.L.: New Directions in Materials Design Science and Engineering. Report of NSF DMR-sponsored workshop held in Atlanta, GA, October 19–21 (1998)

20. Choi, H.-J.: A robust design method for model and propagated uncertainty. Ph.D. Dissertation, G.W. Woodruff School of Mechanical Engineering, Georgia Institute of Technology, Atlanta, GA, USA (2005)

21. Panchal, J.H., Choi, H.-J., Shepherd, J., Allen, J.K., McDowell, D.L., Mistree, F.: A strategy for simulation-based multiscale, multifunctional design of products and design processes. In: ASME Design Automation Conference, Long Beach, CA. Paper Number: DETC2005-85316 (2005)

22. Choi, H.-J., McDowell, D.L., Allen, J.K., Rosen, D., Mistree, F.: An inductive design exploration method for the integrated design of multi-scale materials and products. J. Mech. Des. **130**(3), 031402 (2008)

23. Seepersad, C.C., Fernandez, M.G., Panchal, J.H., Choi, H.-J., Allen, J.K., McDowell, D.L., Mistree, F.: Foundations for a systems-based approach for materials design. In: 10th AIAA/ISSMO Multidisciplinary Analysis and Optimization Conference. AIAA MAO, Albany, NY. AIAA-2004-4300 (2004)

24. Isukapalli, S.S., Roy, A., Georgopoulos, P.G.: Stochastic response surface methods (SRSMs) for uncertainty propagation: application to environmental and biological systems. Risk Anal. **18**(3), 351–363 (1998)

25. Mistree, F., Hughes, O.F., Bras, B.A.: The compromise decision support problem and the adaptive linear programming algorithm. In: Kamat, M.P. (ed.) Structural Optimization: Status and Promise, vol. 150, pp. 251–290. AIAA, Washington (1993)

26. Chen, W.: A robust concept exploration method for configuring complex systems. Ph.D. Dissertation, G.W. Woodruff School of Mechanical Engineering, Georgia Institute of Technology, Atlanta, Georgia (1995)

27. Taguchi, G.: Taguchi on Robust Technology Development: Bringing Quality Engineering Upstream. ASME Press, New York (1993)

28. Choi, H.-J., Austin, R., Shepherd, J., Allen, J.K., McDowell, D.L., Mistree, F., Benson, D.J.: An approach for robust design of reactive powder metal mixtures based on non-deterministic micro-scale shock simulation. J. Comput.-Aided Mater. Des. **12**(1), 57–85 (2005)

29. Panchal, J.H.: A framework for simulation-based integrated design of multiscale products and design processes. Ph.D. Dissertation, G.W. Woodruff School of Mechanical Engineering, Georgia Institute of Technology, Atlanta, GA, USA (2005)

30. Seepersad, C.C.: A robust topological preliminary design exploration method with materials design applications. Ph.D. Dissertation, G.W. Woodruff School of Mechanical Engineering, Georgia Institute of Technology, Atlanta, Georgia (2004)

31. Seepersad, C.C., Kumar, R.S., Allen, J.K., Mistree, F., McDowell, D.L.: Multifunctional design of prismatic cellular materials. J. Comput.-Aided Mater. Des. **11**(2–3), 163–181 (2005)

32. Seepersad, C.C., Allen, J.K., McDowell, D.L., Mistree, F.: Multifunctional topology design of cellular structures. J. Mech. Des. **130**(3), 031404-1-13 (2008)

33. Panchal, J.H., Choi, H.-J., Allen, J.K., McDowell, D.L., Mistree F.: Designing design processes for integrated materials and products realization: a multifunctional energetic structural material example. In: Proceedings of the ASME International Design Engineering Technical Conferences and Computers and Information in Engineering Conference, DETC2006 (2006)

34. Zhou, M., McDowell, D.L.: Equivalent continuum for dynamically deforming atomistic particle systems. Philos. Mag. A **82**(13), 2547–2574 (2002)

35. Muralidharan, K., Deymier, P.A., Simmons, J.H.: A concurrent multiscale finite difference time domain/molecular dynamics method for bridging an elastic continuum to an atomic system. Model. Simul. Mater. Sci. Eng. **11**(4), 487–501 (2003)

36. Chung, P.W., Namburu, R.R.: On a formulation for a multiscale atomistic-continuum homogenization method. Int. J. Solids Struct. **40**, 2563–2588 (2003)

37. Curtarolo, S., Ceder, G.: Dynamics of an inhomogeneously coarse grained multiscale system. Phys. Rev. Lett. **88**(25), 255504 (2002)

38. Kulkarni, Y., Knap, J., Ortiz, M.: A variational approach to coarse-graining of equilibrium and non-equilibrium atomistic description at finite temperature. J. Mech. Phys. Solids **56**, 1417–1449 (2008)

39. Rafii-Tabar, H., Hua, L., Cross, M.: A multi-scale atomistic-continuum modeling of crack propagation in a two-dimensional macroscopic plate. J. Phys. Condens. Matter **10**(11), 2375–2387 (1998)

40. Rudd, R.E., Broughton, J.Q.: Concurrent coupling of length scales in solid state systems. Phys. Status Solidi B **217**(1), 251–291 (2000)
41. Qu, S., Shastry, V., Curtin, W.A., Miller, R.E.: A finite-temperature dynamic coupled atomistic/discrete dislocation method. Model. Simul. Mater. Sci. Eng. **13**(7), 1101–1118 (2005)
42. Cherkaoui, M.: Constitutive equations for twinning and slip in low stacking fault energy metals: a crystal plasticity type model for moderate strains . Philos. Mag. **83**(31–34), 3945–3958 (2003)
43. Svoboda, J., Gamsjäger, E., Fischer, F.D.: Modelling of massive transformation in substitutional alloys. Metall. Mater. Trans. A **37**, 125–132 (2006)
44. Idesman, A.V., Levitas, V.I., Preston, D.L., Cho, J.-Y.: Finite element simulations of martensitic phase transitions and microstructure based on strain softening model. J. Mech. Phys. Solids **53**(3), 495–523 (2005)
45. Needleman, A., Rice, J.R.: Plastic creep flow effects in the diffusive cavitation of grain boundaries. Acta Metall. **28**(10), 1315–1332 (1980)
46. Cocks, A.C.F.: Variational principles, numerical schemes and bounding theorems for deformation by Nabarro-Herring creep. J. Mech. Phys. Solids **44**(9), 1429–1452 (1996)
47. Cleri, F., D'Agostino, G., Satta, A., Colombo, L.: Microstructure evolution from the atomic scale up. Comp. Mater. Sci. **24**, 21–27 (2002)
48. Ghosh, S., Bai, J., Raghavan, P.: Concurrent multi-level model for damage evolution in microstructurally debonding composites. Mech. Mater. **39**(3), 241–266 (2007)
49. Kouznetsova, V., Geers, M.G.D., Brekelmans, W.A.M.: Multi-scale constitutive modelling of heterogeneous materials with a gradient-enhanced computational homogenization scheme. Int. J. Numer. Meth. Eng. **54**(8), 1235–1260 (2002)
50. Kouznetsova, V.G., Geers, M.G.D., Brekelmans, W.A.M.: Multi-scale second-order computational homogenization of multi-phase materials: a nested finite element solution strategy. Comput. Meth. Appl. Mech. Eng. **193**(48/51), 5525–5550 (2004)
51. Vernerey, F., Liu, W.K., Moran, B.: Multi-scale micromorphic theory for hierarchical materials. J. Mech. Phys. Solids **55**(12), 2603–2651 (2007)
52. Zbib, H.M., de la Rubia, T.D., Bulatov, V.: A multiscale model of plasticity based on discrete dislocation dynamics. J. Eng. Mater. Technol. **124**(1), 78–87 (2002)
53. Zbib, H.M., de la Rubia, T.D.: A multiscale model of plasticity. Int. J. Plast. **18**(9), 1133–1163 (2002)
54. Roy, A., Acharya, A.: Size effects and idealized dislocation microstructure at small scales: predictions of a phenomenological model of mesoscopic field dislocation mechanics: II. J. Mech. Phys. Solids **54**, 1711–1743 (2006)
55. Lemaitre, J., Chaboche, J.L.: Mechanics of Solid Materials. Cambridge University Press, Cambridge (1990). ISBN 0521477581
56. McDowell, D.L: Internal state variable theory. In: Yip, S., Horstemeyer, M.F. (eds.) Handbook of Materials Modeling, Part A: Methods, pp. 1151–1170. Springer, The Netherlands (2005)
57. Aifantis, E.C.: The physics of plastic deformation. Int. J. Plast. **3**, 211–247 (1987)
58. Aifantis, E.C.: Update on a class of gradient theories. Mech. Mater. **35**, 259–280 (2003)
59. Beeler, J.R.: Radiation Effects Computer Experiments. North Holland, Amsterdam (1982)
60. Heinisch, H.L.: Simulating the production of free defects in irradiated metals. Nucl. Instrum. Methods B **102**, 47 (1995)
61. Shenoy, M.M., Kumar, R.S., McDowell, D.L.: Modeling effects of nonmetallic inclusions on LCF in DS nickel-base superalloys. Int. J. Fatigue **27**, 113–127 (2005)
62. Shenoy, M.M., Zhang, J., McDowell, D.L.: Estimating fatigue sensitivity to polycrystalline Ni-base superalloy microstructures using a computational approach. Fatigue Fract. Eng. Mater. Struct. **30**(10), 889–904 (2007)
63. Wang, A.-J., Kumar, R.S., Shenoy, M.M., McDowell, D.L.: Microstructure-based multiscale constitutive modeling of γ-γ' nickel-base superalloys. Int. J. Multiscale Comp. Eng. **4**(5–6), 663–692 (2006)
64. Hao, S., Moran, B., Liu, W.-K., Olson, G.B.: A hierarchical multi-physics model for design of high toughness steels. J. Comput.-Aided Mater. Des. **10**, 99–142 (2003)
65. McDowell, D.L.: Simulation-assisted materials design for the concurrent design of materials and products. JOM **59**(9), 21–25 (2007)
66. Adams, B.L., Lyon, M., Henrie, B.: Microstructures by design: linear problems in elastic-plastic design. Int. J. Plast. **20**(8–9), 1577–1602 (2004)
67. Adams, B.L., Gao, X.: 2-point microstructure archetypes for improved elastic properties. J. Comput. Aided Mater. Des. **11**(2–3), 85–101 (2004)
68. Lyon, M., Adams, B.L.: Gradient-based non-linear microstructure design. J. Mech. Phys. Solids **52**(11), 2569–2586 (2004)
69. Kalidindi, S.R., Houskamp, J., Proust, G., Duvvuru, H.: Microstructure sensitive design with first order homogenization theories and finite element codes. Materials Science Forum, vol. 495–497, n PART 1,

Textures of Materials, ICOTOM 14—Proceedings of the 14th International Conference on Textures of Materials, pp. 23–30 (2005)

70. Kalidindi, S.R., Houskamp, J.R., Lyon, M., Adams, B.L.: Microstructure sensitive design of an orthotropic plate subjected to tensile load. Int. J. Plast. **20**(8–9), 1561–1575 (2004)
71. Knezevic, M., Kalidindi, S.R., Mishra, R.K.: Delineation of first-order closures for plastic properties requiring explicit consideration of strain hardening and crystallographic texture evolution. Int. J. Plast. **24**(2), 327–342 (2008)
72. Ganapathysubramanian, S., Zabaras, N.: Design across length scales: a reduced-order model of polycrystal plasticity for the control of microstructure-sensitive material properties. Comput. Meth. Appl. Mech. Eng. **193**(45–47), 5017–5034 (2004)
73. Sankaran, S., Zabaras, N.: Computing property variability of polycrystals induced by grain size and orientation uncertainties. Acta Mater **55**(7), 2279–2290 (2007)
74. Li, D.S., Bouhattate, J., Garmestani, H.: Processing path model to describe texture evolution during mechanical processing. Materials Science Forum, vol. 495–497, n PART 2, Textures of Materials, ICOTOM 14—Proceedings of the 14th International Conference on Textures of Materials, pp. 977–982 (2005)
75. Kuehmann, C.J., Olson, G.B.: Gear steels designed by computer. Adv. Mat. Process. **153**, 40–43 (1998)
76. Kuehmann, C.J., Tufts, B., Trester, P.: Computational design for ultra-high-strength alloy. Adv. Mat. Process. **166**(1), 37–40 (2008)
77. Suh, N.P.: Axiomatic design theory for systems. Res. Eng. Des. **10**(4), 189–209 (1998)
78. Bender, M.: unpublished doctoral research, Northwestern University (2008)
79. Kuehmann, C.J., Olson, G.B.: Computer-aided systems design of advanced steels. In: Hawbolt, E.B. (ed.) Proceedings of the International Symposium on Phase Transformations During Thermal/Mechanical Processing of Steel, pp. 345–356. Metallurgical Society of Canadian Institute of Mining, Metallurgy and Petroleum, Vancouver (1995)
80. Stephenson, T.A., Campbell, C.E., Olson, G.B.: Systems design of advanced bearing steels. In: Richmond, R.J., Wu, S.T. (eds.) Advanced Earth to Orbit Propulsion Technology, vol. 3174, no. 2, pp. 299–307. NASA Conference publication (1992)
81. Campbell, C.E., Olson, G.B.: Systems design of high performance stainless steels I. Conceptual and computational design. J. Comput.-Aided Mater. Des. **7**, 145–170 (2001)
82. Campbell, C.E., Olson, G.B.: Systems design of high performance stainless steels II. Prototype characterization. J. Comput.-Aided Mater. Des. **7**, 171–194 (2001)
83. Carr, S.H., D'Oyen, R., Olson, G.B.: Design of thermoset resins with optimal graded structures. In: Hui, D. (ed.) Proceedings of the 4th International Conference on Composites Engineering, 205 pp. International Community for Composites Engineering (1997)
84. Neubauer, C.M., Thomas, J., Garci, M., Breneman, K., Olson, G.B., Jennings, H.M.: Cement hydration. In: Proceedings of the 10th International Congress on the Chemistry of Cement. Amarkai AB and Congrex Goteborg AB, Sweden (1997)
85. Olson, G.B., Freeman, A.J., Voorhees, P.W., Ghosh, G., Perepezko, J., Eberhart, M., Woodward, C.: Quest for noburnium: 1300C cyberalloy. In: Kim, Y.W., Carneiro, T. (eds.) International Symposium on Niobium for High Temperature Applications, pp. 113–122. TMS, Warrendale, PA (2004)
86. Saha, A., Olson, G.B.: Computer-aided design of transformation toughened blast resistant naval hull steels: part I. J. Comput.-Aided Mater. Des. **14**, 177–200 (2007)
87. Saha, A., Olson, G.B.: Prototype evaluation of transformation toughened blast resistant naval hull steels: part II. J. Comput.-Aided Mater. Des. **14**, 201–233 (2007)
88. Olson, G.B.: Materials design—an undergraduate course. In: Liaw, P.K. (ed.) Morris E. Fine Symposium, pp. 41–48, TMS-AIME, Warrendale PA (1991)
89. Manuel, M.V., McKenna, A.F., Olson, G.B.: Hierarchical model for coaching technical design teams. Int. J. Eng. Ed. **24**(2), 260–265 (2008)
90. Hirsch, P.L., Schwom, B.L., Yarnoff, C., Anderson, J.C., Kelso, D.M., Olson, G.B., Colgate, J.E.: Engineering design and communication: the case for interdisciplinary collaboration. Int. J. Eng. Ed. **17**, 342–348 (2001)
91. McKenna, A.F., Colgate, J.E., Carr, S.H., Olson, G.B.: IDEA: formalizing the foundation for an engineering design education. Int. J. Eng. Ed. **22**, 671–678 (2006)
92. McKenna, A.F., Colgate, J.E., Olson, G.B., Carr, S.H.: Exploring adaptive Expertise as a target for engineering design education. In: Proceedings of the IDETC/CIE, pp. 1–6 (2001)
93. Files, B., Olson, G.B.: Terminator 3: biomimetic self-healing alloy composite. In: Proceedings of the 2nd Internatioal Conference on Shape Memory & Superelastic Technologies, pp. 281–286, SMST-97, Santa Clara CA (1997)

94. Olson, G.B., Hartman, H.: Martensite and life—displacive transformations as biological processes. Proc ICOMAT-82. J. de Phys. **43**, C4-855 (1982)
95. Olson, G.B.: Advances in theory: martensite by design. Mater. Sci. Eng. A **438**, 48–54 (2006)
96. Rajan, K.: Learning from systems biology: an "omics" approach to materials design. JOM **60**(3), 53–55 (2008)

Enthalpy landscapes and the glass transition

John C. Mauro · Roger J. Loucks ·
Arun K. Varshneya · Prabhat K. Gupta

Originally published in the journal Sci Model Simul, Volume 15, Nos 1–3, 241–281.
DOI: 10.1007/s10820-008-9092-2 © Springer Science+Business Media B.V. 2008

Abstract   A fundamental understanding of the glass transition is essential for enabling future breakthroughs in glass science and technology. In this paper, we review recent advances in the modeling of glass transition range behavior based on the enthalpy landscape approach. We also give an overview of new simulation techniques for implementation of enthalpy landscape models, including techniques for mapping the landscape and computing the long-time dynamics of the system. When combined with these new computational techniques, the enthalpy landscape approach can provide for the predictive modeling of glass transition and relaxation behavior on a laboratory time scale. We also discuss new insights from the enthalpy landscape approach into the nature of the supercooled liquid and glassy states. In particular, the enthalpy landscape approach provides for natural resolutions of both the Kauzmann paradox and the question of residual entropy of glass at absolute zero. We further show that the glassy state cannot be described in terms of a mixture of equilibrium liquid states, indicating that there is no microscopic basis for the concept of a fictive temperature distribution and that the glass and liquid are two fundamentally different states. We also discuss the connection between supercooled liquid fragility and the ideal glass transition.

In loving memory of Salvatore M. Mauro. "Give a man a fish and he will eat for a day. Teach him how to fish and he will eat for a lifetime."

J. C. Mauro (✉)
Science and Technology Division, Corning Incorporated, Corning, NY 14831, USA
e-mail: mauroj@corning.com

R. J. Loucks
Department of Physics and Astronomy, Alfred University, Alfred, NY 14802, USA
e-mail: loucks@alfred.edu

A. K. Varshneya
New York State College of Ceramics, Alfred University, Alfred, NY 14802, USA
e-mail: varshneya@alfred.edu

P. K. Gupta
Department of Materials Science and Engineering, Ohio State University, Columbus, OH 43210, USA
e-mail: gupta.3@osu.edu

Keywords Glass · Glass transition · Thermodynamics · Statistical mechanics · Modeling ·
Simulation

1 Introduction

For nearly five millennia, the beauty of glass has captured the hearts and imagination of
mankind [1]. Glass has also proven to be one of the key enablers of modern civilization.
Glass windows have allowed for sunlight to enter homes while protecting their inhabitants
from harsh weather conditions. Glass lenses have restored clear vision to people suffering
from degraded eyesight and enabled the exploration of faraway stars and galaxies and the
study of the tiniest of microorganisms. The invention of the glass light bulb envelope has
illuminated the nighttime world, when previously our eyes strained under dim candlelight.
More recently, the invention of glass fibers for optical communication has revolutionized
the way in which humanity communicates, bringing the dawn of the Information Age and
making neighbors out of people on distant continents. Indeed, the technological versatility
of glass, together with the innovation and creativity of mankind, has brought revolution upon
revolution to our global society. We strongly believe that mankind has just scratched the surface of what is possible with this beautiful and mysterious material which appears solid-like
but is understood to have liquid-like atomic arrangements.[1]

To facilitate future breakthroughs in glass science and technology, it is highly desirable
to have a fundamental physical understanding of the glassy state. Traditionally, the study of
glass has been made difficult by the three "*nons*":

1. Glass is *non*-crystalline, lacking the long-range atomic order found in most solid materials. Unlike crystalline materials, the structure of glass cannot be defined in terms of a
   simple unit cell that is repeated periodically in space.
2. Glass is *non*-equilibrium; hence, the glassy state cannot be described using equilibrium
   thermodynamics or statistical mechanics. The macroscopic properties of a glass depend
   on both its composition and thermal history.
3. Glass is *non*-ergodic, since we observe glass on a time scale that is much shorter than
   its structural relaxation time. As time elapses, ergodicity is gradually restored and the
   properties of a glass slowly approach their equilibrium values [3].

At the heart of these issues lies the glass transition, i.e., the process by which an equilibrium, ergodic liquid is gradually frozen into a nonequilibrium, nonergodic glassy state.
Many models of the glass transition have been proposed previously. The phenomenological
models of Tool [4] and Narayanaswamy [5] are useful from an engineering point of view to
fit experimental relaxation data; however, they do not offer any insights into the underlying
physics of the glass transition, and they cannot offer *a priori* predictions of glass transition range behavior. The free volume model Turnbull, Cohen, and Grest [6,7] is based on a
direct correlation between the fluidity of a supercooled liquid and its free volume; however,
this view has proved to be overly simplistic [8]. The thermodynamic model of Gibbs and
DiMarzio [9] is based on a highly idealized system and is not suitable for studying realistic
glass transitions. Kinetic models such as mode-coupling theory [10,11] show promise for the
modeling of supercooled liquids, but they are not yet capable of reproducing glass transition

---

[1] The alleged liquid-like flow of glass windows in old European churches is an interesting urban legend. It is
now believed [2] that the apparent liquid-like behavior is more due to the way the glass window was installed.
Scientific estimates of time to observe liquid-like behavior are, in essence, part of this publication.

range behavior [12]. The molecular dynamics technique cannot be used to study realistic glass transition range behavior since it is limited by the very short integration time step (about 10^{-15} s) [13]. Excellent reviews of these and other previous glass transition models are provided by Scherer [14], Jäckle [15], Gupta [8], and Varshneya [16].

Recently, much progress has been made in modeling and simulation of the glass transition based on the enthalpy landscape formalism. When combined with nonequilibrium statistical mechanics techniques, the enthalpy landscape approach allows for modeling of the glass transition based solely on fundamental physics, without any empirical fitting parameters. Such a model allows for predictive computation of glass transition range behavior under realistic glass-forming conditions. The enthalpy landscape technique also offers great insights into the nature of the supercooled liquid and glassy states.

Our paper is organized as follows. In Sect. 2 we provide an overview of the glass transition and describe some of its general features. We introduce the concept of an enthalpy landscape in Sect. 3 and present a model of the glass transition based on the enthalpy landscape approach. Several new computational techniques are required for implementation of this model. In Sect. 4 we provide an overview of three of these techniques: an eigenvector-following technique for locating minima and transition points in an enthalpy landscape, a self-consistent Monte Carlo method for computing the degeneracy of these states, and a master equation solver for accessing very long time scales. Finally, in Sect. 5 we discuss some the new understanding that the enthalpy landscape approach has brought to our knowledge of the supercooled liquid and glassy states. In particular, we discuss implications related to glassy entropy, fictive temperature, fragility, and the Kauzmann paradox.

2 The glass transition

The phenomenology of the glass transition can perhaps be best elucidated with the volume-temperature (V-T) diagram depicted in Fig. 1. Consider an equilibrium liquid at point a in the diagram. Upon cooling, the volume of the liquid generally decreases along the path abc. Point b occurs at the melting temperature T_m of the corresponding crystal. At this point, the liquid exhibits an infinitesimally small number of crystal nuclei. The degree of crystallization is governed by nucleation and crystal growth rates in the liquid. In this context, a "nucleus" refers to a precursor to a crystal that lacks a recognizable growth pattern. As shown in Fig. 2, the rates of nucleation and crystal growth are both zero at T_m and in the limit of low temperature. The maximum nucleation and crystal growth rates occur slightly below T_m, corresponding to point c in Fig. 1. If crystallization occurs, the system undergoes a phase change from point c to the crystal line in the V-T diagram. Subsequent cooling of the crystal generally results in a decrease in volume along the path de.

The shaded region around point c in Fig. 1 corresponds to the temperature region where perceptible crystallization can occur. If a liquid is cooled quickly enough through this temperature range to avoid a phase change, it moves into the "supercooled liquid" state along the path bcf. The minimum cooling rate required to avoid crystallization can be determined using a time-temperature-transformation (T-T-T) diagram, as depicted in Fig. 3. The solid "transformation" curve in this figure represents the locus of all points in the temperature-time plane which yield a given crystal concentration (typically 10^{-6} volume fraction) which serves as a threshold for crystal detection. Given a liquid at T_m, the minimum cooling rate to avoid crystallization is determined by the slope of the line which just touches the transformation curve (the dashed line in Fig. 3).

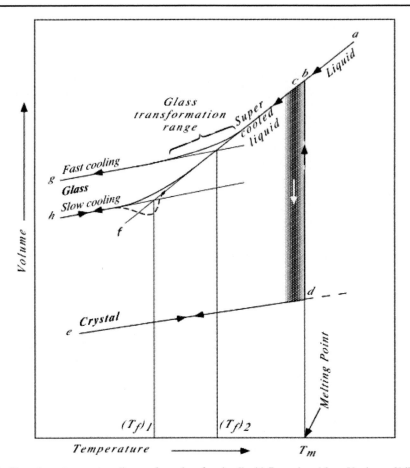

Fig. 1 The volume-temperature diagram for a glass-forming liquid. Reproduced from Varshneya [16]

As a supercooled liquid continues to cool, the molecular motion slows down sufficiently such that the system departs from equilibrium. In other words, the molecules do not have sufficient time to rearrange themselves into the volume characteristic of that temperature and pressure. Further cooling results in a gradual solidification of the system, and the resulting material is a called a "glass." Glass is remarkable in that it has a liquid-like structure but with solid-like behavior. As shown in Fig. 1, the final volume of a glass depends on the cooling rate. A faster cooling rate generally results in a higher volume since the molecules are given less time to relax into a lower energy structure before the onset of viscous arrest.

The smooth curve between the supercooled liquid and glassy regions in Fig. 1 is termed the "glass transition" or "glass transformation" range. It should be emphasized that the transition from the supercooled liquid to glassy states does not occur at a single, well-defined temperature. Rather, the change is gradual and occurs over a range of temperatures. This temperature range depends on both the material under study and the particular cooling path. In other words, glass transition range behavior, and hence the properties of the final glass itself, depend on both composition and thermal history.

Another interesting behavior of glass is that it never retraces its cooling path in the transition range upon reheating. Figure 4 shows a variety of reheating curves for different cooling

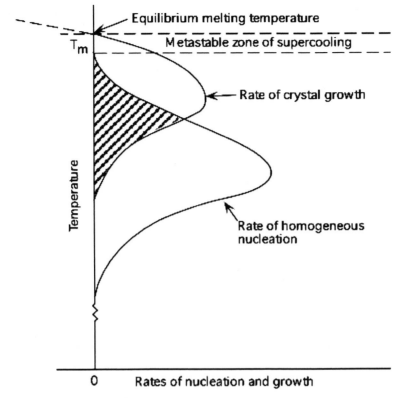

Fig. 2 Nucleation and crystal growth rates with respect to temperature. Reproduced from Varshneya [16]

and reheating rates; many glasses can actually show a decrease in volume as they are reheated through the transition range.

The glass transition is not a phase transition in the thermodynamic sense. Whereas the crystallization of a liquid results in a discontinuity in first-order thermodynamic variables (such as volume and enthalpy), there is no such discontinuity in the case of a glass transition. Instead, the glass transition often involves a rather sharp change in second-order thermodynamic variables such as heat capacity and thermal expansion coefficient. Figure 5 shows the heat capacity of glycerol as it undergoes a glass transition. This change in heat capacity is dramatic, but still not discontinuous. This observation has led some researchers to postulate an "ideal glass transition," in which a system undergoes a discontinuity in second-order thermodynamic variables. Many thermodynamic models of glass assume an ideal glass transition [9], but none has ever been observed in experiment.

Glass is a nonequilibrium, nonergodic state of matter. In fact, the only reason we observe the existence of glass at all is that the relaxation dynamics in glass are much slower than our observation time scale. While glass appears as a rigid, solid material to us humans, on a geologic time scale it is actually relaxing toward the supercooled liquid state. Hence, glass is unstable with respect to the supercooled liquid state. The supercooled liquid is, in turn, metastable with respect to the equilibrium crystal.

Finally, we should comment on the difference between a glass and an amorphous solid. Many researchers use these terms interchangeably, as they are both types of non-crystalline

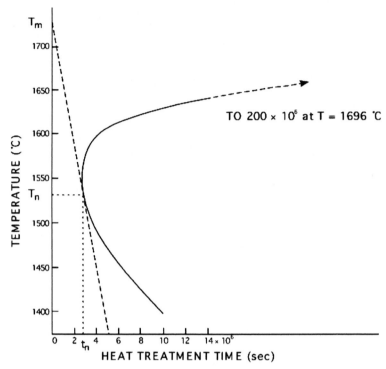

Fig. 3 Time-temperature-transformation (T-T-T) diagram for silica using a crystal volume fraction of 10^{-6}. Reproduced from Varshneya [16]

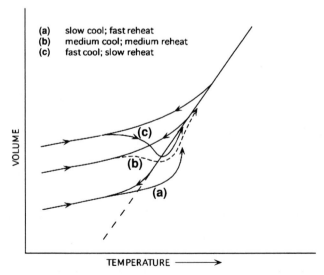

Fig. 4 Various glass reheating curves. Reproduced from Varshneya [16]

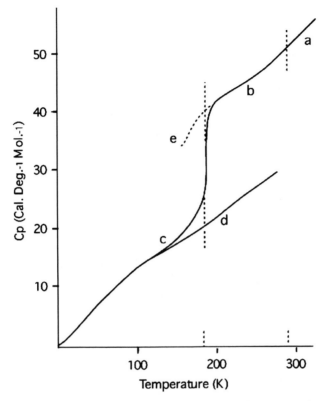

Fig. 5 Heat capacity of glycerol: (a) liquid, (b) supercooled liquid, (c) glass, (d) crystal, and (e) very slowly cooled supercooled liquid. Reproduced from Varshneya [16]

solids. However, glasses and amorphous solids are two thermodynamically distinct classes of material. The short-range order (i.e., the local environment around the atoms) in a glass is the same as that in the corresponding liquid; the short-range order of an amorphous solid is different from that of the liquid. For this reason, an amorphous solid cannot be formed by cooling from a melt. (Amorphous silicon, for example, is prepared by a vapor deposition process.) Furthermore, glasses and amorphous solids exhibit different behaviors upon heating. Whereas a glass undergoes a gradual transition to the liquid state upon heating, an amorphous solid will crystallize, sometimes explosively. Interestingly, glassy silicon has not yet been experimentally realized: liquid silicon is 12-coordinated and always crystallizes on cooling, even for the fastest experimentally realizable cooling rates. Figure 6 illustrates schematically the relative stability of the glassy, supercooled liquid, amorphous, and crystalline states. In this report we are only concerned with glasses and supercooled liquids, not amorphous solids. For a detailed discussion of the differences between a glass and an amorphous solid, we refer the interested reader to the work of Gupta [17].

3 The enthalpy landscape approach

In a pioneering 1969 paper, Goldstein [18] proposed the idea that atomic motion in a supercooled liquid consists of high-frequency vibrations in regions of deep potential energy

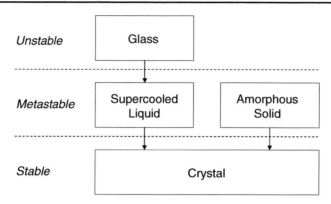

Fig. 6 Schematic diagram illustrating relative thermodynamic stability. The glassy state is unstable with respect to the supercooled liquid, and the supercooled liquid is metastable with respect to the crystal. The amorphous state is metastable with respect to the crystal and not connected to the liquidus path [16]

minima, with less frequent transitions to other such minima. The transport properties of the supercooled liquid, he argued, are linked to the ability of the atoms to flow among these various minima. This idea was subsequently extended in a series of papers by Stillinger and coworkers [19–24]. In this section, we review the concept of potential energy and enthalpy landscapes and present a nonequilibrium statistical mechanical approach for modeling the dynamics of glass-forming systems under isobaric conditions.

3.1 Potential energy landscapes

The potential energy of a system of N interacting particles can be written as

$$U = U\left(\mathbf{r}_1, \mathbf{r}_2, \ldots, \mathbf{r}_N\right) \geq CN, \tag{1}$$

where $\mathbf{r}_1, \mathbf{r}_2, \ldots, \mathbf{r}_N$ are the particle position vectors and C is a constant giving the energy per particle for a perfect crystal at absolute zero. There is no upper bound on potential energy, owing to the high repulsive energy that exists at small separation distances. The potential energy $U\left(\mathbf{r}_1, \mathbf{r}_2, \ldots, \mathbf{r}_N\right)$ is continuous and at least twice differentiable with respect of the configurational coordinates $\mathbf{r}_1, \mathbf{r}_2, \ldots, \mathbf{r}_N$. The partition function of the potential energy landscape is therefore

$$Q = \int_0^L \cdots \int_0^L \int_0^L \exp\left[-\frac{U\left(\mathbf{r}_1, \mathbf{r}_2, \ldots, \mathbf{r}_N\right)}{kT}\right] d^3\mathbf{r}_1 d^3\mathbf{r}_2 \cdots d^3\mathbf{r}_N, \tag{2}$$

where k is Boltzmann's constant, T is absolute temperautre, and L is the length of the simulation cell, which here we assume to be cubic.

In effect, $U\left(\mathbf{r}_1, \mathbf{r}_2, \ldots, \mathbf{r}_N\right)$ represents a continuous multidimensional potential energy "landscape" with many peaks and valleys. The local minima in the potential energy landscape correspond to mechanically stable arrangements of the system's particles. These stable configurations are termed "inherent structures." The number Ω of inherent structures in the limit of large N is given by

$$\ln \Omega \approx \ln\left(N! \sigma^N\right) + aN, \tag{3}$$

where σ is a symmetry factor and $a > 0$ is a constant relating to the number density N/L^3. The first term on the right-hand side of Eq. (3) accounts for the symmetry of the potential energy landscape with respect to $\mathbf{r}_1, \mathbf{r}_2, \ldots, \mathbf{r}_N$, and the second term expresses the exponential increase in the number of distinguishable inherent structures with increasing N [19–24].

In the Stillinger approach, the potential energy landscape is divided into a discrete set of "basins," where each basin contains a single minimum in U, i.e., a single inherent structure. A basin itself is defined to be the set of all coordinates in the $3N$-dimensional configuration space that drains to a particular minimum via steepest descent. With this approach, we can rewrite the partition function as a summation of integrals over each of the individual basins,

$$Q = \sum_{\alpha=1}^{\Omega} \int_{\{\mathbf{r}|\mathbf{r} \in \mathbf{R}_\alpha\}} \exp\left[-\frac{U(\mathbf{r})}{kT}\right] d^{3N}\mathbf{r}, \tag{4}$$

where $d^{3N}\mathbf{r} = d^3\mathbf{r}_1 d^3\mathbf{r}_2 \cdots d^3\mathbf{r}_N$ and $\mathbf{R}_\alpha$ denotes the set of all position vectors in basin α. If we denote the particular configuration of inherent structure α as $\mathbf{r}_\alpha$, the partition function can be rewritten as

$$Q = \sum_{\alpha=1}^{\Omega} \int_{\{\mathbf{r}|\mathbf{r} \in \mathbf{R}_\alpha\}} \exp\left[-\frac{U(\mathbf{r}_\alpha) + \Delta U_\alpha(\mathbf{r})}{kT}\right] d^{3N}\mathbf{r}, \tag{5}$$

where $\Delta U_\alpha(\mathbf{r})$ gives the increase in potential energy at any point in basin α relative to the inherent structure potential, $U(\mathbf{r}_\alpha)$, as shown schematically in Fig. 7. Since $U(\mathbf{r}_\alpha)$ is a constant for any given α, Q becomes

$$Q = \sum_{\alpha=1}^{\Omega} \exp\left[-\frac{U(\mathbf{r}_\alpha)}{kT}\right] \int_{\{\mathbf{r}|\mathbf{r} \in \mathbf{R}_\alpha\}} \exp\left[-\frac{\Delta U_\alpha(\mathbf{r})}{kT}\right] d^{3N}\mathbf{r}. \tag{6}$$

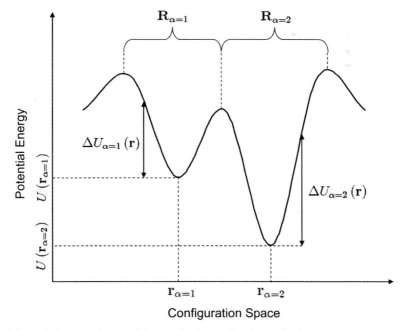

Fig. 7 Schematic diagram of a potential energy landscape showing two basins

We now introduce the normalized coordinates $s_1, s_2, \ldots, s_N$, defined by

$$s = \frac{r}{B_\alpha^{1/3N}},$$ (7)

where the quantity B_α is the volume of basin α in $3N$-dimensional space, given by the integral

$$B_\alpha = \int_{\{r|r \in R_\alpha\}} d^{3N} r.$$ (8)

Since $B_\alpha^{1/3N}$ has units of length, the normalized coordinates s are dimensionless. With this definition of s, the partition function becomes

$$Q = \sum_{\alpha=1}^{\Omega} B_\alpha C_\alpha \exp\left[-\frac{U(r_\alpha)}{kT}\right],$$ (9)

where the integral

$$C_\alpha = \int_{\{s|s \in S_\alpha\}} \exp\left[-\frac{\Delta U_\alpha(s)}{kT}\right] d^{3N} s$$ (10)

depends only on the shape of basin α and not its volume in $3N$-dimensional space. Here, the integration is over the scaled set of position vectors S_α in basin α.

The fundamental assumption of the energy landscape approach is that the partition function Q can be separated into independent configurational and vibrational contributions:

$$Q = Q_{conf} Q_{vib}.$$ (11)

Hence, the energy landscape approach has an implicit (and often overlooked) assumption that the normalized basin shape is a constant ($C_\alpha = C$). In this manner, the configurational and vibrational contributions to the partition function can be written independently as

$$Q_{conf} = \sum_{\alpha=1}^{\Omega} B_\alpha \exp\left[-\frac{U(r_\alpha)}{kT}\right]$$ (12)

and

$$Q_{vib} = \int_{\{s|s \in S\}} \exp\left[-\frac{\Delta U(s)}{kT}\right] d^{3N} s.$$ (13)

Please note that while the basin volume B_α can vary by many orders of magnitude [25], it is quite reasonable to assume that the normalized basin *shape* is a constant.[2] With this assumption, the fast vibrations within a basin are independent of the slower inter-basin transitions. With the partition function written in the separable form of Eq. (11), it follows that all equilibrium thermodynamic properties can be written in terms of separate configurational and vibrational contributions.

---

[2] Each basin must contain a single minimum, and the edges of the basin are a locus of points giving the transition barriers to adjacent basins. The potential energy varies smoothly between the minimum and each of the transition barriers.

3.2 Enthalpy landscapes

While potential energy landscapes are suitable for modeling glass transition range behavior under isochoric conditions, almost all laboratory glass formation occurs under constant pressure and not constant volume conditions. Here we extend our discussion of the previous section to the isothermal-isobaric ensemble, which allows for changes in both particle positions as well as the overall volume of the system. In the isothermal-isobaric ensemble, we work with an enthalpy landscape rather than a potential energy landscape. The enthalpy landscape at zero temperature (i.e., with zero kinetic energy) corresponds to an underlying surface that is sampled by a system at finite temperature under isobaric conditions.

The zero temperature enthalpy landscape of a system of N atoms can be expressed as

$$H = U (\mathbf{r}_1, \mathbf{r}_2, \ldots, \mathbf{r}_N, V) + PV, \tag{14}$$

where V is the volume of the system and the pressure P is constant. The isothermal-isobaric partition function can be written as

$$Y = \int_0^\infty \int_0^{V^{1/3}} \cdots \int_0^{V^{1/3}} \int_0^{V^{1/3}} \exp\left[-\frac{H (\mathbf{r}_1, \mathbf{r}_2, \ldots, \mathbf{r}_N, V)}{kT}\right] d^3\mathbf{r}_1 d^3\mathbf{r}_2 \cdots d^3\mathbf{r}_N dV, \tag{15}$$

where the system is again assumed to be cubic. It is useful to divide the partition function into two separate integrals for volumes below and above $V_{\max}$, which we define to be the volume of the system at which the interaction potentials no longer contribute significantly to the overall enthalpy. In other words,

$$U (\mathbf{r}_1, \mathbf{r}_2, \ldots, \mathbf{r}_N, V_{\max}) \ll PV_{\max}, \tag{16}$$

such that U can be safely ignored for $V \geq V_{\max}$. The partition function can then be written as a sum of interacting and non-interacting contributions:

$$Y = Y_I + Y_{NI}, \tag{17}$$

where

$$Y_I = \int_0^{V_{\max}} \int_0^{V^{1/3}} \exp\left[-\frac{H (\mathbf{r}_1, \mathbf{r}_2, \ldots, \mathbf{r}_N, V)}{kT}\right] d^{3N}\mathbf{r}\, dV \tag{18}$$

and

$$Y_{NI} = \int_{V_{\max}}^\infty \int_0^{V^{1/3}} \exp\left[-\frac{PV}{kT}\right] d^{3N}\mathbf{r}\, dV. \tag{19}$$

The non-interacting integral reduces to a constant in the form of an incomplete gamma function:

$$Y_{NI} = \int_{V_{\max}}^\infty V^N \exp\left[-\frac{PV}{kT}\right] dV \tag{20}$$

$$= \left(\frac{kT}{P}\right)^{N+1} \Gamma\left(N + 1, \frac{PV_{\max}}{kT}\right). \tag{21}$$

Since the enthalpy landscape in the non-interacting regime is strictly a linear function of volume, there are no minima in the landscape and hence no inherent structures. As a result, we need only consider the interacting portion of the partition function.

It is convenient to rewrite the interacting partition function in terms of the length of the simulation cell, $L = V^{1/3}$, such that all coordinates have units of length:

$$Y_I = \int_0^{L_{max}} 3L^2 \int_0^L \exp\left[-\frac{H(\mathbf{r}, L)}{kT}\right] d^{3N}\mathbf{r}dL. \tag{22}$$

Following the approach in the previous section, we divide the continuous enthalpy landscape into a discrete set of basins; each basin contains a single minimum in H with respect to all of the $3N + 1$ coordinates. We can thus rewrite the interacting partition function as a summation of integrals over each of the individual basins,

$$Y_I = \sum_{\alpha=1}^{\Omega} \int_{\{L|L\in\mathcal{L}_\alpha\}} 3L^2 \int_{\{\mathbf{r}|\mathbf{r}\in\mathbf{R}_\alpha(L)\}} \exp\left[-\frac{H(\mathbf{r}, L)}{kT}\right] d^{3N}\mathbf{r}dL, \tag{23}$$

where the set of position vectors $\mathbf{R}_\alpha(L)$ in basin α is a function of L, and $\mathcal{L}_\alpha$ is the set of length values in the basin α. If we denote the particular coordinates of inherent structure α as $\{\mathbf{r}_\alpha, L_\alpha\}$, the partition function can be rewritten as

$$Y_I = \sum_{\alpha=1}^{\Omega} \exp\left[-\frac{H(\mathbf{r}_\alpha, L_\alpha)}{kT}\right] \int_{\{L|L\in\mathcal{L}_\alpha\}} 3L^2 \int_{\{\mathbf{r}|\mathbf{r}\in\mathbf{R}_\alpha(L)\}} \exp\left[-\frac{\Delta H_\alpha(\mathbf{r}, L)}{kT}\right] d^{3N}\mathbf{r}dL, \tag{24}$$

where $\Delta H_\alpha(\mathbf{r}, L)$ gives the increase in enthalpy at any point in basin α relative to the inherent structure enthalpy, $H(\mathbf{r}_\alpha, L_\alpha)$. The volume of basin α in $(3N + 1)$-dimensional space is given by

$$\mathcal{B}_\alpha = \int_{\{L|L\in\mathcal{L}_\alpha\}} \int_{\{\mathbf{r}|\mathbf{r}\in\mathbf{R}_\alpha(L)\}} d^{3N}\mathbf{r}dL \tag{25}$$

We now introduce the normalized positions $\tilde{\mathbf{s}}_1, \tilde{\mathbf{s}}_2, \ldots, \tilde{\mathbf{s}}_N$, defined by

$$\tilde{\mathbf{s}} = \frac{\mathbf{r}}{\mathcal{B}_\alpha^{1/(3N+1)}}, \tag{26}$$

and a normalized length, defined similarly as

$$\tilde{s}_L = \frac{L}{\mathcal{B}_\alpha^{1/(3N+1)}}. \tag{27}$$

With these normalized coordinates, the partition function becomes

$$Y_I = \sum_{\alpha=1}^{\Omega} \mathcal{B}_\alpha^{1+2/(3N+1)} \tilde{C}_\alpha \exp\left[-\frac{H(\mathbf{r}_\alpha, L_\alpha)}{kT}\right], \tag{28}$$

where the integral

$$\tilde{C}_\alpha = 3 \int_{\{\tilde{s}_L|\tilde{s}_L\in\tilde{\mathcal{L}}_\alpha\}} \tilde{s}_L^2 \int_{\{\tilde{\mathbf{s}}|\tilde{\mathbf{s}}\in\tilde{\mathbf{S}}_\alpha(\tilde{s}_L)\}} \exp\left[-\frac{\Delta H_\alpha(\tilde{\mathbf{s}}, \tilde{s}_L)}{kT}\right] d^{3N}\tilde{\mathbf{s}}d\tilde{s}_L \tag{29}$$

depends only on the shape of basin α and not on its volume in $(3N + 1)$-dimensional space. Again, we wish to write the partition function in terms of separate configurational and vibrational contributions,

$$Y_I = Y_{conf}Y_{vib}, \tag{30}$$

which can be accomplished by assuming a constant normalized basin shape ($\tilde{C}_\alpha = \tilde{C}$). With this assumption,

$$Y_{conf} = \sum_{\alpha=1}^{\Omega} \mathcal{B}_\alpha^{1+2/(3N+1)} \exp\left[-\frac{H(\mathbf{r}_\alpha, L_\alpha)}{kT}\right] \qquad (31)$$

and

$$Y_{vib} = 3 \int_{\{\tilde{s}_L | \tilde{s}_L \in \tilde{\mathcal{L}}_\alpha\}} \tilde{s}_L^2 \int_{\{\tilde{s} | \tilde{s} \in \tilde{S}(\tilde{s}_L)\}} \exp\left[-\frac{\Delta H(\tilde{s}, \tilde{s}_L)}{kT}\right] d^{3N}\tilde{s} d\tilde{s}_L. \qquad (32)$$

3.3 Nonequilibrium statistical mechanics

Using the partition functions of Eqs. (11) and (30), we can compute all the equilibrium thermodynamic properties of a system. However, study of the glass transition necessarily involves a departure from equilibrium. A recent model by Mauro and Varshneya [26] combines the inherent structure approach with nonequilibrium statistical mechanics in order to follow the evolution of an arbitrary macroscopic property as a glass-forming system departs from its equilibrium state. Here we present the Mauro-Varshneya model for isobaric conditions.

The dynamics of the system moving between pairs of basins involves the transition states (first-order saddle points) between adjacent basins. The underlying assumption is that while landscape itself is independent of temperature, the way in which the system samples the landscape depends on its thermal energy, and thus the temperature of the system. At high temperatures, the system has ample thermal energy to flow freely between basins; this corresponds to the case of an equilibrium liquid. As the liquid is supercooled, the number of available basin transitions decreases owing to the loss of kinetic energy. Finally, the glassy state at low temperatures corresponds to the system getting "stuck" in a single region of the landscape where the energy barrier for a transition is too high to overcome in laboratory time scales.

If we consider a total of Ω basins in the enthalpy landscape, we can construct an $\Omega \times \Omega$ enthalpy matrix,

$$\mathbf{H} = \begin{pmatrix} H_{11} & H_{12} & \cdots & H_{1\Omega} \\ H_{21} & H_{22} & \cdots & H_{2\Omega} \\ \vdots & \vdots & \ddots & \vdots \\ H_{\Omega 1} & H_{\Omega 2} & \cdots & H_{\Omega\Omega} \end{pmatrix}, \qquad (33)$$

where the diagonal elements ($H_{\alpha\alpha}$) are the inherent structure enthalpies and the off-diagonal elements ($H_{\alpha\beta}$, $\alpha \neq \beta$) are the transition enthalpies from basin α to basin β. The matrix $\mathbf{H}$ is symmetric by construction ($H_{\alpha\beta} = H_{\beta\alpha}$).

In order to capture all "memory" effects that can occur in a glassy system, we have the initial condition of an equilibrium liquid at the melting temperature T_m. From equilibrium statistical mechanics, the initial phase space distribution for any basin α is

$$\left(p_{eq}\right)_\alpha = \frac{1}{Y_{conf}} g_\alpha \exp\left(-\frac{H_{\alpha\alpha}}{kT_m}\right), \qquad (34)$$

where g_α is the degeneracy of basin α, weighted by the basin volume factor $\mathcal{B}_\alpha^{1+2/(3N+1)}$, as in Eq. (31).

As we cool the equilibrium liquid into the supercooled liquid and glassy regimes, the structural relaxation time exceeds the experimental measurement time. We are interested in calculating the phase space distribution of the system as it relaxes, which can be accomplished by writing a master equation for each basin,

$$\frac{d}{dt} p_\alpha(t) = \sum_{\beta \neq \alpha}^{\Omega} K_{\beta\alpha}[T(t)] p_\beta(t) - \sum_{\beta \neq \alpha}^{\Omega} K_{\alpha\beta}[T(t)] p(t), \tag{35}$$

The probabilities satisfy

$$\sum_{\alpha=1}^{\Omega} p_\alpha(t) = 1 \tag{36}$$

for all times t. The rate parameters $K_{\alpha\beta,\beta\alpha}$ are defined parametrically in terms of an arbitrary cooling path, $T(t)$, and form a matrix:

$$\mathbf{K} = \begin{pmatrix} 0 & K_{12} & K_{13} & \cdots & K_{1\Omega} \\ K_{21} & 0 & K_{23} & \cdots & K_{2\Omega} \\ K_{31} & K_{32} & 0 & \cdots & K_{3\Omega} \\ \vdots & \vdots & \vdots & \ddots & \vdots \\ K_{\Omega 1} & K_{\Omega 2} & K_{\Omega 3} & \cdots & 0 \end{pmatrix}. \tag{37}$$

Unlike $\mathbf{H}$, $\mathbf{K}$ is not a symmetric matrix ($K_{\alpha\beta} \neq K_{\beta\alpha}$).

The set of Ω master equations has been defined without any reference to the underlying kinetic model. If we assume the system is always in a state of local equilibrium, the rate of transition from basin α to basin β can be approximated using transition state theory:

$$K_{\alpha\beta}[T(t)] \approx \nu_{\alpha\beta} g_\beta \exp\left[-\frac{(H_{\alpha\beta} - H_{\alpha\alpha})}{kT(t)}\right], \tag{38}$$

where $\nu_{\alpha\beta}$ is the attempt frequency [27,28]. This equation also assumes that there is a single dominant transition barrier between basins α and β. Of course, a more accurate model of $K_{\alpha\beta}$ may be substituted for Eq. (38) without changing the underlying formulation of Eq. (35). The volume of the system can be computed at any time t using the phase space average,

$$V(t) = \sum_{\alpha=1}^{\Omega} V_\alpha p_\alpha(t), \tag{39}$$

where V_α is the volume associated with inherent structure α. Note that this formulation ignores any perturburbation in volume caused by asymmetry of the enthalpy basin along the volume dimension.

4 Simulation techniques

In order to implement the enthalpy landscape approach of Sect. 3, we must have simulation techniques to:

- Map the continuous enthalpy landscape to a discrete set of inherent structures and transition points. This allows construction of the enthalpy matrix in Eq. (33).
- Calculate the inherent structure density of states, which provides us with the degeneracy factors in Eqs. (34) and (38).

– Solve the system of Ω master equations in Eq. (35) for an arbitrary temperature path T (t). This allows for calculation of the evolution of macroscopic properties such as volume and enthalpy on an arbitrary time scale.

In this section, we review simulation techniques for addressing each of these three aspects of the problem. First, we review a split-step eigenvector-following technique for locating inherent structures and transition points in an enthalpy landscape. Next, we provide an overview of a self-consistent Monte Carlo technique for computing inherent structure density of states. Finally, we describe a metabasin technique for solving a large system of master equations on an arbitrary time scale.

4.1 Locating inherent structures and transition points

Several techniques exist for locating inherent structures and transition points in a potential energy landscape [29], and many of these can be extended to isobaric conditions to enable the mapping of enthalpy landscapes. One method of particular interest for mapping enthalpy landscapes is the split-step eigenvector-following technique [30, 31], which includes two steps at each iteration to vary independently the system volume and relative atomic positions. Let us consider the zero temperature enthalpy landscape of an N-particle system:

$$H = U (x_1, x_2, \ldots, x_{3N}, L) + PL^3, \tag{40}$$

where the potential energy U is a function of $3N$ position coordinates, $x_1, x_2, \ldots, x_{3N}$, and the length L of the simulation cell. The pressure P of the system is constant, and we assume a cubic cell volume of $V = L^3$. The enthalpy landscape therefore has a dimensionality of $3N + 1$, minus any constraints.

The eigenvector-following technique proceeds by stepping iteratively through the enthalpy landscape towards either a minimum or a transition point. Let us denote the enthalpy of the system with initial positions x_i^0, where $i = 1, 2, \ldots, 3N$, and initial length L^0 as $H\left(x_i^0, L^0\right)$. We can approximate the enthalpy at a new position $x_i = x_i^0 + h_i$ and new length $L = L^0 + h_L$ using the Taylor series expansion,

$$\begin{aligned}
H\left(x_i, L\right) \approx H\left(x_i^0, L^0\right) &+ \sum_{i=1}^{3N} h_i \left.\frac{\partial H}{\partial x_i}\right|_{x_i=x_i^0, L=L^0} + h_L \left.\frac{\partial H}{\partial L}\right|_{x_i=x_i^0, L=L^0} \\
&+ \frac{1}{2} \sum_{i=1}^{3N} \sum_{j=1}^{3N} h_i h_j \left.\frac{\partial^2 H}{\partial x_i \partial x_j}\right|_{x_{i,j}=x_{i,j}^0, L=L^0} + \frac{1}{2} h_L^2 \left.\frac{\partial^2 H}{\partial L^2}\right|_{x_i=x_i^0, L=L^0} \\
&+ \frac{1}{2} \sum_{i=1}^{3N} h_i h_L \left.\frac{\partial^2 H}{\partial x_i \partial L}\right|_{x_i=x_i^0, L=L^0} + \frac{1}{2} \sum_{i=1}^{3N} h_L h_i \left.\frac{\partial^2 H}{\partial L \partial x_i}\right|_{x_i=x_i^0, L=L^0},
\end{aligned} \tag{41}$$

which can be written in matrix notation as

$$H\left(\mathbf{q}\right) \approx H\left(\mathbf{q}^0\right) + \mathbf{g}^\top \mathbf{h} + \frac{1}{2}\mathbf{h}^\top \mathbf{H}\mathbf{h}. \tag{42}$$

The position vectors are given by

$$
\mathbf{q} = \begin{pmatrix} x_1 \\ x_2 \\ \vdots \\ x_{3N} \\ L \end{pmatrix} ; \mathbf{q}^0 = \begin{pmatrix} x_1^0 \\ x_2^0 \\ \vdots \\ x_{3N}^0 \\ L^0 \end{pmatrix} \tag{43}
$$

and the displacement vector $\mathbf{h} = \mathbf{q} - \mathbf{q}^0$ is

$$
\mathbf{h} = \begin{pmatrix} h_1 \\ h_2 \\ \vdots \\ h_{3N} \\ h_L \end{pmatrix}. \tag{44}
$$

The gradient vector $\mathbf{g}$ and the $(3N + 1) \times (3N + 1)$ Hessian matrix $\mathcal{H}$, evaluated at $\mathbf{q} = \mathbf{q}^0$, are given by

$$
\mathbf{g} = \begin{pmatrix} \dfrac{\partial H}{\partial x_1} \\[2mm] \dfrac{\partial H}{\partial x_2} \\[2mm] \vdots \\[2mm] \dfrac{\partial H}{\partial x_{3N}} \\[2mm] \dfrac{\partial H}{\partial L} \end{pmatrix}_{\mathbf{q}=\mathbf{q}^0} \tag{45}
$$

and

$$
\mathcal{H} = \begin{pmatrix} \dfrac{\partial^2 H}{\partial x_1^2} & \dfrac{\partial^2 H}{\partial x_1 \partial x_2} & \cdots & \dfrac{\partial^2 H}{\partial x_1 \partial x_{3N}} & \dfrac{\partial^2 H}{\partial x_1 \partial L} \\[3mm] \dfrac{\partial^2 H}{\partial x_2 \partial x_1} & \dfrac{\partial^2 H}{\partial x_2^2} & \cdots & \dfrac{\partial^2 H}{\partial x_2 \partial x_{3N}} & \dfrac{\partial^2 H}{\partial x_2 \partial L} \\[3mm] \vdots & \vdots & \ddots & \vdots & \vdots \\[3mm] \dfrac{\partial^2 H}{\partial x_{3N} \partial x_1} & \dfrac{\partial^2 H}{\partial x_{3N} \partial x_2} & \cdots & \dfrac{\partial^2 H}{\partial x_{3N}^2} & \dfrac{\partial^2 H}{\partial x_{3N} \partial L} \\[3mm] \dfrac{\partial^2 H}{\partial L \partial x_1} & \dfrac{\partial^2 H}{\partial L \partial x_2} & \cdots & \dfrac{\partial^2 H}{\partial L \partial x_{3N}} & \dfrac{\partial^2 H}{\partial L^2} \end{pmatrix}_{\mathbf{q}=\mathbf{q}^0} , \tag{46}
$$

respectively.

For potential energy landscapes [32], the Hessian matrix would be symmetric since it would contain second derivatives with respect to position coordinates x_i only. However, for an enthalpy landscape, the Hessian matrix has one additional row and column including derivatives with respect to the simulation cell length L. When undergoing a change in the length of simulation cell, the atomic positions scale according to

$$\frac{\partial x_i}{\partial L} = \frac{x_i}{L}, \tag{47}$$

since any change in length produces a corresponding "stretching" of atomic positions. This relationship leads to

$$\frac{\partial^2 H}{\partial x_i \partial L} = \frac{\partial^2 H}{\partial L \partial x_i} + \frac{1}{L}\frac{\partial H}{\partial x_i}, \tag{48}$$

and therefore the Hessian matrix $\mathcal{H}$ for an enthalpy landscape is not symmetric. This poses a problem when computing the eigenvalues and eigenvectors of $\mathcal{H}$, since an asymmetric matrix can lead to complex eigenvalues. This problem can be overcome by introducing the normalized particle coordinates,

$$\bar{x}_i = \frac{x_i}{L}, \tag{49}$$

which are independent of changes in L. The split-step eigenvector-following technique consists of iteratively stepping through the enthalpy landscape toward a minimum or transition point, where each iteration involves two steps:

1. Step of the simulation box length L while maintaining constant normalized positions $\bar{x}_i$.
2. Step of the positions x_i while maintaining constant L.

In the first step of the iteration, the enthalpy can be written in terms of a Taylor series expansion:

$$H\left(\bar{x}_i^0, L\right) \approx H\left(\bar{x}_i^0, L^0\right) + h_L \left.\frac{\partial H}{\partial L}\right|_{\bar{x}_i=\bar{x}_i^0, L=L^0} + \frac{1}{2}h_L^2 \left.\frac{\partial^2 H}{\partial L^2}\right|_{\bar{x}_i=\bar{x}_i^0, L=L^0}. \tag{50}$$

We can now write a Lagrange function in one dimension,

$$\mathcal{L}_L = -H\left(\bar{x}_i^0, L^0\right) - h_L \left.\frac{\partial H}{\partial L}\right|_{\bar{x}_i=\bar{x}_i^0, L=L^0}$$
$$- \frac{1}{2}h_L^2 \left.\frac{\partial^2 H}{\partial L^2}\right|_{\bar{x}_i=\bar{x}_i^0, L=L^0} + \frac{1}{2}\lambda_L\left(h_L^2 - c_L^2\right), \tag{51}$$

where λ_L is a Lagrange multiplier. Taking the derivative with respect to h_L yields

$$\frac{\partial \mathcal{L}_L}{\partial h_L} = 0 = -\left.\frac{\partial H}{\partial L}\right|_{\bar{x}_i=\bar{x}_i^0, L=L^0} - h_L \left.\frac{\partial^2 H}{\partial L^2}\right|_{\bar{x}_i=\bar{x}_i^0, L=L^0} + \lambda_L h_L. \tag{52}$$

Defining

$$F_L = \left.\frac{\partial H}{\partial L}\right|_{\bar{x}_i=\bar{x}_i^0, L=L^0} = \left(\frac{\partial E}{\partial L} + 3PL^2\right)_{\bar{x}_i=\bar{x}_i^0, L=L^0} \tag{53}$$

and

$$b_L = \left.\frac{\partial^2 H}{\partial L^2}\right|_{\bar{x}_i=\bar{x}_i^0, L=L^0} = \left(\frac{\partial^2 E}{\partial L^2} + 6PL\right)_{\bar{x}_i=\bar{x}_i^0, L=L^0}, \tag{54}$$

we have

$$h_L = \frac{F_L}{\lambda_L - b_L}. \tag{55}$$

The change in enthalpy ΔH_L for such a step h_L is

$$\Delta H_L = \frac{F_L^2 \left(\lambda_L - \frac{b_L}{2}\right)}{(\lambda_L - b_L)^2}. \tag{56}$$

Hence, the sign of the enthalpy change depends on both b_L and the choice of Lagrange multiplier λ_L.

The second step involves changes in the particle positions x_i with a fixed box length L. In this case, the gradient and second derivative terms reduce to

$$\frac{\partial H}{\partial x_i} = \frac{\partial U}{\partial x_i}; \quad \frac{\partial^2 H}{\partial x_i \partial x_j} = \frac{\partial^2 U}{\partial x_i \partial x_j}. \tag{57}$$

The position step vector is given by

$$\mathbf{h}_x = \sum_{i=1}^{3N} \frac{F_i}{\lambda_i - b_i} \mathbf{V}_i, \tag{58}$$

where b_i and $\mathbf{V}_i$ are the eigenvalues and associated eigenvectors of the symmetric $3N \times 3N$ Hessian matrix $\mathcal{H}'$,

$$\mathcal{H}' \mathbf{V}_i = b_i \mathbf{V}_i, \tag{59}$$

and F_i is defined by

$$\mathbf{g} = \sum_{i=1}^{3N} F_i \mathbf{V}_i. \tag{60}$$

Separate Lagrange multipliers λ_i are used for each eigendirection. The change in enthalpy associated with the position step vector $\mathbf{h}$ is

$$\Delta H_x = \Delta E_x = \sum_{i=1}^{3N} \frac{F_i^2 \left(\lambda_i - \frac{b_i}{2}\right)}{(\lambda_i - b_i)^2}. \tag{61}$$

Again, the sign of the enthalpy change in a particular eigendirection $\mathbf{V}_i$ depends on both the eigenvalue b_i and the choice of Lagrange multiplier λ_i.

In order to locate inherent structures and transition points in the enthalpy landscape, we must step iteratively through the landscape, making an appropriate choice of Lagrange multipliers λ_i at each step and for each eigendirection. Since an inherent structure is defined to be a minimum in the enthalpy landscape, when locating an inherent structure we need to choose λ_i values that minimize enthalpy along every eigendirection. As summarized in Fig. 8, the choice of λ_i depends on both the eigenvalue b_i and the gradient F_i along its eigenvector $\mathbf{V}_i$. The maximum step size along each eigenvector is denoted c_i in the figure, and F_{th} and b_{th} denote threshold values below which the slope and curvature are considered zero. By choosing the Lagrange multipliers according to Fig. 8, each step vector $\mathbf{h}$ will bring the system closer to a minimum in enthalpy. Convergence to an inherent structure is achieved when we have reached a minimum along all eigendirections.

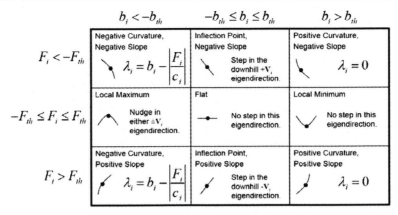

Fig. 8 Choice of Lagrange multipliers for minimization of enthalpy along a given eigendirection

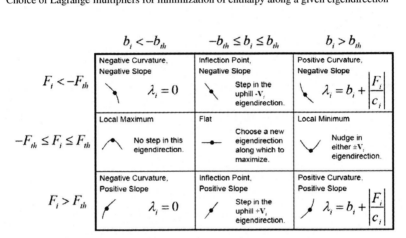

Fig. 9 Choice of Lagrange multipliers for maximization of enthalpy along a given eigendirection

Location of transition points requires a slightly different choice of Lagrange multipliers. Since a transition point is defined as a stationary point where exactly one eigenvalue b_i is negative, here we wish to maximize enthalpy along one eigendirection and minimize enthalpy along all other eigendirections. Figure 8 therefore provides the choice of Lagrange multipliers along all of these eigendirections. However, as we wish to maximize enthalpy along one eigendirection, the choice of Lagrange multipliers in this eigendirection is given in Fig. 9. This ensures enthalpy maximization along one eigendirection while simultaneously minimizing enthalpy along all other eigendirections.

Since many transition points are available from any given inherent structure, the eigenvector-following technique should be applied many times, starting from the same inherent structure but maximizing along different eigendirections. Whenever a transition point is found, the choice of Lagrange multipliers in Fig. 8 can then be used to find the connecting inherent structure on the other side of the transition point. The algorithm can then be applied recursively to locate new transition points from this inherent structure. In this manner, we can construct the enthalpy matrix of Eq. (33), required for simulation of glass transition range behavior.

4.2 Inherent structure density of states

In the previous section we have described a technique for locating inherent structures and transition points in an enthalpy landscape. However, eigenvector-following samples only a small portion of the very large $(3N + 1)$-dimensional enthalpy landscape. A separate calculation is required to ensure that each of these inherent structures and transition points occurs with the correct degeneracy factor. A modified version of the Wang-Landau technique for computing density of states [33] can be used to compute the inherent structure density of states. As described in detail by Mauro et al. [34], the following modifications need to be made to the Wang-Landau technique:

- Extend the technique to enthalpy landscapes, accounting for the coupling of volume variations with atomic displacements.
- Include a minimization after each trial displacement in order to compute the inherent structure density of states rather than that of the continuous landscape.
- Include importance sampling using a Boltzmann weighting factor to sample preferentially the lower enthalpy states, i.e., those that are more likely to be occupied. In other words, we can compute first an inherent structure quench probability, i.e., the probability of occupying a given inherent structure upon instantaneous quenching to absolute zero. The density of states can be computed subsequently from this quench probability.
- As required by Eq. (31), the probability of sampling a particular inherent structure must be weighted by volume of the corresponding basin in $(3N + 1)$-dimensional space.

The technique of Mauro et al. [34] is based on a self-consistent Monte Carlo approach where an initial guess of the inherent structure quench probability is updated as each new state visited with Monte Carlo sampling is either accepted or rejected. In a general form, the quench probability can be represented as a matrix, **g**, in two-dimensional enthalpy-volume space. In this manner, the density of states can be computed for inherent structures with all combinations of enthalpy and volume values. Depending on the system under study, the quench probability matrix could be reduced to a one-dimensional array in either enthalpy or volume space. Here we present the steps of the algorithm in terms of its more general matrix form:

1. Initialize the quench probability matrix, **g**, to unity. The matrix elements g_{ij} cover the full range of enthalpy and volume values of interest. The first index i covers the enthalpy space, and the second index j covers the volume space.
2. Choose a random displacement from a uniform distribution over the configuration space. (The probability of sampling a basin is directly proportional to the volume of that basin.) Any change in volume should be accompanied by a corresponding rescaling of the particle positions according to

$$\frac{\partial \mathbf{r}}{\partial V^{1/3}} = \frac{\mathbf{r}}{V^{1/3}}. \tag{62}$$

3. Each random displacement is followed by a minimization to the inherent structure configuration, which can be accomplished using the eigenvector-following technique of the previous section.
4. The trial displacement is accepted if

$$\text{rand}\,(0,\,1) \leq \frac{g_{ij}}{g_{kl}} \exp\left[-\frac{(H_k - H_i)}{kT}\right], \tag{63}$$

where H_i is the enthalpy of the initial inherent structure and H_k is the enthalpy of the new inherent structure. The matrix elements g_{ij} and g_{kl} correspond to the enthalpy and volume values of the initial and new inherent structures, respectively. T is the temperature of the simulation, and rand $(0, 1)$ is a random number drawn from a uniform distribution between 0 and 1.

5. If the new state is accepted, the quench probability matrix element g_{kl} (corresponding to the new state) is increased by some factor $A > 1$:

$$g_{kl}^{new} = A g_{kl}^{old}. \tag{64}$$

Following Wang and Landau [33], we use $A = e$.

6. If the new state is rejected, the system returns to the previous state and the quench probability matrix element g_{ij} (corresponding to the previous state) is increased by the same factor:

$$g_{ij}^{new} = A g_{ij}^{old}. \tag{65}$$

7. Repeat steps 2–6. The update factor A decreases throughout the simulation according to the procedure of Wang and Landau [33], and convergence is achieved as the update factor approaches unity.

8. The normalized inherent structure density of states, w^{IS}, can then be computed from the quench probabilities using

$$w^{IS}(H, V) = \frac{g(H, V; T) \exp(H/kT)}{\int g(H, V; T) \exp(H/kT) \, dH dV}. \tag{66}$$

4.3 Master equation dynamics

The master equation description of Eq. (35), coupled with the inherent structure and transition point data of the previous two sections, allows for modeling the dynamics of glass-forming systems from the liquid state through the glass transition regime and into the glassy state itself. However, given the exponentially large number of basins computed in Eq. (3), it is impossible to construct a separate master equation for each individual basin. An additional problem is that for most realistic glass-forming systems the transition rates can span over many orders of magnitude. Since the integration time step for solving the system of master equations is limited by the inverse of the fastest transition rate, direct integration of Eq. (35) is often highly inefficient. Moreover, an analytical solution for the system of the master equations must in general assume constant rate parameters; in Eq. (35), the rate parameters change with time as the initially liquid system cools to a glassy state.

Recently, we have described an efficient algorithm for computing the master equation dynamics of systems with highly degenerate states and disparate rate parameters [28] that allows for solution of a reduced set of master equations on the "natural" time scale of the experiment, i.e., the inverse of the cooling rate, rather than being limited by the fastest inter-basin transition rate. Our method consists of two steps: first, incorporate the degeneracy factors into the master equations, and second, partition the enthalpy landscape into "metabasins" and rewrite the master equations in terms of this reduced set of metabasins.

First, we incorporate degeneracy into the master equations. Consider a set of degenerate basins $A = \{1, 2, 3, 4, \ldots, g_A\}$ having the same enthalpy and volume, where g_A is the degeneracy. The total probability of occupying a basin in A is

$$P_A = \sum_{i \in A} p_i = g_A p_{i \in A}. \tag{67}$$

Here, $p_{i\in A}$ is the probability of occupying any of the individual basins i in A. We can write a reduced set of master equations for the degenerate states as

$$\frac{dP_A}{dt} = \sum_{B \neq A} \left(\tilde{K}_{BA} P_B - \tilde{K}_{AB} P_A \right),$$ (68)

where the effective transition rate from set A to set B is

$$\tilde{K}_{AB} = g_B K_{i\in A, j\in B}.$$ (69)

This equation assumes that there is one dominant transition path between any pair of basins.

In the second step, we account for the disparate rate parameters by partitioning the enthalpy landscape into metabasins, which are chosen to satisfy the following two criteria:

1. The relaxation time scale within a metabasin is short compared to the inverse of the cooling rate (i.e., the "observation time scale" over which the temperature can be assumed constant). Hence, the probability distribution within a metabasin follows equilibrium statistical mechanics within the restricted ensemble.
2. The inter-metabasin relaxation time scale is too long to allow for equilibration on the observation time scale.

The inter-metabasin dynamics can be computed based on a reduced set of master equations,

$$\frac{df_\alpha}{dt} = \sum_{\beta \neq \alpha} \left(W_{\beta\alpha} f_\beta - W_{\alpha\beta} f_\alpha \right),$$ (70)

where f_α is the probability of occupying metabasin α. There is one master equation for each metabasin; the $W_{\alpha\beta, \beta\alpha}$ parameters are the effective inter-metabasin transition rates, computed by

$$W_{\alpha\beta} = \frac{n_\beta}{f_\alpha} \sum_{A\in\alpha} K_{A, B\in\beta} P_A.$$ (71)

Thus instead of writing a separate master equation for each basin, we can write a reduced set of master equations with one for each metabasin. At high temperatures the entire system consists of a single metabasin and is governed by equilibrium statistical mechanics. As the system is cooled, the enthalpy landscape is partitioned into metabasins, chosen to satisfy the above two criteria. The procedure for performing such partitioning is beyond the scope of this review. The interested reader is referred to Ref. [28] for details.

By decoupling the inter- and intra-metabasin dynamics, this technique allows for solution of a reduced set of master equations on the "natural" time scale of the experiment, i.e., the observation time scale defined by the inverse of the cooling rate. In other words, the integration time step is governed by the cooling rate rather than by the fastest microscopic transition rate. As a result, the computational time required for computing the system dynamics is largely independent of the fastest inter-basin transition rate.

We have previously used this procedure to simulate the glass transition range behavior of selenium, an elemental glass-former with a chain-like structure [35]. In order to demonstrate the versatility of our model, Fig. 10 shows the molar volume of the selenium system computed with cooling rates covering 25 orders of magnitude. Here, we cool linearly from the melting temperature of selenium (490 K) to room temperature (298 K) with cooling rates ranging from 10^{-12} to 10^{12} K/s. For the very slow cooling rates ($< 10^{-9}$ K/s), the system remains a

Fig. 10 Molar volume of selenium after cooling from the melting temperature (490 K) to room temperature (298 K) with linear cooling rates ranging from 10^{-12} to 10^{12} K/s. For extremely slow cooling rates ($< 10^{-9}$ K/s), the system never departs from the equilibrium supercooled liquid line

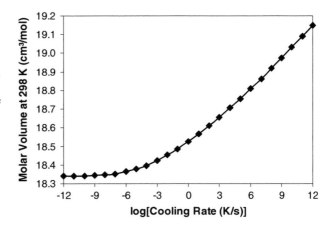

Fig. 11 Volume-temperature diagrams for three selenium systems cooled from the melting temperature (490 K) to room temperature (298 K) at rates of 10^{12}, 1, and 10^{-12} K/s

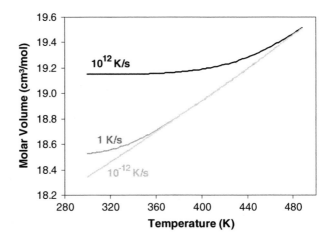

supercooled liquid; glasses form for all of the faster cooling rates. We observe an Arrhenius dependence of molar volume with respect to cooling rate, in excellent agreement with the experimental findings of Moynihan and coworkers [36]. The complete volume-temperature curves for three of the cooling rates (10^{-12}, 1, and 10^{12} K/s) are shown in Fig. 11. Using the master equation technique described above, the computation time is approximately equal for all cooling rates.

These simulations of selenium were performed using potentials based on *ab initio* simulations [35,37]. Hence, the model offers truly predictive calculations of measurable thermodynamic properties without any fitting parameters. As we have demonstrated with selenium, the model is directly applicable to realistic glass-forming compositions. In addition, it incorporates the full thermal history of the system from the equilibrium liquid state through the glass transition. As shown in Fig. 12, this approach can also be used to simulate the long-time ageing behavior of glass. Finally, we note that our model of selenium has been validated against the experimental measurements of Varshneya and coworkers [38,39] for molar volume, thermal expansion coefficient, and viscosity. For more details, please see Ref. [35].

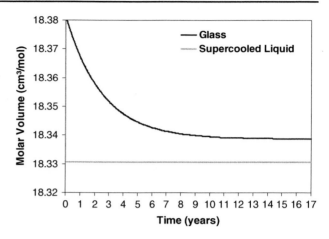

Fig. 12 Long-time ageing of the selenium glass during an isothermal hold at room temperature. The glass undergoes significant relaxation, but not full equilibration, on the time scale of years

5 Nature of the glassy state

The enthalpy landscape approach has enabled a number of significant insights into the nature of the glass transition and the glassy state. In this section we review several of these new insights and discuss their implications for our understanding of glass.

5.1 Continuously broken ergodicity and the residual entropy of glass

One of the most common assumptions in statistical mechanics is that of ergodicity, which asserts the equivalence of time and ensemble averages of the thermodynamic properties of a system. Ergodicity implies that, given enough time, a system will explore all allowable points of phase space. The question of ergodicity is really a question of time scale. At high temperatures, the internal relaxation time scale of a liquid is small compared to an external observation time scale, i.e., $t_{int} < t_{obs}$. When the liquid is cooled, t_{int} increases as the kinetics of the system slow down. The glass transition occurs when the internal relaxation time is equal to the external observation time, i.e., $t_{int} = t_{obs}$. The glassy state itself is necessarily nonergodic, since at low temperatures $t_{int} > t_{obs}$. Thus, the glass transition entails a loss of ergodicity as the ergodic liquid is cooled to a nonergodic glassy state.

In a landmark 1982 paper, Palmer [40] proposed that a glassy system can be described in terms of mutually inaccessible regions (termed "components") that meet the conditions of *confinement* and *internal ergodicity*. Confinement indicates that transitions are not allowed between components on the given observation time scale. The condition of internal ergodicity states that ergodicity holds within a given component. As a result, ergodic statistical mechanics can be applied within each of the individual components, and the overall properties of the nonergodic system can be computed using a suitable average over these components. With this approach, glass is said to exhibit "broken ergodicity." This has important implications related to the entropy of a glassy system. Before proceeding, it is useful to review some common definitions of entropy, following the terminology of Goldstein and Lebowitz [41]:

1. The thermodynamic entropy of Clausius, given by

$$dS = \frac{dE}{T},$$

(72)

where E is internal energy and T is absolute temperature. This definition of entropy is applicable only for equilibrium systems and reversible processes.

2. The Boltzmann entropy,

$$S_B = k \ln \Omega_B, \tag{73}$$

applicable to the microcanonical ensemble, where Ω_B is the volume of phase space (i.e., the number of microstates) visited by a system to yield a given macrostate.

3. The Gibbs or "statistical" entropy,

$$\bar{S} = -k \sum_i p_i \ln p_i, \tag{74}$$

where p_i gives the probability of occupying microstate i, and the summation is over all microstates. Here, the probabilities are computed based on an ensemble average of all possible realizations of a given system. With the assumption of ergodicity, the ensemble-averaged values of p_i are equal to the time-averaged values of p_i.

One particular point of confusion in the glass science community relates to the application of Eq. (72) in conjunction with differential scanning calorimetry (DSC) experiments to compute the entropy of a system on both cooling and heating through the glass transition [16]. This results in two main findings: (a) there is no change in entropy at the laboratory glass transition, and (b) glass has a positive residual entropy at absolute zero. We argue that these results are incorrect since glass is not an equilibrium materials and the glass transition is not a reversible process [42]; hence, Eq. (72) cannot be applied. Moreover, the ergodic formula for Gibbs entropy, which approximately equals the thermodynamic entropy [43], cannot be used since it cannot account for the defining feature of the glass transition, viz., the breakdown of ergodicity. Using the enthalpy landscape approach and the concept of broken ergodicity, the glass transition must entail a *loss* of entropy rather than a "freezing in" of entropy. Furthermore, the entropy of glass at absolute zero is always zero (in the classical limit).

With Palmer's conditions of confinement and internal ergodicity, we can apply the Gibbs definition of entropy within each individual component α:

$$S_\alpha = -k \sum_{i \in \alpha} \frac{p_i}{P_\alpha} \ln \frac{p_i}{P_\alpha}, \tag{75}$$

where the probability of occupying component α, P_α, is the sum of the probabilities of occupying the various microstates within that component:

$$P_\alpha = \sum_{i \in \alpha} p_i. \tag{76}$$

The configurational entropy of the system as a whole is simply a weighted sum of the Gibbs entropies of the individual components:

$$\langle S \rangle = \sum_\alpha S_\alpha P_\alpha = -k \sum_\alpha \sum_{i \in \alpha} p_i \ln \frac{p_i}{P_\alpha}. \tag{77}$$

Equation (77) gives the entropy of the system accounting for the breakdown of ergodicity at the glass transition. If the nonergodic nature of glass were not accounted for, the entropy would be given by direct application of the statistical formula of Eq. (74).

The difference between the statistical entropy and the nonergodic glassy entropy of Eq. (77) is called the "complexity" of the component ensemble [40] and is given by

$$I = \bar{S} - \langle S \rangle = -k \sum_{\alpha} P_{\alpha} \ln P_{\alpha}. \qquad (78)$$

The complexity is a measure of the nonergodicity of a system. At absolute zero, the glass is trapped in a single microstate; here, each microstate is itself a component, and the entropy is necessarily zero. Hence, the complexity adopts its maximum value of $I = \bar{S}$. For a completely ergodic system (e.g., at high temperatures or for long observation times), all microstates are part of the same component, and the complexity is zero. The process of ergodicity-breaking, as occurs at the glass transition, necessarily results in a loss of entropy (and increase in complexity) as the phase space divides into mutually inaccessible component.

Recently, we have generalized the Palmer approach to account for a *continuous*, rather than sudden, breakdown of ergodicity at the glass transition [44] (such as would occur with normal laboratory glass formation using a finite cooling rate). This generalization is accomplished by relaxing the assumption of confinement. Suppose that we make an instantaneous measurement of the microstate of a system at time t'. The act of measurement causes the system to "collapse" into a single microstate i with probability $p_i(t')$. In the limit of zero observation time, the system is confined to one and only one microstate and the observed entropy is necessarily zero. However, the entropy becomes positive for any finite observation time, t_{obs}, since transitions between microstates are not strictly forbidden (except at absolute zero, barring quantum tunneling). The question we seek to answer is, "What is the entropy of the system over the observation window $[t', t' + t_{obs}]$?"

We can answer this question by following the dynamics of a system whose basin is known at t', the beginning of the observation window. Let $f_{i,j}(t)$ be defined as the conditional probability of the system occupying basin j after starting in a known basin i and subsequently evolving for some time t, accounting for the actual transition rates between microstates. The conditional probabilities satisfy

$$\sum_{j=1}^{\Omega} f_{i,j}(t) = 1, \qquad (79)$$

for any initial state i and for all time t. Hence, $f_{i,j}(t_{obs})$ gives the probability of transitioning to basin j after an initial measurement in basin i and evolving through an observation time t_{obs}. The configurational entropy associated with collapsing into initial state i is therefore:

$$S_i = -k \sum_{j=1}^{\Omega} f_{i,j}(t_{obs}) \ln f_{i,j}(t_{obs}). \qquad (80)$$

This represents the entropy of one possible realization of the system and corresponds roughly to Palmer's component entropy of Eq. (75); however, we have not made any assumptions about confinement or partitioning of the enthalpy landscape into components. Note that while the above equation is of the same form as the Gibbs entropy of Eq. (74), the value of $f_{i,j}(t_{obs})$ above gives the probability of the system *transitioning* from an initial state i to a final state j within a given observation time t_{obs}. The probability p_i in the Gibbs formulation represents the ergodic limit of $t_{obs} \to \infty$; hence, p_i represents an ensemble-averaged probability of the system *occupying* a given state and does not account for the finite transition time required to *visit* a state, which may be long compared to the observation time scale.

The expectation value of entropy at the end of the observation window $\left[t', t' + t_{obs}\right]$ is simply the weighted sum of entropy values for all possible realizations of the system:

$$\langle S \rangle = \sum_{i=1}^{\Omega} S_i\, p_i\left(t'\right) = -k \sum_{i=1}^{\Omega} p_i\left(t'\right) \sum_{j=1}^{\Omega} f_{i,j}\left(t_{obs}\right) \ln f_{i,j}\left(t_{obs}\right). \qquad (81)$$

With this approach, there is no need to define components or metabasins of any kind. By considering all possible configurations of the system and the actual transition rates between basins, this approach can be applied to any arbitrary energy landscape and for any temperature path. It is thus suitable for modeling systems in all regimes of ergodicity: fully ergodic, fully confined, and everything in between.

With Eq. (81), the configurational entropy of a system is a function of not only the system itself, but also an external observation time t_{obs} imposed on the system. The imposition of this external observation time creates a kinetic constraint, which causes the breakdown of ergodicity when $t_{int} > t_{obs}$. The implication is that the glassy state only exists when there is an external observer who observes the system on a time scale shorter than the relaxation time scale of the system. If there are two observers who monitor the system on highly disparate time scales, one observer could see a glass while the other sees an ergodic supercooled liquid. Hence, Eq. (81) captures the very essence of what is meant by "glass." With this equation, the entropy of the system is zero both in the limits of $t_{obs} \to 0$ and $T \to 0$. The first case ($t_{obs} \to 0$) is in agreement with Boltzmann's notion of entropy as the number of microstates *visited* by a system to yield a given macrostate [41,45,46]: with zero time the system can visit only a single microstate—no transitions are allowed since there is no time for them to occur. The second case ($T \to 0$) is due to the system being kinetically trapped in one and only one microstate. Since other microstates are never visited, they cannot contribute to the overall entropy of the system [47]. Both of these limits are equivalent in the Palmer model of discontinuously broken ergodicity [40], where each microstate would have its own separate component; no transitions are allowed among the components, so the entropy is necessarily zero.[3]

For any positive temperature, the limit of $t_{obs} \to \infty$ yields an equilibrated system with complete restoration of ergodicity. In the Palmer view [40], this is equivalent to all microstates being members of the same component with transitions freely allowed among all microstates. Both the Palmer approach and our generalization above yield the same result as the Gibbs formulation of entropy in the limit of $t_{obs} \to \infty$, i.e., for a fully ergodic, equilibrated system.

Implementation of the above approach requires calculation of the conditional probability factors $f_{i,j}$, which can be accomplished using the hierarchical master equation approach described in Ref. [44]. Figure 13 plots the computed nonergodic glassy entropy of Eq. (81) and ergodic statistical entropy of Eq. (74) for a simple model landscape described in Ref. [44]. This figure shows that the departure of the nonergodic and ergodic entropies from equilibrium exhibit markedly different behaviors. Using the ergodic statistical formulation of Eq. (74), the glass transition corresponds to a freezing of the basin occupation probabilities and hence a freezing of the entropy. However, when we account for the continuous breakdown of ergodicity, the glass transition results in a smooth loss of entropy as the system becomes confined to smaller regions of phase space. Whereas the ergodic formula predicts a large residual entropy of the glass at absolute zero, the nonergodic formalism of Eq. (81) correctly predicts zero entropy at absolute zero.

---

[3] The fact that glass has zero entropy at absolute zero does not imply that glass adopts a crystalline configuration. Rather, the glass is kinetically trapped in one of many possible non-crystalline configurations.

Fig. 13 Evolution of glassy,
statistical, and equilibrium
entropy values with respect to
temperature for a simple model
landscape using an observation
time is 0.01 s. Glassy entropy is
computed with Eq. (81)
accounting for broken ergodicity,
and statistical entropy is
computed under the ergodic
assumption using Eq. (74)

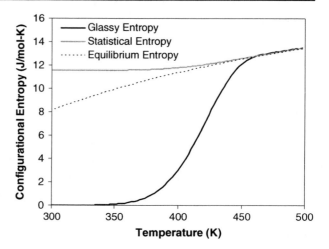

Finally, we note that the concept of continuously broken ergodicity also has important implications beyond entropy, as it play a key role in determining the dynamic properties of glass-forming systems (e.g., viscosity) at and below the glass transition range [35].

5.2 Supercooled liquid fragility

Supercooled liquids can be classified as either "strong" or "fragile," depending on their observed scaling of shear viscosity η with temperature. According to Angell's criterion [48–51], strong liquids exhibit a scaling that is close to the Arrhenius form,

$$\eta = \eta_0 \exp\left(\frac{\Delta H}{kT}\right), \tag{82}$$

where ΔH is an activation barrier to flow and η_0 is a constant. When the logarithm of viscosity is plotted as a function of inverse temperature, as in Fig. 14, strong liquids will have a near-straight-line relationship. Examples of strong liquids include silica (SiO_2) and germania (GeO_2).

Fragile liquids, on the other hand, exhibit a large departure from this straight-line relationship. In this case, the viscosity-temperature relationship can often be described by the Vogel-Fulcher-Tamman (VFT) relation,

$$\eta = \eta_0 \exp\left[\frac{\Delta H}{k\,(T - T_0)}\right], \tag{83}$$

where T_0 is a constant. Examples of fragile liquids include *o*-terphenyl, heavy metal halides, and calcium aluminosilicates.

Stillinger [21] has proposed a direct link between the fragility of a supercooled liquid and the topography of the underlying potential energy or enthalpy landscape. He suggests that the Arrhenius behavior of strong liquids indicates that their energy landscape has a rather uniform roughness, as shown in Fig. 15(a). In contrast, fragile liquids are likely to have a highly non-uniform topography in their potential energy landscape, as indicated in Fig. 15(b). At high temperatures a fragile liquid is able to flow among basins with relatively low activation barriers, indicating the rearrangement of a small number of molecules. As the fragile liquid is supercooled, it samples deeper basins with a greater separation in the energy

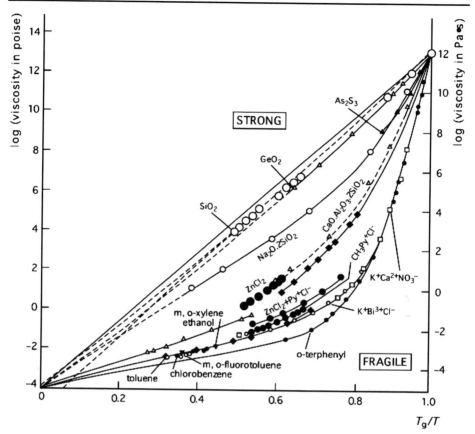

Fig. 14 Variation of viscosity with normalized inverse temperature T_g/T for strong and fragile liquids. Reproduced from Varshneya [16]

landscape. Flow between these basins requires the cooperative rearrangement of many molecules and occurs less frequently than the higher temperature relaxations. Hence, as a fragile liquid is supercooled it encounters continually higher and higher barriers to shear flow.

As noted by Angell [52], a higher value of fragility leads to a sharper, more well-defined glass transition. This is also consistent with Stillinger's picture, since with a higher fragility the effective energy barrier for inter-basin transitions increases more rapidly as the system is cooled through the glass transition.

It is interesting to consider that in the limit of infinite fragility the energy barriers become effectively infinite at the glass transition temperature, prohibiting any structural relaxation and resulting in a perfectly sharp (ideal) glass transition. Such an ideal glass transition would yield a discontinuity in the slopes of the volume and enthalpy curves. An ideal glass transition also implies a complete and discontinuous breakdown of ergodicity at the glass transition temperature.

The subject of an ideal glass transition has been considered previously by Gibbs and DiMarzio [9] and others seeking to develop a thermodynamic model of the glass transition [16]. The appeal of the ideal glass transition is that it represents an ideal second-order Ehrenfest phase transition. With an ideal glass transition, the full thermal history of the glass can be accounted for using just a single order parameter. While this allows for a greatly

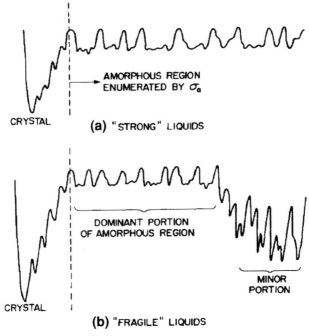

Fig. 15 Schematic representation of the potential energy landscape for (**a**) "strong" and (**b**) "fragile" liquids. Reproduced from Stillinger [21]

simplified picture of the glassy state, an ideal glass transition can occur *only* in the limit of infinite fragility. Hence, a thermodynamic model of the glassy state that uses just a single order parameter is insufficient to capture accurately the full effect of thermal history on glass properties.

We note also that supercooled liquid fragility is well correlated with the magnitude of the change in second-order thermodynamic variables (such as heat capacity and thermal expansion coefficient) during the glass transition. For example, Fig. 16 shows that fragile liquids experience a much greater change in heat capacity during the glass transition than strong liquids. This relationship can be explained easily in the enthalpy landscape framework. Since a more fragile system experiences a greater increase in transition energy barrier as it cools through the glass transition, there is a greater and more sudden loss of ergodicity compared to a less fragile system. As a result, the system becomes confined to a smaller region of phase space, allowing for fewer fluctuations in enthalpy and volume. Since second-order thermodynamic properties are directly related to fluctuations in first-order properties such as enthalpy and volume, it is natural that more fragile liquids would experience a greater change in second-order properties through the glass transition.

5.3 The Kauzmann paradox and the ideal glass transition

One of the decades-old "mysteries of glass science" is that of the Kauzmann paradox. In a landmark paper, Kauzmann [53] plotted the difference in configurational entropy between several supercooled liquids and their corresponding crystalline states. He extrapolated the curves to low temperatures, such as shown in Fig. 17 for glycerol, and found that the entropy

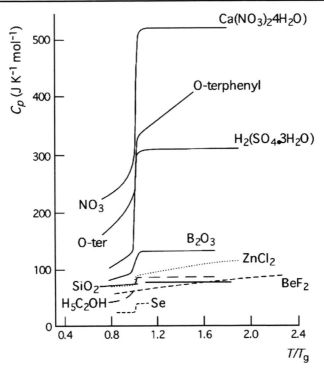

Fig. 16 Heat capacity as a function of normalized temperature, T/T_g, for various strong and fragile liquids. Reproduced from Varshneya [16]

difference becomes zero at a finite temperature T_K, the Kauzmann temperature. Continued extrapolation below T_K would yield negative configurational entropy for the supercooled liquid, in violation of the third law of thermodynamics. Examples of this so-called "Kauzmann paradox" are shown in Fig. 18.

Kauzmann himself proposed a resolution to this paradox, arguing that the energy barrier to crystallization must decrease to the same order as the thermal energy. In this way crystallization would be inevitable at low temperatures, and the issue of negative entropy would be completely meaningless. Figure 19 illustrates this point of view in terms of total configurational plus vibrational entropy. The entropy of an infinitely slowly cooled liquid would meet that of the crystalline state at T_K and then follow it down to absolute zero. Other researchers, including Gibbs and DiMarzio [9], have proposed that the Kauzmann temperature represents a lower limit to the glass transition temperature of a system. An infinitely slowly cooled system would thus achieve the "ground state" of disordered packing at $T = T_K$ and be unable to undergo any further rearrangements of its structure. In this manner, the configurational energy of a system remains at zero between the Kauzmann temperature and absolute zero.

Greater insight into the question of the Kauzmann paradox can be gained using the energy landscape approach. As shown by Stillinger [21] in Fig. 20, it is not possible for the configurational entropy of a supercooled liquid to reach zero at finite temperature, since some transitions are always allowed between basins. The entropy of a supercooled liquid reaches zero only at absolute zero, when it is confined to just a single microstate. Hence, the whole notion of the "Kauzmann paradox" is based on an incorrect extrapolation of the entropy of a supercooled liquid at low temperatures. The correct extrapolation of entropy is as shown

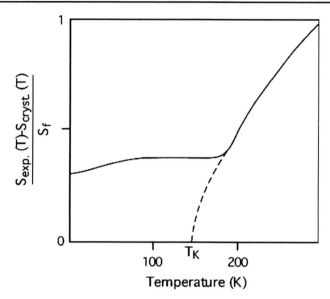

Fig. 17 Entropy of supercooled and glassy glycerol relative to that of the crystal and normalized to the entropy of fusion S_f. Reproduced from Varshneya [16]

Fig. 18 Examples of the Kauzmann paradox, where supercooled liquid entropies apparently extrapolate to negative values at low temperature. Reproduced from Varshneya [16]

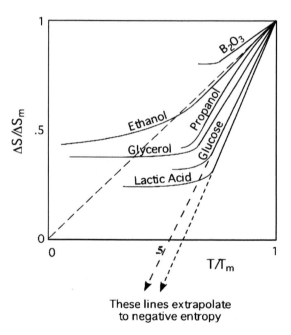

These lines extrapolate to negative entropy

in Fig. 20. This view is supported by the recent experiment of Huang, Simon, and McKenna [54], who measured the heat capacity of poly(α-methyl styrene) at very low temperatures.

While Stillinger successfully debunked the notion of a Kauzmann paradox, he does not dismiss the utility of a Kauzmann temperature at which the entropies of two different phases of a material become equal. In fact, Stillinger extends the concept of a single Kauzmann

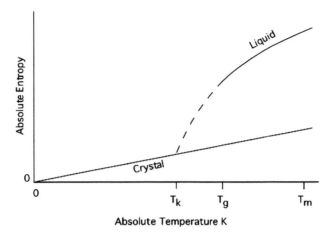

Fig. 19 Absolute entropy of a supercooled liquid and its corresponding crystal. Reproduced from Varshneya [16]

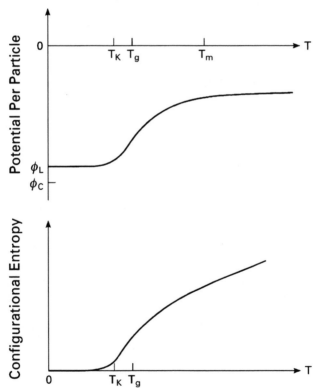

Fig. 20 Extrapolations of potential energy and configurational entropy below the glass transition temperature, T_g, and Kauzmann temperature, T_K, according to the Stillinger model of inherent structures [21]. Reproduced from Varshneya [16]

temperature to multiple "Kauzmann points" in the temperature-pressure plane of a system [55,56]. Consider the Claussius-Clapeyron equation that describes the slope of the melting curve in the temperature-pressure plane:

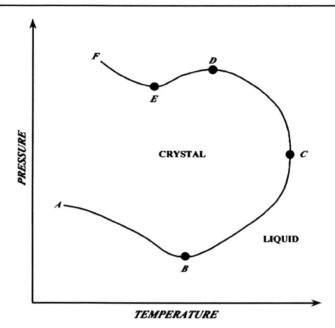

Fig. 21 Example melting curve in the temperature-pressure plane. Reproduced from Stillinger and Debenedetti [56]

$$\frac{d P_m (T)}{dT} = \frac{S_l - S_c}{V_l - V_c}.$$

(84)

Here, $P_m (T)$ is the temperature-dependent melting pressure, and $S_{l,c}$ and $V_{l,c}$ are the molar entropies and volumes of the liquid and crystalline phases, respectively. Figure 21 shows an example melting curve $ABCDEF$ to demonstrate possible melting phenomena. The "normal" melting scenario, in which the molar volume increases upon melting of a crystal, occurs between points B and C. The intervals between C and D and between E and F show the case where the molar volume decreases upon melting of a crystal, while again heat is absorbed. This corresponds to the familiar case of ice melting to form liquid water. The intervals between A and B and between D and E show the perhaps unfamiliar case of inverse melting, in which an isobaric heating of the system causes the liquid to freeze into the crystalline phase. Materials that exhibit inverse melting behavior include the helium isotopes ^{3}He and ^{4}He at low temperatures and the polymeric substance poly(4-methylpentene-1), denoted P4MP1 [56].

Points B, D, and E in Fig. 21 are all points of zero slope in the melting curve:

$$\frac{d P_m (T)}{dT} = 0 \ (B, D, E).$$

(85)

Since the difference in molar volumes, $V_l - V_c$, can never be infinite, these points must have vanishing entropy change:

$$S_l - S_c = 0 \quad (B, D, E).$$

(86)

Hence, points B, D, and E are all "Kauzmann points" occurring naturally in the temperature-pressure plane, and *the existence of these Kauzmann points does not depend upon or necessitate the existence of an ideal glass transition*. Since the chemical potentials of the coexisting liquid and crystal phases are equal, the Kauzmann points must also correspond to points of vanishing enthalpy, $\Delta H = T \Delta S$. Stillinger and Debenedetti [56] show that these Kauzmann points lie on a general "Kauzmann curve,"

$$\left(\frac{dP}{dT}\right)_{\Delta S=0} = \frac{C_{p,l} - C_{p,c}}{T\left(\alpha_l V_l - \alpha_c V_c\right)}, \tag{87}$$

where C_p is heat capacity and α is the isobaric thermal expansion coefficient.

5.4 Fictive temperature and the glassy state

The macroscopic state of a system is defined based on the values of its observable properties. For a glass, these property values are dependent on thermal history. As such, the equilibrium macrostate variables (e.g., temperature and pressure) are not sufficient to describe a glassy state; additional state variables are necessary. In 1932, Tool and Eichlin [4,57] proposed a description of glass in terms of an equivalent equilibrium liquid state at a different temperature. The temperature of this equivalent liquid state was originally termed the "equilibrium temperature" but later became known as the "effective temperature" or, more commonly, the "fictive temperature," denoted T_f. Ideally, the concept of fictive temperature would allow for the difficult glassy problem to be solved in terms of a simpler equilibrium approach. However, Ritland [58] demonstrated in 1956 with his famous "crossover" experiments that two glasses having different thermal histories but the same T_f are actually in two different macroscopic states (as they subsequently relaxed differently). Ritland's result implied that a single additional parameter, T_f, is inadequate to describe the state of a glass. This conclusion has been reinforced recently by simulation studies [59, 60]. However, the question remained as to whether the microscopic state of a glass can be described using a *distribution* of fictive temperatures. In other words, can the nonequilibrium glassy state be described in terms of a mixture of supercooled liquid states? Recently, we have addressed this question in a general sense within the enthalpy landscape framework [61].

Suppose we have a probability distribution of fictive temperatures, $h\left(T_f\right)$, which satisfies:

$$\int_0^\infty h\left(T_f\right) dT_f = 1. \tag{88}$$

For a infinitely fast quench, the glass can be described fully using a single fictive temperature and the fictive temperature distribution is a Dirac delta function:

$$h\left(T_f\right) = \delta\left(T_f - T_{quench}\right). \tag{89}$$

A glass formed in the laboratory with a finite cooling rate would, in principle, have some $h\left(T_f\right)$ distribution with a nonzero width. A more slowly cooled glass would likely have a wider distribution of fictive temperatures, covering the full glass transition range. If the concept of a fictive temperature distribution is meaningful to describe the microscopic state of a glass, then the occupation probabilities p_i of any glass with any thermal history $T\left(t\right)$ could be described using an average of equilibrium states corresponding to the various fictive temperatures. This can be expressed as:

$$p_i\left[T\left(t\right)\right] = \int_0^\infty h\left[T_f, T\left(t\right)\right] p_i^{eq}\left(T_f\right) dT_f. \tag{90}$$

The left hand side of the equation gives the nonequilibrium probability of occupying basin i after evolving through the particular thermal history given by $T\left(t\right)$. The integral on the right hand side of the equation gives a weighted average of equilibrium probabilities for basin occupation at the various fictive temperatures. In other words, the integral provides for an arbitrary mixing of equilibrium states. The weighting function $h\left[T_f, T\left(t\right)\right]$ is the fictive temperature distribution, satisfying Eq. (88), which depends on the thermal history $T\left(t\right)$ of the glass.

If Eq. (90) is valid, then there would be a real physical significance to the concept of fictive temperature distribution since the glassy state could be described in terms of a distribution of equilibrium states. This would enable computation of the thermodynamic properties of a glass using just the physical temperature T and the distribution of fictive temperatures $h\left[T_f, T\left(t\right)\right]$. To test the validity of Eq. (90) we consider the selenium system of Ref. [35] using a linear cooling path to 250 K at a rate of 1 K/min. As the system is cooled, the probabilities $p_i\left(t\right)$ evolve in time according to the coupled set of master equations in Eq. (35).

To make use of Eq. (90), we must first compute the equilibrium occupation probabilities $p_i^{eq}\left(T\right)$ for the various basins. Figure 22(a) plots the equilibrium probabilities of the eleven most significant basins for the temperature range of interest (between 250 and 400 K). Each of these basins corresponds to a different molar volume, as indicated in the legend. As the temperature decreases, the basins with lower molar volume become more favorable. The probabilities of the individual basins display peaks owing to a trade-off of entropy (i.e., degeneracy) and enthalpy effects.

Figure 22(b) plots the basin probabilities for the glass formed by cooling at a rate of 1 K/min. In order to test Eq. (90), we optimize the fictive temperature distribution $h\left[T_f, T\left(t\right)\right]$ at each new temperature T to provide a best fit of the nonequilibrium probabilities in Fig. 22(b) based on a mixture of equilibrium probabilities in Fig. 22(a). At $T = 400$ K, the system is in equilibrium, so $h\left[T_f, T\left(t\right)\right]$ reduces to a Dirac delta function. As the system is cooled, the fictive temperature distribution broadens and lags behind the physical temperature. At low temperatures the distribution becomes frozen. Figure 22(c) shows the result of this optimization, i.e., the best possible fit of the nonequilibrium glassy probabilities based on the right hand side of Eq. (90). Comparing parts (b) and (c) of the figure, it is clear that the glass exhibits a much narrower distribution of basin probabilities p_i compared to the fictive temperature representation. In fact, it is not possible to construct a system with such a narrow distribution of basin probabilities based only on a mixing of equilibrium states.

Figure 23 shows the evolution of the breadth of the basin probability distribution as a function of temperature. Whereas the nonequilibrium probability distribution undergoes a significant narrowing through the glass transition regime, the breadth of the "best fit" fictive temperature-based distribution remains fairly constant. The implication is that the fictive temperature description of the glassy state does not capture accurately the fluctuations in density and enthalpy observed in the glassy state. This is consistent with the well known behavior of second-order thermodynamic properties (e.g., heat capacity and thermal expansion coefficient) through the glass transition range. While these second-order properties show marked changes at the glass transition for a nonequilibrium system, the equilibrium supercooled liquid values undergo no such change [16].

Hence, we conclude that the concept of a fictive temperature distribution has no physical basis for describing the microscopic state of a realistic glass. The fictive temperature representation is rigorous only in the special case of an ideal glass transition, which is never possible in practice. The nonequilibrium glassy state is fundamentally different from the equilibrium liquid state, and the concept of fictive temperature should not be employed by researchers seeking a realistic microscopic description of glass. The inherent inequality

Fig. 22 (**a**) Equilibrium distribution of basin probabilities for supercooled liquid selenium. (**b**) Distribution of basin probabilities for selenium glass prepared with a cooling rate of 1 K/min. (**c**) The best approximation of the above glass using fictive temperature mapping. For clarity, every tenth data point is plotted

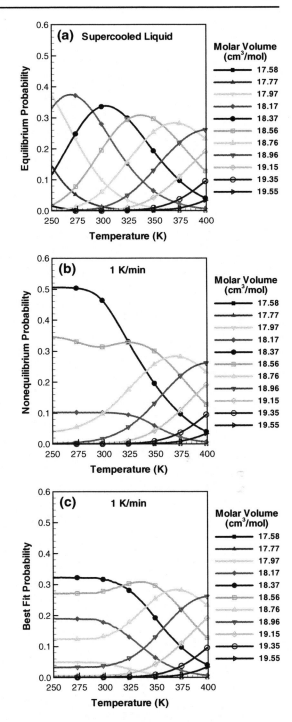

between the glassy and liquid states has been argued previously by Gupta and Mauro [3] based on the nonergodic nature of glass. Here, by showing that the basin probabilities p_i of a glass cannot be represented in terms of a linear combination of equilibrium p_i values, we

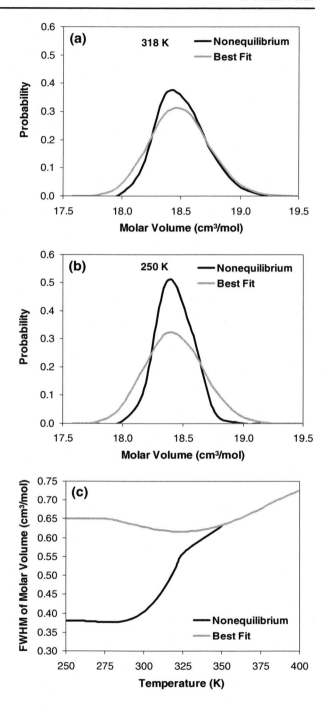

Fig. 23 Computed distribution of molar volume probabilities for selenium glass (black line) and best fit using fictive temperature mapping (gray line) for physical temperatures of (**a**) 318 K and (**b**) 250 K, assuming a cooling rate of 1 K/min. (**c**) Plot of the full width at half maximum (FWHM) of the molar volume probabilities as a function of temperature. The FWHM of the nonequilibrium distribution falls dramatically as the system is cooled through the glass transition

extend this argument to encompass also the nonequilibrium nature of the glassy state. Hence, the glassy state is fundamentally different from the liquid based on both its nonergodic *and* nonequilibrium features.

6 Conclusions

The enthalpy landscape formalism offers a powerful approach for modeling the glass transition range behavior of realistic glass-forming systems on laboratory time scales. Implementation of the enthalpy landscape approach requires a mapping of the continuous landscape to a discrete set of inherent structures and transition states. Several new simulation techniques have been developed to enable this mapping of the landscape and to solve for the dynamics of the system on long time scales. The enthalpy landscape approach offers several important insights into the nature of the glass transition and the glassy state:

- The laboratory glass transition represents a continuous breakdown of ergodicity as the internal relaxation time scale of the system exceeds an external observation time. This breakdown of ergodicity must be accompanied by a loss of entropy as the system becomes confined to a small subset of phase space. The entropy of glass is zero at absolute zero.
- The fragility of a supercooled liquid is directly related to the distribution of transition barriers in the enthalpy landscape. A more fragile liquid leads to a greater and more sudden breakdown of ergodicity at the glass transition. Hence, a more fragile liquid will have a sharper glass transition accompanied by a larger change in second-order thermodynamic properties such as heat capacity and thermal expansion coefficient. A perfectly sharp (i.e., ideal) glass transition can be obtained theoretically in the limit of infinite fragility.
- The Kauzmann paradox is the result of improper extrapolation of the configurational entropy of supercooled liquids at low temperature. In the enthalpy landscape approach, the configurational entropy of a supercooled liquid becomes zero only at absolute zero, and no conflict with the third law of thermodynamics arises. However, the existence of Kauzmann points in the temperature-pressure plane where the entropy of two separate phases is identical is not negated.
- The glassy state cannot be described in terms of a mixture of supercooled liquid states. Hence, there is no microscopic meaning to the concept of a fictive temperature distribution for a laboratory-produced glass. As a nonequilibrium, nonergodic material, glass is in a fundamentally different state than the supercooled liquid.

Acknowledgements The authors take great pleasure in acknowledging the support and encouragement of many colleagues at Corning Incorporated, including Douglas Allan, Roger Araujo, Jitendra Balakrishnan, Dipak Chowdhury, Phong Diep, Gautam Meda, Srikanth Raghavan, and Amy Rovelstad.

References

1. Tait, H.: Five Thousand Years of Glass. British Museum Press, London (1991)
2. Zanotto, E.D., Gupta, P.K.: Do cathedral glasses flow?—additional remarks. Am. J. Phys. **67**(3), 260–262 (1999)
3. Gupta, P.K., Mauro, J.C.: The laboratory glass transition. J. Chem. Phys. **126**(22), 224504 (2007)
4. Tool, A.Q.: Relation between inelastic deformability and thermal expansion of glass in its annealing range. J. Am. Ceram. Soc. **29**(9), 240–253 (1946)
5. Narayanaswamy, O.S.: A model of structural relaxation in glass. J. Am. Ceram. Soc. **54**(10), 491–498 (1971)
6. Turnbull, D., Cohen, M.H.: On the free-volume model of the liquid-glass transition. J. Chem. Phys. **52**(6), 3038–3041 (1970)
7. Cohen, M.H., Grest, G.S.: Liquid-glass transition, a free-volume approach. Phys. Rev. B **20**(3), 1077–1098 (1979)
8. Gupta, P.K.: Models of the glass transition. Rev. Solid State Sci. **3**(3–4), 221–257 (1989)
9. Gibbs, J.H., DiMarzio, E.A.: Nature of the glass transition and the glassy state. J. Chem. Phys. **28**(3), 373–383 (1958)

10. Leutheusser, E.: Dynamical model of the liquid-glass transition. Phys. Rev. A **29**(5), 2765–2773 (1984)
11. Bengtzelius, U., Götze, W., Sjölander, A.: Dynamics of supercooled liquids and the glass transition. J. Phys. C: Solid State Phys. **17**(33), 5915–5934 (1984)
12. Debenedetti, P.G.: Metastable Liquids: Concepts and Principles. Princeton University, Princeton (1996)
13. Kob, W.: Computer simulation of supercooled liquids and glasses. J. Phys.: Condens. Matter **11**(10), R85–R115 (1999)
14. Scherer, G.W.: Relaxation in Glass and Composites. Wiley, New York (1986)
15. Jäckle, J.: Models of the glass transition. Rep. Prog. Phys. **49**(2), 171–231 (1986)
16. Varshneya, A.K.: Fundamentals of Inorganic Glasses, 2nd edn. Society of Glass Technology, Sunderland (2007)
17. Gupta, P.K.: Non-crystalline solids: glasses and amorphous solids. J. Non-Cryst. Solids **195**(1–2), 158–164 (1996)
18. Goldstein, M.: Viscous liquids and the glass transition: a potential energy barrier picture. J. Chem. Phys. **51**(9), 3728–3739 (1969)
19. Stillinger, F.H., Weber, T.A.: Hidden structure in liquids. Phys. Rev. A **25**(2), 978–989 (1982)
20. Stillinger, F.H., Weber, T.A.: Dynamics of structural transitions in liquids. Phys. Rev. A **28**(4), 2408–2416 (1983)
21. Stillinger, F.H.: Supercooled liquids, glass transitions, and the Kauzmann paradox. J. Chem. Phys. **88**(12), 7818–7825 (1988)
22. Debenedetti, P.G., Stillinger, F.H., Truskett, T.M., Roberts, C.J.: The equation of state of an energy landscape. J. Phys. Chem. B **103**(35), 7390–7397 (1999)
23. Debenedetti, P.G., Stillinger, F.H.: Supercooled liquids and the glass transition. Nature **410**(6825), 259–267 (2001)
24. Stillinger, F.H., Debenedetti, P.G.: Energy landscape diversity and supercooled liquid properties. J. Chem. Phys. **116**(8), 3353–3361 (2002)
25. Massen, C.P., Doye, J.P.K.: Power-law distributions for the areas of the basins of attraction on a potential energy landscape. Phys. Rev. E **75**, 037101 (2007)
26. Mauro, J.C., Varshneya, A.K.: A nonequilibrium statistical mechanical model of structural relaxation in glass. J. Am. Ceram. Soc. **89**(3), 1091–1094 (2006)
27. Zwanzig, R.: Nonequilibrium Statistical Mechanics. Oxford University, New York (2001)
28. Mauro, J.C., Loucks, R.J., Gupta, P.K.: Metabasin approach for computing the master equation dynamics of systems with broken ergodicity. J. Phys. Chem. A **111**, 7957–7965 (2007)
29. Wales, D.J.: Energy Landscapes. Cambridge University, Cambridge (2003)
30. Middleton, T.F., Wales, D.J.: Energy landscapes of model glasses. II. Results for constant pressure. J. Chem. Phys. **118**(10), 4583–4593 (2003)
31. Mauro, J.C., Loucks, R.J., Balakrishnan, J.: Split-step eigenvector-following technique for exploring enthalpy landscapes at absolute zero. J. Phys. Chem. B **110**(10), 5005–5011 (2006)
32. Mauro, J.C., Loucks, R.J., Balakrishnan, J.: A simplified eigenvector-following technique for locating transition points in an energy landscape. J. Phys. Chem. A **109**(42), 9578–9583 (2005)
33. Wang, F., Landau, D.P.: Determining the density of states for classical statistical models: a random walk algorithm to produce a flat histogram. Phys. Rev. E **64**, 056101 (2001)
34. Mauro, J.C., Loucks, R.J., Balakrishnan, J., Raghavan, S.: Monte Carlo method for computing density of states and quench probability of potential energy and enthalpy landscapes. J. Chem. Phys. **126**, 194103 (2007)
35. Mauro, J.C., Loucks, R.J.: Selenium glass transition: a model based on the enthalpy landscape approach and nonequilibrium statistical mechanics. Phys. Rev. B **76**, 174202 (2007)
36. Moynihan, C.T., Easteal, A.J., DeBolt, M.A., Tucker, J.: Dependence of the fictive temperature of glass on cooling rate. J. Am. Ceram. Soc. **59**(1–2), 12–16 (1976)
37. Mauro, J.C., Varshneya, A.K.: Model interaction potentials for selenium from ab initio molecular simulations. Phys. Rev. B **71**, 214105 (2005)
38. Sreeram, A.N., Varshneya, A.K., Swiler, D.R.: Molar volume and elastic properties of multicomponent chalcogenide glasses. J. Non-Cryst. Solids **128**(3), 294–309 (1991)
39. Senapati, U., Varshneya, A.K.: Viscosity of chalcogenide glass-forming liquids: an anomaly in the 'strong' and 'fragile' classification. J. Non-Cryst. Solids **197**(2–3), 210–218 (1996)
40. Palmer, R.G.: Broken ergodicity. Adv. Phys. **31**(6), 669–735 (1982)
41. Goldstein, S., Lebowitz, J.L.: On the (Boltzmann) entropy of non-equilibrium systems. Physica D **193**(1–4), 53–66 (2004)
42. Speedy, R.J.: The entropy of glass. Mol. Phys. **80**(5), 1105–1120 (1993)
43. Jäckle, J.: On the glass transition and the residual entropy of glasses. Philos. Mag. B **44**(5), 533–545 (1981)

44. Mauro, J.C., Gupta, P.K., Loucks, R.J.: Continuously broken ergodicity. J. Chem. Phys. **126**, 184511 (2007)
45. Lebowitz, J.L.: Microscopic origins of irreversible macroscopic behavior. Physica A **263**, 516–527 (1999)
46. Lebowitz, J.L.: Statistical mechanics: a selective review of two central issues. Rev. Mod. Phys. **71**(2), S346–S357 (1999)
47. Kivelson, D., Reiss, H.: Metastable systems in thermodynamics: consequences, role of constraints. J. Phys. Chem. B **103**, 8337–8343 (1999)
48. Angell, C.A.: Spectroscopy simulation and scattering, and the medium range order problem in glass. J. Non-Cryst. Solids **73**(1–3), 1–17 (1985)
49. Angell, C.A.: Structural instability and relaxation in liquid and glassy phases. J. Non-Cryst. Solids **102**(1–3), 205–221 (1988)
50. Angell, C.A.: Relaxation in liquids, polymers and plastic crystals—strong/fragile patterns and problems. J. Non-Cryst. Solids **131–133**(1), 13–31 (1991)
51. Angell, C.A., Ngai, K.L., McKenna, G.B., McMillan, P.F., Martin, S.W.: Relaxation in glassforming liquids and amorphous solids. J. Appl. Phys. **88**(6), 3113–3157 (2000)
52. Angell, C.A.: Liquid fragility and the glass transition in water and aqueous solutions. Chem. Rev. **102**, 2627–2650 (2002)
53. Kauzmann, W.: The nature of the glassy state and the behavior of liquids at low temperatures. Chem. Rev. **43**, 219–256 (1948)
54. Huang, D., Simon, S.L., McKenna, G.B.: Equilibrium heat capacity of the glass-forming poly (α-methyl styrene) far below the Kauzmann temperature: the case of the missing glass transition. J. Chem. Phys. **119**(7), 3590–3593 (2003)
55. Stillinger, F.H., Debenedetti, P.G., Truskett, T.M.: The Kauzmann paradox revisited. J. Phys. Chem. B **105**(47), 11809–11816 (2001)
56. Stillinger, F.H., Debenedetti, P.G.: Phase transitions, Kauzmann curves, and inverse melting. Biophys. Chem. **105**(2), 211–220 (2003)
57. Tool, A.Q., Eichlin, C.G.: Variations caused in the heating curves of glass by heat treatment. J. Am. Ceram. Soc. **14**, 276–308 (1931)
58. Ritland, H.N.: Limitations of the fictive temperature concept. J. Am. Ceram. Soc. **39**(12), 403–406 (1956)
59. Giovambattista, N., Stanley, H.E., Sciortino, F.: Cooling rate, heating rate, and aging effects in glassy water. Phys. Rev. E **69**, 050201(R) (2004)
60. Lubchenko, V., Wolynes, P.G.: Theory of aging in structural glasses. J. Chem. Phys. **121**(7), 2852–2865 (2004)
61. Mauro, J.C., Loucks, R.J., Gupta, P.K.: Fictive temperature and the glassy state. Phys. Rev. E. (2008, submitted)

Advanced modulation formats for fiber optic communication systems

John C. Mauro · Srikanth Raghavan

Originally published in the journal Sci Model Simul, Volume 15, Nos 1–3, 283–312.
DOI: 10.1007/s10820-008-9106-0 © Springer Science+Business Media B.V. 2008

Abstract   Choice of modulation format plays a critical role in the design and performance of fiber optic communication systems. We discuss the basic physics of electro-optic phase and amplitude modulation and derive model transfer functions for ideal and non-ideal Mach-Zehnder modulators. We describe the generation and characteristics of the standard nonreturn-to-zero (NRZ) modulation format, as well as advanced formats such as return-to-zero (RZ), carrier-suppressed RZ (CSRZ), duobinary, modified duobinary, differential phase-shift keyed (DPSK), and return-to-zero DPSK (RZ-DPSK). Finally, we discuss the relative merits of these formats with respect to a variety of system impairments.

Keywords   Fiber optics · Modulation formats · Optical communication system · Modeling

1 Introduction and background

Recent advances in glass science and processing technology have enabled the development of low-loss silica fibers for long-haul transmission of optical signals. Various types of modulation formats can be used to encode the data, and the relative merits of these modulation formats is currently a topic of great interest, as modern optical communication systems are characterized by high bit rates and complex channel and network architecture. Modeling and simulation play a critical role in evaluating the impact of modulation format on the resulting system performance and reliability. With modeling it is possible to isolate the effects of different types of system impairments, including linear phenomena (such as dispersion and amplifier noise) and nonlinear phenomena (such as self-phase modulation, cross-phase modulation, and four-wave mixing). Modeling thus provides great insight into the physics of optical communication systems and enables cost-effective optimization of system parameters to maximize performance.

J. C. Mauro (✉) · S. Raghavan
Science and Technology Division, Corning Incorporated, Corning, NY 14831, USA
e-mail: mauroj@corning.com

In this paper, we begin with an overview of the electro-optic effect and then derive transfer functions for electro-optic modulators. We describe in detail how to model each of the following modulation formats: nonreturn-to-zero (NRZ), return-to-zero (RZ) with different duty cycles, chirped RZ (CRZ), carrier-suppressed RZ (CSRZ), duobinary, modified duobinary, differential phase-shift keyed (DPSK), and return-to-zero DPSK (RZ-DPSK). Finally, we discuss the relative merits of these formats with respect to a variety of linear and nonlinear system impairments.

2 Modulation techniques

In the most general sense, modulation involves impressing a prescribed time-dependent waveform upon a continuous-wave (CW) signal. In the context of optical communication systems, the CW signal is typically monochromatic light from a laser with a wavelength suitable for low-loss propagation (around 1300–1600 nm). The CW light is modulated using an electrical signal with a bit rate of 2.5–40 Gb/s.

Optical modulation can be accomplished using a number of different techniques, including direct modulation, electro-absorption modulation, and electro-optic modulation. In this report, we will focus on electro-optic modulation techniques because they yield the highest quality signal and are thus most suitable for generating advanced modulation formats. In particular, we will examine electro-optic phase modulators suitable for generating phase-shift keyed (PSK) formats, and electro-optic amplitude modulators used for generating amplitude-shift keyed (ASK) formats.

2.1 The electro-optic effect

Electro-optic phase and amplitude modulation is based on a phenomenon where the refractive index of a material changes when it is subjected to an applied electric field. This electro-optic effect [1] is caused when the application of a dc or low-frequency electric field distorts the positions, orientations, or shapes of atoms or molecules in a material, thereby resulting in an increase in its refractive index. Generally, the electro-optic effect comes in one of two forms:

1. The refractive index changes linearly with the applied electric field, a form known as the linear electro-optic effect or the Pockels effect; or
2. The refractive index changes quadratically with the applied electric field, known as the quadratic electro-optic effect or the Kerr effect.

Hence, the refractive index of an electro-optic material can be written as

$$n(E) = n_0 + a_1 E + \frac{1}{2} a_2 E^2, \tag{1}$$

where n_0 is the refractive index with no applied field, and $a_{1,2}$ are the expansion coefficients of the refractive index for weak fields. Clearly the second and third terms in Equation (1) correspond to the Pockels and Kerr effects, respectively.

The most common material used for electro-optic modulators is crystalline lithium niobate ($LiNbO_3$), wherein the Kerr effect is negligible and the Pockels effect dominates, i.e., $a_2 \ll a_1$. Thus, we may confine our attention to Pockels media and rewrite Equation (1) as

$$n(E) \approx n_0 - \frac{1}{2} \varrho n_0^3 E, \tag{2}$$

where $\varrho = -2a_1/n_0^3$ is the Pockels coefficient and has typical values in the range of 10^{-12} to 10^{-10} m/V. Other Pockels media include crystalline $NH_4H_2PO_4$, KH_2PO_4, $LiTaO_3$, and CdTe [1].

2.2 Phase modulators

Perhaps the simplest application of the Pockels effect is in the use of electro-optic phase modulators. When a beam of CW light travels through a Pockels medium of length L under an applied electric field E, the light undergoes a phase shift of

$$\varphi = \varphi_0 - \pi \frac{\varrho n_0^3 E L}{\lambda_0}, \tag{3}$$

where λ_0 is the free-space wavelength of the light and $\varphi_0 = 2\pi n_0 L/\lambda_0$. If the electric field is obtained by applying a voltage V across two faces of the Pockels medium separated by distance d, then $E = V/d$ and Equation (3) yields

$$\varphi = \varphi_0 - \pi \frac{V}{V_\pi}. \tag{4}$$

The quantity

$$V_\pi = \frac{d}{L} \frac{\lambda_0}{\varrho n_0^3} \tag{5}$$

is termed the half-wave voltage [1] of the modulator and represents the magnitude of the applied voltage required to induce a phase shift of π.

Thus, one can modulate the phase of a CW optical signal by passing it through a Pockels medium subjected to a varying applied voltage $V(t)$. The key parameter that characterizes an electro-optic phase modulator is the half-wave voltage, V_π, which depends on material properties such as n_0 and ϱ, the shape of the device in terms of its aspect ratio d/L, and the optical wavelength λ_0. Given an input optical signal of $E_0(t)$, the output signal from the phase modulator is

$$E(t) = E_0(t) e^{i\pi V(t)/V_\pi}. \tag{6}$$

If we define a normalized applied voltage,

$$V_0(t) = \frac{V(t)}{V_\pi}, \tag{7}$$

then we can rewrite Equation (6) as

$$E(t) = E_0(t) e^{i\pi V_0(t)}. \tag{8}$$

A schematic diagram of an electro-optic phase modulation is shown in Fig. 1.

2.3 Amplitude modulators

The ability of an electro-optic device to manipulate the phase of an optical field is also key to manipulating its amplitude. Consider, for example, a simple extension of the phase modulator setup using a Mach-Zehnder interferometer, illustrated in Fig. 2. The interferometer splits an input signal $E_0(t)$ into two paths of equal length L. Since the waveguides are composed of

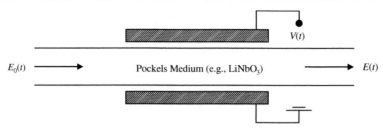

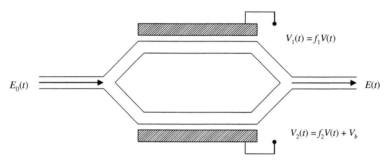

Fig. 1 Schematic diagram of an electro-optic phase modulator

Fig. 2 Schematic diagram of a Mach-Zehnder modulator

LiNbO$_3$ or another Pockels medium, an applied electric field can be used to change the refractive indices of the two arms relative to each other. The recombined field $E(t)$ at the output of the interferometer is simply the superposition of these two fields. If different voltages are applied to the two arms of the interferometer, they experience different changes in refractive index, thereby causing the optical fields to travel at slightly different speeds. Consequently, there may be a phase shift between the two fields when they are recombined at the output of the interferometer, leading to destructive interference and a subsequent decrease in the magnitude of $E(t)$ relative to $E_0(t)$. Thus, a Mach-Zehnder interferometer can be used as an amplitude modulator by controlling the amount of this destructive interference as a function of time.

2.3.1 Mach-Zehnder modulation with an ideal branching ratio

At the input of an ideal Mach-Zehnder modulator, the incident optical power is divided equally between the two arms of the interferometer. In this case, the output optical field is given by

$$E(t) = \frac{1}{2} E_0(t) \left[e^{i\beta_1(t)L} + e^{i\beta_2(t)L} \right], \tag{9}$$

where $\beta_1(t)$ and $\beta_2(t)$ are the phase shifts per unit length induced in the upper and lower arms, respectively. Application of Euler's formula to Equation (9) yields

$$E(t) = \frac{E_0(t)}{2} \{\cos[\beta_1(t)L] + i\sin[\beta_1(t)L] + \cos[\beta_2(t)L] + i\sin[\beta_2(t)L]\}. \tag{10}$$

Employing the following trigonometric identities:

$$\cos[\beta_1(t)L] + \cos[\beta_2(t)L] = 2\cos\left[\frac{\beta_1(t) + \beta_2(t)}{2}L\right]\cos\left[\frac{\beta_1(t) - \beta_2(t)}{2}L\right]; \quad (11)$$

$$\sin[\beta_1(t)L] + \sin[\beta_2(t)L] = 2\sin\left[\frac{\beta_1(t) + \beta_2(t)}{2}L\right]\cos\left[\frac{\beta_1(t) - \beta_2(t)}{2}L\right]; \quad (12)$$

we obtain

$$E(t) = E_0(t)\cos\left[\frac{\beta_1(t) - \beta_2(t)}{2}L\right] \cdot \left\{\cos\left[\frac{\beta_1(t) + \beta_2(t)}{2}L\right]\right.$$
$$\left. + i\sin\left[\frac{\beta_1(t) + \beta_2(t)}{2}L\right]\right\}, \quad (13)$$

which can be simplified as

$$E(t) = E_0(t)\cos\left[\frac{\beta_1(t) - \beta_2(t)}{2}L\right]\exp\left[i\frac{\beta_1(t) + \beta_2(t)}{2}L\right]. \quad (14)$$

Let us assume that the two arms of the interferometer are composed of LiNbO$_3$ or another Pockels medium. Because the electro-optic response is linear, the total phase shift induced in each arm of the interferometer is directly proportional to the applied voltage. Therefore, we can write

$$\beta_1(t)L = V_1(t)\eta \quad (15)$$

and

$$\beta_2(t)L = V_2(t)\eta, \quad (16)$$

where $V_1(t)$ and $V_2(t)$ are the applied voltages in the upper and lower arms of the interferometer, respectively, and η is change in refractive index per unit voltage. Substituting these two equations into Equation (14) yields

$$E(t) = E_0(t)\cos\left[\frac{V_1(t) - V_2(t)}{2}\eta\right]\exp\left[i\frac{V_1(t) + V_2(t)}{2}\eta\right]. \quad (17)$$

We typically drive both $V_1(t)$ and $V_2(t)$ with the same voltage function $V(t)$, but at different magnitudes and with one arm under an additional bias voltage V_b. If f_1 and f_2 are the fractions of $V(t)$ applied to each arm, then we can write

$$V_1(t) = f_1 V(t) \quad (18)$$

and

$$V_2(t) = f_2 V(t) + V_b. \quad (19)$$

Expressing V_b in terms of a bias coefficient ξ and the half-wave voltage V_π, we have

$$V_b = \frac{\xi}{2}V_\pi. \quad (20)$$

We substitute this equation into Equation (19) to obtain

$$V_2(t) = f_2 V(t) + \frac{\xi}{2}V_\pi.$$

$$(21)$$

Since the half-wave voltage is defined as the applied voltage leading to a phase shift of π, we know that

$$\eta V_\pi = \pi. \tag{22}$$

Hence, if one arm of the interferometer is subjected to a voltage V_π relative to the other arm, the output field from the interferometer is zero since it undergoes complete destructive interference.

We now substitute Equations (18), (21), and (22) into our equation for $E(t)$ to obtain

$$E(t) = E_0(t) \cos\left[\pi \frac{f_1 V(t) - f_2 V(t) - \frac{\xi}{2} V_\pi}{2 V_\pi}\right] \exp\left[i\pi \frac{f_1 V(t) + f_2 V(t) + \frac{\xi}{2} V_\pi}{2 V_\pi}\right]. \tag{23}$$

Algebraic manipulation yields

$$E(t) = E_0(t) \cos\left[\frac{\pi}{2}\left((f_1 - f_2)\frac{V(t)}{V_\pi} - \frac{\xi}{2}\right)\right] \exp\left[i\frac{\pi}{2}\left((f_1 + f_2)\frac{V(t)}{V_\pi} + \frac{\xi}{2}\right)\right], \tag{24}$$

and we again define a normalized applied voltage

$$V_0(t) = \frac{V(t)}{V_\pi} \tag{25}$$

such that

$$E(t) = E_0(t) \cos\left[\frac{\pi}{2}\left((f_1 - f_2) V_0(t) - \frac{\xi}{2}\right)\right] \exp\left[i\frac{\pi}{2}\left((f_1 + f_2) V_0(t) + \frac{\xi}{2}\right)\right]. \tag{26}$$

We have now obtained a generalized expression for the output field from a Mach-Zehnder interferometer in terms of f_1 and f_2. However, it more common to specify these values in terms of swing s and chirp α' parameters, where

$$s = f_2 - f_1 \tag{27}$$

and

$$\alpha' = \frac{f_1 + f_2}{f_1 - f_2} = -\frac{f_1 + f_2}{s}. \tag{28}$$

With these definitions, our expression for $E(t)$ becomes

$$E(t) = E_0(t) \cos\left[\frac{\pi}{2}\left(s V_0(t) + \frac{\xi}{2}\right)\right] \exp\left[-i\frac{\pi}{2}\left(s\alpha' V_0(t) - \frac{\xi}{2}\right)\right]. \tag{29}$$

2.3.2 Mach-Zehnder modulation with a non-ideal branching ratio

If a Mach-Zehnder modulator has an ideal branching ratio, then the input optical power is split evenly between the two arms of the interferometer such that half enters the upper arm and half enters the lower arm. Equation (29) is derived under this assumption; however, in reality the branching ratio may not be ideal. Let us consider the generalized case where a fraction a of the input power enters the upper arm and the remaining fraction $1 - a$ of the input power enters the lower arm. The output field from the interferometer in this case is given by

$$E(t) = E_1(t) e^{i\beta_1(t)L} + E_2(t) e^{i\beta_2(t)L} \tag{30}$$

$$= E_0(t) \left[a e^{i\beta_1(t)L} + (1-a) e^{i\beta_2(t)L} \right]. \tag{31}$$

Substituting Equations (15) and (16) into this equation, we obtain

$$E(t) = E_0(t) \left[a e^{i V_1(t)\eta} + (1-a) e^{i V_2(t)\eta} \right]. \tag{32}$$

Next, we substitute in Equations (18) and (21) to get

$$E(t) = E_0(t) \left[a \exp\left[i f_1 V(t) \eta\right] + (1-a) \exp\left[i \left(f_2 V(t) + \frac{\xi}{2} V_\pi \right) \eta \right] \right], \tag{33}$$

and from Equation (22) we have

$$E(t) = E_0(t) \left[a \exp\left[i\pi f_1 \frac{V(t)}{V_\pi}\right] + (1-a) \exp\left[i\pi \left(f_2 \frac{V(t)}{V_\pi} + \frac{\xi}{2} \right)\right] \right]. \tag{34}$$

Using the normalized voltage in Equation (25), we obtain

$$E(t) = E_0(t) \left[a \exp\left[i\pi f_1 V_0(t)\right] + (1-a) \exp\left[i\pi \left(f_2 V_0(t) + \frac{\xi}{2} \right)\right] \right]. \tag{35}$$

Finally, we wish to express our expression for $E(t)$ in terms of swing s and chirp α' parameters, defined in Equations (27) and (28), respectively. From these equations, we can write

$$f_1 = -\frac{s(1+\alpha')}{2} \tag{36}$$

and

$$f_2 = \frac{s(1-\alpha')}{2}. \tag{37}$$

Therefore, we obtain the following expression for the output optical field given a non-ideal power branching ratio:

$$E(t) = E_0(t) \left\{ a \exp\left[-i\frac{\pi}{2} s(1+\alpha') V_0(t)\right] \right. $$
$$\left. + (1-a) \exp\left[i\frac{\pi}{2} \left[s(1-\alpha') V_0(t) + \xi\right]\right] \right\}. \tag{38}$$

2.3.3 Calculation of extinction ratio

A Mach-Zehnder modulator with a non-ideal branching fraction ($a \neq \frac{1}{2}$) leads to pulses with a finite extinction ratio X, defined by

$$X = \frac{I_{\max}}{I_{\min}}, \tag{39}$$

where $I_{\max}$ is the maximum intensity of output field (corresponding to a "one" bit) and $I_{\min}$ is the minimum intensity (corresponding to a "zero" bit). With an ideal Mach-Zehnder modulator, the intensity of a "zero" bit is $I_{\min} = 0$, so X is infinite. A finite extinction ratio means that the output intensity of a "zero" bit is some finite value $I_{\min} > 0$, a condition which necessarily degrades the performance of the system.

In order to calculate extinction ratio, we must first determine values for I_{max} and I_{min} in a non-ideal Mach-Zehnder modulator. From Equation (38), we see that the normalized output intensity from such a modulator is

$$I(t) = \left| \frac{E(t)}{E_0(t)} \right|^2 \tag{40}$$

$$= \left| a \exp\left[-i\frac{\pi}{2}s\,(1+\alpha')\,V_0(t) \right] + (1-a) \exp\left[i\frac{\pi}{2}\left[s\,(1-\alpha')\,V_0(t) + \xi \right] \right] \right|^2 \tag{41}$$

$$= a^2 + (1-a)^2 + 2a\,(1-a)\cos\left[\frac{\pi}{2}s\,(1+\alpha')\,V_0(t) \right.$$
$$\left. + \frac{\pi}{2}\left[s\,(1-\alpha')\,V_0(t) + \xi \right] \right] \tag{42}$$

$$= 2a^2 - 2a + 1 + 2a\,(1-a)\cos\left[\pi s\,V_0(t) + \frac{\pi}{2}\xi \right]. \tag{43}$$

It follows that

$$I_{max} = 2a^2 - 2a + 1 + 2a\,(1-a) \tag{44}$$

$$= 1 \tag{45}$$

and

$$I_{min} = 2a^2 - 2a + 1 - 2a\,(1-a) \tag{46}$$

$$= 4a^2 - 4a + 1 \tag{47}$$

$$= (2a-1)^2. \tag{48}$$

Therefore, extinction ratio can be expressed as a function of a as

$$X(a) = \frac{1}{(2a-1)^2}. \tag{49}$$

Please note that $X(a) \to \infty$ for the ideal case (when $a = \frac{1}{2}$). Also, it is common to express extinction ratio in terms of decibel units:

$$X_{dB}(a) = -10\log_{10}\left[(2a-1)^2 \right]. \tag{50}$$

Given an extinction ratio X, the effective branching fraction a can be calculated by

$$a = \frac{1}{2}\left(1 + \frac{1}{\sqrt{X}} \right), \tag{51}$$

or

$$a = \frac{1}{2}\left(1 + 10^{-X_{dB}/20} \right). \tag{52}$$

2.3.4 Chirp induced by Mach-Zehnder modulation

Besides extinction ratio, another important parameter governing system performance is the chirp of the optical pulses. Generally speaking, chirp degrades system performance by broadening the optical spectrum of the signal without adding any new information. The chirp of an optical signal, $\alpha(t)$, is defined in terms of a ratio between the phase and intensity modulations [2–4],

$$\alpha(t) = \frac{-\dfrac{\partial\phi(t)}{\partial t}}{\dfrac{1}{2I(t)}\dfrac{\partial I(t)}{\partial t}}. \tag{53}$$

Assuming an ideal Mach-Zehnder modulator where the output optical field is governed by Equation (29), the phase of the signal is given by

$$\phi(t) = -\frac{\pi}{2}\left(s\alpha'V_0(t) - \frac{\xi}{2}\right). \tag{54}$$

Hence, we can calculate the phase modulation as

$$\frac{\partial \phi(t)}{\partial t} = -\frac{\pi}{2}\frac{\partial}{\partial t}\left(s\alpha'V_0(t) - \frac{\xi}{2}\right) \tag{55}$$

$$= -\frac{\pi}{2}s\alpha'\frac{\partial V_0(t)}{\partial t}. \tag{56}$$

Equation (29) also gives us

$$I(t) = [E_0(t)]^2\cos^2\left[\frac{\pi}{2}\left(sV_0(t) + \frac{\xi}{2}\right)\right], \tag{57}$$

such that

$$E(t) = \sqrt{I(t)}\exp[i\phi(t)]. \tag{58}$$

We can calculate the chirp, $\alpha(t)$, by taking the time derivative of $I(t)$:

$$\frac{\partial I(t)}{\partial t} = 2E_0(t)\frac{\partial E_0(t)}{\partial t}\cos^2\left[\frac{\pi}{2}\left(sV_0(t) + \frac{\xi}{2}\right)\right]$$
$$-\pi s[E_0(t)]^2\frac{\partial V_0(t)}{\partial t}\cos\left[\frac{\pi}{2}\left(sV_0(t) + \frac{\xi}{2}\right)\right]\sin\left[\frac{\pi}{2}\left(sV_0(t) + \frac{\xi}{2}\right)\right]. \tag{59}$$

Therefore,

$$\frac{1}{2I(t)}\frac{\partial I(t)}{\partial t} = \frac{1}{E_0(t)}\frac{\partial E_0(t)}{\partial t} - \frac{\pi}{2}s\frac{\partial V_0(t)}{\partial t}\tan\left[\frac{\pi}{2}\left(sV_0(t) + \frac{\xi}{2}\right)\right], \tag{60}$$

and it follows that

$$\alpha(t) = \frac{\frac{\pi}{2}s\alpha'\frac{\partial V_0(t)}{\partial t}}{\frac{1}{E_0(t)}\frac{\partial E_0(t)}{\partial t} - \frac{\pi}{2}s\frac{\partial V_0(t)}{\partial t}\tan\left[\frac{\pi}{2}\left(sV_0(t) + \frac{\xi}{2}\right)\right]}. \tag{61}$$

Under CW conditions for the input optical signal, i.e., $E_0(t) = E_0$, Equation (61) reduces to

$$\alpha(t) = -\alpha'\cot\left[\frac{\pi}{2}\left(sV_0(t) + \frac{\xi}{2}\right)\right]. \tag{62}$$

This expression shows that chirp can be minimized by setting $\alpha' = 0$. From Equation (28), we see that $\alpha' = 0$ can be obtained by setting $f_1 = -f_2$, i.e., by driving both arms of the interferometer by electrical voltage functions $V_{1,2}(t)$ of the same amplitude but opposite sign. In some less expensive Mach-Zehnder modulators, voltage is applied to only one arm of the interferometer; these "single-drive" modulators necessarily lead to chirped output pulses and generally poorer system performance.

3 Modulation formats

In the previous section, we have defined several parameters which control the characteristics of the output pulses from a Mach-Zehnder modulator, viz.,

- Normalized drive voltage, $V_0(t)$;
- Chirp factor, α';
- Bias, ξ;
- Swing, s; and
- Extinction ratio, X_{dB}.

In this section, we describe how these parameters can be adjusted to generate optical signals with various modulation formats. Since different modulation formats have different advantages and disadvantages, choice of modulation format is an important consideration when designing optical communication systems.

3.1 Nonreturn-to-zero (NRZ)

The most basic of all modulation formats is nonreturn-to-zero (NRZ). Figure 3 shows an NRZ signal generated using an ideal Mach-Zehnder modulator. The four plots in Fig. 3 show different aspects of the same signal. Figure 3(a) shows a plot of optical power versus time to show the form of the actual bit sequence. Figure 3(b) shows the same information displayed as an "eye diagram," which represents a superposition of all bits in the signal on top of each other; a larger eye "opening" indicates less noise or distortion and therefore a higher quality signal. Note that each bit slot occupies a time of 0.1 ns; hence the bit rate of the system is (1 bit)/(0.1 ns) = 10 Gb/s. Figure 3(c) shows the phase of the optical signal as a function of time; in the case of ideal NRZ modulation, all pulses have identical phase. (The phase of the "zero" bits is plotted as zero since here there is no light for which to measure the phase.) Finally, Fig. 3(d) shows the optical spectrum of the NRZ signal, where we note sharp peaks at multiples of the bit rate.

With the NRZ modulation format, each "one" pulse occupies an entire bit slot, and the signal intensity does not drop to zero between successive "one" bits. As shown in Fig. 3(a), the only time when the signal intensity changes between the "one" and "zero" levels is when the bit sequence itself changes between these states. Since rising or falling between these two intensity levels takes some finite amount of time governed by the electronics that generate the RF signal, we would like to describe the shape of the NRZ intensity modulation by a function that is the same in any bit slot. Since the intensities of "one" bits do not return to zero at the edges of the bit slots, adjacent NRZ pulses can be expressed as a sum:

$$V_{0,n}(t) = \sum_{j=n-1}^{n+1} a_j S(t), \tag{63}$$

where $V_{0,n}(t)$ is the driving function for the nth pulse in a bit stream, a_j is the bit value (zero or one), and $S(t)$ is the shape function. Mathematically, we can write $S(t)$ as

$$S(t) = \begin{cases} \frac{1}{2}\left(1 + \sin\left(\frac{\pi t}{T_R}\right)\right), & -\frac{T_R}{2} < t < \frac{T_R}{2} \\ 1, & \frac{T_R}{2} < t < T_B - \frac{T_R}{2} \\ \frac{1}{2}\left(1 - \sin\left(\frac{\pi}{T_R}(t - T_B)\right)\right), & T_B - \frac{T_R}{2} < t < T_B + \frac{T_R}{2} \end{cases}, \tag{64}$$

where T_B is the bit period and T_R is the rise time (and fall time) of the pulse. The above functional form ensures that half the rising portion of the pulse falls in the preceding bit period and half the falling portion falls in the succeeding bit period with the cross-over value of $1/2$ occurring at the bit boundary. This ensures that successive ones (e.g., the "111" bit pattern) are exactly compensated in value by neighboring bits and thus yield a constant intensity.

In order to preserve the logic of the bit sequence encoded in the $V_0(t)$ electrical drive, "0" bits in the electrical domain should correspond to "0" bits in the optical domain, and "1" electrical bits should correspond to "1" optical bits. This is accomplished by driving the modulator at $V_0(t) = V_\pi$ for "0" bits (to achieve destructive interference) and $V_0(t) = 0$ for "1" bits (to achieve constructive interference); hence, for NRZ modulation we require $\xi = 2$ and $s = -1$. Chirp factor and extinction ratio are ideally $\alpha' = 0$ and $X_{dB} = \infty$, respectively, but other values can still be used to generate (somewhat degraded) NRZ modulation.

3.2 Return-to-zero (RZ)

Return-to-zero (RZ) modulation is accomplished by sending an NRZ-modulated optical signal into a second Mach-Zehnder modulator with a sinusoidal driving function:

$$V_0(t) = A \cos\left(2\pi \nu_c t - \pi\theta\right), \tag{65}$$

where A is the amplitude (as a fraction of V_π), ν_c is the frequency of the RZ drive, and θ is the phase offset as a fraction of π. Since the NRZ signal going into the RZ modulator has

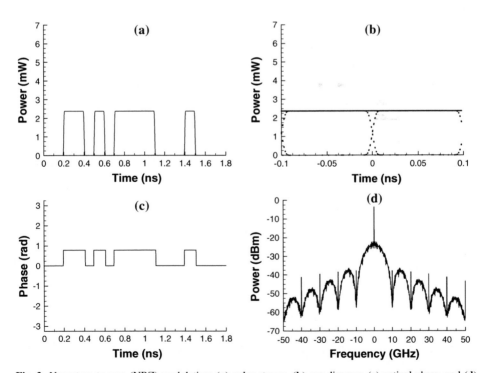

Fig. 3 Nonreturn-to-zero (NRZ) modulation: (**a**) pulse stream, (**b**) eye diagram, (**c**) optical phase, and (**d**) optical power spectrum

already been encoded with the appropriate sequence of ones and zeros, the RZ modulator does not need a separate electrical input for the desired bit stream.

RZ modulation takes three popular forms: RZ with 50% duty cycle, RZ with 33% duty cycle, and carrier-suppressed RZ (CSRZ) with 67% duty cycle. Duty cycle refers to the ratio of the full-width at half-maximum (FWHM) of the pulse intensity to the time duration of the entire bit slot:

$$d_c = \frac{T_{FWHM}}{T_B}. \tag{66}$$

We will discuss the methods of generating each form of RZ in the subsections below, but first let us specify the properties that pulses must satisfy in order to be classified as ideally "return-to-zero":

1. $E(t)$ should be periodic with period equal to T_B;
2. $E(t)$ should be zero at the bit period boundaries, $t = 0$ and $t = T_B$;
3. $E(t)$ should reach its maximum value at the center of the bit period, $t = T_B/2$; and
4. $E(t)$ should not have any other local maxima or minima, i.e., $dE(t)/dt \neq 0$, except at the minima, $t = 0$ and T_B, and at the maximum, $t = T_B/2$.

3.2.1 RZ with 50% duty cycle

We generate RZ pulses with 50% duty cycle using the following drive parameters for the RZ Mach-Zehnder modulator:

$\alpha' = 0$	$\xi = 1$	$s = 1/2$
$v_c = B$	$A = 1$	$\theta = 0$

where B is the bit rate of the incident NRZ signal. In order to explain the origin of these parameters, let us consider the illustration in Fig. 4. The upper curve shows the transfer function of the Mach-Zehnder modulator, with 100% transmittance at $V_0(t) = 0$ and 0% transmittance at $V_0(t) = V_\pi$. The lower curve shows the electrical drive behavior used to generate RZ 50% signals. The drive voltage varies sinusoidally in time, as indicated by Equation (65), with a frequency of $v_c = B$ and an amplitude of $sA = 1/2$ of the half-wave voltage. (While we have chosen $A = 1$ and $s = 1/2$, any combination of s and A that satisfies $sA = 1/2$ will yield the same results.) Finally, the bias (dc) voltage is $V_b = V_\pi/2$, so by Equation (20) we have $\xi = 1$. The phase $\theta = 0$ is chosen to align the RZ modulator drive with the input NRZ signal, and we have chosen an ideal chirp value of $\alpha' = 0$.

Assuming an ideal Mach-Zehnder modulator, the output RZ pulses will be described by

$$E(t) = E_0(t) \cos\left[\frac{\pi}{2}\left(sV_0(t) + \frac{\xi}{2}\right)\right] \exp\left[-i\frac{\pi}{2}\left(s\alpha'V_0(t) - \frac{\xi}{2}\right)\right] \tag{67}$$

$$= E_0(t) \cos\left[\frac{\pi}{2}\left(sA\cos(2\pi v_c t - \pi\theta) + \frac{\xi}{2}\right)\right] \exp\left[-i\frac{\pi}{2}\left(s\alpha'V_0(t) - \frac{\xi}{2}\right)\right], \tag{68}$$

where $E_0(t)$ now represents the input NRZ optical signal. Substituting the parameter values from above, we obtain

$$E(t) = E_0(t) \cos\left[\frac{\pi}{4}\cos(2\pi Bt) + \frac{\pi}{4}\right] \exp\left[i\frac{\pi}{4}\right]. \tag{69}$$

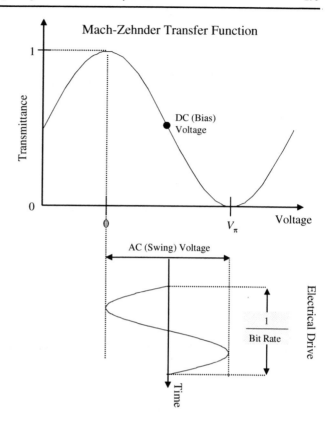

Fig. 4 Drive settings for the generation of RZ pulses with 50% duty cycle

We can use this equation to confirm that these pulses have a duty cycle of 50%. The duty cycle is determined by the full-width at half maximum of the pulse intensity as a percentage of the entire bit period. When the intensity is at half its value, the field amplitude reaches $1/\sqrt{2}$ of its maximum value. Thus, we first determine the time during the bit interval when the field amplitude reaches $1/\sqrt{2}$ of its maximum value, and we find from Equation (69) that

$$\cos\left(\frac{\pi}{4}\cos(2\pi Bt) + \frac{\pi}{4}\right) = \frac{1}{\sqrt{2}} = \cos\left(\pm\frac{\pi}{4}\right). \tag{70}$$

This implies that $\cos(2\pi Bt) = 0$, which happens at $t = \pm 1/(4B)$. Thus the duty cycle is given by

$$d_c = \frac{\frac{1}{4B} - \left(-\frac{1}{4B}\right)}{\frac{1}{B}}\,(100\%) = 50\%. \tag{71}$$

Figure 5 shows output from the RZ modulator. The time and eye diagrams show that the pulse intensity returns to zero between consecutive "1" bits; hence, unlike NRZ, all RZ pulses have exactly the same pulse shape, independent of the neighboring bit values. As with NRZ, the phases of the RZ pulses are identical and chirp-free under ideal modulation. Figure 5(d) shows the optical spectrum of the RZ signal, which is much broader than that of NRZ due to the reduced width of the RZ pulses.

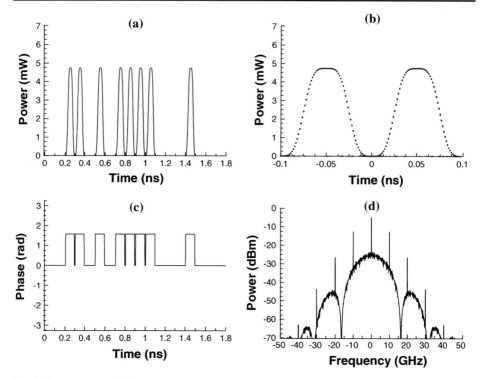

Fig. 5 Return-to-zero (RZ) modulation with 50% duty cycle: (**a**) pulse stream, (**b**) eye diagram, (**c**) optical phase, and (**d**) optical power spectrum

Figure 6 shows the RZ 50% modulation format using non-ideal NRZ and RZ Mach-Zehnder modulators, both with an extinction ratio of $X_{dB} = 13$ dB. Here, we see several drawbacks of non-ideal modulation, viz.,

1. Since the "0" level of the NRZ modulator does not reach zero intensity, Fig. 6(a) shows that both the "1" and "0" levels are modulated by the RZ modulator, leading to the eye-closure penalty shown in Fig. 6(b);
2. Due to the non-ideal branching ratio, the pulses are necessarily chirped, as shown in Fig. 6(c); and
3. The non-ideal optical spectrum in Fig. 6(d) is much broader than that in Fig. 5(d), thereby increasing cross-talk with any neighboring channels.

3.2.2 RZ with 33% duty cycle

In order to generate RZ pulses with 33% duty cycle, we use the following Mach-Zehnder drive parameters:

$\alpha' = 0$	$\xi = 0$	$s = -1$
$v_c = B/2$	$A = 1$	$\theta = 0$

which are illustrated in Fig. 7. Here, we see that the drive signal is biased about the maximum transmission point of the modulator and the swing voltage is twice as large as in the RZ 50% case. Also, since the period of the RZ 33% drive voltage is twice that of RZ 50%, one RZ 33%

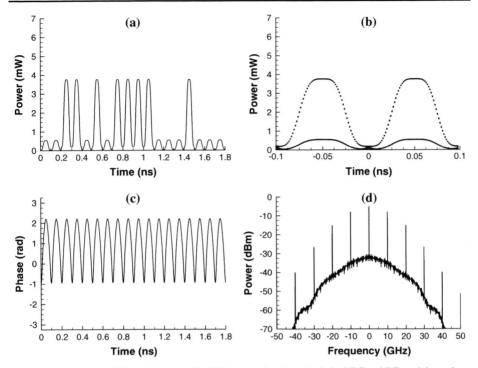

Fig. 6 Return-to-zero (RZ) modulation with 50% duty cycle, where both the NRZ and RZ modulators have a 13 dB extinction ratio: (**a**) pulse stream, (**b**) eye diagram, (**c**) optical phase, and (**d**) optical power spectrum

pulse is generated for each half-period of the RZ 33% sinusoidal drive. In other words, one pulse is generated during each transition between the nodes of the Mach-Zehnder transfer function.

Mathematically, we can write the optical field of the RZ 33% signal as

$$E(t) = E_0(t) \cos\left[\frac{\pi}{2}\cos(\pi Bt)\right]. \tag{72}$$

Therefore, the duty cycle can be calculated as

$$\cos\left[\frac{\pi}{2}\cos(\pi Bt)\right] = \frac{1}{\sqrt{2}} = \cos\left(\pm\frac{\pi}{4}\right); \tag{73}$$

$$\Rightarrow \cos(\pi Bt) = \pm\frac{1}{2}; \tag{74}$$

$$\Rightarrow t = \frac{1}{3B} \text{ and } \frac{2}{3B}, \tag{75}$$

such that

$$d_c = \frac{\frac{2}{3B} - \frac{1}{3B}}{\frac{1}{B}}(100\%) = 33\%. \tag{76}$$

Figure 8 shows the properties of the RZ 33% modulation format. Note that because of its shorter duty cycle, RZ 33% has a much larger eye opening than both NRZ and RZ 50%. Like ideal NRZ and RZ 50%, ideal RZ 33% pulses all have the same phase. Finally, comparing

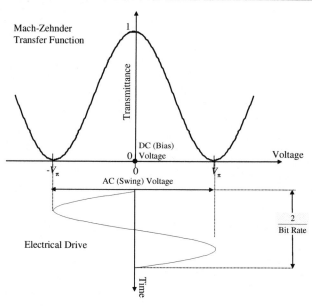

Fig. 7 Drive settings for the generation of RZ pulses with 33% duty cycle

the optical spectra of the RZ 50% and RZ 33% signals, we note that while the initial lobe centered at $f = 0$ is larger for RZ 33% than for RZ 50%, the outer lobes are suppressed much faster for RZ 33%. This is because the RZ 33% pulse shape is much closer to an ideal Gaussian shape than RZ 50%.

3.2.3 Carrier-suppressed RZ (CSRZ) with 67% duty cycle

Carrier-suppressed RZ (CSRZ) is similar to standard RZ except that the phase of the pulses alternates by π every consecutive bit period. The Mach-Zehnder drive parameters for this case are:

$\alpha' = 0$	$\xi = 2$	$s = -1$
$\nu_c = B/2$	$A = 1$	$\theta = 1/2$

Figure 9 shows that, as with RZ 33%, the sinusoidal drive voltage has a frequency half that of RZ 50% and swings at twice the amplitude. However, unlike RZ 33%, the CSRZ drive is biased at the minimum transmittance point of the Mach-Zehnder transfer function. During each half-period, the voltage swings from minimum to maximum transmittance and then back to minimum transmittance again. However, since this swing is in the opposite direction for consecutive bit periods, adjacent pulses are π-phase shifted from each other.

Using the drive parameters above, the amplitude of the CSRZ signal is given by

$$|E(t)| = |E_0(t)| \sin\left[\frac{\pi}{2} \sin(\pi B t)\right]. \tag{77}$$

Hence, the duty cycle can be calculated by

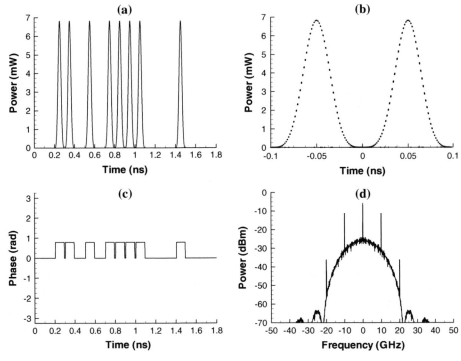

Fig. 8 Return-to-zero (RZ) modulation with 33% duty cycle: (**a**) pulse stream, (**b**) eye diagram, (**c**) optical phase, and (**d**) optical power spectrum

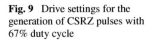

Fig. 9 Drive settings for the generation of CSRZ pulses with 67% duty cycle

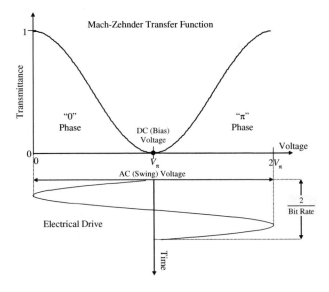

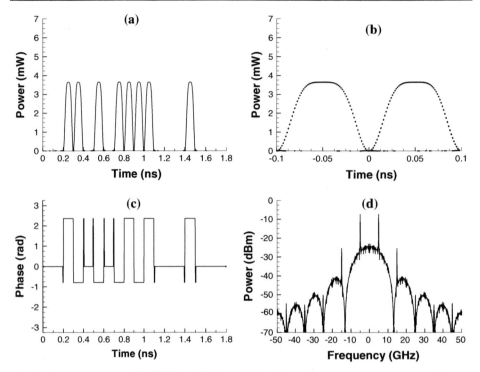

Fig. 10 Carrier-suppressed RZ (CSRZ) modulation with 67% duty cycle: (**a**) pulse stream, (**b**) eye diagram, (**c**) optical phase, and (**d**) optical power spectrum

$$\sin\left[\frac{\pi}{2}\sin\left(\pi Bt\right)\right] = \frac{1}{\sqrt{2}} = \sin\left(\frac{\pi}{4}\right) \text{ or } \sin\left(\frac{3\pi}{4}\right); \tag{78}$$

$$\Rightarrow \sin\left(\pi Bt\right) = \frac{1}{2} = \sin\left(\frac{\pi}{6}\right) \text{ or } \sin\left(\frac{5\pi}{6}\right); \tag{79}$$

$$\Rightarrow t = \frac{1}{6B} \text{ or } \frac{5}{6B}; \tag{80}$$

Therefore,

$$d_c = \frac{\frac{5}{6B} - \frac{1}{6B}}{\frac{1}{B}}(100\%) = 67\% \tag{81}$$

Figure 10 shows the properties of the CSRZ 67% modulation format. In Fig. 10(c), we see that the phase alternates by π every bit period. Over a long bit sequence, this leads to destructive interference of the carrier frequency: note in Fig. 10(d) that there is no peak at $f = 0$. We will see later in this report that carrier-suppression helps to reduce the interactions between neighboring pulses and thus improve signal quality.

3.2.4 Chirped RZ (CRZ)

Thus far, we have only considered RZ modulation with $\alpha' = 0$. If a less expensive single-drive modulator is used, then the chirp factor becomes $\alpha' = \pm 1$, and the pulses are chirped according to Equation (61). Figure 11 shows chirped RZ (CRZ) pulses with 50% duty cycle,

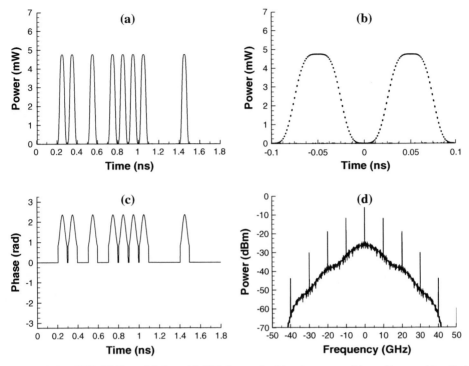

Fig. 11 Chirped RZ (CRZ) modulation with 50% duty cycle: (**a**) pulse stream, (**b**) eye diagram, (**c**) optical phase, and (**d**) optical power spectrum

generated with the same parameters as RZ 50% except that $\alpha' = 1$. This figure shows that the pulse power as a function of time is exactly the same for CRZ 50% as for RZ 50%. However, the phase varies within the time span of each pulse, and hence the spectrum is significantly broader. Although the chirp of the pulses can be used to counteract fiber dispersion [5], it generally leads to increased cross-talk penalty and lower overall performance.

We should also note that CRZ 33% can be generated using the same parameters as RZ 33%, except for $\alpha' \neq 0$. If $\alpha' \neq 0$ with CSRZ 67% parameters, pulses are chirped with opposite sign in adjacent bit periods. In this case, the modulation format is known as alternate chirp RZ (ACRZ), and the signal is no longer carrier-suppressed.

3.3 Duobinary

We have seen that alternating the phase between adjacent bit periods can lead to a carrier-suppressed optical spectrum, as in CSRZ. The duobinary modulation format combines a more intelligent kind of phase control with an NRZ pulse shape. A duobinary signal is generated by first encoding an input electrical bit sequence into a three-level electrical waveform [6]. This three-level waveform, with the "0" bits having a value of "0" and the "1" bits having a value of "1" or "−1", is then used as the driving function for a Mach-Zehnder modulator. The change in sign between the "1" and "−1" states reflects itself as a π phase shift between sequences of "one" bits in the output optical signal.

Given an input binary sequence a_n, the duobinary encoded sequence is generated as follows:

1. Invert the bits.

$$b_n = 1 - a_n \tag{82}$$

2. Generate an auxiliary bit sequence.

$$c_n = (b_n + c_{n-1}) \bmod 2 \tag{83}$$

3. Encode the duobinary sequence.

$$d_n = c_n + c_{n-1} - 1 \tag{84}$$

For example, an input sequence of

$$0, 0, 1, 1, 0, 1, 0, 1, 1, 1, 1, 0, 0, 0, 1, 0, 0, 0$$

is encoded as

$$0, 0, -1, -1, 0, 1, 0, -1, -1, -1, -1, 0, 0, 0, 1, 0, 0, 0.$$

Otherwise, the Mach-Zehnder drive parameters are the same as for NRZ.

An example duobinary signal is shown in Fig. 12. Since there are no phase shifts between directly adjacent "1" bits, the NRZ electrical waveform converts into an NRZ optical signal. This is in contrast to modified duobinary (discussed in the next subsection), where the phase shifts of consecutive "1" bits cause the optical signal to become RZ, even when the input electrical waveform is NRZ in shape. Figure 12(c) shows that with duobinary modulation the phase of consecutive groups of "1" bits is shifted by π. As shown in Fig. 12(d), this results in a narrow optical spectrum without any sharp peaks.

3.4 Modified duobinary

Modified duobinary, also known as alternate mark inversion (AMI), is an improved version of duobinary where the phase alternates by π between every pair of "1" bits, regardless of how many "0" bits, if any, separate them [6,7]. As with duobinary, modified duobinary involves generating a three-level electrical waveform using an input bit sequence a_n and a counter sequence c_n. This electrical waveform is then used as the driving function for a Mach-Zehnder modulator to generate a two-level optical signal with "1" bits of alternating phase.

With modified duobinary, a counter sequence c_n begins with a value of "0" and then toggles state at each "1" bit in the input bit sequence a_n. For example:

$a_n =$	0 1 1 0 0 1 0 0 1 1 1 0 0 1 0 0
$c_n =$	0 1 0 0 0 1 1 1 0 1 0 0 0 1 1 1

Then, a three-level electrical waveform b_n is generated by combining a_n and c_n through the logical procedure shown below.

a_n	c_n	$a_n \times c_n$	$\overline{c_n}$	$a_n \times \overline{c_n}$	$b_n = (a_n \times c_n) - (a_n \times \overline{c_n})$
1	1	1	0	0	1
1	0	0	1	1	-1
0	1	0	0	0	0
0	0	0	1	0	0

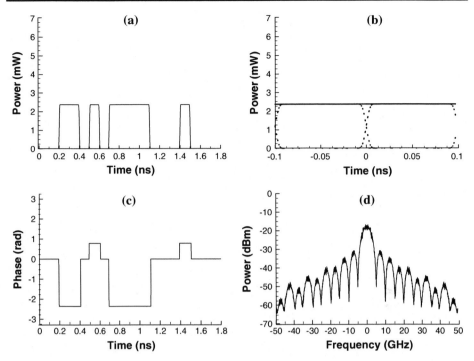

Fig. 12 Duobinary modulation: (**a**) pulse stream, (**b**) eye diagram, (**c**) optical phase, and (**d**) optical power spectrum

When this electrical waveform is converted to optical amplitude modulation using a Mach-Zehnder interferometer, the positive "1" bits yield optical "1" bits with a zero phase shift, and the negative "1" bits yield optical "1" bits with a π phase shift. Please note that no two adjacent "1" bits are in phase in the resulting optical signal. Thus, although modified duobinary is typically driven with an NRZ electrical waveform, adjacent "1" bits in the output optical signal behave as if they were in an RZ sequence.

Consider the example modified duobinary signal in Fig. 13 and note the return-to-zero shape of the optical pulses. The phase plot in Fig. 13(c) shows that each "1" bit is π phase-shifted from its two neighboring "1" bits, regardless if there are any "0" bits in between. Finally, the optical spectrum plot in Fig. 13(d) shows that modified duobinary, because of its intelligent phase encoding, yields a very high degree of carrier suppression, much greater even than CSRZ (Fig. 10).

3.5 Differential phase-shift keyed (DPSK)

Thus far, we have only discussed amplitude-shift keyed (ASK) formats, where the bit sequence is encoded in the amplitude of the optical signal. Alternatively, phase-shift keyed (PSK) formats encode information in the phase of the optical signal. The most important PSK formats for fiber optic communication systems are differential phase shift keyed (DPSK), and return-to-zero DPSK (RZ-DPSK).

As with the duobinary formats, DPSK formats require encoding of the input bit sequence [8]. Given an input bit sequence a_n, an auxiliary bit sequence b_n is generated by $b_n =$

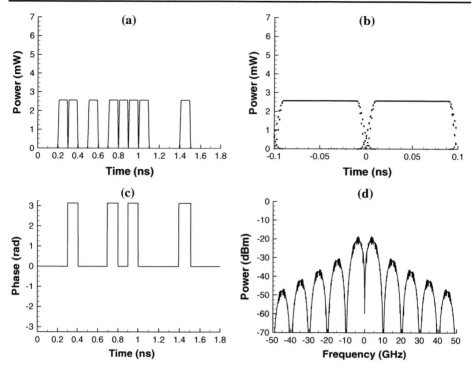

Fig. 13 Modified duobinary modulation: (**a**) pulse stream, (**b**) eye diagram, (**c**) optical phase, and (**d**) optical power spectrum

$\overline{(a_n \oplus b_{n-1})}$, where $\oplus$ denotes exclusive OR. The auxiliary sequence is one bit longer than the input sequence and always begins with $b_0 = 1$. This auxiliary sequence is then encoded into the phase of a CW optical stream using an electro-optic phase modulator, where "0" bits in b_n are encoded as a zero phase shift and "1" bits in b_n are encoded as a π phase shift. The example below illustrates the encoding procedure.

$a_n =$		0	1	1	0	0	1	0	0	1	1	1	0	0	1	0	0
$b_{n-1} =$		1	0	0	0	1	0	0	1	0	0	0	0	1	0	0	1
$b_n =$	1	0	0	0	1	0	0	1	0	0	0	0	1	0	0	1	0
$\phi_n =$	π	0	0	0	π	0	0	π	0	0	0	0	π	0	0	π	0

However, using a standard photodetector we can only detect optical power, not phase. Therefore, in order to detect the DPSK signal, we must first convert the phase modulation to amplitude modulation.

Decoding is accomplished by passing the phase-modulated signal through a Mach-Zehnder interferometer just before the receiver. One arm of the interferometer has a time delay of one bit period, T_B, with respect to the second arm. Hence, at the output of the interferometer the phase data, ϕ_n, interferes with a bit-shifted copy of itself, ϕ_{n-1}, resulting in either constructive or destructive interference. Constructive interference (if ϕ_n and ϕ_{n-1} are of the same phase) produces a "1" bit, whereas destructive interference (if ϕ_n and ϕ_{n-1} are of opposite phase) produces a "0" bit. The following example demonstrates this procedure.

$\phi_n =$		π	0	0	0	π	0	0	π	0	0	0	0	π	0	0	π	0
$\phi_{n-1} =$	π	0	0	0	π	0	0	π	0	0	0	0	π	0	0	π	0	
$a_n =$		0	1	1	0	0	1	0	0	1	1	1	0	0	1	0	0	

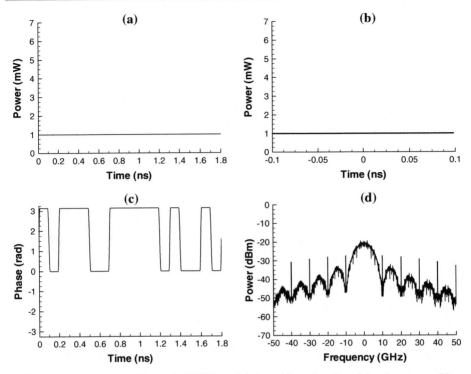

Fig. 14 Differential phase-shift keyed (DPSK) modulation without decoding: (**a**) pulse stream, (**b**) eye diagram, (**c**) optical phase, and (**d**) optical power spectrum

Notice that the decoded signal a_n is exactly the same as the input bit sequence above.

A DPSK-encoded signal is shown in Fig. 14. Here, we see that whereas the optical power remains constant in time, information is encoded in the phase of the signal. After decoding using a Mach-Zehnder interferometer, the amplitude-modulated signal is shown in Fig. 15. At this point, the signal is almost identical to duobinary, except for the occurrence of sharp peaks between adjacent bits slots. These peaks form because of the finite time taken to switch between the 0 and π phase states in the phase modulation process. When a phase shift occurs between two adjacent bits, the phase becomes $\pi/2$ at the bit boundary. Therefore, when decoding occurs, the $\pi/2$ phase at one bit boundary can align with the $\pi/2$ phase at the adjacent bit boundary, leading to constructive interference and the appearance of a sharp spike in the resulting amplitude-modulated signal.

The quality of the received DPSK signal can be improved by using a second photodetector in a setup known as the "balanced receiver." When the phase-modulated signal interferes with the bit-shifted copy of itself in the decoding process, the two signals combine via a directional coupler [9]. The constructive interference propagates through one output port of the coupler, while the destructive interference is outputted at the second port. Hence, "0" bits at the destructive port appear with "1" intensity. Two photodetectors are used: one to measure the constructive output and one to measure the destructive output. Finally, the current from the destructive photodetector is subtracted from the constructive current such that "0" bits have a current of "−1." This effectively doubles the eye opening of DPSK signals and improves the receiver sensitivity by 3 dB [10, 11].

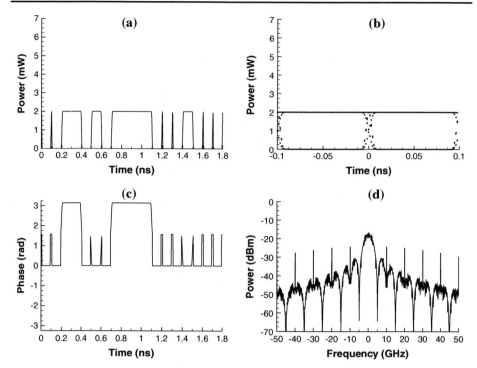

Fig. 15 Differential phase-shift keyed (DPSK) modulation after decoding: (**a**) pulse stream, (**b**) eye diagram, (**c**) optical phase, and (**d**) optical power spectrum

3.6 Return-to-zero DPSK (RZ-DPSK)

The return-to-zero DPSK (RZ-DPSK) modulation format is generated by sending a DPSK-encoded optical signal through a Mach-Zehnder modulator set for generation of RZ 50%, RZ 33%, or CSRZ 67% pulses. As shown in Fig. 16 for RZ-DPSK 33%, all information is still encoded in the phase of the optical signal. The optical intensity takes on the shape of the RZ pulses, but unlike ASK RZ formats there are no bit periods with "0" intensity. Figure 16(d) shows that the RZ-DPSK 33% spectrum is very similar to that of RZ 33%, except that the RZ-DPSK format contains no peaks in the spectrum.

Decoding of RZ-DPSK signals is accomplished in the same manner as for standard DPSK. As shown by the decoded signal in Fig. 17, RZ-DPSK has two main advantages over DPSK. Firstly, the eye opening is much greater because of its reduced duty cycle. Secondly, since the intensity of the signal goes to zero at the bit boundaries, there are no spikes between bits as observed in standard DPSK.

4 Impact on system performance

Various properties of the optical fiber can lead to penalties that deteriorate system performance:

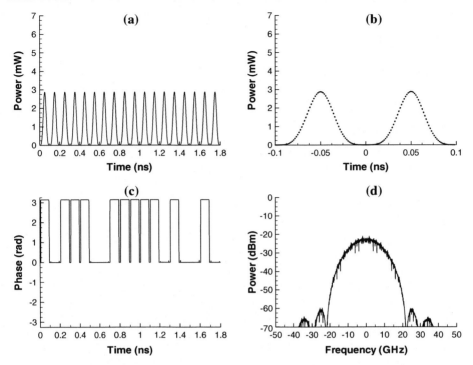

Fig. 16 Return-to-zero DPSK (RZ-DPSK) modulation with 33% duty cycle, without decoding: (**a**) pulse stream, (**b**) eye diagram, (**c**) optical phase, and (**d**) optical power spectrum

1. Attenuation due to absorption and Rayleigh scattering makes it necessary to amplify the signal periodically. Amplification necessarily introduces noise in the system that degrades the signal quality.
2. The finite low-loss bandwidth of the fiber (and limitations on transmitter and receiver wavelengths) often force adjacent optical channels to partially overlap. Data from one signal interferes with data from the adjacent signal, thereby creating linear cross-talk penalty.
3. Chromatic dispersion and polarization mode dispersion lead to distortion and patterning of optical pulses as they broaden and shift temporally.
4. Nonlinear optical effects such as self-phase modulation (SPM), cross-phase modulation (XPM), and four-wave mixing (FWM) generate noise and distortion in the signal.

Because of their different temporal, spectral, and phase characteristics, different modulation formats display different tolerances to these penalties. In this section, we will give a brief overview describing the behavior of the various modulation formats with respect to each of the above impairments. But before we proceed, we should take care to describe how system performance is quantified.

The performance of each channel can be quantified in terms of the Q parameter, which has different definitions depending on the experimental measurement and underlying assumptions about the nature of the noise distribution. The value of Q is commonly measured in one of two ways. Experimentally, a bit error rate tester (BERT) can be used to detect the exact number of errors in a very long bit sequence after transmission through the system. The Q parameter is then obtained from the bit error rate (BER) by inverting [12]

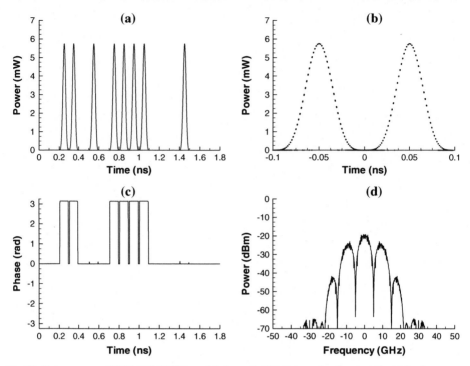

Fig. 17 Return-to-zero DPSK (RZ-DPSK) modulation with 33% duty cycle, after decoding: (**a**) pulse stream, (**b**) eye diagram, (**c**) optical phase, and (**d**) optical power spectrum

$$BER = \frac{1}{2}\text{erfc}\left(\frac{Q}{\sqrt{2}}\right) \approx \frac{\exp\left(-Q^2/2\right)}{Q\sqrt{2\pi}}, \tag{85}$$

where Q is given in linear units. Note that the approximation holds for large values of Q. Also, it is common to express Q in terms of dB units by $Q_{dB} = 10\log_{10} Q$.

The second method for obtaining Q involves measuring the mean and the standard deviation values of the "1" and "0" levels using an eye diagram produced by a digital communication analyzer (DCA). In this case, Q in linear units is given by

$$Q = \frac{I_1 - I_0}{\sigma_1 + \sigma_0}, \tag{86}$$

where $I_{1,0}$ and $\sigma_{1,0}$ refer to the mean and the standard deviation values of the "1" and "0" levels, respectively. A larger Q value indicates a lower BER and hence better performance. Note, however, that Equation (86) assumes Gaussian statistics for the noise distribution, an assumption that is often violated in modern communication systems [13–15].

4.1 Amplified spontaneous emission (ASE) noise

Optical amplification has significantly increased the capacity and reach of fiber optic communication systems. The primary drawback of such amplification is the unavoidable generation of amplified spontaneous emission (ASE) noise, which leads to degradation of the optical signal to noise ratio (OSNR) and decreased Q. In fact, all current optical communication systems are fundamentally limited by ASE noise.

ASE noise can be classified into two components: signal-spontaneous beat noise and spontaneous-spontaneous beat noise [16–18]. Signal-spontaneous beat noise represents a beating between the signal and noise power at the receiver and contributes primarily to the σ_1 term in Equation (86). Spontaneous–spontaneous noise refers to the noise power beating with itself and is hence a much smaller effect. If we assume $I_0 \approx 0$ and $\sigma_0 \approx 0$, then the Q factor can be approximated as

$$Q \approx \frac{I_1}{\sigma_1}. \tag{87}$$

Since the signal-spontaneous noise contribution, σ_1, increases proportionally with I_1 for a given average signal power, we can conclude that the ASE contribution to Q is independent of eye opening. Hence, the only way to increase tolerance to ASE noise is to increase the OSNR of the signal, i.e., increase the average channel power. Since average channel power is limited by nonlinear optical effects such as SPM, XPM, and FWM, the performance of various modulation formats with respect to ASE is intimately connected to its tolerance of nonlinear impairments. In other words, a higher tolerance to nonlinear impairments means that more power can be launched into the fiber, thereby improving the OSNR and increasing the Q factor.

4.2 Fiber nonlinearities

Fiber nonlinearities can be classified into many different types, including self-phase modulation (SPM), cross-phase modulation (XPM), and four-wave mixing (FWM) [19]. The dominant nonlinear optical penalty depends on the design of the system. In dense wavelength-division multiplexed (DWDM) systems, channels are typically spaced close together, e.g., 50 GHz channel spacing for 10 Gb/s systems or 100 GHz channel spacing for 40 Gb/s systems. Cross-channel effects such as XPM and FWM typically dominate in 10 Gb/s DWDM systems, but SPM can also be an issue. Typically, nonlinear penalties can be suppressed by spreading out the power of the signal in the spectral domain. Hence, reduced duty cycle formats such as RZ 33% and RZ-DPSK 33% display greater tolerance to nonlinear effects than spectrally narrow formats such as NRZ [20–22].

The spectrum of a 40 Gb/s signal is four times as broad as the same signal at 10 Gb/s, so cross-channel nonlinearities are greatly suppressed. However, the various frequencies within a single channel can mix with each other via intrachannel nonlinear effects, viz., intrachannel cross-phase modulation (iXPM) and intrachannel four-wave mixing (iFWM) [23]. Typically, iXPM is the dominant nonlinear penalty in fibers with low chromatic dispersion, e.g., nonzero-dispersion-shifted (NZDSF) fiber with $D = 4$ ps/nm/km, and iFWM dominates in fibers with high chromatic dispersion, e.g., standard single-mode fiber (SMF) with $D = 17$ ps/nm/km.

Whereas iFWM is sensitive to the phase of pulses, iXPM depends only on intensity. A shorter duty cycle leads to significant suppression of iXPM penalty, so RZ 33% and RZ-DPSK 33% are good choices if this impairment is dominant (i.e., for NZDSF systems). On the other hand, iFWM can be suppressed by alternating the phases of adjacent bits. CSRZ 67% provides some suppression of iFWM penalty [24], but the maximum performance can be obtained through the intelligent phase control of modified duobinary [25–27].

4.3 Linear cross-talk

Another problem associated with DWDM systems is the overlap of the spectra of adjacent channels, leading to linear cross-talk penalty. The most effective way to suppress this penalty is to use a spectrally compact format such as NRZ, duobinary, modified duobinary, or DPSK. In general, as the duty cycle of the pulses decreases, the linear cross-talk penalty can be expected to increase [28–30].

However, it is also important to account for the specific shapes of the pulses. As mentioned previously, RZ 33% offers greater suppression of its outer spectral lobes compared to RZ 50% because of its near-Gaussian shape. Hence, although RZ 33% has a shorter duty cycle than RZ 50%, it also experiences less linear cross-talk penalty [31].

4.4 Chromatic dispersion

Ideally, the net chromatic dispersion at the end of a system is zero since dispersion-compensating fiber (DCF) is used to counteract the dispersion of the transmission fiber. However, temperature fluctuations can lead to nonzero dispersion at the receiver, and hence it is important to consider the impact of a finite net residual dispersion on the performance of various modulation formats. In general, dispersion acts to broaden pulses in the time domain; therefore, short duty cycle formats such as RZ 33% have a distinct advantage in that they have more room to broaden before encroaching on pulses in the neighboring bit slots [32,33]. NRZ and duobinary have especially poor tolerance to dispersion since their duty cycle is effectively 100%.

DPSK and RZ-DPSK formats have perhaps the worst tolerance to dispersion of all formats since they are very sensitive to any phase distortions. Also, dispersion can act to convert phase modulation to amplitude modulation, thereby generating noise in the received signal.

4.5 Polarization mode dispersion (PMD)

Polarization mode dispersion (PMD) occurs due to random birefringence in optical fibers, which creates a differential group delay (DGD) between the two principal states of polarization. The DGD manifests itself as a random pulse distortion and acts to degrade the Q of the system. In general, shorter duty cycle formats can withstand more DGD distortion and thus yield a higher Q value. However, the greatest tolerance to PMD has been displayed by DPSK and RZ-DPSK formats [34].

PMD is generally not a limiting impairment in today's optical fiber due to the use of fiber spinning and other techniques that minimize birefringence [35].

5 Conclusions

In summary, we have discussed the basic physics of how electro-optic modulators work and derived the transfer functions for ideal and non-ideal Mach-Zehnder modulators. We have described the application of Mach-Zehnder and phase modulators to generate a variety of modulation formats. Different modulation formats offer their own advantages and disadvantages with respect to various fiber impairments, and choice of modulation format depends largely on the type of system being implemented. Short duty cycle formats such as RZ 33% and RZ-DPSK 33% offer a high tolerance to most nonlinear and dispersion penalties, but they are less spectrally compact than NRZ, duobinary, and DPSK. Because of their alternating phase behavior, CSRZ 67% and modified duobinary give advantage in high bit rate SMF systems where ifWM is the dominant impairment.

Acknowledgements We would like to acknowledge the valuable assistance and support of a number of researchers at Corning Incorporated, including Sergey Ten, Yihong Mauro, Dipak Chowdhury, Shiva Kumar, and Jan Conradi.

References

1. Saleh, B., Teich, M.: Fundamentals of Photonics, Chap. 18. John Wiley & Sons, New York (1991)
2. Koyama, F., Iga, K.: Frequency chirping in external modulators. J. Lightwave Technol. **6**(1), 87–93 (1988)
3. Djupsjöbacka, A.: Residual chirp in integrated-optic modulators. IEEE Photon. Technol. Lett. **4**(1), 41–43 (1992)
4. Fishman, D.A.: Design and performance of externally modulated 1.5-μm laser transmitter in the presence of chromatic dispersion. J. Lightwave Technol. **11**(4), 624–632 (1993)
5. Mu, R.-M., Yu, T., Grigoryan, V.S., Menyuk, C.R.: Dynamics of the chirped return-to-zero modulation format. J. Lightwave Technol. **20**(1), 47–57 (2002)
6. Haykin, S.: Communication Systems, 4th edn. John Wiley & Sons, New York (2001)
7. Cheng, K.S., Conradi, J.: Reduction of pulse-to-pulse interaction using alternative RZ formats in 40-Gb/s systems. IEEE Photon. Technol. Lett. **14**(1), 98–100 (2002)
8. Proakis, J.G.: Digital Communications, 3rd edn, Chap. 5. McGraw-Hill, New York (1995)
9. Ghatak, A., Thyagarajan, K.: An Introduction to Fiber Optics, Chap. 17. Cambridge University Press, (1998)
10. Zhu, B., Nelson, L.E., Stulz, S., Gnauck, A.H., Doerr, C., Leuthold, J., Grüner-Nielsen, L., Pedersen, M.O., Kim, J., Lingle, R.L. Jr..: High spectral density long-haul 40-Gb/s transmission using CSRZ-DPSK format. J. Lightwave Technol. **22**(1), 208–214 (2004)
11. Gnauck, A.H., Liu, X., Wei, X., Gill, D.M., Burrows, E.C.: Comparison of modulation formats for 42.7-Gb/s single-channel transmission through 1980 km of SSMF. IEEE Photon. Technol. Lett. **16**(3), 909–911 (2004)
12. Agrawal, G.P.: Fiber-Optic Communication Systems, 2nd edn. John Wiley & Sons, New York (1997)
13. Georges, T.: Bit error rate degradation of interacting solitons owing to non-Gaussian statistics. Electron. Lett. **31**(14), 1174–1175 (1995)
14. Chandrasekhar, S., Liu, X.: Experimental study on 42.7-Gb/s forward-error-correction performance under burst errors. IEEE Photon. Technol. Lett. **20**(11), 927–929 (2008)
15. Lobanov, S., Raghavan, S., Downie, J., Mauro, Y., Sauer, M., Hurley, J.: Impact of uncompensated dispersion on non-Gaussian statistics in duobinary transmission. Opt. Eng. **46**, 010501 (2007)
16. Olsson, N.A.: Lightwave systems with optical amplifiers. J. Lightwave Technol. **7**(7), 1071–1082 (1989)
17. Marcuse, D.: Derivation of analytical expressions for the bit-error probability in lightwave systems with optical amplifiers. J. Lightwave Technol. **8**(12), 1816–1823 (1990)
18. Kumar, S., Mauro, J.C., Raghavan, S.: Impact of modulation format and filtering on the calculation of amplified spontaneous emission noise penalty. J. Opt. Comm. **25**, 945–953 (2004)
19. Agrawal, G.P.: Nonlinear Fiber Optics, 2nd edn. Academic Press, San Diego (1995)
20. Hayee, M.I., Willner, A.E.: NRZ versus RZ in 10-40-Gb/s dispersion-managed WDM transmission systems. IEEE Photon. Technol. Lett. **11**(8), 991–993 (1999)
21. Breuer, D., Petermann, K.: Comparison of NRZ- and RZ-modulation format for 40-Gb/s TDM standard-fiber systems. IEEE Photon. Technol. Lett. **9**(3), 398–400 (1997)
22. Suzuki, M., Edagawa, N.: Dispersion-managed high-capacity ultra-long-haul transmission. J. Lightwave Technol. **21**(4), 916–929 (2003)
23. Kumar, S., Mauro, J.C., Raghavan, S., Chowdhury, D.Q.: Intrachannel nonlinear penalties in dispersion-managed transmission systems. IEEE J. Sel. Top. Quant. Electron. **8**(3), 626–631 (2002)
24. Appathurai, S., Mikhailov, V., Killey, R.I., Bayvel, P.: Suppression of intra-channel nonlinear distortion in 40 Gbit/s transmission over standard single mode fibre using alternate-phase RZ and optimised pre-compensation. Proc. Euro. Conf. Opt. Comm. Paper Tu3.6.5 (2003)
25. Forzati, M., Mårtensson, J., Berntson, A., Djupsjöbacka, A., Johannisson, P.: Reduction of intra-channel four-wave mixing using the alternate-phase RZ modulation format. IEEE Photon. Technol. Lett. **14**(9), 1285–1287 (2002)

26. Kanaev, A.V., Luther, G.G., Kovanis, V., Bickham, S.R., Conradi, J.: Ghost-pulse generation suppression in phase-modulated 40-Gb/s RZ transmission. J. Lightwave Technol. **21**(6), 1486–1489 (2003)

27. Mauro, J.C., Raghavan, S., Ten, S.: Generation and system impact of variable duty cycle α-RZ pulses. J. Opt. Comm. **26**, 1015–1021 (2005)

28. Bosco, G., Carena, A., Curri, V., Gaudino, R., Poggiolini, P.: On the use of NRZ, RZ, and CSRZ modulation at 40 Gb/s with narrow DWDM channel spacing. J. Lightwave Technol. **20**(9), 1694–1704 (2002)

29. Hoshida, T., Vassilieva, O., Yamada, K., Choudhary, S., Pecqueur, R., Kuwahara, H.: Optimal 40 Gb/s modulation formats for spectrally efficient long-haul DWDM systems. J. Lightwave Technol. **20**(12), 1989–1996 (2002)

30. Downie, J.D., Ruffin, A.B.: Analysis of signal distortion and crosstalk penalties induced by optical filters in optical networks. J. Lightwave Technol. **21**(9), 1876–1886 (2003)

31. Mauro, J.C., Raghavan, S., Rukosueva, M., Stefanini, C., Ten, S.: Impact of OADMs on 40 Gb/s transmission with alternative modulation formats. Proc. Euro. Conf. Opt. Comm. Paper We4.P.93 (2003)

32. Belahlou, A., Bickham, S., Chowdhury, D., Diep, P., Evans, A., Grochocinski, J.M., Han, P., Kobyakov, A., Kumar, S., Luther, G., Mauro, J.C., Mauro, Y., Mlejnek, M., Muktoyuk, M.S.K., Murtagh, M.T., Raghavan, S., Ricci, V., Sevian, A., Taylor, N., Tsuda, S., Vasilyev, M., Wang, L.: Fiber design considerations for 40 Gb/s systems. J. Lightwave Technol. **20**(12), 2290–2305 (2002)

33. Hodžić, A., Konrad, B., Petermann, K.: Alternative modulation formats in $N \times 40$ Gb/s WDM standard fiber RZ-transmission systems. J. Lightwave Technol. **20**(4), 598–607 (2002)

34. Xie, C., Möller, L., Haunstein, H., Hunsche, S.: Comparison of system tolerance to polarization-mode dispersion between different modulation formats. IEEE Photon. Technol. Lett. **15**(8), 1168–1170 (2003)

35. Chen, X., Li, M.-J., Nolan, D.A.: Spun fibers for low PMD: understanding of fundamental characteristics. Dig. LEOS Summer Topical Meet. 123–124 (2003)

Computational challenges in the search for and production of hydrocarbons

John Ullo

Originally published in the journal Sci Model Simul, Volume 15, Nos 1–3, 313–337.
DOI: 10.1007/s10820-008-9095-z © Springer Science+Business Media B.V. 2008

Abstract   The exploration and production of oil and natural gas facing unprecedented demands for a secure energy supply worldwide is continuing a long trend to develop and adopt new technologies to help meet this challenge. For many oilfield technologies mathematical modeling and simulation have played a truly enabling role throughout their development and eventually their commercial adoption. Looking ahead, the vision of data-driven "intelligent" oilfields designed and operated using simulations to reach higher recovery factors is becoming a reality. Very little of this vision would be possible let alone make sense without the capability to move information across several simulation domains. We will examine several successes of modeling and simulation as well as current limitations which need to be addressed by new developments in theory, modeling, algorithms and computer hardware. Finally, we will mention several fundamental issues affecting oil recovery for which increased understanding is needed from new experimental methods coupled to simulation.

Keywords   Modeling and simulation · Oil and gas · Seismic imaging · Porous media · Sensors

1 Introduction

The Oil & Gas Exploration & Production (E&P) industry is facing many challenges to satisfy a growing worldwide demand for a secure energy supply for many decades to come. These include developments in harsher and more complex environments (deep water and Arctic) to locate and produce hydrocarbons, a greater emphasis on exploring for and producing unconventional resources (heavy oil, tar sands, oil shale and coal-bed methane) and a need for more real time data to manage production and risk. Increasing the quantity and quality of information about the reservoir and the capital assets deployed will enable oilfield operations with dramatically improved safety, reliability, and effectiveness as measured by both ultimate

J. Ullo (✉)
Schlumberger-Doll Research, One Hampshire St, Cambridge, MA 02141, USA
e-mail: ullo@boston.oilfield.slb.com

S. Yip & T. Diaz de la Rubia (eds.), *Sci Model Simul*, DOI: 10.1007/978-1-4020-9740-9_17

recovery factor and financial profitability. To achieve this game-changing "illumination" of the entire enterprise including its geological and capital assets, simulations of many kinds are being employed with greater sophistication. These simulations naturally span the R&D landscape with a focus on new materials and sensor systems as enabling technologies for the operational part of the enterprise. However, some of the heaviest computing and simulation applications reside on the operational side where imaging and measurements of subsurface rock and fluid properties are made over spatial scales ranging from mm to km. All of this feeds into hydrocarbon reservoir simulation as a guide to more optimal strategies for oil and gas extraction.

But big challenges still remain. More accurate measurements of rock and fluid properties must extend farther from wells to cover the entire reservoir. Critical subsurface infrastructures must be instrumented to provide real-time information about their state of health to trigger replacement or maintenance at the most cost-effective, yet still safe, times. We look forward to sensing systems that are able to penetrate away from wells within the reservoir without additional drilling. Enormous amounts of data must be combined and processed intelligently to focus the operator's attention on the salient bits for the decisions at hand.

Within this context, we will examine a selection of current practices and opportunities for exploiting large scale and multi-scale modeling and simulation—first starting with the big picture associated with geophysical imaging and inversion, then moving into subsurface measurements made from wells and finally to the design of enabling technologies such as new nano sensors, networks and materials which will take us down to the scales of atomic dimensions. Overall, we hope to portray that simulation of one kind or another is found in: 1) the daily operation of the entire enterprise, 2) the R&D of new services and understanding of fundamental processes to aid oil recovery and 3) the provision of new enabling technologies.

2 Part I. The big picture—geophysical imaging and inversion

2.1 Seismic imaging for exploration and production

Historically seismic surveying has been one of the most important techniques used in the oil industry to identify new subsurface sources of hydrocarbons. This continues today, but we also observe important extensions to the production domain using so-called time-lapse seismic imaging. Basically, seismic imaging is the oilfield analog to the commonly used MRI technique in medicine. Generating high resolution images of the earth's subsurface is essential for minimizing the substantial financial risks associated with exploration. Seismic imaging has undergone a number of paradigm shifts in capability over several decades, and each has brought forth the need to acquire, transport, store, and process greater amounts of raw data. Today a typical 3-D survey may acquire 100's of Terabytes of data which need to be subjected to multifaceted mathematical processing to generate the images which geologists and geophysicists can then interpret for the presence of hydrocarbon deposits.

A typical marine seismic survey is often acquired by a purpose built ship that systematically cruises over an area of interest and fires air guns that send powerful sound waves into the ocean. These waves propagate through the water and down through sub-seafloor layers of sandstone, shale, salt, and other materials, producing echoes that return to the surface. The ship may tow a dozen or more cables, each up to 10 km long, carrying thousands of hydrophones that measure the minute pressure waves of the returning reflected wavefield.

Several thousand square kilometers, sometimes with a quite irregular surface footprint, may be covered by a survey.

Duration of this process is often measured in months for acquisition and possibly a good fraction of a year for processing. The industry has traditionally been a rapid adopter of the latest computing, data storage and visualization technologies for decades. Today the industry can claim ownership of some of the largest computing centers on earth. A look at the top machines [46] as of this writing reveals several devoted to seismic processing, and all of them consisting of several thousand computing nodes adding up to several hundreds of Teraflops of peak performance.

However, despite already large computational resources, higher levels of performance are being demanded from increased levels of market activity and a need for higher fidelity images. The latter is a direct result of having to image evermore complex geologies. Higher resolution images reduce risk of increasingly expensive decisions. This requires an even larger density in the amount of data acquired and the application of more rigorous mathematical techniques to create the images. Data volumes and mathematical approximations have historically traded off one another depending on computational and human costs. Often if the geologies were relatively simple, these tradeoffs could be managed. This is changing with data volumes continuing to increase at high rates due to single-sensor technology [37], which allows higher density spatial sampling, and a desire for fewer physics and mathematical approximations in the processing algorithms.

To create an image the energy in the reflected wavefield measured on the surface must be extrapolated mathematically to points in the subsurface where the reflected energy originated from the original down going wavefield. This process called migration locates reflection points in the subsurface which tend to accumulate along loci associated with rapid changes in acoustic impedance which are often associated with a geological boundary or a tectonic disruption of the geology. This produces an image of impedance changes that serves as a proxy for the geometric changes in geology. Most algorithms in commercial practice today start with the Helmholtz acoustic wave equation to mathematically extrapolate energy (in reverse time) back to points in the subsurface where the reflected energy originated. An example is shown in Fig. 1a which is a 2-D vertical slice from a 3-D volume of image data. Solving the wave equation in three dimensions starts by creating a 3-D grid of acoustic wave velocity values that spans the surveyed volume of ocean and sub-seafloor earth. The dimensionality of the problem could easily range from one to five thousand intervals in each of the X, Y lateral directions and one thousand in Z depth direction to achieve 10 m resolution. In addition, the time discretization could range from four to eight thousand samples. At each time step the pressures of one or more sound waves present are assigned to each grid point. In the past many seismic imaging codes created 3-D maps of the subsurface by extrapolating using the one-way wave equation, i.e., seismic waves traveling in just one direction. Physically, this approximation accounts only for propagation of reflected energy upward from subsurface points to the surface.

One way migration techniques were generally satisfactory for simple layer-like geologies and could scale with computer power to deal with ever larger survey data sets. However, the computational load for seismic imaging continues to grow for several reasons:

1. The quest for higher resolution images has led to greater use of single sensor imaging which now can employ 30,000 (and growing) sensors at a time to achieve denser spatial sampling of the wavefields. The challenge of this massive increase in data is further magnified by growing use of computationally intensive methods to determine iteratively better subsurface velocity distributions to extrapolate surface data to the right points in

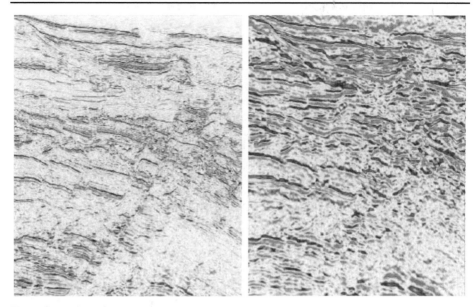

Fig. 1 Comparison of new high resolution seismic images (a-left) with conventional images (b-right). Note many fine scale features associated with geological faulting are better resolved on the left (Graphic courtesy of WesternGeco)

    depth. It is clear that these procedures have led to dramatic improvements in resolution over previous experience. Even to an untrained eye the single sensor image in Fig. 1a is far easier to interpret for subtle subsurface structure than what was available before (Fig. 1b).

2. By now it has become normal to extrapolate using the two-way wave equation. In two-way methods the wave propagation from the air guns to subsurface reflection points is also accounted for. A good image is obtained when upward reflected energy from points in the subsurface is consistent with downward propagating energy arriving at the same points from the surface. This has aided hydrocarbon discoveries in 3-D geologies with overburden and large geologic lateral heterogeneities such as salt domes and basaltic bodies that literally can mask the oil deposits (Fig. 2) unless we properly account for all seismic wave propagating directions in the subsurface.

3. More and more we wish to interpret seismic data not just in terms of the wiggles of a reflected wavefield, but that same data inverted for underlying reservoir properties that influence the sound propagation. In Fig. 3 we have an example where seismic waveform data have been inverted for acoustic impedances [8]. Additionally, this has been done for two separate times using time lapse seismic, and Fig. 3 actually shows the impedance changes that in this case are interpreted as water (in red) encroaching on a planned well path. This caused a change in the planned well path to stay above the water level and eliminated a potentially costly mistake. With accurate well log data for local calibration other reservoir properties can be estimated such as types of rock (lithology), elastic moduli, pressure, density, fluid saturations and fracture fields.

4. The relevant wave equation for seismic wave propagation is ultimately the full elastic equation which includes conversions to shear waves and the possibility of anisotropic wave propagation effects. This is only being done currently for research studies and some small pilot projects and is not yet a common commercial practice.

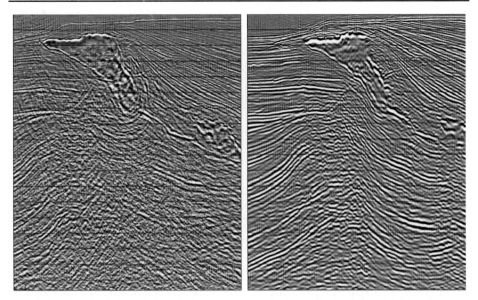

Fig. 2 Wave equation based imaging (right) exposes greater detail beneath subsurface salt bodies (upper center of the image) compared to more approximate methods (left) that limit tracking of energy that has traveled at high angles to reach the region of interest. Many sedimentary structures below such bodies can hold trapped hydrocarbons (Graphic courtesy of WesternGeco)

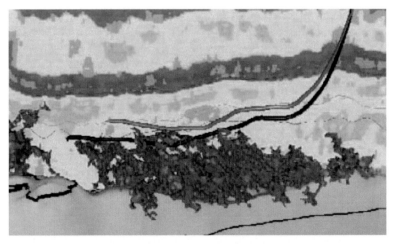

Fig. 3 Time lapse differences in acoustic impedance in a local seismic data volume show encroachment of water (red) upward past a barrier that was thought to be sealing from below [8]. The original well plan (black) was deemed to be too close to the water for efficient production. The new well plan (gray) is shown just above the water in the producing zone [8]

5. Long term the prospect of doing global inversion is a high priority. This refers to automated inversion of multiple physics data reflecting underlying earth model properties. Besides surface seismic this could include borehole-to-surface and cross borehole acoustic data and data from other imaging measurement modalities such as electromagnetic surveys which will be discussed further below because of their complementarity to acoustic techniques.

2.2 Towards inversion for reservoir properties

Formally inversion refers to estimating earth properties or more rigorously their probability distribution functions (pdfs) from calibrated seismic impedance maps by fitting synthetic seismic waveforms to measured (processed) seismic data. Contributions to the pdfs come from both data and model uncertainties. A good introduction of the subject is given by Tarantola [45], and several useful articles are given by [10,38,43]. An objective is to derive subsurface spatial properties with such reliability so as to reduce or eliminate the need for drilling appraisal wells. Eventually, we wish to link processing and inversion to bring seismic more and more into the daily reservoir engineering domain.

Most inversion problems are numerically performed by iterating on the parameters of forward models to a point where an optimal match is made between the predicted and measured seismic waveform data. When this happens the resulting features and properties of the earth model expose in principle the additional information content in the seismic wavefield that we are seeking through inversion. In mathematical terms Gauss Newton iteration is often the method of choice today. In general we wish to invert a forward partial differential equation $S\mathbf{u} = \mathbf{f}$ where the $\mathbf{f}$ represents the sources (acoustic or electromagnetic). The resulting spatially and time dependent wavefields $\mathbf{u}(\mathbf{x},t)$ are linked to $\mathbf{f}$ via the forward operator S. Often either finite difference or finite element discretizations may be used resulting in a matrix equation of the form $[S]\,\underline{\mathbf{u}} = \underline{\mathbf{f}}$ with the matrix $[S]$ usually nonlinearly dependent of the underlying model parameters and now $\underline{\mathbf{u}}$ and $\underline{\mathbf{f}}$ are column vectors.

The optimization problem is posed as a minimization of the data residuals

$$\delta d_i = u_i - d_i, \quad i = (1, 2, \dots n) \tag{1}$$

where the δd_i expresses the mismatch between the calculated u_i and the measured data d_i. The index i goes over all time samples at each surface receiver location for each source.

Commonly, a least-squares technique is used where we seek to minimimize the L_2 norm of the data residuals

$$E\left(\mathbf{p}\right) = \frac{1}{2}\delta\mathbf{d}^t\delta\mathbf{d}^* \tag{2}$$

where $\mathbf{p}$ respresents the vector of model parameters which is to be optimized by minimizing $E(\mathbf{p})$. Expanding $E(\mathbf{p})$ in a Taylor series and retaining terms up to second order yields

$$E\left(\mathbf{p} + \delta\mathbf{p}\right) = E\left(\mathbf{p}\right) + \delta\mathbf{p}^t\nabla_p E\left(\mathbf{p}\right) + 1/2\left(\delta\mathbf{p}^t H\delta\mathbf{p}\right) \tag{3}$$

where $\mathbf{H}$ is the approximate Hessian matrix having elements (i,j) given by,

$$H = \partial^2 E\left(\mathbf{p}\right)/\partial p_i \partial p_j \tag{4}$$

To update the parameter space vector $\mathbf{p}$ from one iteration (k) to the next (k+1) we minimize the quadratic form of the misfit to get

$$p^{(k+1)} = p^{(k)} - \mathbf{H}^{-1}\nabla_p E\left(\mathbf{p}\right) \tag{5}$$

The iterations proceed until some chosen convergence criterion is met.

We are mindful that such inverse methods can be risky from non uniqueness of the results. To meet this challenge substantial use of auxiliary data is made to constrain the inversion process. Also, due to high computational costs, many approximate forward modeling techniques have served as engines to drive the iterative inversion. No doubt there is growing interest in exploiting large finite difference models for practical applications, whereas up to now they have been used sparingly for that purpose. Even a modest survey block might require at least a billion spatial mesh cells and a thousand time points making each iterative forward simulation very expensive in time without access to large computing resources. Key to any success is the parametrization of the models in such a way that the degrees of freedom are limited to the order of 100–1000. A brute force voxel-based approach where we have to update and optimize the parameters for each of one billion mesh cells would be impossible. Fortunately, there are often many sources of prior information that can be exploited to vastly reduce the parameter space that has to be searched to find a best model fit. Prior information about geological boundaries can be extracted from image data, and they can be supplemented by models of geological processes that create the architectures and properties in the earth. Well log data can be used to constrain properties at or near well locations, and geostatistical methods can extend that information deeper away from the wells. Furthermore, the industry has ample expertise in linking elastic properties (seismic wave propagation) to petrophysical properties (reservoir) based on laboratory measurements of rock cores. The inverse problem computational size is still formidable, but it is becoming more tractable for large commercial data sets especially on new computing architectures that are in sight.

2.3 Evolution to 4-D (time lapse) seismic and reservoir simulations

There is little doubt that the emergence of practical 3-D seismic imaging in the early 90's led to significant increases in exploration efficiency and rapidly became a standard practice. In the meantime time-lapse or 4-D seismic has begun to emerge as a value-added proposition for managing efficient production and recovery processes [36]. Seismic reflectivity is a function of both static and dynamic reservoir properties. As dynamic properties such as pressure and fluid saturations change due to reservoir production, seismic responses change as well, opening up the possibility of using observed seismic wavefield changes as quantitative measures of reservoir property changes. This concept, while not new, has taken some time to reach a degree of practicality and has benefitted from several step changes in acquisition technology and data processing methods.

A major achievement of 4-D acquisition is the replication of the exact conditions of a baseline survey so that differences observed between the surveys are dominated by changes due to reservoir production and the injection of fluids to push the oil out. In other words, the results should repeat in areas where there are no expected changes in geophysical properties. This has been accomplished in marine environments through many innovations in acquisition employing steerable sources and receivers so as to place source and receiver data locations in the repeat surveys close to the original locations. This requires remarkable computer driven real time control of locations under dynamical conditions induced by wind, currents, tides and physical obstacles. Also in real time, new data are processed to monitor signal to noise characteristics compared to the baseline data and allow decisions to repeat parts of the data acquisition that fall outside established QC standards.

Such control of acquisition allows more reliable interpretation of changes in seismic amplitudes and relative acoustic impedances that are associated with changes in fluid saturations and pressures within the reservoir. However, the greatest impact of this will come if the data are ready to be interpreted along with reservoir simulation data in a matter of days after the acquisition. This too is benefitting from the aforementioned QC controls and the advanced computing capabilities on board to do at least some low level image processing that is good enough for a first look at changes in relative acoustic impedances. At this stage these changes can be reliable qualitative indicators of where fluid fronts are moving. Comparing the images to predictions from the reservoir simulators either validates the reliability of the existing

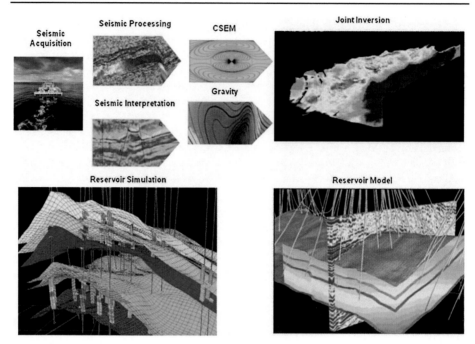

Fig. 4 The vision of the seismic-to-simulation integrated workflow clockwise from upper left. (*Upper left*)—the seismic portion showing the steps of acquisition, data processing to produce detailed images of the subsurface and subsequent interpretation of the images into geological boundaries and fault lines. (*Upper center*)—Acquisition of several kinds of complementary data ranging from well logs (e.g. gravity) to electromagnetic surveys. (*Upper right*)—Joint inversion of single and multiple data sets to derive reservoir properties over the data volume. (*Lower right*)—Creation of the reservoir model showing use of modern visualization techniques to compare seismic data with key model features. (*Lower left*)—Creation of a cellular model to transform the reservoir model into a complex 3-D grid for simulation of subsurface flows. The workflow comprises several large scale computational steps including real time acquisition QC and initial field processing of imaging data from both seismic and electromagnetic methods, subsequent data processing to refine the images using more computationally intense methods, joint inversions of the data using well log measurements as constraints, and finally mapping the reservoir properties into a simulation model. Production rates and fluid movements from simulation are tracked against actual time lapse fluid changes of the reservoir leading to model improvements with better predictive power. Today these workflows can be done approximately, but we expect the continuity of this feedback loop to improve dramatically in coming years

reservoir model or more likely is used to feedback refinements to the reservoir model so that predictions become more reliable. In this way a sophisticated feedback loop (Fig. 4) from seismic to reservoir simulation is set up which over time increases the predictive value of the simulations and decreases the risk of costly drilling decisions often associated with locating by-passed hydrocarbons.

Beyond this compute-intensive real time application, the data will continue to be processed using the sophisticated migration techniques discuss above and subjected to more elaborate interpretations using new inversion methods. The objective is fuller examination of detailed changes in seismic properties and linking these to a greater number of reservoir variables. This can go on for months, but again advances in automation, computing power and algorithmic improvements are making these analyses available in shorter times or allowing more rigorous analyses which would not have been attempted before. Either way we expect over time to achieve greater risk reduction and a positive impact on eventual recovery rates.

2.4 Inversion of seismic and electromagnetic wavefields

The problem of imaging sedimentary structures obscured by high-velocity layers, such as carbonate, basalt or salt, using conventional seismic techniques is well known (Fig. 2). When this problem is encountered in offshore areas, marine electromagnetic data can provide valuable alternative and complementary constraints on the structure. Marine controlled-source electromagnetic (CSEM) imaging in the frequency domain uses an electric dipole source to transmit a low frequency narrow band signal to an array of fixed seabed receivers or to a towed array just off the seafloor [14,42]. Overall, the geometric features of the acquisition are quite similar to seismic surveys. The measurements are guided by the fact that the amplitude and phase features of the detected electric fields are dominated by components that penetrated the seafloor and the underlying sediments which are more resistive than the column of seawater. The method also relies on the large resistivity contrasts between hydrocarbon bearing sediments and those saturated with saline fluids. Thus the returning signals contain information of subsurface resistivity structures. Since electromagnetic energy preferentially propagates in resistive media, conductive structures can be detected and delineated.

The feasibility of mapping hydrocarbon reservoir structures using this technique have been explored through numerous modeling and simulation studies. These studies generally follow the inversion procedures described above specifying subsurface models which range from simple 1-D planar geometries [14,42] to complex 3-D geological models [20] with parametrizations resembling those likely to be encountered in actual practice. Finite difference forward models for the electromagnetic field equations are used to generate the synthetic data.

An example (Fig. 5) which is based on an actual field data set demonstrates the model-based inversion procedures using a 2.5-D forward modeling approach [2]. The data were acquired over a portion of a field using twenty three electric dipole transmitters along a 13 km line at a depth around $z = 300$ m. Details of the acquisition and problem domain are given in Fig. 5. The initial model (Fig. 5a) used to start the inversion consisted of an air layer, a water layer and a sea floor layer. The inversion results are shown in Fig. 5b which also shows the depth of the reservoir estimated from seismic denoted by a dashed-line. We observe that the depth of the reservoir is accurately estimated along with its broad outline. This was for a 2-D acquisition data set with a 2.5-D forward modeling package. This can easily be done on current pc computing nodes, but scaling to large 3-D data sets with source receiver combinations similar to modern day seismic acquisitions suggests the electromagnetic inversion problem will rival that of seismic inversion in size.

Looking ahead we expect increased use of joint seismic and electromagnetic data to raise the value of geophysical inversion. We have seen that inverted seismic wavefields yield information directly on acoustic impedance maps which can be used to delineate geological boundaries as well as elastic parameters, porosity and fluid types using suitable local calibrations from well log data. Resistivity maps, on the other hand, yield information on fluid boundaries and can be used with accurate log data to yield information on fluid types, relative saturations and porosity. Thus both types of data supplement and complement each other, suggesting that a simultaneous inversion of combined data would likely result in improved estimation of flow parameters for input into reservoir simulation models and simulations. This could be done for multiple configurations of measurements, e.g., borehole-to-borehole, borehole-to-surface and surface-to-surface. Recent reported work [12] on simultaneous inversions already lends some confirmation to this expectation. We can also foresee, generalizing the concept of 4-D seismic discussed above to include electromagnetic time-lapse data to provide almost continuous monitoring of the reservoir fluid pathways. This begs the

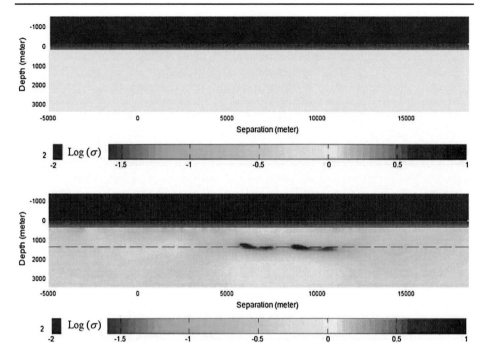

Fig. 5 Inversion of an EM field data set with 45 receiver units employed along a 24 km line at a depth around $z = 320$ m. The transmitter fundamental was 0.25 Hz. The inversion domain ranged from $x = -7$ km to $x = 19$ km and $z = 0.4$ km to $z = 3.5$ km and was discretized into 260 by 62 grid cells with sizes of 100 m by 50 m. (**a**) The initial model showing only the sea column (blue), sea bed (orange) and a uniform subsurface (yellow). (**b**) The inverted data showing the reservoir in a position consistent with the depth derived from seismic data (dashed line)

question - can better knowledge of these pathways be used to assist economic ultimate recovery? We believe the answer is yes. While the cost of this data acquisition and the inversion computational load will initially be high, we can expect continued improvements in hardware and algorithmic efficiencies. However, the value of more precise exploitation of reservoirs leading to improved recovery rates could be huge.

The prospects for computer hardware over the next five years show multicore processor speeds growing beyond a TFlop and no doubt greater access to Petaflop multi-processor computing platforms. Combined with algorithmic improvements, transformations of reservoir geophysics will continue and likely will produce:

– Much shorter imaging workflows to accomplish what we do today (10 months goes to 2 months or even days) which will certainly improve the matching of 4-D seismic imaging to reservoir engineering time scales potentially producing a game changer for reservoir recovery.
– Far more automated velocity field determinations to support accurate 3-D depth imaging (months to days).
– Multi-component imaging of fully elastic wavefields.
– New generations of sound wave inversion methods for reservoir properties using the elastic wave equation.
– Joint inversion of 3-D surface and borehole seismic data for reservoir properties.
– Joint inversion of 3-D seismic and 3-D electromagnetic data for reservoir properties.

- Vastly improved accuracy of reservoir simulations from geophysical data-driven workflows that constrain detailed flow pathways.
- Improved uncertainty quantification and control driving better decision making.

3 Part II. Formation evaluation—pore scale fundamentals of oil recovery

3.1 Borehole measurements

Once a well is drilled the industry has developed over the past 80 years many sophisticated measurement instruments for determining important physical properties of the rock/fluid formations along a well path. These instruments are often linked together and lowered to the bottom of a well on the end of a wireline cable. The entire assembly is then systematically pulled to the surface while making continuous measurement readings along the way as a function of depth (the log). The wireline serves not only as a tether but also is the means whereby raw data is transmitted to surface computers and processed in real time to allow local and remote experts to observe the properties of the formations penetrated by the well.

These measurements focus on properties such as locations of geological boundaries, porosity, formation electrical resistivities, oil vs water ratios of the fluids confined in the pore systems, fluid permeabilities, rock mineral constituents, formation fluid chemical compositions, rock elastic parameters, pressure and temperature. The modalities include electric current injection, electromagnetic waves, nuclear gamma ray radiation, neutron radiation, coupled neutron to gamma ray radiation, acoustic waves, nuclear magnetic resonance and infrared spectroscopy. We will not endeavor here to describe in detail each of these measurements. Interested readers can consult several general references, e.g., [15,24,29].

Over the past twenty-five years similar measurements have been designed and incorporated into the structural components of drill strings in order to make these measurements as a function of depth during the drilling phase of the well. This has been an extraordinary feat of mechanical engineering in addition to managing the measurement physics. These logging while drilling (LWD) techniques have enabled measurements to be made very close in time to the actual penetration which has enormous benefits for mitigating the effects of formation alteration from the drilling process.

Research and engineering development of these modern logging measurements has benefited greatly from modeling and simulation. Throughout the past three decades this effort has evolved from 1-D models based on simple, but essential, physics to rigorous 3-D time-dependent numerical models using finite difference and finite element approaches for wave propagation [1,20] and Monte Carlo techniques for nuclear radiation transport [32]. In all cases we are now able to simulate the details of the measurement instrument, its placement in a well and the details of the surrounding rock formation. In fact systematic studies of the almost infinite variety of formation characteristics would not be possible today without the capabilities to simulate their imprint on measurement responses.

3.2 Drilling and geosteering

A good example of these modeling and simulation capabilities is associated with recent practices in horizontal well drilling. Starting about 15 years ago it has become common to drill wells horizontally, and today this extended-reach drilling is reaching distances up to 20 km from the original vertical wellbore. Environmentally, this practice becomes attractive by substantially reducing the surface footprint of drilling and production facilities. However, it also has significant economic advantages. Whereas traditional vertical wells may only penetrate several 10's of meters of producing zones, a horizontal well suitably drilled to reach a reservoir target could make contact with several kilometers of a single producing zone or even intersect multiple producing zones in different isolated reservoirs.

To accomplish this geosteering, which has obvious analogies to modern medical micro-surgery practices, the well must be drilled with real time steering information. Early LWD logging measurements commonly measuring resistivity close to the drill bit were used to locate the reservoir zone and subsequently steer horizontally using a touch and feel technique. If the drill bit penetrated the top or bottom layer boundaries of the reservoir zone, the resistivity measurement would sense an immediate decrease in resistivity and indicate to the driller to steer down or up. While this practice had some success, it suffered from not being able to see farther to avoid exiting the production zone in the first place and the directional information proved to be quite crude.

Now new deep directional electromagnetic measurements exist that allow well placement by real time mapping of distances to geological boundaries [3,27]. Besides novel new sensors, the method rests on being able to use multiple data in real time to perform a model-based parametric inversion to translate the data into a local layer structural map thereby obtaining measurement distances to nearby boundaries as well as bed resistivities. Various levels of model sophistication can be supported again in real time, but none of this would be possible without the continued evolution of efficient forward modeling numerical algorithms, data handling, graphics and computer hardware to match the real time constraints of the application.

Figure 6 shows the results of this process applied to a well with a complex vertical extent of the reservoir layer some 40 feet thick extending over a horizontal interval of some 2600 feet. If only seismic imagery were used to "steer" the well, a trajectory following the cyan line would have been followed. As a result of the new directional measurement with real time model-based inversion the path denoted by the red line was followed showing excellent steering within the reservoir zone. One can also see evidence of smaller-scale intrabedding shale layers (dark color), and it is remarkable that the well trajectory was able to avoid them and preferentially penetrate the yellow color-coded hydrocarbon bearing sands.

3.3 Porescale physics

3.3.1 Introduction

Carbonate reservoirs represent roughly 60% of the remaining conventional hydrocarbon reserves. A deep understanding of these rocks that promotes greater recovery is therefore vital to our continued dependence on hydrocarbons for many decades throughout the 21st century.

Carbonate rocks are well known to have highly heterogeneous hydraulic properties that make delineating recoverable reserves and producing them uniquely difficult. Their pore space is often partially filled with brine of varying pH and salinity. These fluids have been reacting with the rock matrix minerals for eons and are typically in states of disequilibrium today. Adding hydrocarbons with a wide range of viscosity and chemical behavior further complicates the story. As a result surface effects (roughness, chemistry and wettability) can have profound influences on recovery method choices and their efficiencies. Often to enhance production in carbonate reservoirs, fluids (water, produced gas, steam, CO_2, synthetics) are

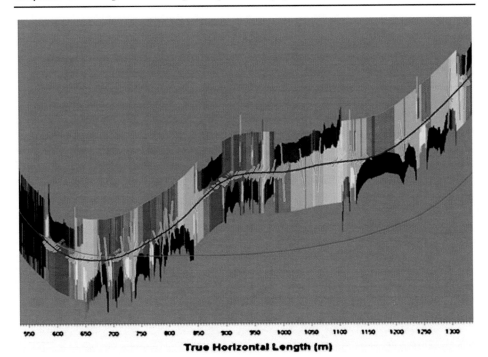

True Horizontal Length (m)

Fig. 6 Visualization of a complex well through a 2600 foot horizontal reservoir section (approximately 40 feet thick) which is highly faulted. In this case staying in the reservoir also meant navigating fault planes across which reservoir surfaces were displaced from one another. The cyan line was the planned trajectory from seismic data while the red line was the actual trajectory. Numerous intrabedded shale layers are shown in the darker shades. These also had to be navigated in such a way to maximize contact with the lighter layers which contained hydrocarbons. Figure reproduced with permission from [3]

injected into the formation. The success or failure of these efforts is strongly influenced by the multi-scale heterogeneity of the reservoir down to the rock-fluid interfacial dynamics. In simple terms, whether the rock is oil or water wet can result in very different outcomes from a production enhancement program such as a water flood. Characterizing and understanding how this heterogeneity from pore to reservoir spatial scales determines production is an ongoing and difficult technical challenge.

3.3.2 What do we want to achieve?

Characterizing the multi-scale fluid flow properties of carbonate reservoirs has typically been highly empirical, contributed to in an ad hoc sense by many types of physical measurements and often not transferable. To get the hard oil and gas out and hence raise recovery rates, we are up against significant gaps in our understanding of the role of surface physics and chemistry governing flow in hydrocarbon bearing porous media. Mineral-water interfaces have been under intense study outside the oil industry for several decades driven by interest in natural geochemical processes and a whole host of environmental concerns related to hazardous wastes. Currently in the oil industry there is a growing appreciation that more focus is needed on fundamental understanding of processes at rock-fluid interfaces. Lack of understanding may indeed be the most important limitation to growing recovery rates significantly beyond where best practice is now.

Even for flow conditions over a "planar" mineral surface (no pore walls yet to contend with) complexity over length scales will start from atomic-scale structure at ideal crystallographic mineral/fluid terminations and move into topologies associated with defects, micro-cracks, impurities, inorganic and organic coatings and even biofilms. This is the scale where chemical bond breaking and formation will be most important, and it is the scale where wetting behavior will be established. The next level will be concerned with the flow boundary layer next to the mineral surface which is often suggested to be less than 10 nm thick [49]. Beyond that we get into bulk flow properties and the details at the interface can serve to establish proper boundary conditions for study of flows at larger scales. As pore walls are added, the situation can change remarkably as the surface to volume ratio increases to a point where the interfacial characteristics could significantly perturb or even dominate the local flow properties.

Rephrasing this into specific goals, we want to:

1. Improve our basic understanding of multi-scale heterogeneity effects on all our measurements to better link macro hydraulic models at the reservoir (production) scale to those at meso scales (borehole or near borehole) and then to micro scales (pore level) and finally to nano scales (rock/fluid interface) and vice versa. This represents a formidable multi-scale problem spanning 12 orders of magnitude.
2. Extend our knowledge of physics and chemistry associated with flows at mineral/pore fluid interfaces, especially with respect to chemical reactions, factors affecting wettability in the presence of multiple fluid species and chemical interventions associated with Enhanced Oil Recovery (EOR) techniques.
3. Develop new laboratory techniques for item 2 capable of direct observation and characterization of mineral/fluid wetting layers and their alteration.
4. Extend and apply this new knowledge base to unconventional sources of oil & gas such as heavy oils, shale oil, tar sands and gas hydrates.
5. Extend this work to unique issues relevant to long term CO_2 sequestration.

While progress has been made in many of these areas of rock physics and chemistry, much of this progress has been compartmentalized among the different domains of hydrology, contaminant cleanup, earth science, petroleum engineering, etc. An understanding of carbonate rocks will require an integrated understanding of their rock/fluid interfacial physics and chemistry over multiple spatial and time scales starting from atomistic/molecular behaviors.

It is this last point that needs some elaboration with respect to simulation. In order to predict the transport properties linked to spatial/temporal scales at the rock/fluid interface several nontrivial issues immediately arise:

- How do me measure geometric features at the interfaces and at what spatial resolution
- How is geometry to be represented
- What geometric parameters are necessary for which properties
- Is there a minimum set of geometrical parameters
- What is the relevant surface chemistry at the interfaces
- What are appropriate models for surface defects and their chemical influence
- How are the interfacial properties transformed into up-scaled boundary conditions.

Each will have to be studied concurrently and systematically integrated to make useful progress. Simulation will have to play a strong role throughout.

3.3.3 Porescale simulations

So far approaches to this problem have paid most attention to predicting permeabilities using microstructural models. Early on, geometric information focused on porosity and specific surface areas which were all that were needed for simple empirical models [5], but now detailed geometries are derived from analyses of rock thin sections using optical imagery [22] and x-ray microtomographic measurements [4,17] which are reaching spatial resolutions approaching micron and even sub-micron scales.

Modeling has progressed from simple heuristic models [6] to geometrically complicated models needing large scale numerical simulations which are commonly based on network models and lattice Boltzmann (LB) techniques. Network models which are derived from microscopic properties of real porous media continue to have some success [7,48]. In these models idealized 3-D networks of pores connected by throats are constructed in such a way to preserve measured geometric and topological information for a particular porous medium. This information includes size distributions for pores and throats and their coordination derived from microtomographic data. At first the network elements were taken to be circular, but this has now moved on to square and triangular elements for more realistic modeling of wetting states. Fluid movement through these networks proceeds using displacement rules based on capillary pressures between network elements and gravity. These models yield reasonable predictions of fluid transport properties for single and multi-phase flows, but usually this success is highly dependent on conditioning the model parameters to experimental data. Transferability of model predictions to a wider range of rock systems is still problematical and often depends on getting a good geometric match to the system of interest. Also the inclusion of wettability distributions on permeability is still an open question.

Beyond network models most attention is now being given to 3-D LB methods [31,39]. LB methods are uniquely powerful for being able to represent complex pore geometries [4,17] and boundary conditions if they can be properly derived from smaller-scale dynamics that exist at rock/pore fluid interfaces. It is no surprise that a major drawback of LB methods is computational expense, but LB methods are highly parallel, and they will benefit from the continued evolution of massively parallel computing. More problematical will be describing realistic microscopic boundary conditions conditioned by nanoscopic details at rock/fluid interfaces. This will require unique new measurements to probe rock/fluid physical (electrostatic) and chemical (reaction) interactions at the scale of nanoscopic topography (roughness). For example, it is thought that precipitation of certain heavy hydrocarbon components (asphaltenes) [9,33] can alter the nature of the wetting fluid at the rock/fluid interfaces resulting in changes of macroscopic hydrological properties such as the relative permeabilities of water and oil. This can have sometimes a very detrimental effect on oil recovery, but still little is understood about the general nature of these processes. We note that these issues, which in detail are somewhat unique to the oil industry, are still reminiscent of challenges in the general field of pore scale reactive transport studies which have highlighted long term difficulties associated with physical and chemical scale-dependent heterogeneities in the systems studied [44].

On the experimental side we see availability of high resolution x-ray microtomographic images [41] no longer solely reliant on large 3-D synchrotron radiation facilities. More portable measurement systems are becoming available which will greatly increase the diversity of the data sets (Figs. 7 [4], 8 [40]) and challenge the generality of the simulation models. There are still important open issues on how to transform the 3-D image data to a form suitable for inclusion in simulations models whether they be LB or otherwise. We still don't know how limiting the resolution of the 3-D image data is for prediction of transport properties

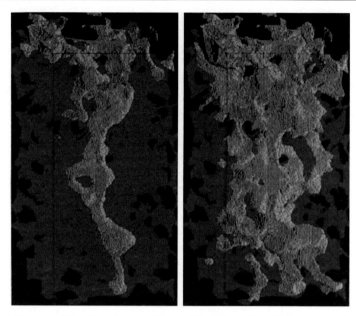

Fig. 7 Simulation showing invasion of a nonwetting fluid (red) from above displacing a wetting fluid (blue). The left panel shows initial breakthrough and the right panel shows the nonwetting fluid in steady state. The original tomographic image is of a Fountainebleau sandstone with a voxel edge of 7.5 μm. Reproduced from [4] with permission

and correspondingly how big the computational domains must be for highly heterogeneous pore systems. Thus validation of LB techniques even for single-phase fluid flows still has a ways to go, but here we can expect good progress.

The situation becomes even more challenging as we move into multi-phase and multi-component flows with chemical reactions. Recently LB techniques were used to simulate reactive transport for a model pore system [21] with multiple fluid components and minerals taking account of both homogeneous reactions among fluid species and heterogeneous reactions between the mineral surfaces and the fluid components. Homogeneous reactions were included as "simple" source/sink terms for each possible chemical reaction with reaction rates for each species specified as a function of concentrations. At interfaces mineral reactions were handled via boundary conditions accounting for diffusion of solute species to and from the mineral surface. Overall, the methods produced reasonable results, but the challenges for real systems will be supplying suitable rate constants for all applicable reactions, scaling to larger models to include additional pore heterogeneity and validating predictions against experiments.

We expect that LB simulations of complex (multi-phase, multi-component, reactive) flows in porous media will be done at micron or sub-micron resolution with upscaling to larger volumes to aid linkage to macro properties. Important interfacial effects due to boundary flows and chemical reactions operating at the nano scale must come in as boundary conditions at the micro scale. Resolving these effects involves a different area of both experiment and simulation. Fortunately, new optical imaging techniques now allow us to directly observe what is taking place at or near mineral surfaces subjected to controlled flows. Laser scanning microscopy allows us to image flow profiles with submicron resolution. Likewise electron microscopy, atomic force microscopy and techniques like vertical scanning interferometry

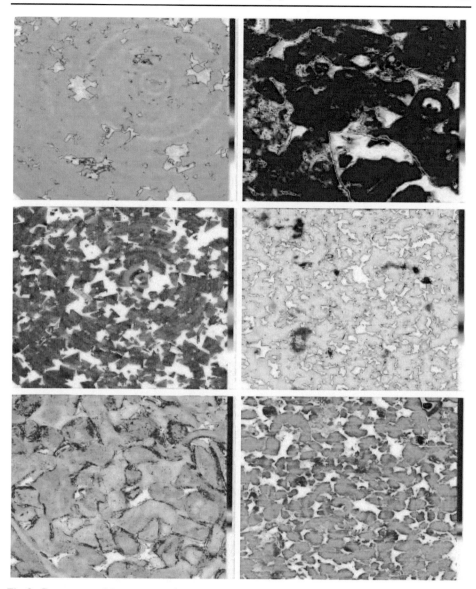

Fig. 8 Current state of the art tomographic images of six rocks at 2.85 μm resolution [40]

(VSI) [26,30] are being successfully applied to image mineral surfaces allowing direct quantitative measurements of microscopic changes in the mineral surface topography due to dissolution/precipitation processes with nearly angstrom resolution in the direction normal to the surface. This capability to observe mineral/fluid interfacial behavior will also serve to complement use of ab initio computational methods like density functional theory molecular dynamics to simulate the structure and dynamics at interfaces involving thousands of atoms over sufficient times to determine average mineral-water interface properties. If necessary for the study of long time scales, kinetic Monte Carlo techniques [51] may have to be invoked as well.

Finally, introducing hydrocarbon phases also adds the complexity of mixed wettability. Microscopic studies of mixed wettability in systems with multiple fluid phases are still in their infancy [31,48], but fundamental understanding of interfacial physics and chemistry surrounding mixed wettability at mineral surfaces will be essential for innovations leading to better recovery rates. A particular issue, mentioned earlier, will be understanding and managing the wettability altering role of heavier hydrocarbon fractions should they precipitate out from solution and deposit on mineral surfaces.

This surface deposition problem is quite widespread. Heavy organics such as paraffin, resin, and asphaltenes exist in petroleum fluids throughout the world. These compounds are usually found in solution, but some can precipitate out of solution and deposit on mineral surfaces due to changes in temperature, pressure or chemical composition. When this happens in the formation, severe impairment of porous flow can ensue, but they can also precipitate in wells and other flowlines leading in many cases to total blockage which must undergo expensive remediation. Asphaltene precipitation and deposition often lies at the center of these problems. Given the industry trends to develop and produce hydrocarbons with heavier asphaltic fractions, one can expect this problem to become more prevalent. In fact some urgency may be called for since these so-called heavy oils are more and more associated with very expensive operations, e.g., deep water.

While the asphaltene problem is not new and in fact has been studied for many decades, the understanding of the precipitation/deposition mechanisms from a fundamental standpoint remains poor. The result has been a strong tendency for the industry to rely on phenomenological models at best and at worst to ignore the problem all together until it is too late. Several difficulties that impede investigation of these heavy oil fractions are worth noting:

1. Asphaltenes are a whole class of different compounds with varying molecular weights which may act differentially with respect to precipitation properties. The chemical structure of these compounds have been under study using all applicable physical methods including IR, NMR, ESR, mass spectrometry, electron microscopy and so on. Only recently has there been some agreement on their molecular weight distributions and structures [33].
2. The mechanisms of precipitation are believed to be highly dependent on their molecular weights and the existence of other constituents [9]. For example, the paraffins are believed to mediate the larger molecular weight compounds coming out of solution and flocculating to form larger particles. These larger particles can have a strong affinity to solid surfaces due to polar groups associated with the asphaltene molecules themselves.
3. Resins, if present, may act to stabilize the flocculates by adsorbing onto their surfaces enabling the asphaltene particles to remain in colloidal suspension. Thus we can postulate a general condition whereby lighter asphaltene particles are dissolved in oil, others are stabilized by resins as flocculates in a suspended colloidal state and still others (possibly the heaviest) are precipitated out and adhering to the rock mineral surface. Thus the onset of precipitates and their potential for damage is a dynamic interplay of various constituents and their physical environments.

Ultimately, we want to understand asphaltene chemistry and dynamics sufficiently to avoid precipitation in the first place, but a next level of understanding will necessitate fundamental studies of asphaltic particles adhering to mineral surfaces and how interventions can be designed to eliminate or mitigate impairment to hydrocarbon flows and recovery.

4 Part III. Simulation for new enabling technologies in the oil and gas industry

4.1 Introduction

We have discussed how modern simulation tools are becoming indispensable aids for monitoring production from oil reservoirs and developing future production strategies. A new level of data-driven hydrocarbon reservoir management is being achieved from advances in multiphysics modeling and simulation, algorithms exploiting massive parallelism, and processes for continuous use of dynamic well-bore and seismic data. Looking ahead, the possibility of combining large scale reservoir simulations with ubiquitous monitoring sensors embedded in reservoir-fields, e.g., permanent downhole sensors and seismic sensors anchored at the seafloor, could provide a new level of symbiotic feedback between measured data and model predictions. This is sometimes referred to as the "intelligent oilfield". While this vision is not yet reality, we can forecast several enabling technologies that will help it become so.

4.2 Illuminating the oilfield with new sensor systems

New micro/nano sensors and network systems are one of the pillars of this outlook, and one can see them massively deployed as embedded elements in the subsurface infrastructure of tubulars, valves and even in the steel and cement used to mechanically support well structures. In this scenario possibilities abound for increased monitoring of the mechanical integrity and safety of the system (recall structures will be deployed in a hostile environment for many years). This is a strategy similar to what the aerospace industry is following with respect to monitoring critical flight elements and structures. In addition, these sensors could yield a wealth of additional information on production behavior within the well and quite possibly behavior in the immediate vicinity of the well.

A longer term scenario foresees sensor systems that could be injected into the reservoir after which they may migrate to locations from which they will "report" properties such as temperature, pressure and maybe fluid content at their position with an accuracy not obtainable by any other method known today. The value of this distributed information coupled back to reservoir simulation could be another enormous step in data-driven reservoir management. The components (sensing, power and communication elements) of these systems will likely be at the nanoscale opening an opportunity to leverage much of the work on nanoscale devices currently ongoing, but often driven by biotechnology and biomedical applications. No doubt the extreme environments of the oilfield will provide new challenges for practical devices.

One expects that these devices could be based on a number of well known nano building blocks such as nanowires, nanotubes, quantum dots and wells, as so on. Studies of these devices rely heavily on a blend of experimental and simulation efforts. Simulations must deal with multi-scale phenomena beginning with atomic structure at lead contacts, to quantum descriptions of non-equilibrium electronic transport, to interfaces with power and communication components (which will likely be nano-sized), and finally to interfaces to the macro world. These are daunting challenges, but the smallness of the sensors combined with their demonstrated high sensitivities for detecting mechanical alteration and chemical species makes them attractive for distributed sensing applications.

For active sensing, predicting and controlling current in these devices as a function of their environment is essential, but is complicated by quantum phenomena such as tunneling, confined dimensions, and fluctuations that come into play. Structural details at interfaces will

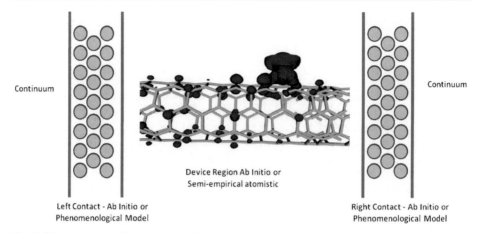

Fig. 9 Conceptual view of the simulation of a nanowire device (a carbon nanotube is shown with an attached molecule). A multi-scale description includes ab initio, atomistic, semi-empirical and continuum models

play a role since free surfaces along conduction paths will undergo reconstruction possibly with passivating species, and heterointerfaces will often involve charge transport across two different materials systems. For example, contacts with multi-moded metals will bring a highly conductive material in thermal equilibrium with a sparsely moded active channel that operates out of thermal equilibrium.

In Fig. 9 we illustrate the multi-scale nature of the problem ranging from a quantum mechanical description of the conduction channel and nearby lead structures to a microscopic macroscopic connection with the outside. If one can specify the atomic configuration and Hamiltonian [H] of the device connected to the two contacts, it is possible to calculate the transport properties using several methods, the most popular being the nonequilibrium Green's function (NEGF) approach [13,25]. This is usually done self-consistently with an electrostatics calculation based on the Poisson equation and can include many parasitic effects such as phonon scattering. The computational challenge depends on the physics included in the Hamiltonian. Some semiconductor nanodevices can be simulated using an effective mass Hamiltonian in conjunction with a finite difference description of the Schrodinger equation. Another level of rigor is reached using atomistic tight binding approaches which will scale with the number of atomic orbitals per atom. For many semiconductors sp^3d^5s* orbital sets are often used leading to matrices that could easily reach 10^9 elements and beyond depending on the size of the device. This could make direct inversions to calculate the energy dependent Green's functions in the NEGF method quite expensive if not for some elegant approaches [25] to invert for the Green's functions at reduced cost. At the highest level one can still resort to density functional theory, but with some constraint on the number of atoms in the active device. How predictions from these different approximations compare amongst themselves is still not well established [23,50]. Furthermore, comparisons against experimental data are often clouded by variable experimental controls [11]—so we can expect their use to guide development of new sensors to remain semi-quantitative for some time, but still useful.

4.3 Computational materials

The oil industry uses a lot of different materials to accomplish its business and in many situations these materials are intended for use in the rather extreme environments of the subsurface in terms of temperatures, pressures and risk of chemical attack. The most common structural materials include steels, specialty stainless steels and cements. Polymer composites though desirable because of weight are used in a more limited sense (often because of temperature limitations and strength). Ceramics are most often used as protective coatings or in situations with severe wear, but have not found general use because of classical tradeoffs of hardness vs brittleness. Current trends in subsurface operational environments point to more situations where we will become materials limited. There is also a growing need to place far more infrastructure into the subsurface to better manage the overall extraction of hydrocarbons over long periods of time (several decades). This introduces more severe durability requirements on materials (wear, strain fatigue, corrosion resistance) in addition to improving the usual mechanical requirements of strength and toughness. All this suggests that new materials are desired to operate efficiently, and their development time must become much shorter. A partial list of desirable materials would include:

1) New specialty stainless steels with less costly components and manufacture and with enhanced strength and toughness and resistance to corrosion, a difficult set of tradeoffs.
2) New high temperature polymer composites whose mechanical properties ideally one day could match those of current generation stainless steels.
3) Tougher ceramics without compromising their hardness.
4) New classes of cements which are lighter and more resistant to cracking and chemical attack over long periods of time (many decades).
5) In all cases materials that have some capability to repair themselves if they become damaged.

This is a tall order, but we can expect that advances from such things as combinatorial analyses of composite properties [47] and multi-scale simulation [18,51] will make it easier to identify interesting candidates which could be developed in shorter times. We'll elaborate on the case for high temperature composites to make the point.

4.3.1 High temperature polymer composites

A common approach to new composites today is the use of well dispersed nano-sized fillers that can significantly improve mechanical behavior and other properties at macro scales. Traditional composite property models which work well for large phase domains may no longer apply. New theories [18] have to take account of particle sizes (especially particle specific surface areas) and the interfacial layers between particles and the matrix. Because we lack theories that describe the mechanics of nanointerfaces and how these are manifested at larger scales (new constitutive laws), simulations will have to play a role in advancing the state of the art in order to avoid the time costly trial and error approaches of the past. We can envision that ab initio calculations will be required at and near interfacial boundaries to study structures as well as the dynamics of relative displacements of the two phases when larger stresses are applied potentially disrupting bonding that may be present. This can best be handled with density functional theory molecular dynamics since tailoring chemical bonding at the interface is an important strategy for maximizing load transfers [16]. Of course we would like this information to transition into a classical molecular dynamics simulation over a larger volume perhaps with appropriate reactive force fields as long as bond breaking is still an issue and then on to more mesoscopic simulations and finally to finite element methods.

To our knowledge the full multi-scale problem has not been fully integrated and tested on multiple systems. Instead a large patchwork of results based on different modeling assumptions seems to exist. Continuum mechanics approaches [52] treating carbon nano-tubes (CNTs) as beams, thin shells or solids with cylindrical shapes have been tried to get a handle on overall load behavior. However, they are of little value to understanding details of interfacial interactions with the polymer matrix. Various MD simulations have been attempted with some success [18]. Force fields have been based on fully atomistic potentials to more coarse-grain potentials often used with dissipative particle dynamics DPD which extends classical MD methods to larger systems and longer time scales. Molecular dynamics was employed to calculate the elastic properties of an effective fiber [34] consisting of a CNT surrounded by polymer in a cylindrical shape. A micromechanics homogenization model with this effective fiber was then used to calculate the properties of composites with different CNTs, lengths and volume fractions. Comparisons with experimental data were reasonable. In another example a full MD approach [19] was employed for two CNT polymer com-posites. A surprising result was that the calculated elastic moduli were in rough agreement with the simple classical rule of mixtures. Further work is needed to see if this is a general correlation or not. At a larger scale a continuum Boundary Element Model (BEM) [28] has been used to simulate a CNT reinforced composite containing 16,000 straight rigid fibers. The effective Young's modulus of this model compared closely to an MD result. Interest-ingly, the rigid fiber approximation is not limiting since the elasticity of the fibers and more realistic interface conditions based on MD CNT composite results could be included in the BEM.

There is little doubt that polymer nanocomposites simulation is a growing field lever-aging computational power, theory and algorithmic improvements in line with studies of other materials systems. Designing, producing and optimizing new composites with prop-erties amenable to the extreme environments of the oil industry is a formidable challenge, but it also is one shared throughout materials science which gives cause for hope. Even the few simulation results mentioned above which involve some linking of information over different length and time scales is testimony that useful progress is happening toward com-putational materials design. However, we would agree that validated methods on how to transfer information from the molecular level to the mesoscale is the critical bottleneck for realizing the potential of computational materials for industrial applications [35]. We also believe the benefits are worth the effort.

5 Summary

By many indications the petroleum industry is entering a new phase in which demand for oil & gas continues to reach unprecedented levels due to global growth. We can expect this to continue with minor fluctuations due to economic cycles. At the same time the age of easy oil is nearing an end. Two outcomes of this new scenario are already being observed. The first is greater emphasis on recovery rates from existing resources raising them from today's average values of around 30–35% to greater than 60–70%, and the second will be increased develop-ment of unconventional resources, e.g., heavy oils, bitumen, tar sands, shale oil, gas hydrates and coalbed methane. For the time being there are still plenty of hydrocarbon molecules, but the easy ones to produce in terms of today's technology and economics are reaching shorter supplies probably over the next couple of decades. Reactions to both trends will place new demands on increased fundamental understanding and technologies.

With the above macro-economic scenario serving as a context, we have discussed several of application areas with particular focus on the role of modeling and simulation. These applications were deliberately chosen from the upstream part of the oil industry which deals with the subsurface, which historically has had the greatest risk. That risk has spawned a chain of notable technical innovations, and we need this to continue.

The enterprise is already becoming data-driven in the sense that continuous new data reflecting static and dynamic properties of the oil reservoir are being acquired and interpreted, and the results are being compared to reservoir simulator predictions and then used to guide updating the underlying simulator models. This dynamic feedback loop will continue over the lifetime of the asset starting with appraisal where the models are most uncertain to late development where they have matured and help focus on by-passed oil and gas. Precision management of reservoir drainage is a key activity that helps raise recovery factors. Both seismic and electromagnetic imaging are used to illuminate the reservoir at the largest scales. The processing of those data into images are among the largest computational tasks done anywhere today, and we still haven't exhausted the physics rigor. In a time-lapse sense we want the image data to become more real time to support drilling decisions which cannot wait months while data are processed. Novel short cuts are used today to meet this requirement, but at the sacrifice of physics.

Beyond imagery the same data is being inverted to derive other reservoir properties. Three-dimensional inversion over the scale of the reservoir is still impractical using finite difference based methods to solve the forward wave propagation problems. However, algorithmic advances and the growing power of computing platforms now reaching PetaFlop scale are making this more possible with fewer physics approximations. Logs acquired during drilling or via wireline after drilling produce the most accurate data which can be used to constrain inversion. But logs only measure at or near well penetrations and must be supplemented by geological models to fill in between wells. Thus uncertainties in reservoir model descriptions between wells are still major risks. This can have a detrimental impact on horizontal drilling through large sections of a thin reservoir. We saw an example where a directional EM measurement while drilling supplemented by real time inversion is now able to pinpoint the reservoir upper and lower boundaries with impressive results over just using seismic imagery to guide the process. This enables well placement with far greater reservoir contact and improved recovery.

At the smallest scales approaching atomistic we covered three areas where modeling and simulation supplemented by modern tools capable of nanoscopic examinations will likely impact oilfield technology. New materials, especially polymer composites, are interesting from a strength-to-weight point of view if they can be also engineered to withstand subsurface environments for long time. Likewise, new sensors with micro to nano footprints may someday be in the offering to provide ubiquitous sensing within and from wellbores. There is even the possibility that tiny devices could perhaps penetrate the formation and act either as smart tracers or remote sensors of reservoir properties. While some of these ideas are clearly long term we have to be mindful that in other application domains (medicine, aerospace, electronics, environment, etc.) work is going on with an impressive blend of experimental and modeling and simulation efforts that can be leveraged for oilfield applications. Finally, interfacial physics and chemistry at nano scales can play dominant roles influencing how reservoir fluids flow in porous media. We made the argument that increasing recovery may indeed rely on improving understanding of interfacial transport processes and how they link to macro transport properties. The industry due to lack of tools and economic drive has ignored the nanoscopic scale in favor of more microscopic to mesoscopic scales as starting points. These limitations are being removed. New tools, including simulation, for probing rock mineral/fluid interfaces provide a compelling opportunity to gain insight on some basic road blocks affecting recovery.

Acknowledgements The author would like to thank Tarek Habashy of Schlumberger-Doll Research for providing the electromagnetic inversion example shown in Fig. 5 and through his work also Statoil and EMGS for the use of their field data and permission to show the results. Further the author would like to thank Anthony Curtis and Anthony Cooke of Schlumberger WesternGeco for help in preparing Fig. 3 and Statoil for permission to show the results. He also thanks Paine Mccown and Dominic Louden both of Schlumberger WesterGeco for useful discussions of other examples. He thanks Raphael Altman, Paolo Ferraris and Fabricio Filardi of Schlumberger Data and Consulting Services for permission to use the example in Fig. 6. Finally, he thanks Seungoh Ryu of Schlumberger-Doll Research for his work in rendering the images shown in Fig. 8 and for permission to show them and Prof Sidney Yip of MIT for his gentle encouragement and advice in preparing the manuscript.

References

1. Abubakar, A., van den Berg, P.M., Habashy, T.M.: An integral equation approach for 2.5-dimensional forward and inverse electromagnetic scattering. Geophys. J. Int. **165**, 744–762 (2006)
2. Akubakar, A., Habashy, T.M., Drushkin, V.L., Knizhnerman, L., Alumbaugh, D.: Two-and-half dimensional forward and inverse modelling for interpretation of low-frequency electromagnetic measurements. to be published in Geophysics (2008)
3. Altman, R., Ferris, P., Filardi, F.: Latest generation horizontal well placement technology helps maximize production in deep water turbidite reservoirs. In: SPE 108693. Veracruz (2007)
4. Auzerais, F., et al.: Transport in sandstone: a study based on three dimensional microtomography. Geophys. Lett. **23**, 705–708 (1996)
5. Bear, J.: Dynamics of Fluids in Porous Media. Dover, Mineola (1972)
6. Blair, S.A., Berge, P.A., Berryman, J.G.: Using two point correlation functions to characterize microgeometry and estimate permeabilities of sandstones and porous glass. J. Gepphys. Res. **101**, 20359–20375 (1996)
7. Blunt, M.J.: Flow in porous media-pore-network models and multiphase flow. Curr. Opin. Colloid Interface Sci. **6**, 197–207 (2001)
8. Boutte, D.: Seismic-to-simulation arrives. Hart's E&P J (July 2007)
9. Branco, V., Mansoori, G.A., Xavier, L., Park, S.J., Manafi, H.: Asphaltene flocculation and collapse from petroleum fluids. J. Petrol. Sci. Eng. **32**, 217–230 (2001)
10. Burstedde, C., Ghattas, O.: Algorithmic strategies for full waveform inversion: 1D experiments. In: Society of Exploration Geophysicists, San Antonio, 1913–1917 (2007)
11. Chen, Z., Appenzeller, J., Knoch, J., Lin, Y.-M., Avouris, P.: The role of metal-nanotube contact in the performance of carbon nanotube field-effect transistors. Nano Lett. **5**, 1497–1502 (2005)
12. Colombo, D., De Stefano, M.: Geophysical modeling via simultaneous joint inversion of seismic, gravity, and electromagnetic data: application to prestack depth imaging. Leading Edge **26**, 326–331 (2007)
13. Datta, S.: Quantum Transport: Atom to Transister. Cambridge University Press, Cambridge (2005)
14. Eidesmo, T., et al.: Sea bed logging. A new method for remote and direct identification of hydrocarbon-filled layers in deepwater areas. First Break **20**, 144–152 (2002)
15. Ellis, D.V., Singer, J.M.: Well Logging for Earth Scientists. Springer Verlag (2007)
16. Frankland, S.J.V., Harik, V.M.: Simulation of carbon nanotube pull-out when bonded to a polymer matrix. In: Materials Research Society Proceedings, vol. 740, paper I12.1, 1-6 (2003)
17. Friedrich, J.T., DiGiovanni, A.A., Noble, D.R.: Predicting macroscopic transport properties using microscopic image data. J. Geophys. Res. **111**, 1–14 (2006)
18. Glotzer, S., Paul, W.: Molecular and mesoscale simulation methods for polymer materials. Annu. Rev. Mater. Res. **32**, 401–436 (2002)
19. Griebel, M., Hamaekers, J.: Molecular dynamics simulations of the elastic moduli of polymer carbon nanotube composites. Comput. Methods Appl. Mech. Eng. **193**, 1773–1788 (2004)
20. Habashy, T.M., Abuakar, A.: A general framework for constraint minimization for inversion of electromagnetic measurements. Progr. Electromagn. Res. **46**, 265–312 (2004)
21. Kang, Q., Lichtner, P.C., Zhang, D.: Lattice-Boltzmann pore-scale model for multicomponent reactive transport in porous media. J. Geophys. Res. **111**, B05203 (2006)
22. Keehm, Y., Nur, A.: Permeability from thin sections: 3D reconstruction and lattice-Boltzmann simulation. Geophys. Res. Lett. **31**, 1–4 (2004)
23. Khairul, A., Lake, R.K.: Leakage and performance of zero-Schottky-barrier carbon nanotube transistors. J. Appl. Phys. **98**, 064307 (2005)
24. Kleinberg, R., Clark, B.: Physics in oil exploration. Phys. Today **48**, 48–53 (2002)

25. Lake, R., Klimeck, G., Bowen, R.C., Jovanovic, D.: Single and multiband modeling of quantum electron transport through layered semiconductor devices. J. Appl. Phys. **81**, 7845–7869 (1997)
26. Lasaga, A.C., Luttge, A.: Variation of crystal dissolution rate based on a dissolution stepwave model. Science **291**, 2400–2404 (2001)
27. Li, Q., et al.: New directional electromagnetic tool for proactive geosteering and accurate formation evaluation while drilling. In: SPWLA 46th Annual Logging Symposium. New Orleans (2005)
28. Liu, Y.J.: Large scale modeling of carbon nanotube composites by a fast multipole boundary element method. Comput. Mater. Sci. **34**, 173–187 (2005)
29. Luthi, S.M.: Geological Well Logs: Their Uses in Reservoir Modeling. Springer Verlag (2001)
30. Luttge, A., Winkler, U., Lasaga, A.C.: Interferometric study of the dolomite dissolution: a new conceptual model for mineral dissolution. Geochim. Cosmochim. Acta **67**, 1099–1116 (2003)
31. Martys, N., Chen, H.: Simulations of multicomponent fluids in complex three dimensional geometries by the lattice Boltzmann method. Phys. Rev. E **53**, 743–750 (1996)
32. Mendoza, A., Preeg, W., Torres-Verdin, C., Alpak, C.: Monte Carlo modeling of nuclear measurements in vertical and horizontal wells in the presence of mud-filtrate invasion and salt mixing. In: Society of Petrophysicists and Well Log Analysts 46th Annual Logging Symposium, New Orleans (2005)
33. Mullins, O., Sheu, E.Y., Hammami, A., Marshall, A.G. (eds.): Asphaltenes, Heavy oils, and Petroleomics. Springer Science, New York (2007)
34. Odegard, G.M., Gates, T.S., Wise, K.E., Park, C., Siochi, E.J.: Constitutive modeling of carbon nanotube re-inforced polymer composites. Compos. Sci. Technol.**63**, 1671–1687 (2003)
35. Oden, J.T., et al.: Revolutionizing engineering science through simulation. Report of the National Science Foundation Blue Ribbon Panel on Simulation-Based Engineering Science, National Science Foundation (2006)
36. Osdal, B., et al.: Mapping the fluid front and pressure buildup using 4D data on Norne field. Leading Edge **9**, 1134–1141 (2006)
37. Pickering, S.: Q-reservoir: advanced seismic technology for reservoir solutions. Oil Gas North Africa Mag. Jan–Feb 2003, 26–30 (2003)
38. Pratt, R.G., Shin, C., Hicks, G.J.: Gauss–Newton and full Newton methods in frequency-space seismic waveform inversion. Geophys. J. Int. **133**, 341–362 (1998)
39. Rothman, D.H., Zaleskki, S.: Lattice-Gas Cellular Automata. Cambridge University Press (1997)
40. Ryu, S.: Personnal Communication (2008)
41. Sham, T.K., Rivers, M.L.: A brief overview of synchrontron radiation in low temperature geochemistry and environmental science. Rev. Mineral. Geochem. **49**, 117–147 (2002)
42. Sinha, M.C., MacGregor, L.M.: Use of marine controlled-source electromagnetic sounding for sub-basalt exploration. Geophys. Prospecting **48**, 1091–1106 (2000)
43. Sirgue, L., Pratt, R.G.: Efficient waveform inversion and imaging: a strategy for selecting temporal frequencies. Geophysics **69**, 231–248 (2004)
44. Steefel, C.I., MacQuarrie, K.T.B.: Approaches to modeling reactive transport in porous media. Rev. Minerol. **34**, 83–125 (1996)
45. Tarantola, A.: Inverse Problem Theory. Elsevier Science B.V., Amsterdam (1987)
46. Top500. http://www.top500.org/.
47. Tweedie, C.A., Anderson, D.G., Langer, R., Van Vliet, K.J.: Combinatorial materials mechanics: high-throughput polymer synthesis and nanomechanical screening. Adv. Mater. **17**, 2599–2604 (2005)
48. Valvatne, P.H., Blunt, M.J.: Predictive pore scale modeling of two phase flow in mixed wet media. Water Resour. Res. **40**, 1–21 (2004)
49. Van der Veen, J.F., Reichert, H.: Structural ordering at the solid–liquid interface. Mater. Res. Soc. Bull. **29**, 958–962 (2004)
50. Venugopal, R., Ren, Z., Datta, S., Lundstrom, M.S., Jovanovic, D.: Simulating quantum transport in nanoscale MOSFETs: real vs mode space approaches. J. Appl. Phys. **92**, 3730–3739 (2002)
51. Voter, A.F., Montalenti, F., Germann, T.C.: Extending the time scale in atomistic simulation of materials. Annu. Rev. Mater. Res. **32**, 321–346 (2002)
52. Wong, E.W., Sheehan, P.E., Lieber, C.: Nanobeam mechanics: elasticity, strength and toughness of nanorods and nanotubes. Science **277**, 1971–1975 (1997)

Microscopic mechanics of biomolecules in living cells

Fabrizio Cleri

Originally published in the journal Sci Model Simul, Volume 15, Nos 1–3, 339–362.
DOI: 10.1007/s10820-008-9104-2 © Springer Science+Business Media B.V. 2008

Abstract The exporting of theoretical concepts and modelling methods from physics and mechanics to the world of biomolecules and cell biology is increasing at a fast pace. The role of mechanical forces and stresses in biology and genetics is just starting to be appreciated, with implications going from cell adhesion, migration, division, to DNA transcription and replication, to the mechanochemical transduction and operation of molecular motors, and more. Substantial advances in experimental techniques over the past 10 years allowed to get unprecedented insight into the elasticity and mechanical response of many different proteins, cytoskeletal filaments, nucleic acids, both in vitro and, more recently, directly inside the cell. In a parallel effort, also theoretical models and computational methods are evolving into a rather specialized toolbox. However, several key issues need to be addressed when applying to life sciences the theories and methods typically originating from the fields of condensed matter and solid mechanics. The presence of a solvent and its dielectric properties, the many subtle effects of entropy, the non-equilibrium thermodynamics conditions, the dominating role of weak forces such as Van der Waals dispersion, hydrophobic interactions, and hydrogen bonding, impose a special caution and a thorough consideration, up to possibly rethinking some basic physics concepts. Discussing and trying to elucidate at least some of the above issues is the main aim of the present, partial and non-exhaustive, contribution.

Keywords Biomolecules · Mechanical properties · Configurational entropy · Molecular dynamics · Jarzynski identity

From the point of view of a materials scientist, living tissues exhibit rather extraordinary properties compared to both standard structural materials, such as metals or ceramics, and functional materials, such as semiconductors or ionic crystals. On the one hand, living tissues have specific values of mechanical properties that would superficially rule them out of advanced engineering: their Young's moduli are typically within a few MPa, i.e. orders of

F. Cleri (✉)

Institut d'Electronique, Microélectronique et Nanotechnologie, Université des Sciences et Technologies de Lille, Av. Poincaré, BP 60069, 59652 Villeneuve d'Ascq, France

e-mail: fabrizio.cleri@univ-lille1.fr

magnitude smaller than even the softer polycarbonate ($Y \sim 2.5\,\mathrm{GPa}$); their typical fracture resistance under tension is in the range from a few kPa (animal skin, muscle fibers) to a few MPa (cortical bone, with a maximum of $\sim 100\,\mathrm{MPa}$ for limb bones) [1, 2], while plastics such as polypropylene or polystyrene have values of a few tens MPa, and metals are in the range of thousands of MPa. Their thermal performances are substantially restricted from below by the freezing and, from above, by the boiling temperature of water at ambient pressure; however, already at temperatures above $40°\mathrm{C}$ most metabolic functions of many living organisms will be endagered, and most tissue cells will die well below $50°\mathrm{C}$. As far as electronic properties are concerned, ion mobility along a neural axone is so slow that it would take years for a neurotransmitter to cover the distance from the brain to the limbs by pure diffusion, whereas the typical commuting times of a neural synapse are in the range of msec, i.e. at least six orders of magnitude slower than some outdated VLSI circuit from the late 90s.

However, a closer examination of biological materials reveals astounding surprises. For example, abalone shell which is just 97% calcium carbonate, turns out to be 3,000 times more resistant to fracture than pure calcium carbonate [3], thanks to a molecular-scale "glue" component that holds together individual mineralized fibrils. Another well-known example is spider silk which, although having a Young's modulus more than ten times smaller than Kevlar (the best artificial fibre commercially available), is largerly better than Kevlar as far as failure strain (10 times larger), and resilience (more than twice) [4]. On the other hand, despite being quite slow at pure number-crunching computing, the human brain can outperform by orders of magnitude any present digital computer in such tasks as template matching, pattern recognition and associative memory, with moreover a ridiculously small size of about 1.5 liters, mostly of water, and a power consumption of about 23 W, compared to IBM's Blue Gene supercomputer which covers $250\,\mathrm{m}^2$ with its 106,000 computing nodes, made of 320 kg of silicon drawing nearly 2 MW.

In the above examples—just a few representative ones from an overwhelmingly long list— the key feature that defies any simplistic material scientists' intuition about biological materials is the presence of a tightly interconnected hierarchy of microstructures, over all the length scales. Figs. 1a and 1b show examples of bone and muscle microstructures, respectively, with their whole complexity growing from the molecular level all the way up to the macroscopic living tissue. At every length scale the constituent materials appear to adopt a well-organized spatial microstructure, which results from the evolutionary design of specialized molecular factories, capable of assembling kilograms of matter according to a precise and functionally optimized layout. For a mechanical engineer willing to model such structures by, e.g., finite-element three-dimensional calculus, the structures shown in Fig. 1 are as breathtaking as a gothic cathedral.

However, as physicists and materials scientists, we can give a possibly helpful contribution to the understanding of biological materials provided we try to keep a rather general perspective, aiming at understanding the underlying organization and processing principles of living microstructures, rather than focussing on all the possible details, no matter how important they could be for the proper functioning of a cell or tissue. (A typical, maybe extreme, example is that when a biologist and a physicist are asked how many different types of neurons there are in the cerebral cortex, the biologist answer is 'ten million' and the physicist says 'two' [5].)

Self-organized microstructures at all length scales are widespread in every biological context, from the molecular level and up. Cells are the basic units, coming in a wide variety of shapes and sizes. The overall organization of such hierachical microstructures is aimed at obtaining the maximum performances with a minimum of material, by adopting composite and gradient material structures. The response of cells to external and intrinsic stimulation is

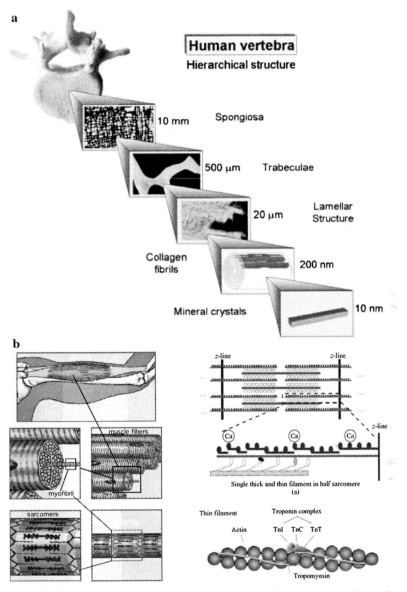

Fig. 1 **a** Hierarchical structure of a human cancellous bone such as that found, e.g. in vertrebrae or limbs. The cancellous structure is filled with a highly porous structure of spongy appearance (*spongiosa*). The *trabeculae*, or hanging bridges forming this cellular material are typically 200 µm wide and made of a composite of collagen and calcium phosphate mineral. This composite has a *lamellar structure*, where each lamella consists of layers of parallel mineralised *collagen fibrils*. Individual collagen fibrils have a diameter of a few hundred nanometers, while the individual reinforcing *mineral particles* have a thickness of only a few nanometers. **b** (Left) Hierarchical structure of a muscle, starting from the sarcomere, then assembling into myofibrils, which make up the muscle fibers. (Right) Molecular structure of the sarcomere: the z-lines limit the contraction, which is activated by the gliding of interdigitated thick (myosin) and thin (actin) filaments. Myosin heads can bind to specific sites on actin (see crosshatched example) and then rotate to stretch the extensible region. Rotated heads generate a net force between thick and thin filaments. Regulatory units (tropomyosin) bind calcium (Ca)

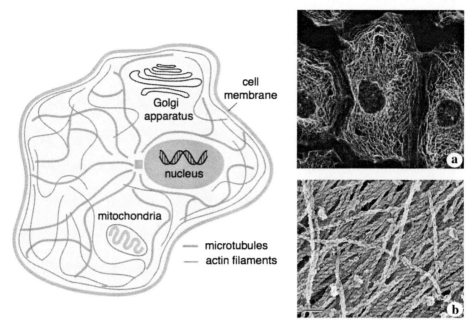

Fig. 2 (Left) Simplified scheme of a living cell, underscoring the relationship between the overall shape taken by the lipid membrane, and the mechanical scaffold represented by the cytoskeleton. The latter has at least two different microstructural components, microtubules and filaments (actin is one kind of structural protein), differing in their intrinsic length scale and mechanical properties. In general, microtubules are thicker and more rigid than filaments, and are organized around a central node (centrosome), while actin filaments are more similar to a dense network spreading all over the cytoplasmic fluid that fills up the whole cell, colored in light yellow. (Right) **a** Cytoskeleton of epithelial cells highlighted via a fluorescent stain attached to keratin filaments (image by W. W. Franke). **b** Scanning electron micrography of part of an actin filament bundle from a REF-52 fibroblast after decoration with the protein S1 (skeletal myosin subfragment 1) (from Ref. [6])

determined by complex, interconnected chains of biochemical reactions, defying any endeavour aimed at drawing solid borders between the various levels: molecular, supramolecular (individual cell components and structures), genetic (scale of the cell nucleus), transcriptional, post-transcriptional and regulatory (single-cell level), metabolic (from tissues up to whole organs).

While all living tissues are generally formed by assemblies of cells, the outer wall of most cells is made by a very soft and deformable membrane: a double-shell of well-ordered lipid molecules, having the mechanical consistency of a soft gel. However, cell deformations as large as 100% are easily observed, e.g., red blood cells of about 7–8 μm fitting into capillary vases of about 1–3 μm diameter. In fact, besides its very limited elastic capability of just a few % maximum strain, the main mechanical response of the cell resides in its ability to continuously supplement new membrane material, taken from some special reservoir, while adjusting its shape by restructuring the internal cytoskeleton (Fig. 2).

Clearly, living tissue microstructures have plenty of hidden secrets that should greatly stimulate the imagination of materials scientists. Study and observation of such microstructures is complex, since there is no single tool, either experimental or theoretical, capable of covering all the various levels of organization of the biological materials. While structures down to about a μm in size are accessible to conventional light and fluorescence microscopy, higher resolution down to the single-molecule level can be achieved by other probes, such as confocal, total internal-reflection (TIRF), or resonant energy-transfer (FRET) laser microscopy, scanning (SEM) or transmission (TEM) electron microscopy, x-ray diffraction (XRD) or small-angle x-ray scattering (SAXS), and by a variety of spectroscopic techniques, from nuclear magnetic resonance (NMR), to Fourier-transform infrared (FTIR) spectroscopy, and so on.

In this contribution, I wish to focus on the issues raised by the study of the mechanical deformation of living cells and of their molecular constituents, by means of computer modelling. Far from being exhaustive, my modest (and possibly too naive) purpose is that of addressing a few general issues about the design and use of theoretical models, from a physics and materials science point of view, aimed at linking up our rather well-assessed knowledge in solid mechanics and molecular physics to the domain of cell biology.

1 Cell mechanics and adhesion

The microscopic mechanics of molecules, including proteins and nucleic acids, can be crucial to understanding several fundamental biological processes, e.g., the manner in which cells sense mechanical forces or deformation, and transduce mechanical signals into alterations in such biological processes as cell growth, differentiation and movement, protein secretion, etc. Still in its infancy, this emerging field of mechanics aims to investigate such issues as: how the structural and mechanical properties of DNA (deoxyribonucleic acid), RNA (ribonucleic acid) and proteins under stretching, twisting, bending and shearing conditions, affect DNA—protein, RNA—protein and protein–protein interactions [7]; what function the deformation of proteins and nucleic acids does have in DNA condensation, gene replication, transcription and regulation [8]; how mechano-chemical coupling works in enzymes as nanomachines, how use of DNA and proteins as a component of nanosystems, and the attendant interface considerations [9] can be understood in terms of the deformation and mechanics of biomolecules.

At the level of a single cell, microrheological measurements show that the cytoplasm is a viscous fluid, permeated by a gel-like network (the cytoskeleton, see Fig. 2) with a rather large mesh size, to the point that particles of up to about 50 nm in diameter can freely diffuse through the cell. The viscosity of the cytoplasm is about 10–100 mPas, i.e. about 10–100 times that of pure water, reflecting the high concentration of proteins in the fluid (about 20% weight in a typical cell). The modulus of elasticity of the actin network in the cell typically has a value in the range 0.1–1 MPa [10].

From a structural mechanics point of view, a tissue can be considered as a compact structure, endowed with mathematical continuity. In other words, a tissue is seen as a continuous pattern of cells, possibly evolving in number, shape and size, but generally smooth and, in mathematical terms, convex, i.e. containing no holes, tears or discontinuities.

Mechanical stresses can be transmitted between cells, via either direct cell-cell contacts, or cell to the extracellular matrix (ECM) contacts. In either case, the physical link is not a flat surface between the respective membranes, but is rather represented by a chain of transmembrane proteins which ensure the solid contact among membrane portions belonging to adjacent cells. Transmembrane proteins may link, via a complex microstructure of secondary proteins, either directly to the cytoskeleton, or to the ECM. Therefore, in order to link the mechanical deformation of a single cell to the mechanical response of a tissue-like aggregate of cells, one must necessarily consider also the issue of *cell adhesion*.

Cell adhesion is a central property of biological tissues, at the origin of numerous important properties of living organisms. For example, cell adhesion represents the most direct way of coordinating the synaptic connectivity in the brain, promoting the contact between

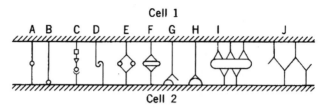

Fig. 3 Examples of some non-covalent association of molecules leading to intercellular bonds between two hypothetical cell membranes: (**A**) two receptors linked by a bivalent ligand, (**B**) a receptor on cell 1 binding to a ligand on cell 2, (**C**) an enzyme-substrate bond, (**D**) a (hypothetical) bond between identical receptors, (**E**) antibodies (or glycoproteins) bridged by two bivalent antigens (or lectins), (**F**) two antibodies bridged by a multivalent antigen, (**G**) antibody binding singly to antigen, (**H**) antibody binding multiply to antigen, (**I**) multiple antibodies bridged by a multivalent antigen, (**J**) complementary antibody bonds. (Adapted from Ref. [20].)

dendrites and axons [11]. In a different context, pathologies in genes controlling cell–cell adhesion are among the main causes of cancerous cells spreading, since the under expression of adhesive proteins increases cell motility and, on the other hand, the over expression of different adhesive proteins facilitates the wandering behavior of tumoral cells [12]. Notably, from the point of view of a biophysicist, the adhesion of a biological cell to another cell, or to the ECM scaffold, involves complex couplings between biochemistry, structural mechanics and surface physics/chemistry. As such, it may constitute one of the very few occasions in which a strictly physical mechanism has a direct connection with biological and genetic phenomena.

At a microscopic, molecular level, adhesion takes place through association and dissociation of non-covalent bonds between very large transmembranal proteins. Different kinds of associations are observed (Fig. 3), with either single or multiple non-covalent bonds being formed between similar or complementary receptors on the proteins on each side. It is to be noted that membrane proteins are rather free to diffuse in the plane of the lipid layer, therefore non-covalent association rates will depend on the diffusivity, especially when the binding rate is much faster than the rates of encounter and formation or dissolution of the receptor-ligand complex. In particular, the kinetics of reaction between surface-bound reactants leads to much lower 2D diffusion and rates, compared to the 3D diffusive features of ligands in solution. Moreover, the association/dissociation rates change considerably under mechanical stress.

The association within, and across a portion of the membrane, of a large number of proteins, enzymes, lipids, proteoglycans etc., during adhesion to the substrate or to a nearby cell, leads to the formation of the so-called focal adhesion complex (Fig. 4). Such a structure (a true *microstructure*, to the eyes of a materials scientist) is indeed a specialized organelle, and may involve dozens of different molecular species, while covering a large extension of the membrane surface [13]. Moreover, the protein–protein and protein-ligand interactions are dynamic, namely the focal adhesion complexes between the cell membrane and cytoskeletal proteins are formed and continuously adjusted in shape and size, as a function of time, stress conditions and elastic properties of the substrate [14]. Recent experiments suggested for the first time the possibility of a direct connection between mechanical forces at the cell surface, and biochemical mechanisms of genetic expression [14–16]. For example, Discher [15] could highlight a direct role of the Youngs' modulus of the substrate in promoting the correct striation of myoblast pre-fibers, starting from stem cells attached onto plastic substrates of varying elasticity.

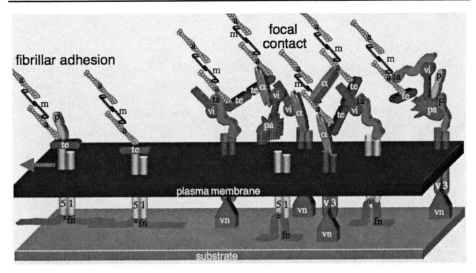

Fig. 4 Hypothetical molecular model of focal contacts and fibrillar adhesions. Since substrate-attached vitronectin forms a rigid matrix, $\alpha_v\beta_3$ integrin remains immobile despite the applied contraction force. In contrast, $\alpha_5\beta_1$ integrin is bound to a relatively soft fibronectin matrix and thus translocates centripetally owing to the actomyosin-driven pulling. The translocation of the fibronectin receptor can also stretch the fibronectin matrix and promote fibrillogenesis. Abbreviations: a, actin; α, α-actinin; F, FAK; fn, fibronectin; m, myosin II; P, parvin/actopaxin; pa, paxillin; ta, talin; te, tensin; vi, vinculin; vn, vitronectin; 51, $\alpha_5\beta_1$ integrin; v3, $\alpha_v\beta_3$ integrin. (adapted from Ref. [13])

However, despite that chemical signals involved in mechanical sensing are starting to be identified, the initial sensor of mechanical force is not yet known. It is generally thought that eliciting a mechanical response in a cell requires at least two steps [17]: (1) a structure which is deformed by the applied force, for example the unfolding of some protein domain, in one of the focal adhesion proteins, and (2) one or more elements that transmit the information to the 'target', for example a transcription initiation site in the cell nucleus.

Just by looking at a simplified scheme of the cell adhesion microstructure (see again Fig. 4), it may appear surprising that some materials scientist could even conceive of trying to describe such a complexity in terms of a homogeneized continuum model, as simple as, e.g., a diffusion-like equation [18]. Nevertheless, thermodynamics models only slightly more sophisticated than a diffusion equation have today become largely accepted biophysical theories [19,20], capable of outlining some basic concepts in cell adhesion. The Bell theory of reversible bond formation and breaking [20] is but one outstanding example of the possibilities offered by the exporting of physical concepts into the domain of biology.[1] With a synthesis capability which may appear extraneous, or even misleading, to the biologists, and despite their only apparent simplicity, such theories are trying to provide a general framework for what would be otherwise an overwhelming complexity, devoid of any common interpretative features.

Since this is a review and position paper rather than a scientific communication, I wish to stress once more that, in order to make true impact in biological sciences as a physicist, one should resist the temptation of following the immediate path of a strict interpretation of

---

[1] It may be worth noting that, before producing his famous theory about cell adhesion, G. I. Bell had already amply made his career as a nuclear physicist, including a world-respected book about nuclear reactors theory, written together with S. Glasstone.

individual experiments, i.e., having ultimately in mind the possible clinical and therapeutic outcomes of a series of specific molecular or supramolecular processes. The latter is, indeed, the business of biomedical and clinical research, and certainly there is people out there much better trained at it, than a physicist or a materials scientist. I believe that our guiding idea should always be that of finding general principles, even among the great complexity of life sciences, aiming at connecting and incorporating in a more general framework the existing, fundamental theories of non-living matter. As suggested by Buiatti in their papers about the concept of a "living state of matter" (see [21, 22] and refs. therein), which are in turn based on a philosophical and episthemologic background linking ideas from I. Prigogine, M. Eigen, and others, present-day physics should be viewed as the theory of only a part of the Universe (the non-living one), and the eventual inclusion of life sciences into a futuristic "theory of everything" should be attempted in the spirit of *completing* physics into a new science with an enlarged basis of principles, rather than attempting to reduce the whole of biology to just some 'special kind' of physics.

In modern physics, going beyond even the simplest theoretical descriptions necessarily requires the use of numerical simulations. Models for such a complex phenomenology as cell adhesion and mechanics should be capable of starting from the most elementary mechanisms of protein-ligand interaction, and proceed upwards in both the length and time scales, all the way up to the level of the elasto-plastic deformations of the cell membrane and cytoskeletal scaffold. To this end, a combination of computational methods must be used, ranging from Molecular Dynamics (MD), at the most microscopic scale, to micromechanical models, in which the molecular degrees of freedom are progressively coarse-grained, to continuum or semi-continuum (microstructured) models, in other words by adopting a *multiscale* computational strategy [23, 24].

To be specific, a molecular-level study of cell adhesion and mechanical deformation from the point of view of computer simulations poses numerous daunting challenges, both methodological and conceptual. To list just the most outstanding ones:

– the description of non-covalent interactions, such as Van de Waals, hydrophobic-hydrophilic, and hydrogen-bonding forces, is still an unsolved problem, even within the most advanced electron density functional theories;
– the deformation rates applicable in MD simulations are orders of magnitude faster than in the real experiments and, as a result, the peak forces are overestimated, and irreversible work is injected in the computer simulation;
– moreover, the too fast MD deformation rates artificially smooth out force fluctuations, which would allow to sample the space of conformationally-independent deformation modes, therefore failing to represent the inherent dynamics and stress-force dependence of actual atomic-force microscopy (AFM) or laser-tweezer pulling experiments;
– linked to the previous point, is the issue of how to correlate the microscopic simulation of mechanical loading at the atomic scale, to the meso- and macroscopic deformation state of cell membranes, and tissues;
– fundamental questions still remain about the most adequate statistical mechanical treatment to describe the entropic contribution to the unfolding forces, since the current methods based, e.g., on steered-molecular dynamics (see below) involve fixing the end-points of molecules, thereby arbitrarily restricting the configuration space sampling.

An excellent review of open problems in the application of micromechanics models to cell mechanics has been recently presented by Bao (see Ref. [25] and references therein). It is not my intention to recapitulate published material and, moreover, the present contribution is not intended as an exhaustive review article, but rather as a perspective paper. In the remainder of this article I will therefore focus on the interplay between protein deformation and cell adhesion, and mechanical behavior of living tissues, and I will try to address the above questions concerning molecular-scale and multi-scale modelling of the mechanical deformation of proteins from a rather general, qualitative standpoint, going from single-molecule to more complex systems.

2 Modelling molecules inside cells

In the past ten years, it has become clear that studying the deformation of molecules (nucleic acids, proteins) inside cells could be the key to understand mechano-chemical coupling, and how mechanical forces could play a role in regulating cell division, shape, spreading, crawling etc (see for example Ref. [26]). Internal mechanical forces $\mathbf{F}_\alpha$, generated e.g. by motor proteins, induce conformational changes in a protein that can be understood by studying the behavior of the total free energy of the system:

$$G = U + \sum_k \mu_k c_k - \sum_\alpha \mathbf{F}_\alpha \cdot \mathbf{x}_\alpha - TS \tag{1}$$

where U is the internal energy, T the temperature, S the entropy, and the $\mathbf{x}_\alpha$ represent vectors of generalized displacements. 'Total' here means that one cannot consider the protein as an isolated object, but must account for the presence of the surrounding solvent, ions, small molecules (e.g., sugars), and any other adjacent, deformable bodies (cell membrane, ligands, other proteins), each one indicated by its concentration c_k and chemical potential μ_k. By assuming quasi-equilibrium conditions under the applied forces $\mathbf{F}_\alpha$, the conformational changes can be determined by the condition:

$$\frac{\partial G}{\partial \mathbf{x}_\alpha} = 0, \quad \forall \alpha \tag{2}$$

The different contributions to U and S upon variations of the atomic positions by $\mathbf{x}_\alpha$ are rather difficult to assess. Besides covalent, or *bonding*, force terms, arising from bond-stretching, bond-bending, dihedral forces etc., and affecting the internal coordinates of the molecule, U may contain *non-bonding* contributions, such as electrostatic and Van der Waals, hydrogen bonding, hydration, hydrophobic, steric and brownian contributions, affecting the translational and rotational degrees of freedom, and the overall 3D conformation of the molecule. However, even more subtle and crucial is the role of the entropy, S, for biological macromolecules. In fact, aside from a usually negligible vibrational entropy term, an important contribution to G in proteins and long polymers arises from *configurational*, or 'conformational' entropy (the first name being preferable, to emphasize the connection with the microstates of a statistical ensemble) which, in turn, is usually negligible when considering the condensed phases of matter.

Configurational entropy is a measure of the contribution to the free energy arising from all the possible geometrical arrangements of a polymer chain, compatible with a given set of macroscopic thermodynamical constraints. It is related to the amount of free volume surrounding the object, and this is the reason why it is usually negligible in the high-density, condensed states of matter. Intuitively, a polymer chain in solution completely folded into a globular state, may have a huge number, Ω, of nearly-equivalent geometrical conformations, all corresponding to the same values of energy, enthalpy, temperature, volume, etc. The same chain, once unfolded into an almost rectilinear string, has but a very small number, $\Omega \sim 1$, of equivalent configurations. The configurational entropy corresponding to either state is, by definition, $S_{conf} = -k_B \log \Omega g$ Thereby, the entropy of a randomly coiled polymer, or a denatured protein, is much larger that of a protein folded to its native state. Indeed, at the very extreme, one may say that there exists only one conformation of a protein, $\Omega \equiv 1$ and $S_{conf} = 0$, that makes it fully effective for the function to which it is destined. The variation in configurational entropy is probably the single largest contribution to the stability of the aminoacid chain, and a similar, albeit non exclusive, entropy role is thought to be involved in the stabilization of nucleic acid chains. At variance with the world of pure mechanics, the world of flexible polymers in solution (which includes proteins and nucleic acids) is governed by the competition between attractive interactions and unfavorable entropy terms. Describing the evolution of such a system under the action of external mechanical forces cannot be limited to the minimization of mechanical energy, but must take into account the configurational degrees of freedom, i.e., the changes in the *multiplicity* of configurations belonging to a same set of values of energy. This is what is concisely expressed by the concept of entropy. In the remainder of this Section, I wish to elaborate briefly on the relationship between MD simulations and configurational entropy, in context of the elasticity of living polymers at null or small applied forces.

The general features of a chemical reaction, including, e.g., biomolecular association into a complex, $A + B \Leftrightarrow AB$, association of a soluble ligand to a surface binding site, $L + S \Leftrightarrow L : S$, or the two-state folding/unfolding of a protein, $U \Leftrightarrow F$, are customarily described in terms of a standard reaction free-energy, ΔG, and the relative equilibrium constant, K, for the reaction:

$$\Delta G = -RT \, \log K \tag{3}$$

A first problem arises when considering single molecule experiments. As the presence of the molar gas constant R suggests, the above expression is valid for an *ensemble* of equivalent objects in the thermodynamic limit. One should ask whether is it still feasible to extrapolate the use of a similar description, when an *individual* object or reaction is being monitored?

In the standard connection between classical MD and thermodynamics, the ergodic hypothesis is assumed, namely that a "sufficiently long" time observation of the system evolution can represent a statistically meaningful ensemble average. This, in turn, allows to connect the statistical average values of estimated physical quantities to thermodynamic parameters, such as ΔG and K. It turns out that such an assumption is readily verified in (and, indeed, has been instrumental for the understanding of) a wide range of condensed-phase systems. This is due to the fact that in such cases, and at relatively low temperatures and pressures, the free energy is practically dominated by the internal energy contribution, a quantity which is easily accessible to MD calculations; at non-zero pressure or stress the main contribution to the free energy is embodied in the enthalpy, again directly accessible to MD calculations by means of several extended methods to include also external thermodynamic and mechanical constraints.

However, in a single-molecule experiment, e.g., observation of a protein folding to its native state, or pulling a protein from one end with an AFM or tweezer apparatus, the system must be able to explore the entire phase space unconstrained, and the sampling of the intermediate configurations along the minimum free-energy path must be distributed according to the right multiplicity, i.e., the partition functions of the reactants and the complex. The classical expression for the equilibrium constant reads like [27]:

$$K \propto \frac{Z_{N,AB} Z_{N,0}}{Z_{N,A} Z_{N,B}} \tag{4}$$

with $Z_{N,x}$ the statistical partition functions for: $x = AB$ the complex, $x = A$, $x = B$ the reactants (all at dilution $1/N$ in the solvent), and $x = 0$ the pure solvent, respectively; the partition function of the species X in N solvent molecules is proportional to the configurational integral,

$$Z_{N,X} \propto Q_{N,X} = \int e^{-\beta U(\mathbf{r}_X, \mathbf{r}_N)} d\mathbf{r}_X d\mathbf{r}_N \tag{5}$$

The appearance of the ratio of the partition functions in Eq. (4) can be understood since, during the experiment, the system goes from the initial to the final state, and the difference between the two free energies (logarithms of K) turns into the ratio between the configurational integrals.

The free energy difference between initial and final state can be evaluated by the method of thermodynamic integration (Kirkwood 1935), in practice by performing a set of equilibrium Molecular Dynamics simulations of the two extreme states, and interpolating by a numerical approximation [28–30]:

$$\Delta G_{AB} = \int \frac{\partial G(\lambda)}{\partial \lambda} d\lambda = \int \left\langle \frac{\partial U(\lambda, \mathbf{r}_A, \mathbf{r}_B, \mathbf{r}_{AB}, \mathbf{r}_N)}{\partial \lambda} \right\rangle_{I(r_{AB}) \approx 1} d\lambda \tag{6}$$

The averaging condition $I(r_{AB}) \approx 1$ indicates that the configurations over which the integral (6) is to be evaluated are those for which the ligand is at, or near the binding site (those in which A and B are far apart giving a negligible contribution). This is true *also for the initial state*, when the complex AB is not yet formed, and the ligand is in solution. In other words, the ligand is surrounded by the solvent, but it is always close to the binding site: this avoids to wrongly compare calculations of the initial and final states having different values of the chemical potentials. Such a condition can be implemented during the MD simulation, e.g., by a hard wall, the $I(r_{AB})$ being represented by a step function, equal to 1 for r_{AB} smaller than some cut-off distance and equal to 0 elsewhere; or more effectively by a confining harmonic potential [31,32].[2]

It should be noted that all the free-energy methods rest on the central idea [28] that the rate of change of the system (i.e., the variation of the parameter λ in Eq. (6) above) is sufficiently *slow* to represent an ideally adiabatic transformation: if this is the case, the system follows a constant-entropy path. If, on the other hand, the perturbation rate is too fast, the final state will depend on the initial conditions, and the calculated free energy will not be a unique function of time. In practical cases, this means that the minimum change in λ at each step can be so small that the computation time becomes prohibitive.

In MD calculations one gets access to the internal energy and enthalpy of the system, but the entropy must be separately added from a model calculation.[3] The simplest approach may involve the use of the classical Sackur-Tetrode equation for the translational entropy of the ideal gas, plus the ideal gas rotational entropy, and some approximated expression for the mixing entropy (see [27] and Refs. therein). Otherwise, an explicit calculation of the configurational integral Q can be carried out for each degree of freedom separatedly (all being expressed in terms of the internal coordinates q_i), and subsequent mixing by a correlation function, usually of Gaussian form [33,34]. It can be shown, after some algebric manipulation

---

[2] Note that the free energy ΔG_{AB} as computed, must be corrected by a term describing the transfer of the ligand from the solvent to the gas phase.

[3] Since the Newton's equations of motion conserve the total hamiltonian as the integral of motion, MD simulations are performed at constant energy. Therefore, the entropy is not constant during the simulation and cannot be calculated explicitly.

and approximations, that such a method requires only knowledge of the covariance matrix:

$$\sigma_{ij} = \langle (q_i - \langle q_i \rangle) \cdot (q_j - \langle q_j \rangle) \rangle \tag{7}$$

The individual averages and fluctuations of each internal coordinate are readily available in a MD simulation, therefore to obtain the configurational entropy it is just necessary to add the calculation of a determinant.

An improved method to obtain a close upper bound for the configurational entropy has been introduced by Schlitter in 1993 [35,36]. The heuristic Schlitter's method works in 'natural' Cartesian coordinates, instead of the more cumbersome internal coordinates, with the advantage of allowing inclusion of translational and rotational degrees of freedom. Moreover, it is shown to converge to the correct quantum and classical limits when the temperature goes to zero or to infinity, respectively.

3 Mechanical loading of single molecules

The considerations in the previous Section concerned the spontaneous evolution of a molecular system in solution, going from a well defined initial state to a well defined final state. In a mechanical deformation experiment (and simulation), instead, the system is artificially driven to an unknown final state by applying some external constraint, typically a force at one extremity of a molecule. Compared to a macroscopic mechanical loading, in which the sample is deformed in a smooth and non ambiguous way, the kind of loading applied to a molecular system during an AFM or tweezer pulling experiment has an intrinsic molecular, or 'nanoscale' character. When observed at the nanoscopic scale, the pulling of a molecule follows a rather noisy, fluctuating path, with a well defined average value and direction of the force, but the instantaneous movement being continuously biased by the Brownian motion of the surrounding solvent.[4] The values of the forces applied by the AFM or tweezer apparatus are in the same range, or smaller, than the binding forces holding together the molecule in its 3-dimensional shape, and comparable to the $k_B T$ of thermal fluctuations. As a result, the deformation path is rather erratic in time and space, and follows the overall free energy changes of the molecule, the solvent, and the pulling apparatus, all working on comparable energy, time and length scales.

Dissociation under the action of an external force is a non-equilibrium process, characterized by a 'rupture strength' (the inverse of the binding affinity) which is not a constant, but rather depends on the rate of force application and the duration of loading. The 'off' rate, v, in the already cited model of Bell [20] is the product of a bond vibration frequency, ω, and the likelihood of reaching a transition state x_t, characterized by a 'reduced' energy barrier $(E_b - Fx_t)$:

$$v \approx \omega \exp\left[-(E_b - Fx_t)/k_B T\right] \tag{8}$$

The above equation shows that the dissociation rate increases with the applied force as $v \approx v_0 \exp(F/F_t)$, with $F_t = k_B T/x_t$ a scale factor for the force at the transition point. Of course, the complexity of biomolecular systems is far too large to be captured by a single parameter, x_t, and more refined models are required to describe, e.g., brittle vs. ductile rupture, entropic elasticity, competition between hydrogen bonds and van der Waals forces, and so on (see below, in the concluding Section). However, Bell's theory points out the importance of mechanical forces in biology: in fact, in the presence of non-covalent association, even

---

[4] Quantum motion and uncertainty effects are not a concern for the large masses of biological molecules.

an infinitesimal force if applied for a long enough time can cause breakage. The question therefore is: can MD simulations reproduce the actual physics of the mechanical deformation and rupture of a single molecule?

The practical answer to this problem, over the past ten years or so, has been the method called "steered Molecular Dynamics", or SMD [37–39]. SMD refers, in fact, to a collection of methods intended to apply a force to a molecular-scale complex, such as a bound protein-ligand ensemble, and monitor the subsequent unbinding, or unfolding process. A first SMD method prescribes to restrain the ligand to a point x_0 in space by an external (e.g., harmonic) potential, $U = k(x - x_0)^2/2$, and subsequently let the point x_0 to move at a constant velocity v during the simulation. The resulting external force is $F = k(x_0 + vt - x)$, the ligand being pulled by an harmonic spring of constant stiffness k. Alternatively, the point x_0 can be held fixed at some distance well from the equilibrium position, and the spring constant can be slowly increased in time, as $k = \alpha t$. In this case, the resulting force is equal to $F = \alpha t(x_0 - x)$. Other variations of the basic methods involve the application of a constant force or torque, to a whole set of atoms. Application of the SMD method requires to prescribe a path, i.e. a series of directions of the force vector (which of course can also be a constant, resulting in a rectilinear path), thereby restricing arbitrarily the region of the phase space compatible with the given thermodynamic constraints. This is coherent with the observation that the procedure described by the SMD algorithm is, indeed, a non-equilibrium process. As such, the SMD method has been employed in a rather large range of cases, such as dissociation of biotin from avidin and streptavidin [37], unbinding of retinal from bacteriorhodopsin [40], release of phospate from actin [41], stretching of immunoglobulin domains in various molecules [42,43], glycerol conduction through the aquaglyceroporin [44], the helix-coil transition of deca-alanine [45], elasticity of tropocollagen [46], and more.

The SMD methods seems to stand on a physical contradiction. The evolution of the molecular complex under an external force is clearly a non equilibrium, dissipative process, but the simulation aims nevertheless at finding a final equilibrium state, characterized by some thermodynamic potential, and estimating free energy differences with respect to the initial state. The so called Jarzynski identity is invoked to solve this contradiction [47–49]:

$$\langle \exp(-W(t_s)/k_B T) \rangle_{\Gamma_0} = \exp(-\Delta G/k_B T) \tag{9}$$

The puzzling result summarized by the above equation, in only apparent violation of the Second Law of thermodynamics, is that the ensemble average of the exponential of the work performed on the system during the "switching time" t_s (clearly a non-equilibrium transition between an initial state Γ_0 and some final state) is equivalent to the exponential of the free energy difference between the same two states, i.e. a quantity independent on the time trajectory, or phase-space path. In other words, the identity allows to obtain equilibrium quantities from a collection of non-equilibrium processes. The importance of Eq. (9) for the case of single-molecule experiments is obvious: one does not need a statistical ensemble in order to derive the free energy of a single process, it is just sufficient to perform many irreversible experiments on copies of the same system.

The Jarzynski identity has been debated in the literature and, even if it is currently accepted on theoretical grounds [50,51], yet is rests on rather stringent practical conditions. In a non-equilibrium process the work is path-dependent, and the ensemble average in Eq. (9) must be computed for a collection of time trajectories, with a whole distribution of initial Γ_0. The limited size of the trajectory sample, and the fact that only a few, rare events contribute importantly to the Boltzmann average, make the use of the identity quite inaccurate in principle.

Moreover, the final state should be uniquely identifiable, and be practically the same for any choice of Γ_0, which is not always the case.

In practice, however, the results of SMD simulations often seem to achieve a good qualitative comparison with experimental results, for example when comparing the force-displacement curves on a relative scale [52], and several successful examples have been already reported, in which the simulation helped identify crucial details of the molecular processes during a single-molecule nanomechanics experiment [53]. In general, the absolute values of the forces observed for, e.g., the unfolding of a molecular domain under an applied force in SMD are substantially larger than those observed in experiments. This is expected, because of the high stretching velocity used in simulations, which in turn is linked to the very small time-step ($\Delta t \approx 10^{-15}$ sec) necessary to ensure a numerically stable integration of the atomic equations of motion. Moreover, the fluctuations of the forces in constant-velocity SMD simulations depend on the stretching velocity and spring constant used, and thereby the displacement path followed, and possibly the final state reached, is a function of the same parameters. Advocates of the SMD method concede that "the smaller the velocity used, the less drastic the perturbation of the system is, and more details along the reaction coordinate are likely to be captured" and that "(too) fast pulling may alter the nature of the protein's elastic response" [53]. Indeed, even the slowest stretching velocities used in simulations (of the order of 10^{-4} Å per time-step) are orders of magnitude faster than those used in equivalent AFM stretching experiments, a situation which is common to all nanomechanics computational studies performed by MD [54]. Althought the *a posteriori* comparison with AFM or tweezers experiments can increase confidence in the computational results, for example by looking at the hierarchy of mechanical stability of the various unfolding steps, and by comparing on a relative scale the peak-to-peak displacement distances and force ratios, yet some deep questions about the dependence of the monitored force on the stretching velocity, and about the actual dependence on the Γ_0 sample size in Eq. (9), remain unsolved. Notably, both questions are bound only by computational power limitations, and not by fundamental theoretical restrictions. It is, moreover, worth noting that most experiments, albeit far slower than simulations, are rarely slow enough to work in a truly reversible regime. I will further elaborate on this point in the following Section.

With respect to the considerations in the preceding Section, about the role of entropy and the need for an unbiased sampling of the configuration space, it is clear that a non-equilibrium process like the one described by SMD cannot but have a strong dependence on the initial conditions. The computation of the binding affinity (i.e., binding free energy), moreover in presence of subtle, non-covalent interactions such as the desolvation of ligand and receptor binding site, and the attending changes in conformational free energy upon binding, involves a near-compensation of entropy and enthalpy, with a conformational reorganization spanning large length and time scales. Therefore computer simulations should be particularly careful in this respect, at least by allowing an adequate sampling of the initial conditions, and by restarting the simulation from intermediate points along the SMD trajectory ('pinned' simulations, in the language of Harris et al. [55]). The latter procedure may reveal, for example, instability towards different final states. To some extent, the rapid pulling experiment mimicked by SMD amounts to travel 'over the hills' of the energy landscape, once the first minimum is escaped, with only occasionally looking at some of the downslopes leading to intermediate states. A more careful (i.e., slower) exploration of the landscape should be considered, e.g. by a rare-events search method (see also the final Section). In any case, the role of configurational entropy, qualitatively translating in the distribution of multiplicity of nearly-equivalent paths across a given barrier, cannot be overlooked [55], and it should be included by one of the methods described in the previous Section.

4 Nanomechanics of living polymers

Although DNA has been, and still is, the object of intense and thorough experimental and theoretical studies under any possible respect, the details and biological relevance of the mechanical response of DNA under applied forces (either external, or arising from inside the cell) are still unclear in many aspects. It is well known that DNA functions, replication, transcription, etc., rely on the matching with other nucleic acids and associated proteins. However, local and global deformations of the DNA in its coiled and supercoiled state can alter this matching, and thereby its functions. Moreover, there are clear evidences that it would be impossible to describe specific protein-DNA interactions entirely by a simple 'recognition code', based on direct chemical contacts of amino acid side chains to bases [56]. Even when taking the 3D arrangement of protein-DNA contacts into account [57,58], the observed specificity cannot always be explained. In general, complexation free energies also depend on the deformation required to distort both the protein and the DNA binding site into their 3D structure in the complex. In this way, sequence-dependent structure and the deformability of DNA contribute to sequence-specific binding, the effect called 'indirect readout'. Up to date, we do not yet have a clear picture of how mechanical forces are applied to DNA inside cells and, more generally, we lack understanding of how mechanical forces acting on DNA may contribute to regulation and control of gene replication and transcription.

Notably, DNA under tension offers a relatively simple and biologically relevant example of a living polymer for which one can develop rather elegant theoretical expressions of the free energy G, to be compared to experimental data, and a possibly useful test to check direct molecular-scale SMD simulations. When pulled along its axis by an external force F (Fig. 5), double-stranded DNA (or B-DNA) is firstly unwrapped from its supercoiled state, without

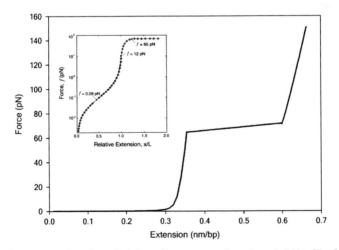

Fig. 5 Schematic representation of a typical force-distance curve for polymeric DNA. The flat region of the force-distance curve at 65 pN is commonly known as the B-S plateau. The DNA unbinds at forces of $\sim$150 pN (depending upon experimental details such as the pulling rate). The inset shows on a log-log scale two sets of experimental data [57,58] at small forces, up to the B-S plateau.

appreciable change in enthalpy. In this force range ($F < 0.08$ pN) configurational entropy is the main contribution to the observed elasticity [59]. For $0.08 < F < 0.12$ pN, the bending rigidity of the chain starts contributing, i.e., the local chain curvature is slowly changed by the applied force: this involves rearranging a number of hydrogen bonds, and the π-stacking of the bases, thereby giving also an enthalpic contribution to the elasticity. The well-known freely-jointed chain (FJC) and worm-like chain (WLC) models of random polymers can describe quite well the experimental data in the above two small-force regimes [60,61]. Both models assume the polymer chain to be formed by inextensible segments, implying that the force diverges as the total chain length approaches the uncoiled length L.

At larger forces, the length of the double-twisted chain becomes longer than its uncoiled length, meaning that in this force range single covalent bonds start to be stretched and bent [62,63], thereby making the FJC or WLC models unapplicable. Upon reaching a force of $\sim$65 pN, DNA elongates abruptly at nearly constant force up to some critical length (see Fig. 5), after which it stiffens again and rapidly breaks apart at a force of about $\sim$150 pN. Two physical pictures have been developed to describe this overstretched, or 'plateau' state. The first proposes that strong forces induce a phase transition to a molten state consisting of unhybridized single strands. The second picture introduces an elongated hybridized phase called S-DNA, and the plateau is a phase transformation at (nearly) constant stress, from the B-DNA (the ordinary, right-handed double helix) to S-DNA. Little thermodynamic evidence exists to discriminate directly between these competing pictures [64].

Several direct simulations of DNA extension by SMD have been performed in the past, with a moderate degree of success (see, e.g., [65,66]). Basically, such simulations observed the plateau in the force-extension curve for short fragments, around 12 bp, with features of the B-S transition. However, the plateau is experimentally observed only for longer fragments (>30 bp), and interpretation of the experimental results in terms of the B-S transition has been therefore subject of debate. Since the experiments are not slow enough to be reversible, the plateau should be more probably an evidence of force-induced melting. This is a good example of the careless use of simulations inducing misleading information. As clearly indicated by Harris et al. [55], the neglect of entropic contributions in the SMD simulations was at the origin of the wrong interpretation.

The most current interpretation of the mechanical response of short DNA fragments under external forces is that the two strands detach due to a rare high-energy fluctuation which brings the system above the thermal barrier. The breaking force is therefore correlated to the spontaneous dissociation rate $\nu(0)$ (i.e., the frequency at zero applied force), rather than to the equilibrium free energy of binding, as [67]:

$$F = \frac{k_B T}{x_t} \ln \frac{R x_t}{\nu(0) k_B T} \tag{10}$$

The SMD simulations can provide a value of the transition state x_t, at which the force is maximum and the free energy goes through zero. As shown in Ref.[55], the estimate of $x_t = 50$ Å, obtained from earlier SMD simulations including the solvent but without explicit account for the configurational entropy of the stretched DNA, underestimates the experimental force by a factor of $\sim$3.5. However, when the entropy is calculated according to, e.g., the Schlitter's heuristic formula, the total free energy $\Delta G = \Delta U_{DNA} + \Delta G_{solv} - T \Delta S_{DNA}$ goes through zero at $x_t = 18$ Å, and the SMD calculated force is in very good agreement with the experimental value.

The simulations carried out on the DNA double helix served as an example to attack other similar problems, such as the study of mechanical strenght of collagen, a triple helix held together by hydrogen bonds, similar in some respects to the DNA structure [46,68].

Also in this case, the first SMD simulations have neglected the role of entropy in the free energy of deformation. The work by Buehler and Wong [46] points out, indeed, an important role of entropic elasticity. However their finding is indirect, simply based on the observation of a steady elongation at nearly constant energy in the first phase of the deformation. The questions about the efficacity of the plain SMD procedure, in the absence of an adequate averaging over Γ_0, and of a repeated sampling of the non-equilibrium trajectories by restarting the simulations at mid-points, continue to hold.

However, the SMD simulations have been able to reveal important features the unfolding or unbinding process in many cases for which a direct experimental validation has been possible. Not surprisingly, in all such cases the enthalpic component was the main responsible for the observed behavior. Several experiments of forced protein unfolding by means of optical tweezers and AFM have shown a characteristic sawtooth pattern [69–74]. Such a behavior is reminiscent of the unfolding of DNA, but occuring in a many-steps fashion, compared to the single-step action of DNA. Since adhesion proteins, such as the VCAM or NCAM, cytoskeletal proteins such as spectrin, and other proteins, such as titin or avidin, have a multidomain structure, the most obvious interpretation is that each subsequent branch of the sawtooth corresponds to the repeated unfolding of one single domain. However, as SMD simulations helped revealing, the situation is not always so simple and clean.

A more complex multidomain unfolding behavior has been demonstrated in the case of the human heart muscle titin [72], for which Marszalek et al. were able to identify the microscopic mechanism responsible for the deviation from the WLC curves of the individual branches of the sawtooth pattern. A peculiar intermediate state occurring during the unfolding was identified by SMD simulations, connected with the extra resistance to entropic unfolding imparted by a patch of six hydrogen bonds. The enthalpy of binding, the force range and critical extension predicted by SMD nicely matched the experimental data obtained from AFM single-molecule extension, and the authors of that study were even able to give the experimental counter proof, by engineering a mutant protein without the six H-bonds, which did not show any deviation from the purely entropic behavior of the WLC chain.

5 Perspectives: physics, mechanics, and the multiscale modelling of biomolecules

The development of a new discipline necessarily involves a critical revision of the existing concepts, as well as the introduction of new ideas. Combining biology and physics, and notably mechanics, can be a worthwhile effort provided we can go beyond a simple cut-and-paste of concepts from either one of the fields to the other. To deal with biological systems, physics—and mechanics in particular—have to face the challenge of adapting to working conditions which are extraneous to their traditional culture. For example, the presence of the solvent with its own chemistry and subtle entropic effects, leading to hydrophobic and hydrophilic forces, and to the still somewhat puzzling 'depletion' force. Notably, changes in the ion concentration in the solvent can influence the structural stability of a protein or nucleic acid: for example, the persistence length of DNA was found to decrease with the inverse of the ionic strength [75], i.e., an increase in ion concentration leads to more curved and bent DNA strands. However, the stretching modulus K increases with ion concentration, in contradiction with the classical elasticity theory that predicts K to be, instead, proportional to the persistence length.

Mechanochemical coupling in living cells involves, on the one hand, the way in which forces and movement are sensed by the cells, and how these are transduced into biochemical signals and, ultimately, genetic transcription [15,16]. From the opposite point of view, it also involves the way in which mechanical forces can be generated inside cells starting from chemical events, such as the activation of movement by molecular motors (see, e.g., [76–78]). While the coupling between mechanics and chemistry may be embodied, in principle, in the thermodynamics description of Eq. (1), in practice the microscopic mechanisms involved in the transduction of chemical into mechanical signals, and vice versa, defy such a simple picture. The description of a molecular motor cannot stop at the conventional mechanical relationships between forces and displacements, torques and angular displacements, power, energy and momentum, but must take into account the randomness of Brownian motion, since the level of forces and displacements involved, in the range of a few pN times a few Å, is comparable to the energy of thermal fluctuations $k_B T$. For example, Brownian forces from the thermal fluctuation of water molecules can assist breakage of hydrogen bonds along the backbone of a protein, as a sort of background random force. Even more importantly, it has become clear only in recent years that 'rectified' Brownian motion (Fig. 6) could be responsible for the displacement of motor proteins along the cytoskeletal filaments [77,79], the average velocities observed being too large to be explained by simple diffusional displacement based on the classical Fick's law. Indeed, both the microtubules and the actin filaments in the cytskeleton show a 'polarity' at the molecular level, which should be at the basis of

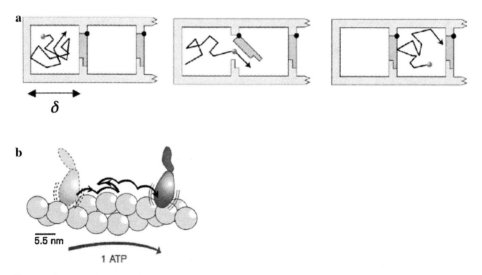

Fig. 6 a Conceptual model of rectified Brownian motion. Consider a particle diffusing in one dimension with diffusion coefficient D. According to Einstein's relation, the mean time it takes a particle to diffuse from the origin, $x = 0$, to the point $x = \delta$, is $\tau = \delta^2/2D$. Now, suppose that a domain extending from $x = 0$ to $x = L$ is subdivided into $N = L/\delta$ intervals, and that each boundary is a "ratchet": the particle can pass freely through a boundary from the right, but having once passed it cannot go back (i.e., the boundary is absorbing from the right, but reflecting from the left). The actual physico-chemical mechanism of the ratchet, schematized in the figure by the door and lock mechanism, depends on the situation; for example, the particle may be prevented from reversing its motion by a polymerizing fiber to its left. The time to diffuse a distance $L = N\delta$ is simply $N\tau = N\delta^2/2D = L\delta/2D$. The average velocity of the particle is $v = L/T$, so the average speed of a particle that is 'ratcheted' at intervals δ is $v = 2D/\delta$. **b** The currently accepted model for the rectified Brownian motion of myosin V along the cytoskeletal actin filaments. The mechano-chemical action provided by the binding and burning of one ATP molecule is necessary to detach the myosin protein from the actin, thereby favouring the reversible clicking of the two globular heads. When the protein is weakly bound to its substrate, the role of thermal fluctuations becomes dominant. From this point the molecule can move to the next stable site, where the ADD is released and the molecule is bound again to the actin. For the myosin, the elementary step of ~35 nm results from several 'ratchet' jumps of 4–5 nm each, helped by thermal fluctuations

the rectification mechanism, i.e., allowing the molecular motors to distinguish between a forward and a backward direction. The case of rectified Browian motion of motor proteins and other biological systems also holds a deeper interest for a physicist [80], since it involves the question of generating order from disorder, thereby touching at fundamental issues about energy conservation and the Second Law of thermodynamics. It turns out that realizing in the laboratory a rectified Brownian motor, or 'thermal ratchet', is an extremely complicated task, while biology naturally offers a host of examples [81], in which many fundamental issues about Brownian motion can be carefully scrutinized and tested.

The mixing of continuum mechanics and statistical mechanics has been invoked in order to improve the simplified physical picture of bond breaking in molecules provided by Bell's theory. Eq. (8) does not include a dependence of the pulling speed on the pulling force. However, during a SMD simulation the maximum rupture force does show a strong dependence on the speed at which the free end of the molecule is pulled, i.e., the fracture is strain-rate dependent. In order to obtain such an information, Ackbarow and Buehler [82] considered that the pulling speed σ could be linearized as the product of the rate ν from Eq. (8) (a statistical mechanics concept) times the amplitude of the displacement at the transition (a continuum mechanics concept), $\sigma = \nu \cdot x_t$. Under this assumption, the product can be inverted to obtain F as a function of the speed, resulting in a logarithmic dependence, $F \propto \ln \sigma$. Apparently, the SMD data can be roughly fitted to $\ln \sigma$, however the values of the fitting parameters E_b and x_t as obtained lend to ambiguity, since the barrier height, $E_b = 5.59 \, \text{kcal} \, \text{mol}^{-1}$, is about 1/3 of the experimental value, and the transition state corresponds to a very small elongation $x_t = 0.185 \, \text{Å}$. While the idea of linearizing the force-velocity relationship (a typical physicist's solution to the problem) may be, indeed, a bold one, the result is not entirely satisfactory.

It is a general statement that extending the ideas and methods of physics to biology does not amount to simply trying to rephrase biological phenomena according to the laws of physics, and especially equilibrium thermodynamics. The use of the Jarzynski identity in single-molecule experiments is a possible example of ways to extend the current domain of physics in order to deal with life sciences. In fact, it represents one first case in which a physical concept has been proven with the help of biology. Representing a bridge between equilibrium and non-equilibrium thermodynamics, a subject that still largely escapes direct physical experimentation, the Jarzynski identity Eq. (9) implies $\langle W \rangle \geqslant \Delta G$ (which, besides, makes it clear that the identity does not violate the Second Law). Notably, for the identity to hold only the average value of W must be restricted, while there can be some non-equilibrium trajectories for which $W < \Delta G$ Such fluctuations are negligible in macroscopic systems, however they can be occasionally realized in a molecular system. When this is the case, their statistical contribution to the Boltzmann integral is very large, since the exponent becomes positive, and a few of such contributions cancel out the effect of many trajectories with $W > \Delta G$. This is the way by which the Jarzynski identity can recover the full ΔG, in the limit of a large number of trajectories. 'Large' means practically infinite for a macroscopic system, but it can be reasonably small to be realized during single-molecule experiments, and this was proven to be indeed the case [83]. Moreover, it is also interesting to note that the recently introduced method of *metadynamics*, aiming at finding rare events along a set of pre-defined reaction coordinates during a MD simulation [84], has been proven to give an unbiased estimate of the free energy of the phase-space explored by the molecular system during the MD trajectory [85]. Since the metadynamics method amounts to exploring the phase-space by a properly constructed collection of non-equilibrium paths according to a history-dependent bias potential, there seems to be a singular correspondence between such an approach and the outcomes of the Jarzynski identity. The implications of such a connection, as well as the possible advantages or complementarities with respect to SMD, are yet to be explored.

Many fundamentally important processes in biology are inherently *multiscale*. Biological processes such as protein folding, nucleic acid packaging and membrane remodeling, that evolve over mesoscopic up to nearly macroscopic length and time scales, are intimately coupled to chemistry and physics by means of atomic and/or molecular level dynamics (e.g. fluctuations in sidechain conformation, or lipid diffusion). Consequently, it is not surprising that many diverse computational methodologies need to be developed for the physical modeling of biological processes, with varying degrees of resolution.

On the one hand, atomic-scale models, such as SMD, will remain a powerful tool for investigating biological structure and dynamics over nanosecond time and nanometer length scales, with femtosecond and Angstrom-level resolution. However, due to the inherent limitations of the computational power, only lower-resolution coarse-grained models could eventually provide the capability for investigating the longer time- and length-scale dynamics that are critical to many biological processes. The level of model detail needs to match the given question and the available computer resources, ranging from quantum mechanical and classical all-atom MD and SMD simulations, to united-atom models, in which some atoms are grouped together in subunits (typically the hydrogens in groups CH_2, CH_3, NH_3, etc.), or residue-based methods, in which only the backbone of the molecule is explicitly described, to even coarser-scale methods in which only the side chain, or only the main $\alpha - C$ of the aminoacid sequence are explicitly modeled, up to mesoscopic and semi-continuum models, in which many residues are grouped together, the solvent is described as a dielectric continuum, and the cell membrane is described as a viscoelastic medium (see Ref. [23], in particular the contribution by Ayton et al. on p. 929, and references therein).

Multiscale techniques have emerged as promising tools to combine the efficiency of coarse-grained simulations with the details of all-atom simulations, for the characterization of a broad range of molecular systems. Recent work has focused on the definition of strategies that combine different resolutions in different regions of the space during a single simulation [86,87]. For instance, this idea has been applied to a system of small molecules where some parts of the space use the all-atom representation and the rest of the space uses a coarse-grained representation [86]. Multiple resolution simulations have also been used to study membrane-bound ion channels by coarse graining the lipid and water molecules, while using an all-atom representation for the polypeptide ion channel [88]. In the context of protein simulation, a similar idea has been applied to represent parts of the protein, such as the active site, with all-atom detail while using a coarse-grained model for the rest of the system [87]. Additional multiscale strategies for biomolecular systems focus on changing the whole system resolution during the same simulation. One of the first applications in this area used a simplified protein model as a starting point to evaluate the folding free energy of the corresponding all-atom model [89]. Candidate structures obtained from coarse-grained simulations can be used as initial configurations for all-atom simulations, allowing for a larger sampling of the protein conformations than using all-atom simulations alone. For example, we recently used this strategy to study adhesion of artificial polypeptide sequences onto inorganic surfaces [90], also coupling low-resolution folding methods based on simple chemical and electrostatic criteria and implicit solvent representation, to all-atom MD with explicit solvent. Another idea, known as 'resolution exchange' or 'model hopping', allows jumping between different levels of structural detail in order to cross energy barriers [91]. Also, a coarse-grained variant of the SMD method has been introduced, by using a continuum representation that allows to efficiently conduct many runs of steered Langevin dynamics simulations [92]. Coarse-graining is of course more effective when the system can be considered homogeneous over some length scale, as in the case of lipid bilayers [93]. For this case, the coarse-graining method of dissipative particle dynamics (DPD) has been successfully employed [94]; some extension of DPD to proteins has also been recently attempted [95].

The underlying assumption in the definition of multiscale techniques for biomolecular simulation is that it is possible to reliably and efficiently move between coarse-grained representations and all-atom models. In this respect, the coarse-grained model used must be realistic enough so that the molecular structures being sampled will implicitly represent relevant atomic-scale conformations of the molecule. Rigorous mathematical procedures, such as renormalization group theory, have yet to be applied to the general definition of coarse-grained models. Therefore, an evaluation of the reliability of coarse-grained models is usually obtained by comparison to experimental data. There have been no thorough studies yet, on whether going back from coarse-grained to all-atom protein structures could distort the thermodynamic properties of the corresponding ensembles of structures. Coupling the coarse-grained and atomistic-level systems involves bridging of information across various length and time scales, the end goal ultimately being that of integrating the different resolutions of the system into a single, unified, multiscale simulation methodology [96–99] The development of new theories and computational methodologies for connecting the disparate spatial and temporal scales relevant to cellular processes remains, arguably, one of the most significant challenges for the physical and mechanical modeling of complex biological phenomena.

References

1. Doran, C.F., McCormack, B.A.O., Macey, A.: A simplified model to determine the contribution of strain energy in the failure process of thin biological membranes during cutting. Strain **40**, 173–179 (2004)
2. Feng, Z., Rho, J., Han, S., Ziv, I.: Orientation and loading condition dependence of fracture toughness in cortical bone. Mat. Sci. Eng. C **11**, 41–46 (2000)
3. Fantner, G.E., Hassenkam, T., Kindt, J.H., Weaver, J.C., Birkedal, H., Cutroni, J.A., Cidade, G.A.G., Stucky, G.D., Morse, D.E., Hansma, P.K.: Sacrificial bonds and hidden length dissipate energy as mineralized fibrils separate during bone fracture. Nat. Mater. **4**, 612–616 (2005)
4. Elices, M., Pérez-Rigueiro, J., Plaza, G.R., Guinea, G.V.: Finding inspiration in argiope trifasciata spider silk fiber. JOM J. **57**, 60–66 (2005)
5. Toulouse, G.: Perspectives on neural network models and their relevance to neurobiology. J. Phys. A Math. Gen. **22**, 1959–1960 (1989)
6. Svitkina, T.M., Borisy, G.G.: Correlative light and electron microscopy of the cytoskeleton of cultured cells. Meth. Enzym. **298**, 570–576 (1998)
7. Rudnick, J., Bruinsma, R.: DNA-protein cooperative binding through variable-range elastic coupling. Biophys. J. **76**, 1725–1733 (1999)
8. Wang, J., Su, M., Fan, J., Seth, A., McCulloch, C.A.: Transcriptional regulation of a contractile gene by mechanical forces applied through integrins in osteoblasts. J. Biol. Chem. **277**, 22889–22895 (2002)
9. Chen, Y., Lee, S.-H., Mao, C.: A DNA nanomachine based on a duplex-triplex transition. Angew. Chem. Int. Ed. **43**, 5335–5338 (2004)
10. Satchey, R.I., Dewey, C.F.: Theoretical estimates of mechanical properties of the endothelial cell cytoskeleton. J. Biophys. **71**, 109–118 (1996)
11. Dean, C., Dresbach, T.: Neuroligins and neurexins: linking cell adhesion, synapse formation and cognitive function. Trends Neurosci. **29**, 21–29 (2006)
12. Wijnhoven, B.P.L., Dinjens, W.N.M., Pignatelli, M.: E-cadherin-catenin cell-cell adhesion complex and human cancer. Br. J. Surg. **87**, 992–1005 (2000)
13. Zamir, E., Geiger, B.: Molecular complexity and dynamics of cell-matrix adhesions. J. Cell Sci. **114**, 3577–3579 (2001)

14. Balaban, N.Q., Schwarz, U.S., Riveline, D., Goichberg, P., Tzur, G., Sabanay, I., Mahalu, D., Safran, S., Bershadsky, A., Addadi, L., Geiger, B.: Force and focal adhesion assembly: a close relationship studied using elastic micropatterned substrates. Nat. Cell Biol. **3**, 466–472 (2001)
15. Discher, D.E., Janmey, P., Wang, Y.: Tissue cells feel and respond to the stiffness of their substrate. Science **310**, 1139–1143 (2005)
16. Evans, E.A., Calderwood, D.: Forces and bond dynamics in cell adhesion. Science **316**, 1148–1153 (2007)
17. Janmey, P.A., Weitz, D.A.: Dealing with mechanics: mechanisms of force transduction in cells. Trends Biochem. Sci. **29**, 364–370 (2004)
18. Shenoy, V.B., Freund, L.B.: Growth and shape stability of a biological membrane adhesion complex in the diffusion-mediated regime. Proc. Natl. Acad. Sci. USA **102**, 3213–3218 (2005)
19. Steinberg, M.: Reconstruction of tissues by dissociated cells. Science **141**, 401–408 (1963); see also Steinberg, M.: Adhesion in development: an historical overview. Dev. Biol. **180**, 377–388 (1996)
20. Bell, G.I.: Models for the specific adhesion of cells to cells. Science **200**, 618–627 (1978)
21. Buiatti, M., Buiatti, M.: The living state of matter. Riv. Biol. Biol. Forum **94**, 59–82 (2001)
22. Buiatti, M., Buiatti, M.: Towards a statistical characterisation of the living state of matter. Chaos Sol. Fract. **20**, 55–66 (2004)
23. de Pablo, J.J., Curtin, W.A. (guest eds.): Multiscale modeling in advanced materials research—challenges, novel methods, and emerging applications. MRS Bull. **32**(11) (2007)
24. Buehler, M.: Nature designs tough collagen: explaining the nanostructure of collagen fibrils. Proc. Natl. Acad. Sci. USA **103**, 12285–12290 (2006)
25. Bao, G.: Mechanics of biomolecules. J. Mech. Phys. Sol. **50**, 2237–2274 (2002)
26. Lecuit, T., Lenne, P.-F.: Cell surface mechanics and the control of cell shape, tissue patterns and morphogenesis. Nat. Rev. Mol. Cell Biol. **8**, 633–644
27. Gilson, M.K., Given, J.A., Bush, B.L., McCammon, A.: The statistical-thermodynamic basis for computation of binding affinities: a critical review. Biophys. J. **72**, 1047–1069 (1997)
28. Frenkel, D., Smit, B.: Understanding Molecular Simulation, Chap. 7. Academic Press, New York (2006)
29. McCammon, J.A., Harvey, S.C.: Dynamics of Proteins and Nucleic Acids. Cambridge University Press, Cambridge (1987)
30. Aiay, R., Murcko, M.: Computational methods for predicting binding free energy in ligand-receptor complexes. J. Med. Chem. **38**, 4953–4967 (1995)
31. Hermans, J., Shankar, S.: The free-energy of xenon binding to myoglobin from molecular-dynamics simulation. Isr. J. Chem. **27**, 225–227 (1986)
32. Roux, B., Nina, M., Pomes, R., Smith, J.C.: Thermodynamic stability of water molecules in the bacteriorhodopsin proton channel: a molecular dynamics free energy perturbation study. Biophys. J. **71**, 670–681 (1996)
33. Karplus, M., Kushick, S.: Method for estimating the configurational entropy of macromolecules. Macromolecules **14**, 325–332 (1981)
34. Di Nola, A., Berendsen, H.J.C., Edholm, O.: Free energy determination of polypeptide conformations generated by molecular dynamics simulations. Macromolecules **17**, 2044–2050 (1984)
35. Schlitter, J.: Estimation of absolute and relative entropies of macromolecules using the covariance matrix. Chem. Phys. Lett. **215**, 617–621 (1993)
36. Schaefer, H., Mark, A.E., van Gunsteren, W.F.: Absolute entropies from molecular dynamics simulations trajectories. J. Chem. Phys. **113**, 7809–7817 (2000)
37. Izrailev, S., Stepaniants, S., Balsera, M., Oono, Y., Schulten, K.: Molecular dynamics study of unbinding of the avidin-biotin complex. Biophys. J. **72**, 1568–1581 (1997)
38. Izrailev, S., Stepaniants, S., Isralewitz, B., Kosztin, D., Lu, H., Molnar, F., Wriggers, W., Schulten, K.: Steered molecular dynamics. In: Deuflhard, P., Hermans, J., Leimkuhler, B., Mark, A., Skeel, R.D., Reich, S. (eds.) Algorithms for Macromolecular Modelling, Lecture Notes in Computational Science and Engineering, Springer-Verlag, New York (1998)
39. Evans, E., Ritchie, K.: Dynamic strength of molecular adhesion bonds. Biophys. J. **72**, 1541–1555 (1997)
40. Isralewitz, B., Izrailev, S., Schulten, K.: Binding pathway of retinal to bacterio-opsin: a prediction by molecular dynamics simulations. Biophys. J. **73**, 2972–2979 (1997)
41. Wriggers, W., Schulten, K.: Stability and dynamics of G-actin: back-door water diffusion and behavior of a subdomain 3/4 loop. Biophys. J. **73**, 624–639 (1997)
42. Lu, H., Schulten, K.: Steered molecular dynamics simulation of conformational changes of immunoglobulin domain I27 interprete atomic force microscopy observations. Chem. Phys. **247**, 141–153 (1999)
43. Paci, E., Karplus, M.: Unfolding proteins by external forces and temperature: the importance of topology and energetics. Proc. Natl. Acad. Sci. USA **97**, 6521–6526 (2000)
44. Jensen, M.O., Park, S., Tajkhorshid, E., Schulten, K.: Energetics of glycerol conduction through aquaglyceroporin GlpF. Proc. Natl. Acad. Sci. USA **99**, 6731–6736 (2002)

45. Park, S., Khalili-Araghi, F., Tajkhorshid, E., Schulten, K.: Free energy calculation from steered molecular dynamics simulations using Jarzynski's equality. J. Chem. Phys. **119**, 3559–3566 (2003)
46. Buehler, M.J., Wong, S.Y.: Entropic elasticity controls nanomechanics of single tropocollagen molecules. Biophys. J. **93**, 37–43 (2007)
47. Jarzynski, C.: Nonequilibrium equality for free energy differences. Phys. Rev. Lett. **78**, 2690–2693 (1997)
48. Jarzynski, C.: Entropy production fluctuation theorem and the nonequilibrium work relation for free energy differences. Phys. Rev. E **60**, 2721–2726 (1997)
49. Crooks, G.E.: Path-ensemble averages in systems driven far from equilibrium. Phys. Rev. E **61**, 2361–2366 (2000)
50. Cuendet, M.A.: The Jarzynski identity derived from general Hamiltonian or non-Hamiltonian dynamics reproducing NVT or NPT ensembles. J. Chem. Phys. **125**, 144109 (2006)
51. Rodinger, T., Pomés, R.: Enhancing the accuracy, the efficiency and the scope of free energy simulations. Curr. Opin. Struct. Biol. **15**, 164–170 (2005)
52. Isralewitz, B., Gao, M., Schulten, K.: Steered molecular dynamics and mechanical functions of proteins. Curr. Opin. Struct. Biol. **11**, 224–230 (2001)
53. Sotomayor, M., Schulten, K.: Single-molecule experiments in vitro and in silico. Science **316**, 1144–1148 (2007)
54. Cleri, F., Phillpot, S.R., Wolf, D., Yip, S.: Atomistic simulations of materials fracture and the link between atomic and Continuum length scales. J. Amer. Cer. Soc. **81**, 501–516 (1998)
55. Harris, S.A., Sands, Z.A., Laughton, C.A.: Molecular dynamics simulations of duplex stretching reveal the importance of entropy in determining the biomechanical properties of DNA. Biophys. J. **88**, 1684–1691 (2005)
56. Matthews, B.: No code for recognition. Nature **335**, 294–295 (1988)
57. Suzuki, M., Brenner, S., Gerstein, M., Yagi, N.: DNA recognition code of transcription factors. Protein Eng. **8**, 319–328 (1995)
58. Pabo, C., Nekludova, L.: Geometric analysis and comparison of protein-DNA interfaces: why is there no simple code for recognition?. J. Mol. Biol. **301**, 597–624 (2000)
59. Bustamante, C., Marko, J.F., Siggia, E.D., Smith, S.: Entropic elasticity of lambda-phage DNA. Science **265**, 1599–1600 (1994)
60. Doi, M., Edwards, S.F.: The Theory of Polymer Dynamics. Oxford University Press, Oxford, UK (1986)
61. Marko, J.F., Siggia, E.D.: Bending and twisting elasticity of DNA. Macromolecules **27**, 981–987 (1994)
62. Baumann, C.G., Bloomfield, V.A., Smith, S.B., Bustamante, C., Wang, M.D., Block, S.M.: Stretching of single collapsed DNA molecules. Biophys. J. **78**, 1965–1978 (2000)
63. Strick, T.R., Allemand, J.F., Bensimon, D., Croquette, V.: Stress-induced Structural transitions in DNA and proteins. Ann. Rev. Biophys. Biomol. Struct. **29**, 523–542 (2000)
64. Whitelam, S., Pronk, S., Geissler, P.L.: There and (slowly) back again: entropy-driven hysteresis in a model of DNA overstretching. Biophys. J. **94**, 2452–2469 (2008)
65. Konrad, M.W., Bolonick, J.I.: Molecular dynamics simulation of DNA stretching is consistent with the tension observed for extension and strand separation and predicts a novel ladder structure. J. Am. Chem. Soc. **118**, 10989–10994 (1996)
66. MacKerell, A.D., Lee, G.U.: Structure, force, and energy of a double-stranded DNA oligonucleotide under tensile loads. Eur. Biophys. J. **28**, 415–426 (1999)
67. Strunz, T., Oroszlan, K., Guntherodt, H.J., Henger, M.: Model energy landscapes and the force-induced dissociation of ligand-receptor bonds. Biophys. J. **79**, 1206–1212 (2000)
68. in't Veld, P.J., Stevens, M.J.: Simulation of the mechanical strength of a single collagen molecule. Biophys. J. **95**, 33–39 (2008)
69. Rief, M., Gautel, M., Oesterhelt, F., Fernandez, J.M., Gaub, H.: Reversible unfolding of individual titin immunoglobulin domains by AFM. Science **276**, 1109–1112 (1997)
70. Kellermayer, M.S.Z., Smith, S.B., Granzier, H.L., Bustamante, C.: Folding-unfolding transitions in single titin molecules characterized with laser tweezers. Science **276**, 1112–1116 (1997)
71. Oberhauser, A.F., Marszalek, P.E., Erickson, H.P., Fernandez, J.M.: The molecular elasticity of the extracellular matrix protein tenascin. Nature **393**, 181–185 (1998)
72. Marszalek, P.E., Lu, H., Li, H., Carrion-Vazquez, M., Oberhauser, A.F., Schulten, K., Fernandez, J.M.: Mechanical unfolding intermediates in titin modules. Nature **402**, 100–103 (1999)
73. Carl, P., Kwok, C.H., Manderson, G., Speicher, D.W., Discher, D.E.: Forced unfolding modulated by disulfide bonds in the Ig domains of a cell adhesion molecule. Proc. Natl. Acad. Sci. USA **98**, 1565–1570 (2001)
74. Bhasin, N., Carl, P., Harper, S., Feng, G., Lu, H., Speicher, D.W., Discher, D.E.: Chemistry on a single protein, vascular cell adhesion molecule-1, during forced unfolding. J. Biol. Chem. **279**, 45865–45874 (2004)

75. Baumann, C.G., Smith, S.B., Bloomfield, V.A., Bustamante, C.: Ionic effects on the elasticity of single DNA molecules. Proc. Natl. Acad. Sci. USA **94**, 6185–6190 (1997)
76. Alberts, B., Bray, D., Lewis, J., Raff, M., Roberts, K., Watson, J.D.: Molecular Biology of the Cell. Garland, New York (1994)
77. Dean Astumian, R.: Thermodynamics and kinetics of a brownian motor. Science **276**, 917–922 (1997)
78. Walker, M.L., Burgess, S.A., Sellers, J.R., Wang, F., Hammer, J.A., Trinick, J., Knight, P.J.: Two-headed binding of a processive myosin to F-actin. Nature **405**, 804–807 (2000)
79. Mather, W.H., Fox, R.F.: Kinesin's biased stepping mechanism: amplification of neck linker zippering. Biophys. J. **91**, 2416–2426 (2006)
80. Huxley, A.F.: Muscle structure and theories of contraction. Prog. Biophys. Biophys. Chem. **7**, 255–318 (1957); Huxley, A.F.: Muscular contraction—review lecture. J. Physiol. (London) **243**, 1–43 (1974)
81. Fox, R.F.: Rectified brownian movement in molecular and cell biology. Phys. Rev. E **57**, 2177–2203 (1998)
82. Ackbarow, T., Buehler, M.J.: Superelasticity, energy dissipation and strain hardening of vimentin coiled-coil intermediate filaments: atomistic and continuum studies. J. Mater. Sci. **42**, 8771–8787 (2007)
83. Liphardt, J., Dumont, S., Smith, S.B., Tinoco Jr., I., Bustamante, C.: Equilibrium information from non-equilibrium measurements in an experimental test of Jarzynski's equality. Science **296**, 1832–1835 (2002)
84. Laio, A., Parrinello, M.: Escaping free energy minima. Proc. Natl. Acad. Sci. USA **99**, 12562–12566 (2002)
85. Bussi, G., Laio, A., Parrinello, M.: Equilibrium free energies from nonequilibrium metadynamics. Phys. Rev. Lett. **96**, 090601 (2006)
86. Praprotnik, M., Delle Site, L., Kremer, K.: Adaptive resolution molecular-dynamics simulation: changing the degrees of freedom on the fly. J. Chem. Phys. **123**, 224106 (2005)
87. Neri, M., Anselmi, C., Cascella, M., Maritan, A., Carloni, P.: Coarse-grained model of proteins incorporating atomistic detail of the active site. Phys. Rev. Lett. **95**, 218102 (2005)
88. Shi, Q., Izvekov, S., Voth, G.A.: Mixed atomistic and coarse-grained molecular dynamics: simulation of a membrane-bound ion channel. J. Phys. Chem. B **110**, 15045–15048 (2006)
89. Fan, Z.Z., Hwang, J.K., Warshel, A.: Using simplified protein representation as a reference potential for all-atom calculations of folding free energy. Theor. Chem. Acc. **103**, 77–80 (1999)
90. Popoff, M., Cleri, F., Gianese, G., Rosato, V.: Docking of small peptides to inorganic surfaces. Eur. Phys. J. E (2008) (to appear)
91. Lyman, E., Ytreberg, F.M., Zuckerman, D.M.: Resolution exchange simulation. Phys. Rev. Lett. **96**, 028105 (2006)
92. Klimov, D.K., Thirumalai, D.: Native topology determines force-induced unfolding pathways in globular proteins. Proc. Natl. Acad. Sci. USA **97**, 7254–7259 (2000)
93. Marrink, S.J., de Vries, A.H., Mark, A.E.: Coarse grained model for semiquantitative lipid simulations. J. Phys. Chem. B **108**, 750–760 (2004)
94. Shillcock, J.C., Lipowsky, R.: Equilibrium structure and lateral stress distribution of amphiphilic bilayers from dissipative particle dynamics simulations. J. Chem. Phys. **117**, 5048–5061 (2002)
95. Chen, Q., Li, D.Y., Oiwa, K.: The coordination of protein motors and the kinetic behavior of microtubule—a computational study. Biophys. Chem. **129**, 60–69 (2007)
96. Ayton, G.S., Noid, W.G., Voth, G.A.: Multiscale modeling of biomolecular systems: in serial and in parallel. Curr. Opin. Struct. Biol. **17**, 192–198 (2007)
97. Kmiecik, S., Kolinski, A.: Characterization of protein-folding pathways by reduced-space modeling. Proc. Natl. Acad. Sci. USA **104**, 12330–12335 (2007)
98. Heath, A.P., Kavraki, L.E., Clementi, C.: From coarse-grain to all-atom: toward multiscale analysis of protein landscapes. Proteins Struct. Funct. Bioinfo. **68**, 646–661 (2007)
99. Miao, Y., Ortoleva, P.J.: Viral structural transitions: an all-atom multiscale theory. J. Chem. Phys. **125**, 214901 (2006)

Enveloped viruses understood via multiscale simulation: computer-aided vaccine design

Z. Shreif · P. Adhangale · S. Cheluvaraja · R. Perera · R. Kuhn · P. Ortoleva

Originally published in the journal Sci Model Simul, Volume 15, Nos 1–3, 363–380.
DOI: 10.1007/s10820-008-9101-5 © Springer Science+Business Media B.V. 2008

Abstract   Enveloped viruses are viewed as an opportunity to understand how highly organized and functional biosystems can emerge from a collection of millions of chaotically moving atoms. They are an intermediate level of complexity between macromolecules and bacteria. They are a natural system for testing theories of self-assembly and structural transitions, and for demonstrating the derivation of principles of microbiology from laws of molecular physics. As some constitute threats to human health, a computer-aided vaccine and drug design strategy that would follow from a quantitative model would be an important contribution. However, current molecular dynamics simulation approaches are not practical for modeling such systems. Our multiscale approach simultaneously accounts for the outer protein net and inner protein/genomic core, and their less structured membranous material and host fluid. It follows from a rigorous multiscale deductive analysis of laws of molecular physics. Two types of order parameters are introduced: (1) those for structures wherein constituent molecules retain long-lived connectivity (they specify the nanoscale structure as a deformation from a reference configuration) and (2) those for which there is no connectivity but organization is maintained on the average (they are field variables such as mass density or measures of preferred orientation). Rigorous multiscale techniques are used to derive equations for the order parameters dynamics. The equations account for thermal-average forces, diffusion coefficients, and effects of random forces. Statistical properties of the atomic-scale fluctuations and the order parameters are co-evolved. By combining rigorous multiscale techniques and modern supercomputing, systems of extreme complexity can be modeled.

Keywords   Enveloped viruses · Structural transitions · All-atom multiscale analysis · Multiscale computation · Liouville equation · Langevin equations

Z. Shreif · P. Adhangale · S. Cheluvaraja · P. Ortoleva (✉)
Center for Cell and Virus Theory, Department of Chemistry,
Indiana University, Bloomington, IN 47405, USA
e-mail: ortoleva@indiana.edu

R. Perera · R. Kuhn
Department of Biological Sciences, Purdue University,
West Lafayette, IN 47907, USA

1 Introduction

Deriving principles of microbial behavior from laws of molecular physics remains a grand challenge. While one expects many steps in the derivation can be accomplished based on the classical mechanics of an N-atom system, it is far from clear how to proceed in detail due to the extreme complexity of these supra-million atom systems. Most notably, molecular dynamics (MD) codes are not practical for simulating even a simple bionanosystem of about 2 million atoms (e.g. a nonenveloped virus) over biologically relevant time periods (i.e. milliseconds or longer). For example, the efficient MD code NAMD, run on a 1024-processor supercomputer [1], would take about 3000 years to simulate a simple virus over a millisecond; the largest NAMD simulation published to date is for a ribosome system of approximately 2.64 million atoms over few nanoseconds only [2].

We hypothesize that a first step in the endeavor to achieve a quantitative, predictive virology is to establish a rigorous intermediate scale description. Due to their important role in human health, complex structure, and inherent multiscale nature, enveloped viruses provide an ideal system for guiding and testing this approach. Experimental evidence suggests that an enveloped virus manifests three types of organization:

- an outer protein net that can display several well-defined structures; for each structure, proteins maintain long-lived connectivity with specific nearest-neighbor units (Fig. 1);
- a sea of membranous material below the protein net; this subsystem consists of phospholipid molecules whose nearest-neighbors are continuously changing but for which there is long-lived structure on-the-average. Also, biological membranes display liquid-crystal transitions [3] either autonomously or as promoted by proteins traversing the membranous subsystem; and
- genomic RNA or DNA complexed with proteins in which there is long-lived connectivity between nucleotides but which, as evidenced by cryo-electron microscopy and X-ray diffraction data, often lack well-defined structure [4].

We hypothesize that these three types of organization and the interplay of their stochastic dynamics is the essence of principles governing the structural transitions and stability of enveloped viruses.

Developing a quantitative understanding of enveloped viruses is of great importance for global health. Human pathogenic enveloped viruses include Dengue (Fig. 1) and HIV. Understanding the mechanisms underlying virus entry and hijacking of host cell processes is a main step in preventing the often fatal virus infections. The aim is not only to be able to attack the virus in question (or simply prevent its docking proteins from binding to the host cell receptors), but also to use viruses for therapeutic delivery. Since viruses have a natural ability to find and penetrate host cells, using them as a means to deliver genes, drugs, and other therapeutic agents holds great promise in medical advancement. Enveloped viruses provide a natural choice as they are able to entirely fuse inside the cell before delivering their payload [7]. Our strategy is to develop a predictive whole-virus model that serves as a basis of a computer-aided antiviral vaccine and drug design capability. Furthermore, this predictive model could be a key element of a system for forecasting the potential pandemic of a given strain via an assessment of computer-generated mutants, and similarly for testing the effects of a library of potential antiviral drugs.

To achieve these practical goals, and to arrive at a fundamental understanding of complex bionanosystems, we suggest starting from the laws of molecular physics. These, as considered here, are Newton's equations for an N-atom system. For an enveloped virus, N is on the order of 10^8. However, the conceptual framework within which one discusses viral phenomena does

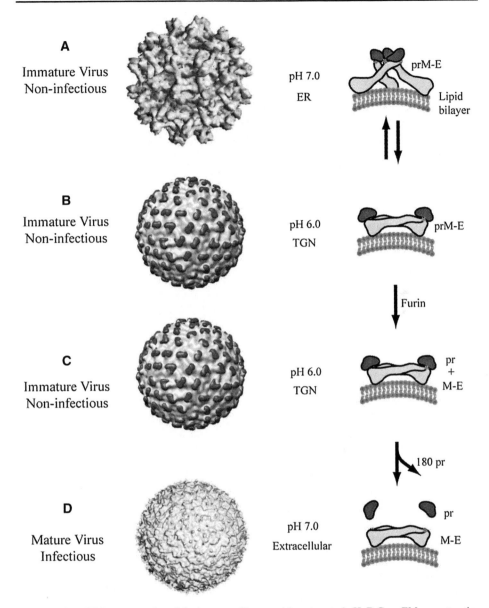

Fig. 1 **A** Cryo-EM reconstruction of the immature Dengue virion at neutral pH. **B** Cryo-EM reconstruction of the immature virion at low pH. **C** Cleavage of the prM protein into its 'pr' peptide and M protein by the host endoprotease, furin. **D** The cryo-EM reconstruction of the mature virion. From Refs. [5,6]

not involve keeping track of the positions and momenta of all the atoms. Nonetheless, an all-atom description is required to derive the principles of enveloped viral behavior from laws of molecular physics. Processes involved in viral behavior include the 10^{-14} second atomic vibration and collisions and the millisecond or longer overall structural transitions. In addition, various size scales are involved: The scale of the nearest-neighbor atom distance (a few angstroms) to the diameter of the enveloped virus and a closely associated aqueous

Fig. 2 Order parameters characterizing nanoscale features affect the relative probability of the atomistic configurations which, in turn, mediates the forces driving order parameter dynamics. This feedback loop is central to a complete multiscale understanding of nanosystems and the true nature of their dynamics. This loop is also the schematic workflow of our AMA/ACM computational strategy.

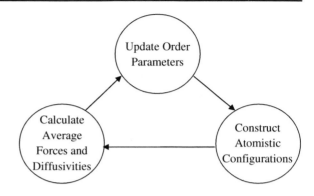

layer (several thousand angstroms). As the short scale phenomena affect the larger scale ones, and conversely (see Fig. 2), viruses have a strongly multiscale character.

Multiscale techniques have been discussed extensively in the literature [8–31]. These studies start with the Liouville equation and arrive at a Fokker-Planck or Smoluchowski type equation for the stochastic dynamics of a set of slowly evolving variables (order parameters). Of particular relevance to the present study are recent advances [20–31] wherein it was shown one could make the hypothesis that the N-atom probability density is a function of the $6N$ atomic positions and momenta both directly and, via a set of order parameters, indirectly. It was shown that both dependencies could be reconstructed when there is a clear separation of timescales, and that such an assumed dual dependence is not a violation of the number ($6N$) of classical degrees of freedom. Technical advances were also introduced which facilitated the derivation of the stochastic equations of the order parameter dynamics and allowed for an atomistic description of the entire system (i.e. within the host medium and the nanostructures). Furthermore, it was shown in earlier work [30,31] how to transcend the conceptual gap between continuum and all-atom theories; starting from the more fundamental all-atom formulation and, via a deductive multiscale analysis, equations coupling field variables with selected facets of the all-atom description were derived.

For the present study, we suggest that a mixed continuum/all-atom approach is a natural framework for achieving an integrated theory of enveloped viruses. As there are several types of molecular organization to account for, we introduce distinct types of order parameters to characterize them. Details on these order parameters are provided in Sects. 2 and 3. Common features of both types of order parameters are as follows:

- they are expressed in terms of the all-atom state (i.e. the $6N$ positions/momenta of the N atoms) in terms of which the basic laws of molecular physics are posed;
- they evolve on timescales relevant to virology (i.e. microseconds or longer, and not 10^{-14} seconds as for atomic vibration/collisions);
- they characterize the nanoscale features of interest to virology (e.g. outer protein net structure, genomic structure, and liquid-crystal order in the membranous subsystem); and
- they form a complete set, i.e. they do not couple to other slow variables not considered in the analysis.

In Sect. 4, we develop a multiscale theory of viral dynamics and, in the process, clarify the need for the above criteria on the nature of order parameters. We derive rigorous equations for their stochastic dynamics, notably structural fluctuations and transitions. In Sect. 5, we discuss our multiscale simulation method and present results for the STMV virus to illustrate the feasibility of our approach. Applications and conclusions are discussed in Sect. 6.

2 Order parameters for connected structures

Order parameters for subsystems of connected atoms have been constructed as generalized Fourier amplitudes [23,24]. They represent the degree to which a structure is a deformation of a reference configuration (e.g. a cryoTEM reconstruction). For Fourier analysis one uses sine and cosine basis functions. In our approach, other basis functions are introduced as follows.

The system is embedded in a volume Ω. Orthogonal basis functions $U_{\underline{\ell}}\left(\vec{r}\right)$ for point $\vec{r}$ in Ω with triplet labeling index $\underline{\ell}$ are introduced. The basis functions are periodic if computations are carried out using periodic boundary conditions to approximate a large system by minimizing boundary effects or to handle Coulomb forces. According to our earlier method [32], a virus and its microenvironment are embedded in a 3-D space where a point $\vec{r}$ is considered a displacement from an original point $\vec{r}^{\,0}$ in the undeformed space. The deformation of space taking any point $\vec{r}^{\,0}$ to the position after deformation $\vec{r}$ and the basis functions are used to introduce a set of order parameters $\vec{\Phi}_{\underline{\ell}}$ via

$$\vec{r} = \sum_{\underline{\ell}} \vec{\Phi}_{\underline{\ell}} U_{\underline{\ell}}\left(\vec{r}^{\,0}\right). \qquad (2.1)$$

As the $\vec{\Phi}_{\underline{\ell}}$ change, a point $\vec{r}^{\,0}$ is moved to $\vec{r}$ and thus, to connect the $\vec{\Phi}_{\underline{\ell}}$ to the physical system, the nanostructure embedded in the space is deformed. The $\vec{\Phi}_{\underline{\ell}}$ are interpreted to be a set of vector order parameters that serve as the starting point of an AMA (All-Atom Multiscale Analysis) approach. In what follows, we show how to use (2.1) for a finite set of basis functions and the associated order parameters, and prove that the latter are slowly evolving for appropriately chosen basis functions.

Each atom in the system is moved via the above deformation by evolving the $\vec{\Phi}_{\underline{\ell}}$. However, given a finite truncation of the $\underline{\ell}$-sum, there will be residual displacement of individual atoms above that due to the continuous deformation generated by the order parameters. Denoting the residual of atom i as $\vec{\sigma}_i$, we write its position $\vec{r}_i$ as

$$\vec{r}_i = \sum_{\underline{\ell}} \vec{\Phi}_{\underline{\ell}} U_{\underline{\ell}}\left(\vec{r}^{\,0}\right) + \vec{\sigma}_i. \qquad (2.2)$$

The size of $\vec{\sigma}_i$ can be controlled by the choice of basis functions, the number of terms in the ℓ sum, and the way the $\vec{\Phi}_{\underline{\ell}}$ are defined. Conversely, imposing a permissible size threshold for the residuals allows us to determine the number of basis functions needed to minimize the $\vec{\sigma}_i$, and hence order parameters, to include in the analysis.

A concrete definition of the order parameters is developed as follows. Define the mass-weighted mean-square residual S via

$$S = \sum_{i=1}^{N} m_i \left|\vec{\sigma}_i\right|^2 \Theta_i^c, \qquad (2.3)$$

where N is the total number of atoms in the system, m_i is the mass of atom i, and Θ_i^c is one when atom i belongs to the connected subsystem and zero otherwise. With (2.2), this yields

$$S = \sum_{i=1}^{N} m_i \Theta_i^c \left| \vec{r}_i - \sum U_\ell \left(\vec{r}_i^{\,0} \right) \vec{\Phi}_\ell \right|^2. \tag{2.4}$$

We assert that the optimal order parameters are those which minimize S, i.e. those containing the maximum amount of information so that the $\vec{\sigma}_i$ are, on the average, the smallest. Thus, we obtain the relationship between $\vec{\Phi}_\ell$ and $\Gamma_r = \left\{ \vec{r}_1, \vec{r}_2, \ldots, \vec{r}_N \right\}$ via minimizing S with respect to the $\vec{\Phi}_\ell$ keeping Γ_r constant [22]. This implies

$$\sum_{\ell'} B_{\ell\ell'} \vec{\Phi}_{\ell'} = \sum_{i=1}^{N} m_i \vec{r}_i U_\ell \left(\vec{r}_i^{\,0} \right) \Theta_i^c, \tag{2.5}$$

$$B_{\ell\ell'} = \sum_{i=1}^{N} m_i U_\ell \left(\vec{r}_i^{\,0} \right) U_{\ell'} \left(\vec{r}_i^{\,0} \right) \Theta_i^c. \tag{2.6}$$

Orthogonality of the basis functions implies that the B matrix is nearly diagonal. Hence, the order parameters can easily be computed numerically in terms of the atomic positions by solving (2.5).

To proceed, we must be more precise regarding the normalization of the basis functions. For the function U_ℓ with $\ell = (0, 0, 0)$, we take $U_0 = 1$. Thus, B_{00} is the total mass of the atoms in the connected structure. Furthermore, if the B matrix is diagonal, one can show that $\vec{\Phi}_0$ is the center-of-mass (CM). From earlier studies [20–29], this implies that $\vec{\Phi}_0$ is slowly varying. If $\varepsilon = m/B_{00}$ for typical atomic mass m, then the eigenvalues of B are large, i.e. $O\left(\varepsilon^{-1}\right)$.

A necessary condition for a variable to satisfy a Langevin equation (i.e. to be an order parameter in our multiscale formulation) is that it evolves on a timescale much longer than that of the vibration/collisions of individual atoms. It can be shown that for diagonal B matrix $d\vec{\Phi}_0/dt = \varepsilon \vec{\Pi}_0$, where $\vec{\Pi}_\ell$ is the conjugate momentum of $\vec{\Phi}_\ell$. Taking $d\vec{\Phi}_\ell/dt = \varepsilon \vec{\Pi}_\ell$, and applying Newton's equations ($d\vec{\Phi}_\ell/dt = -\mathcal{L}\vec{\Phi}_\ell$ for Liouville operator $\mathcal{L}$), yields

$$\varepsilon \sum_{\ell'} \vec{\Pi}_{\ell'} B_{\ell\ell'} = \sum_{i=1}^{N} \vec{p}_i U_\ell \left(\vec{r}_i^{\,0} \right) \Theta_i^c. \tag{2.7}$$

Inclusion of m_i in the above expressions gives the order parameters the character of generalized CM variables.

While the $\vec{\Phi}_\ell$ qualify as order parameters from the above perspective, they do not suffice as a way to characterize the aqueous microenvironment or the membranous material of the enveloped virus. This follows because the molecules in these two subsystems do not maintain nearest-neighbor connectivity. Thus, a second type of order parameters is required for enveloped virus modeling, as developed in the next section.

3 Order parameter fields for disconnected subsystems

For disconnected systems, one needs an all-atom/continuum multiscale (ACM) approach [31]. The starting point of our ACM theory is a set of field variables that change across the system. These field variables must be related to the atomistic description to achieve a rigorous formulation. The membranous material of an enveloped virus contains a subsystem composed of continuously changing molecules. Application of ACM theory to such systems is illustrated as follows. Let the membranous continuum be comprised of N_t types of molecules labeled $k = 1, 2, \ldots, N_t$. Each type is described by a mass density field variable Ψ_k at spatial point $\vec{R}$:

$$\Psi_k \left(\vec{R}, \Gamma_r \right) = \sum_{i=1}^{N} m_i \delta \left(\vec{R} - \varepsilon_d \vec{r}_i \right) \Theta_i^k, \tag{3.1}$$

where δ is the Dirac delta function centered at $\vec{0}$, ε_d is a smallness parameter, and $\Theta_i^k = 1$ when atom i belongs to a molecule of type k in the disconnected subsystem, and zero otherwise. As these field variables are intended to be coherent in character, $\vec{R}$ is scaled such that it undergoes a displacement of about one unit as several nanometers are traversed. In contrast, the $\vec{r}_i$ undergo a displacement of about one unit as a typical nearest-neighbor interatomic distance for a condensed medium (a few angstroms) is traversed. Thereby, it is natural to scale $\vec{r}_i$ to track the fine-structural details in the system. With this, we let ε_d be the ratio of the typical nearest-neighbor interatomic distance to the size of a nanocomponent (e.g. a viral capsomer). This length scale ratio characterizes the multiscale nature of the enveloped virus. As $\varepsilon_d \ll 1$, it provides a natural expansion parameter for solving the equations of molecular physics, i.e. the Liouville equation for the N-atom probability density. Newton's equations imply

$$\frac{d\Psi_k}{dt} = -\mathcal{L}\Psi_k = -\varepsilon_d \vec{\nabla} \vec{G}_k \left(\vec{R}, \Gamma \right) \equiv \varepsilon_d J_k \left(\vec{R}, \Gamma \right) \tag{3.2}$$

$$\vec{G}_k \left(\vec{R}, \Gamma \right) = \sum_{i=1}^{N} \vec{p}_i \delta \left(\vec{R} - \varepsilon_d \vec{r}_i \right) \Theta_i^k, \tag{3.3}$$

where $\vec{G}_k$ is the momentum density of molecules of type k, $\vec{\nabla}$ is the $\vec{R}$-gradient, J_k is the divergence (defined by (3.2)), $\vec{p}_i$ is the momentum of atom i, and $\Gamma = \left\{ \Gamma_r, \vec{p}_1, \vec{p}_2, \ldots, \vec{p}_N \right\}$ is the set of $6N$ atomistic state variables. For quasi-equilibrium conditions, the average momentum $\langle \vec{p}_i \rangle$ is small. Thus, the momenta of the atoms in the expression for $\vec{G}_k$ tend to cancel each other. This suggests that $\vec{G}_k$ are of order $O\left(\varepsilon_d^0 \right)$, and thus Ψ_k are slowly evolving, at a rate of $O(\varepsilon_d)$.

Order parameter fields like Ψ_k are indexed by $\vec{R}$ which varies continuously across the system. Thus, with Γ_r dependency being understood, we sometimes use the notation $\Psi_k(\vec{R})$ to reflect this parameterization, i.e. to label Ψ_k as the $\vec{R}$-associated order parameter, much like $\vec{r}_i$ is the i-associated position variable (although $\vec{r}_i$ is not an order parameter as it has predominantly 10^{-14} second timescale dynamics). That for each molecule type k there is a Ψ_k for every point $\vec{R}$ in the system suggests there is an uncountable infinity of slow field variables, $\Psi_k(\vec{R})$. Finally, in order to connect the smallness parameter, ε, of Sect. 2 with that of this section, ε_d, we suggest that ε_d is proportional to ε, and thus take $\varepsilon_d = \varepsilon$ for simplicity.

4 Multiscale integration for enveloped virus modeling

Integration of the multiple types of order parameters (Sects. 2 and 3) needed for enveloped virus modeling is achieved in a self-consistent fashion as follows. We hypothesize that the N-atom probability density ρ for the composite virus/microenvironment system has multiscale character and can thus be rewritten to express its dependency on both the set of atomic positions and momenta Γ and the order parameters. The reduced probability density W is defined as a function of the set of order parameters describing the connected subsystem $\vec{\Phi} = \vec{\Phi}_\ell$ for all ℓ included, and a functional of the set of order parameter fields $\underline{\Psi}(\vec{R}) = \{\Psi_1(\vec{R}), \dots, \Psi_{N_t}(\vec{R})\}$ describing the disconnected subsystem. By definition, W takes the form

$$
W\left[\vec{\Phi}, \underline{\Psi}, t\right] = \int d^{6N}\Gamma^* \prod_\ell \delta\left(\vec{\Phi}_\ell - \vec{\Phi}_\ell^*\right) \prod_{k=1}^{N_t} \Delta\left(\Psi_k - \Psi_k^*\right) \rho\left(\Gamma^*, t\right), \qquad (4.1)
$$

where Δ is a continuum product of δ-functions for all positions $\vec{R}$, Γ^* is the N-atom state over which integration is taken, and $\vec{\Phi}_\ell^*$ and Ψ_k^* are the order parameters for state Γ^*. With this and the fact that the N-atom probability density ρ satisfies the Liouville equation $\partial\rho/\partial t = \mathcal{L}\rho$, W is found to satisfy the conservation equation

$$
\frac{\partial W}{\partial t} = -\varepsilon \sum_{k=1}^{N_t} \int \frac{d^3 R}{v_c} \frac{\delta}{\delta\Psi_k\left(\vec{R}\right)} \int d^{6N}\Gamma^* \prod_\ell \delta\left(\vec{\Phi}_\ell - \vec{\Phi}_\ell^*\right)
$$

$$
\times \prod_{k=1}^{N_t} \Delta\left(\Psi_k - \Psi_k^*\right) J_k^*\left(\vec{R}\right) \rho\left(\Gamma^*, t\right)
$$

$$
-\varepsilon \sum_\ell \frac{\partial}{\partial\vec{\Phi}_\ell} \int d^{6N}\Gamma^* \prod_\ell \delta\left(\vec{\Phi}_\ell - \vec{\Phi}_\ell^*\right) \prod_{k=1}^{N_t} \Delta\left(\Psi_k - \Psi_k^*\right)\vec{\Pi}_\ell^*\rho\left(\Gamma^*, t\right) \qquad (4.2)
$$

where v_c is the minimal volume for which it is reasonable to speak of a field variable, and the superscript * for any variable indicates evaluation at Γ^*.

We hypothesize that to reflect the multiscale character of the system, ρ should be written in the form

$$
\rho\left(\Gamma, \vec{\Phi}, \underline{\Psi}, t_0, \underline{t}; \varepsilon\right). \qquad (4.3)
$$

The time variables $t_n = \varepsilon^n t$, $n = 0, 1, \dots$ are introduced to track processes on timescales $O\left(\varepsilon^{-n}\right)$ for t_n. The set $\underline{t} = \{t_1, t_2, \dots\}$ tracks the slow processes of interest to viral dynamics, i.e. much slower than those on the 10^{-14} second scale of atomic vibration/collisions. In contrast, t_0 tracks the fast atomistic processes. The ansatz (4.3) is not a violation of the number ($6N$) of degrees of freedom, but a recognition that ρ depends on Γ in two ways (i.e. both directly and, via $\vec{\Phi}$ and $\underline{\Psi}$, indirectly).

With this and the discrete and field order parameters, the chain rule implies the Liouville equation takes the multiscale form

$$\frac{\partial \rho}{\partial t} = (\mathcal{L}_0 + \varepsilon \mathcal{L}_1) \, \rho \tag{4.4}$$

$$\mathcal{L}_0 = -\sum_{i=1}^{N} \left(\frac{\vec{p}_i}{m_i} \frac{\partial}{\partial \vec{r}_i} + \vec{F}_i \frac{\partial}{\partial \vec{p}_i} \right) \tag{4.5}$$

$$\mathcal{L}_1 = \mathcal{L}_\Phi + \mathcal{L}_\Psi \tag{4.6}$$

$$\mathcal{L}_\Phi = -\sum_{\ell} \vec{\Pi}_\ell \frac{\partial}{\partial \vec{\Phi}_\ell} \tag{4.7}$$

$$\mathcal{L}_\Psi = -\sum_{k=1}^{N_t} \int \frac{\mathrm{d}^3 R}{\upsilon_c} J_k \left(\vec{R} \right) \frac{\delta}{\delta \Psi_k \left(\vec{R} \right)} \tag{4.8}$$

The operator $\mathcal{L}_1$ involves derivatives with respect to $\vec{\Phi}$ and functional derivatives with respect to Ψ at constant Γ, and conversely for $\mathcal{L}_0$. By mapping the Liouville problem to a higher dimensional descriptive variable space (i.e. $6N$ plus the number of variables in $\vec{\Phi}$ and the function space of the order parameter fields $\underline{\Psi}$), our strategy as suggested by our earlier studies [20–31] is to solve the Liouville equation in the higher dimensional representation, and then use the solution to obtain an equation of stochastic $\vec{\Phi},\underline{\Psi}$-dynamics.

The development continues with the perturbation expansion

$$\rho = \sum_{n=0}^{\infty} \varepsilon^n \rho_n, \tag{4.9}$$

and examining the multiscale Liouville equation at each order in ε. We hypothesize the lowest order behavior of ρ is slowly varying in time since the phenomena of interest vary on the millisecond or longer, and not the 10^{-14} second time scale. Thus, we assume the lowest order solution ρ_0 is independent of t_0 and, furthermore, is quasi-equilibrium in character.

To O $\left(\varepsilon^0 \right)$, the above assumptions imply $\mathcal{L}_0 \rho_0 = 0$ so that ρ_0 is in the null space of $\mathcal{L}_0$ but is otherwise unknown. We determine ρ_0 by adopting an entropy maximization procedure with canonical constraint of fixed average energy as discussed earlier in the context of nanosystems [24]; this is equivalent to taking the system to be isothermal. With this, we obtain

$$\rho_0 \left[\Gamma; \vec{\Phi}, \underline{\Psi}, \underline{t} \right] = \hat{\rho} \left[\Gamma; \vec{\Phi}, \underline{\Psi} \right] W_0 \left[\vec{\Phi}, \underline{\Psi}, \underline{t} \right] \tag{4.10}$$

$$\hat{\rho} = \frac{e^{-\beta H}}{Q \left[\vec{\Phi}, \underline{\Psi} \right]}, \tag{4.11}$$

where β is the inverse temperature, H is the Hamiltonian,

$$H \left(\Gamma \right) = \sum_{i=1}^{N} \frac{p_i^2}{2m_i} + V \left(\Gamma_r \right), \tag{4.12}$$

for N-atom potential V, and Q is the partition function which is a function of $\vec{\Phi}$ and a functional of Ψ given by

$$Q\left[\vec{\Phi}, \Psi\right] = \int d^{6N}\Gamma^* \prod_{\ell} \delta\left(\vec{\Phi}_{\ell} - \vec{\Phi}_{\ell}^*\right) \prod_{k=1}^{N_t} \Delta\left(\Psi_k - \Psi_k^*\right) e^{-\beta H^*}. \tag{4.13}$$

To O (ε), the multiscale Liouville equation implies

$$\left(\frac{\partial}{\partial t_0} - \mathcal{L}_0\right)\rho_1 = -\frac{\partial \rho_0}{\partial t_1} + \mathcal{L}_1 \rho_0. \tag{4.14}$$

This admits the solution

$$\rho_1 = -\int_0^{t_0} dt_0' e^{\mathcal{L}_0\left(t_0 - t_0'\right)} \left\{\frac{\partial \rho_0}{\partial t_1} - \mathcal{L}_1 \rho_0\right\}, \tag{4.15}$$

where the initial first order distribution was taken to be zero as suggested earlier [25,26] to ensure that the system is initially in equilibrium. As a consequence, the final stochastic equation is closed in W.

Inserting (4.6), (4.7), (4.8), and (4.10) in (4.15) yields

$$\rho_1 = -t_0 \hat{\rho}\frac{\partial W_0}{\partial t_1} - \hat{\rho}\int_0^{t_0} dt_0' e^{\mathcal{L}_0\left(t_0 - t_0'\right)} \left\{\sum_{\ell} \vec{\Pi}_{\ell}\left(\frac{\partial}{\partial \vec{\Phi}_{\ell}} - \beta\left\langle \vec{f}_{\ell}\right\rangle\right)\right.$$
$$\left. + \sum_{k=1}^{N_t} \int \frac{d^3 R}{v_c} J_k\left(\vec{R}\right)\left(\frac{\delta}{\delta \Psi_k\left(\vec{R}\right)} - \beta\left\langle h_k\left(\vec{R}\right)\right\rangle\right)\right\} W_0 \tag{4.16}$$

where the thermal-average forces are given by

$$\left\langle \vec{f}_{\ell}\right\rangle = -\frac{\partial F}{\partial \vec{\Phi}_{\ell}} \tag{4.17}$$

$$\left\langle h_k\left(\vec{R}\right)\right\rangle = -\frac{\partial F}{\partial \Psi_k\left(\vec{R}\right)} \tag{4.18}$$

and F is the free energy related to Q via $Q = e^{-\beta F}$.

Using the Gibbs hypothesis, imposing the condition that ρ_1 be finite as $t_0 \to \infty$, and using the fact that the thermal-averages of J_k and $\vec{\Pi}_{\ell}$ are zero (since the weighing factor $\hat{\rho}$ is even in the $\vec{p}_i$, while J_k and $\vec{\Pi}_{\ell}$ are odd in them), we find W_0 to be independent of t_1. With this, (4.16) becomes

$$\rho_1 = -\hat{\rho} \int_0^{t_0} dt_0' e^{\mathcal{L}_0\left(t_0 - t_0'\right)} \left\{ \sum_\ell \vec{\Pi}_\ell \left(\frac{\partial}{\partial \Phi_\ell} - \beta \left\langle \vec{f}_\ell \right\rangle \right) \right.$$
$$\left. + \sum_{k=1}^{N_t} \int \frac{d^3 R}{v_c} J_k\left(\vec{R}\right) \left(\frac{\delta}{\delta \Psi_k\left(\vec{R}\right)} - \beta \left\langle h_k\left(\vec{R}\right) \right\rangle \right) \right\} W_0 \qquad (4.19)$$

Inserting (4.10) and (4.19) in the conservation equation (4.2) yields

$$\frac{\partial W}{\partial t} = \varepsilon^2 \sum_\ell \sum_{\ell'} \frac{\partial}{\partial \vec{\Phi}_\ell} \vec{D}_{\ell\ell'}^{\vec{\Phi}} \left(\frac{\partial}{\partial \vec{\Phi}_{\ell'}} - \beta \left\langle \vec{f}_{\ell'} \right\rangle \right) W$$
$$+ \varepsilon^2 \sum_{k=1}^{N_t} \sum_\ell \int \frac{d^3 R}{v_c} \frac{\partial}{\partial \vec{\Phi}_\ell} \vec{D}_{\ell k} \left(\frac{\delta}{\delta \Psi_k\left(\vec{R}\right)} - \beta \left\langle h_k\left(\vec{R}\right) \right\rangle \right) W$$
$$+ \varepsilon^2 \sum_{k=1}^{N_t} \sum_\ell \int \frac{d^3 R}{v_c} \frac{\delta}{\delta \Psi_k\left(\vec{R}\right)} \vec{D}_{k\ell} \left(\frac{\partial}{\partial \vec{\Phi}_\ell} - \beta \left\langle \vec{f}_\ell \right\rangle \right) W$$
$$+ \varepsilon^2 \sum_{k=1}^{N_t} \sum_{k'=1}^{N_t} \int \frac{d^3 R\, d^3 R'}{v_c\, v_c} \frac{\delta}{\delta \Psi_k\left(\vec{R}\right)} D_{kk'}^{\Psi} \left(\frac{\delta}{\delta \Psi_{k'}\left(\vec{R}'\right)} - \beta \left\langle h_{k'}\left(\vec{R}'\right) \right\rangle \right) W$$
$$(4.20)$$

where $\vec{D}_{\ell\ell'}^{\vec{\Phi}}$, $\vec{D}_{\ell k}$, $\vec{D}_{k\ell}$, and $D_{kk'}^{\Psi}$ are diffusion coefficients defined as

$$\vec{D}_{\ell\ell'}^{\vec{\Phi}} = \int_{-\infty}^{0} dt_0 \vec{\Pi}_\ell e^{-\mathcal{L}_0 t_0} \vec{\Pi}_{\ell'} \qquad (4.21)$$

$$\vec{D}_{\ell k}\left(\vec{R}\right) = \int_{-\infty}^{0} dt_0 \vec{\Pi}_\ell e^{-\mathcal{L}_0 t_0} J_k\left(\vec{R}\right) \qquad (4.22)$$

$$\vec{D}_{k\ell}\left(\vec{R}\right) = \int_{-\infty}^{0} dt_0 J_k\left(\vec{R}\right) e^{-\mathcal{L}_0 t_0} \vec{\Pi}_\ell \qquad (4.23)$$

$$D_{kk'}^{\Psi}\left(\vec{R}, \vec{R}'\right) = \int_{-\infty}^{0} dt_0 J_k\left(\vec{R}\right) e^{-\mathcal{L}_0 t_0} J_{k'}\left(\vec{R}'\right) \qquad (4.24)$$

As in Ref. [31], there are symmetry rules relating the cross-diffusion coefficients. The set of Langevin equations equivalent to this generalized Smoluchowski equation (4.20), provides a practical approach to the simulation of enveloped virus systems as outlined in the next section.

5 Multiscale computations and the NanoX platform

The all-atom multiscale approach of the previous section can be implemented as a platform for the simulation of nanosystems. In the flow chart of Fig. 3 it is seen how order parameters are co-evolved with the statistics of the atomic fluctuations in our NanoX platform. Interscale coupling as in Fig. 2 is manifested through the thermal-average forces and diffusion coefficients constructed via short-time ensemble/molecular dynamics computations in the indicated modules. We have implemented this Fig. 3 workflow, creating the NanoX simulator.

At this writing, NanoX is built on the order parameters of Sect. 2. For it to be practical, the diffusion coefficients and thermal-average forces must not be excessively demanding on CPU time. In Fig. 4a we show an example of an order parameter trajectory for the STMV (nonenveloped) virus in vacuum. This system has been studied elsewhere using classic MD [33]. Our simulation began from the crystal cryo-structure which was equilibrated for 1 ns at various temperatures. The order parameter trajectory (here we choose the z component of the 001 mode) is shown for the first 20 ps and is seen to change slowly except at the beginning where the system escapes from its potential energy-minimized unphysical structure. The velocity of the order parameters are obtained by differentiating equation (2.1). After thermalization, the order parameters hardly change for several picoseconds (Fig. 4b) whereas the velocity (not shown) appears to fluctuate about zero. These fluctuations are not highly correlated in time as shown by the rapid decay of the velocity auto-correlation function in Fig. 4c. This demonstrates that only short MD runs are required to calculate the diffusion coefficients via (4.21). The behavior of the auto-correlation function was studied at various temperatures to check consistency with the notion that the diffusion coefficient decreases with temperature. The auto-correlations were normalized by dividing by their starting values which are provided in Table 1. If the velocity autocorrelation function does not approach zero exponentially, then the friction-dominated Smoluchowski limit of Sect. 4 is not valid and the order parameter velocities should be added to the list of order parameters, i.e. a Fokker-Planck limit is appropriate [24,25]. This does not appear to be the case in this system. To add artificial intelligence, NanoX will automatically determine if a Smoluchowski or

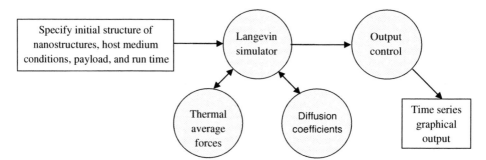

Fig. 3 Schematic NanoX workflow indicating that thermal-average forces and diffusion coefficients are computed "on the fly" since they co-evolve with the order parameters describing overall features of a bionanosystem

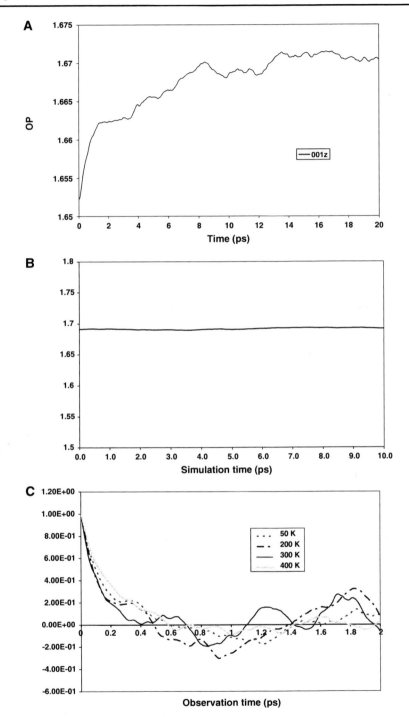

Fig. 4 **A** Order parameter 001z as a function of time. **B** Order parameter 001z as a function of time after equilibration (300 K). **C** Normalized order parameter velocity auto-correlation function for component 001z as a function of observation time for different temperatures

Table 1 Initial order parameter velocity auto-correlation function for component 001z at different temperatures	Temperature (K)	Initial order parameter velocity auto-correlation function
	50	1.49E-09
	200	7.08E-09
	300	9.33E-09
	400	2.08E-08

Fokker-Planck approach is required. In a similar way, NanoX will include additional order parameters if necessary, a capability enabled by our automated order parameter generation scheme of Sect. 2. This feature gives our approach a degree of self-consistency; for example, if the order parameter velocity autocorrelation functions are poorly behaved, then additional order parameters are needed to complete the theory.

The thermal-average forces are calculated using two methods. In the first method, the integral in (4.17) is calculated by a Monte Carlo approach by sampling a fixed order parameter ensemble. Random sampling generates unfavourable configurations and yields high energy configurations with negligible Boltzmann weight. Rather, we use a sampling method whereby only order parameter components orthogonal to the fixed ones are varied. We use this high frequency order parameter ensemble to calculate order parameter velocities and forces. In the second method, a short MD run is performed and Gibbs hypothesis is used to calculate the averages. This approximation is valid since, as discussed above, the order parameters vary slowly in time and the only limitation is the sampling of configuration space by the MD trajectory. We hypothesize that a necessary condition for self-consistency is that the order parameter velocity auto-correlation time is much less than the characteristic time of evolution of the order parameters as driven by the thermal-average forces. Once the thermal-average forces and diffusion coefficients are calculated, the feedback loop in Fig. 3 underlying our multiscale approach is accounted for via the flowchart. The system evolution is carried out via a sequence starting with short MD runs followed by a Langevin evolution timestep for the order parameters. This requires the thermal-average forces and diffusion constants. After the Langevin evolution timestep, the atomic configurations are regenerated by using equation (2.2). Constrained energy minimization and annealing of high energy contacts generated by this "fine graining" is needed to ensure sensible atomic scale configurations. This completes the evolution cycle of Fig. 3. We have implemented this workflow for the STMV virus in our NanoX simulator.

The calculation of the thermal-average forces, diffusion constants, and the energy minimization are all CPU intensive requiring optimizations at different stages. Our simulations were run on the 768 node IBM JS21 cluster at Indiana University using NAMD and we built our simulation modules to effectively utilize existing NAMD resources without introducing unnecessary overhead. Based on preliminary estimates, the thermal-average force module takes about 30 minutes to sample and analyze 2000 configurations, a 10 ps NAMD run for STMV in vacuum takes 15 minutes on 64 processors. Further reductions in CPU time will be possible with greater integration into the NAMD source code.

6 Applications and conclusions

Because of the impracticality of straightforward all-atom models, phenomenological approaches have mainly been used for systems biology. These approaches require recalibration with each new application, a main impediment to progress in computer simulations of biological systems. This difficulty is compounded by the complexity of the systems of interest, leading to ambiguities created by over-calibration, i.e. arriving at the right answer for the wrong reason. To transcend this impediment, we derived principles of microbiology from the laws of molecular physics as outlined for bionanosystems as in Sects. 2 to 4.

To demonstrate that this is possible with extreme complexity, we focused on enveloped viruses. We integrated several types of order parameters into a self-consistent framework and showed that this can be accomplished via rigorous multiscale analysis. This approach overcomes computational difficulties associated with the great number of atoms involved, and accounts for the existence of subsystems with distinct types of physics. Enveloped viruses with their self-assembly, maturation scenario, structural transitions, and richness of phenomena appear to be a prime candidate.

Many systems can be modeled if a self-consistent ACM theory was developed. Examples include

- local sites such as receptors or pores in a cell membrane;
- an enveloped virus involving a composite of a lipid membrane and protein/DNA or RNA structures;
- a nanocapsule with its surrounding cloud of leaking drug molecules;
- an intrabacterial energy-storage granule; and
- a nanodroplet with embedded macromolecule(s).

The fundamental science of these systems and their potential importance for medicine and nanotechnology make the development of ACM approaches of great interest [31]. For example, an ACM approach could provide the basis of a computer-aided strategy for the design of antiviral vaccines or nanocapsules for the delivery of drugs, genes, or siRNA to diseased cells.

Open questions about enveloped viruses include the following.

- What is the structure of the genome-protein core complex deep within the phospholipid subsystem, i.e. is it well organized or highly fluctuating?
- Does the interaction of the phospholipid subsystem with the outer protein net, traversing protein or the genome-protein core, induce liquid-crystal order in the phospholipid subsystem?
- What factors restrict the size, structure, and stability of the ghost particles that are devoid of the genome-protein core?
- What are the ranges of pH, salinity, concentrations of selected ions, and other conditions in the microenvironment that favor a given structure of the protein outer net (see Fig. 1)?
- What is the effect of an applied electric field, or stress applied through an AFM tip, on viral structure and stability?
- Can the virus be grown around, or injected with, a magnetic or fluorescent nanoparticle probe?
- Does the structural transition in Dengue's outer protein net (Fig. 1) involve bond cleavage or formation?
- Can chemical labels pass through the outer protein net and selectively bond to the genome-protein inner core and provide structural information on the core structure [24]?

An integrated multiple order parameter model to address these questions is suggested in Fig. 5.

We propose a model composed of four order parameters accounting for each of the subsystems of Fig.5. Starting from the aqueous microenvironment, we proceed inward to the

Fig. 5 Schematic four subsystem model: AM (aqueous microenvironment), PN (outer protein net), PL (phospholipid zone), and CP (inner genome-protein core particle)

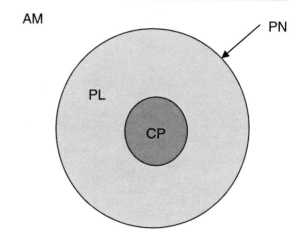

genome-protein core. The aqueous microenvironment is described by a set of order parameter fields Ψ_q^{aq} specifying the mass-weighted position-orientation density for the water molecules (as in Sect. 3). The outer protein net is considered to be connected over the time period of the structural transition. Thus, it is described via a set of structural order parameters $\vec{\Phi}_\ell^{PN}$ from the set of order parameters introduced in Sect. 2 that specify the CM, orientation, and details of the conformation of the outer protein net. The membranous zone contains at least two major components (e.g. phospholipids and Glycoproteins) denoted by A and B here. These are characterized by the order parameter fields Ψ_A and Ψ_B giving the spatial distribution of their CM. The genome-protein core is, for simplicity here, assumed to consist of one connected object described by a set of order parameters, as in Sect. 2. More complex models would also include multiple parts of the core, each of which is connected internally, which require their individual set of order parameters and, therefore, could be suited to study core self-assembly/maturation within the overall assembled enveloped system. Proteins bridging the outer protein net and the core could also be accounted for via such structural order parameters. Finally, order parameter fields could be used to track the exchange of small molecules between the microenvironment and the various inner subsystems. In the integrated model, one solves the Langevin equations as outlined in Sect. 5, but for all the order parameters. All order parameters in the model are coupled in two ways. The thermal-average force for any one order parameter depends on others, thereby accounting for a variety of thermodynamic interactions. Furthermore, the diffusion coefficients for the order parameters provide cross-frictional effects among them. Those associated with the field variables introduce nonlocal effects so that the thermal-average force at $\vec{R}$ affects order parameter fields at $\vec{R}'$.

While the solution of the Langevin equations presents no major computational challenges, the construction of thermal-average forces and diffusion coefficients does. However, recent results of Sect. 5 suggest that the correlation functions for our order parameters have a characteristic time of around a picosecond. Thus, only short MD simulations are needed to estimate them.

A simple system for testing the multiscale model of Sect. 4 is to use the structural transitions in the outer protein net of Dengue virus (Fig. 1). In particular, Dengue "ghosts" consist only of an outer protein net and inner phospholipid material, i.e. they are devoid of the inner protein-genomic core assembly. Thus, the model of a ghost consists of the protein-net, as described via the set of discrete structural parameters introduced in Sect. 2, while the surrounding aqueous medium and inner phospholipid subsystems can be described via order parameter fields. With this, we conclude that with the integration of rigorous multiscale analysis and supercomputing, complex bionanosystems can be modeled, principles of microbiology can be derived, and practical benefits for nanotechnology and biomedicine can be achieved.

Acknowledgements This project was supported in part by the National Institute of Health (NIBIB), the National Science Foundation (CCRC Program), and Indiana University's college of Arts and Sciences through the Center for Cell and Virus Theory.

References

1. Phillips, J.C., Braun, R., Wang, W., Gumbart, J., Tajkhorshid, E., Villa, E., Chipot, C., Skeel, R.D., Kale, L., Schulten, K.: Scalable molecular dynamics with NAMD. J. Comput. Chem. **26**, 1781–1802 (2005)
2. Sanbonmatsu, K.Y., Tung, C.S.: High performance computing in biology: multimillion atom simulations of nanoscale systems. J. Struct. Biol. **157**, 470–480 (2007)
3. Stewart, G.T.: Liquid crystals in biology. I. Historical, biological and medical aspects. Liquid. Cryst. **30**, 541–557 (2003)
4. Zhang, Y., Kostyuchenko, V.A., Rossman, M.G.: Structural analysis of viral nucleocapsids by subtraction of partial projections. J. Struct. Biol. **157**, 356–364 (2007)
5. Zhang, Y., Zhang, W., Ogata, S., Clements, D., Strauss, J.H., Baker, T.S., Kuhn, R.J., Rossmann, M.G.: Conformational changes of the flavivirus E glycoprotein. Structure **12**, 1607–1618 (2004)
6. Zhang, Y., Corver, J., Chipman, P.R., Zhang, W., Pletnev, S.V., Sedlak, D., Baker, T.S., Strauss, J.H., Kuhn, R.J., Rossman, M.G.: Structures of immature flavivirus particles. EMBO J. **22**, 2604–2613 (2003)
7. Klasse, P.J., Bron, R., Marsh, M.: Mechanisms of enveloped virus entry into animal cells. Adv. Drug. Deliv. Rev. **34**, 65–91 (1998)
8. Chandrasekhar, S.: Stochastic problems in physics and astronomy. Rev. Mod. Phys. **15**, 1–89 (1943)
9. Bose, S., Ortoleva, P.: Reacting hard sphere dynamics: Liouville equation for condensed media. J. Chem. Phys. **70**, 3041–3056 (1979)
10. Bose, S., Ortoleva, P.: A hard sphere model of chemical reaction in condensed media. Phys. Lett. A **69**, 367–369 (1979)
11. Bose, S., Bose, S., Ortoleva, P.: Dynamic Padé approximants for chemical center waves. J. Chem. Phys. **72**, 4258–4263 (1980)
12. Bose, S., Medina-Noyola, M., Ortoleva, P.: Third body effects on reactions in liquids. J. Chem. Phys. **75**, 1762–1771 (1981)
13. Deutch, J.M., Oppenheim, I.: The concept of Brownian motion in modern statistical mechanics. Faraday Discuss. Chem. Soc. **83**, 1–20 (1987)
14. Shea, J.E., Oppenheim, I.: Fokker-Planck equation and Langevin equation for one Brownian particle in a nonequilibrium bath. J. Phys. Chem. **100**, 19035–19042 (1996)
15. Shea, J.E., Oppenheim, I.: Fokker-Planck equation and non-linear hydrodynamic equations of a system of several Brownian particles in a non-equilibrium bath. Phys. A **247**, 417–443 (1997)
16. Peters, M.H.: Fokker-Planck equation and the grand molecular friction tensor for combined translational and rotational motions of structured Brownian particles near structures surface. J. Chem. Phys. **110**, 528–538 (1998)
17. Peters, M.H.: Fokker-Planck equation, molecular friction, and molecular dynamics for Brownian particle transport near external solid surfaces. J. Stat. Phys. **94**, 557–586 (1999)
18. Coffey, W.T., Kalmykov, Y.P., Waldron, J.T.: The Langevin Equation with Applications to Stochastic Problems in Physics Chemistry and Electrical Engineering. World Scientific Publishing Co, River Edge (2004)
19. Ortoleva, P.: Nanoparticle dynamics: a multiscale analysis of the Liouville equation. J. Phys. Chem. **109**, 21258–21266 (2005)
20. Miao, Y., Ortoleva, P.: All-atom multiscaling and new ensembles for dynamical nanoparticles. J. Chem. Phys. **125**, 044901 (2006)
21. Miao, Y., Ortoleva, P.: Viral structural transitions: an all-atom multiscale theory. J. Chem. Phys. **125**, 214901 (2006)
22. Shreif, Z., Ortoleva, P.: Curvilinear all-atom multiscale (CAM) theory of macromolecular dynamics. J. Stat. Phys. **130**, 669–685 (2008)

23. Miao, Y., Ortoleva, P.: Molecular dynamics/OP eXtrapolation (MD/OPX) for bionanosystem simulations. J. Comput. Chem. (2008). doi:10.1002/jcc.21071
24. Pankavich, S., Miao, Y., Ortoleva, J, Shreif, Z., Ortoleva, P.: Stochastic dynamics of bionanosystems: multiscale analysis and specialized ensembles. J. Chem. Phys. **128**, 234908 (2008)
25. Pankavich, S., Shreif, Z., Ortoleva, P.: Multiscaling for classical nanosystems: derivation of Smoluchowski and Fokker-Planck equations. Phys. A **387**, 4053–4069 (2008)
26. Shreif, Z., Ortoleva, P.: Multiscale derivation of an augmented Smoluchowski. Phys. A (2008, accepted)
27. Shreif, Z., Ortoleva, P.: Computer-aided design of nanocapsules for therapeutic delivery. Comput. Math. Methods Med. (2008, to appear)
28. Pankavich, S., Shreif, Z., Miao, Y., Ortoleva, P.: Self-assembly of nanocomponents into composite structures: derivation and simulation of Langevin equation. J. Chem. Phys. (2008, accepted)
29. Pankavich, S., Ortoleva, P.: Self-assembly of nanocomponents into composite structures: multiscale derivation of stochastic chemical kinetic models. ACS Nano (2008, in preparation)
30. Pankavich, S., Ortoleva, P.: Multiscaling for systems with a broad continuum of characteristic lengths and times: structural transitions in nanocomposites. (2008, in preparation)
31. Shreif, Z.,Ortoleva, P.: All-atom/continuum multiscale theory: application to nanocapsule therapeutic delivery. Multiscale Model. Simul. (2008, submitted)
32. Jaqaman, K., Ortoleva, P.: New space warping method for the simulation of large-scale macromolecular conformational changes. J. Comput. Chem. **23**, 484–491 (2002)
33. Freddolino, P.L., Arkhipov, A.S., Larson, S.B., McPherson, A., Schulten, K.: Molecular dynamics simulations of the complete satellite tobacco mosaic virus. Structure **14**, 437–449 (2006)

Computational modeling of brain tumors: discrete, continuum or hybrid?

Zhihui Wang · Thomas S. Deisboeck

Originally published in the journal Sci Model Simul, Volume 15, Nos 1–3, 381–393.
DOI: 10.1007/s10820-008-9094-0 © Springer Science+Business Media B.V. 2008

Abstract   In spite of all efforts, patients diagnosed with highly malignant brain tumors (gliomas), continue to face a grim prognosis. Achieving significant therapeutic advances will also require a more detailed quantitative understanding of the dynamic interactions among tumor cells, and between these cells and their biological microenvironment. Data-driven computational brain tumor models have the potential to provide experimental tumor biologists with such quantitative and cost-efficient tools to generate and test hypotheses on tumor progression, and to infer fundamental operating principles governing bidirectional signal propagation in multicellular cancer systems. This review highlights the modeling objectives of and challenges with developing such *in silico* brain tumor models by outlining two distinct computational approaches: discrete and continuum, each with representative examples. Future directions of this integrative computational neuro-oncology field, such as *hybrid multiscale multiresolution* modeling are discussed.

Keywords   Brain tumor · Agent-based model · Cellular automata · Continuum · Multi-scale

Abbreviations

ABM	Agent-based model
CA	Cellular automata
EGFR	Epidermal growth factor receptor
ECM	Extracellular matrix
GBM	Glioblastoma
MRI	Magnetic resonance imaging
PLCγ	Phopholipase Cγ
ROI	Region of interest

Z. Wang · T. S. Deisboeck (✉)
Complex Biosystems Modeling Laboratory, Harvard-MIT (HST) Athinoula A. Martinos Center for Biomedical Imaging, Massachusetts General Hospital-East, 2301, Building 149, 13th Street, Charlestown, MA 02129, USA
e-mail: deisboec@helix.mgh.harvard.edu

2D   Two-dimensional
3D   Three-dimensional

1 Introduction

There are two basic types of brain tumors, i.e. primary tumors and secondary or metastatic brain tumors. Primary brain tumors arise in the brain, and here most often from its supporting astrocytes or glia cells (hence the terminology 'astrocytoma' or 'glioma') and generally do not spread outside the brain tissue; on the contrary, metastatic brain tumors originate elsewhere in the body such as in the lung or skin before disseminated cancer satellite cells spread also to the brain. Carcinogenesis in the brain, much like elsewhere in the body, is a complex multistage process that originates from genetic changes, distortion of the cell cycle, and loss of apoptosis [1], and proceeds to angiogenesis, and extensive local infiltration and invasion [51]. In the United States, for the year 2007 alone, it was estimated that there were 20,500 new cases of (both primary and secondary) brain tumors, and 12,740 deaths related to this disease [37]. Brain tumors are still relatively insensitive to conventional cancer treatments, including radiation and chemotherapy [36]. Despite advances in recent targeted anticancer therapies, the clinical outcome in treating malignant brain tumors remains disappointing [60] with less than 30% of recurrent glioblastoma (GBM; the most aggressive form of gliomas) patients surviving without further progression six months after treatment [9]. This is mainly a result of the tumor's extensive infiltrative behavior, its rapid development of treatment resistance due to its inherent genetic and epigenetic heterogeneity, and the difficulties the so called blood-brain barrier poses for delivery of therapeutic compounds [7,43].

Cellular and microenvironmental factors along with the underlying processes at the molecular level act as regulators of tumor growth and invasion [24,27]. Tumor cells bi-directionally communicate with their microenvironment: they not only respond to various external cues but also impact the environment by e.g. producing (auto- and paracrine) signals and degrading the neighboring tissue with proteases [34]. However, despite a vast amount of qualitative findings, conventional cancer research has made few gains in exploring the *quantitative* relationship between these very complicated intra- and intercellular signaling processes and the behavior they trigger on the microscopic and macroscopic scales [54]. It is here where we and others argue that *systems biology* [42] can provide useful insights, which may eventually promote the development of new cancer diagnostic and therapeutic techniques. While still in its beginning, systems biology has so far focused primarily on the single-cell level [2]. However, the usefulness of computational modeling and simulation, combined with experiment, is being increasingly recognized for exploring the dynamics at a multi-cell or tissue level of a variety of biological systems within a temporal, spatial and physiological context [8].

To date, computational modeling works have produced preliminary quantifications of the links between cell-cell and cell-extracellular matrix (ECM) interactions, cell motility, and local concentration of cell substrates. Already, this sprawling interdisciplinary field draws increasing attention from an experimental and clinical as well as pharmaceutical perspective [14,21,31]. A better understanding of the inherent complexity of these cancer systems requires intensified interdisciplinary research in which the next iteration of innovative computational models, informed by and continuously revised with experimental data, will play an important role of guiding experimental interpretation and design in going forward [29]. Here, we discuss first objectives of and challenges with modeling brain tumors mathematically and computationally and then briefly review some recent developments in using two distinct *in silico*[1] approaches.

2 In silico brain tumor modeling: objectives & challenges

As for other *in silico* oncology efforts, the main objective of modeling brain tumors is to design and develop powerful simulation platforms capable of (**1**) providing a realistic mathematical representation and systematic treatment of the complexity of experimentally observed cancer phenomena, across the scales of interest and within its distinct biological context, (**2**) generating experimentally testable hypotheses to guide wet-lab research and help evaluate the algorithms, and finally (**3**) integrating any number of distinct data qualities (e.g., serum markers, genomics, phospho-proteomics and magnetic resonance images) into these modeling algorithms in an effort to predict tumor growth patterns eventually also in a *patient-specific* context. To achieve such ambitious goals, a computational brain tumor model should be able to (**i**) quantitatively clarify the characteristics of a set of various basic cancer phenotypes (e.g., proliferation, invasion, and angiogenesis) at different scales, and (**ii**) to assess the impact of the microenvironmental cues on these cell phenotypes. Finally, we argue that (**iii**) an advanced brain tumor model should eventually be extended to the molecular level in that it explicitly includes the combinational effects of oncogenes [15] and tumor suppressor genes [11] on the aforementioned microscopic phenotypes.

There are several key challenges confronting a computational tumor biologist in developing any such model. These include: **1) Selection of modeling scale(s)**. Choosing the appropriate scale is the first critical step, usually guided by both the data available and the area of expertise of the investigator. Also, if a model is designed to be composed of different scales, then how to *link* these scales in a way supported by data is another non-trivial step. For example, GBM cells exhibit a variety of point mutations (*molecular level*) [35] that can affect microvascular remodeling (*microscopic level*) which in turn impacts tumor size, shape, and composition (*macroscopic level*) [33]. To date, while some brain tumor modeling studies have dealt with the interaction of processes between cellular and macroscopic levels (for a recent review, see [54]), only very few works made an attempt to quantitatively establish the relationship between the molecular and cellular levels. **2) Level of complexity versus computational cost**. Generally, it holds that the more detailed a model, the more parameters are involved and thus the higher the computational 'cost' of running the algorithm. As such, for the time being, it is a compromise between the biological complexity to be represented and the computational resources this would require. Given the ever increasing amount of data available, *scalability* becomes an issue of paramount interest when deciding on the applicability of any such *in silico* tool in a clinical setting. **3) Tumor boundary definition**. Defining the degree of diffuse invasion of tumor cells into the surrounding brain tissue remains difficult regardless of advancements in conventional medical imaging [69]. While some algorithms have made progress on translating tumor and surrounding tissue information from patient imaging data to the coordinate system of the models with finite element methods [16,49], there is still a long way to go towards accurately predicting where and when what number of the currently still invisible but surely existent mobile tumor cells spread into the adjacent healthy brain tissue.

---

[1] *In silico* refers to experiments carried out entirely using a computer as opposed to being conducted in a wet lab environment (see [47] for a brief review on differences between *in silico* and *in vitro* or *in vivo* studies).

Available computational models have addressed these challenges in one form or another. The next section will detail current approaches with a focus on briefly reviewing some significant findings of representative models developed in the past few years, and highlight some research groups active at the forefront of this interdisciplinary field.

3 Computational modeling approaches

Two major types of modeling strategies currently exist in the computational tumor modeling community: discrete and continuum approaches. *Discrete* models can explicitly represent individual cells in space and time and easily incorporate biological rules (based on data or assumptions), such as defining cell-cell and cell-matrix interactions involved in both chemotaxis and haptotaxis for instance. However, these models are limited to relatively small numbers of cells due to the compute intense nature of the method, and as a result a typical discrete model is usually designed with a sub-millimeter or even lower domain size [70]. In contrast, *continuum* models, by describing e.g. extracellular matrix or the entire tumor tissue as continuum medium rather than at the resolution of individual cells, are able to capture larger-scale volumetric tumor growth dynamics at comparatively lesser computational cost. As a trade-off, continuum models lack sensitivity to small fluctuations or oscillatory behaviors of a tumor system at a smaller segment, such as tumor angiogenetic sprout branching [3]. That is a significant shortcoming as in some cases such small changes can be the leading cause in driving a nonlinear complex biosystem to a different state [10]. In the following, we will introduce the two approaches in more detail.

3.1 Discrete modeling

The two main, related discrete modeling methods extensively used in this context are *cellular automata* (CA) and *agent-based model* (ABM). A generic CA is a collection of cells on a grid of specified shape that synchronously evolves through a number of discrete time steps, according to an identical set of rules (applied to each single cell) based on the states of neighboring cells [71]. The grid can be implemented in any finite number of dimensions, and neighbors are a selection of cells relative to a given cell. In contrast, ABM asynchronously models phenomena as dynamical systems of interactions among and between agents and their environments [12,32]. An agent is any autonomous component that can interact or communicate with other components. Each biological cell is often represented as an agent in an ABM, and indeed ABM is the natural extension of CA. Because of the asynchronous characteristic, the ease of implementation and the richness of detail one can expect in exploring biosystem dynamics, ABM is an appealing choice for the simulation of tumors like glioma where the behavior and heterogeneity of the interacting cells cannot be safely reduced to some averaged, stylized or simple mechanism [66].

For instance, GBM growth dynamics in a three-dimensional (3D) environment have been successfully predicted using a CA model driven by four microscopic parameters (referring to cell-doubling time, nutritional needs of growth-arrested cells, nutritional needs of dividing cells, and effects of mechanical confinement pressure) [39,40]. This model was then used as the basis for a follow-up study to analyze a heterogeneous tumor by introducing a distinct subpopulation of tumor cells that exhibit a growth advantage [38]. The results showed that changes even in a small subpopulation may lead to a drastically altered tumor growth behavior, suggesting that prognosis based on the assumption of a homogeneous tumor cell population can be markedly inaccurate. With a CA approach to study the effects of surgery plus chemotherapy on the evolution of a homogeneous and more realistic heterogeneous GBM mass, it was found that the spatial distribution of chemotherapeutic resistant cells is an important indicator of persistence and continued tumor growth [55]. One clinical implication gained from this study is that the shape of the reoccurring tumor may depend on the rate at which chemotherapy induces mutations. Since these previous iterations made oversimplifying assumptions on tumor vascular and angiogenesis, a recent two-dimensional (2D) CA simulation tool [30] considered the processes of vessel co-option, regression and angiogenesis in tumor growth; it enabled the researchers to study the growth of a primary neoplasm from a small mass of cells to a macroscopic tumor mass, and to simulate how mutations affecting the angiogeneic response subsequently impact tumor development.

To investigate highly malignant brain tumors as *complex, dynamic,* and *self-organizing biosystems* [20], Deisboeck and co-workers have been focusing on the development of ABMs simulating tumor properties *across* multiple scales in time and space. First, the spatio-temporal expansion of virtual glioma cells in a 2D microscopic setup and the relationship between rapid growth and extensive tissue infiltration were investigated [44,45]. These earlier works reported a phase transition leading to the emergence of two distinct spatio-temporal patterns: a) a small number of larger tumor cell clusters exhibiting rapid spatial expansion but shorter overall lifetime of the tumor system, and b) many small clusters with longer lifetime but the tradeoff of a slower velocity of expansion, depending on different implicit chemotactic search strategies. Subsequently, by incorporating a molecular scale in the form of a simplified representation of the epidermal growth factor receptor (EGFR) signaling pathway (important for epithelial cancers in general, and for highly malignant brain tumors in particular [46]), the model was extended to capture tumor growth dynamics to a degree of any specific pathway component [5,6]. Some intriguing, testable hypotheses have been generated in terms of how molecular profiles of individual glioma cells impact the cellular phenotype and how such single-cell decisions can potentially affect the dynamics of the entire tumor system. Most recently, an explicit cell cycle description was introduced to the model and brain tumor growth dynamics were examined in a 3D context with a more complicated ECM representation at the microscopic scale [72]. Together, these works have provided a computational paradigm for simulating brain tumors from the molecular scale up to the cellular level and beyond. It should be noted that in these works some environmental parameters, such as growth factors, nutrient, and oxygen tension, were expressed with a continuum term. Another contribution of the works by Deisboeck and co-workers is that, based on available data [23,48], they propose employing, as an example, an EGFR-downstream protein, phospholipase $C\gamma$ (PLCγ), to determine two phenotypic traits, i.e. cell proliferation and migration, by comparing the rate of change of its molecular-level concentration with a predefined threshold. That is, a glioma cell becomes eligible to 1) migrate if the range of change of PLCγ exceeds the threshold, and 2) proliferate if the range of change of PLCγ is below that set threshold, yet above a noise threshold. More generic, the change in the concentration of a pathway component over time is calculated with a continuum element, i.e., according to the following differential equation:

$$\frac{dXi}{dt} = \alpha Xi - \beta Xi \tag{1}$$

where Xi represents the concentration level of the ith pathway component, and α and β are the reaction rates of producing and consuming Xi, respectively. Figure 1 shows a series of simulation results produced by the model [6], explaining how tumor growth dynamics at the cellular level can be related to alterations at the molecular level. This algorithm is flexible so that it can accommodate the governing, physical requirements of other cancer types, such as non-small cell lung cancer [67], which demonstrates the versatility of this design concept.

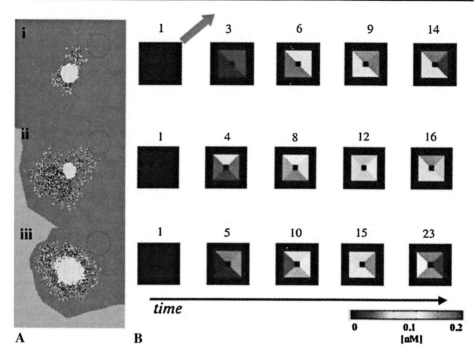

Fig. 1 (a) 2D cross-section of a tumor spheroid, for three different human glioma cell lines, from top to bottom: D-263 MG, D-247 MG, and D-37 MG. Each simulation was terminated when the first tumor cell reached the edge of a (red) nutrient source (representing an orthograde cut cerebral blood vessel) located in the north-east quadrant of the lattice. (b) Polarization of the molecular concentration profiles of the EGFR-pathway downstream component PLCγ in the first cell, at five consecutive time points. A qualitatively similar PLCγ polarization pattern emerges in the three cell lines as higher concentrations of PLCγ eventually accumulate in the apical part of the cell that faces nutrient abundance. Adapted from [6]

It is noteworthy that some efforts employ techniques analogous to ABM to study the clinical level of brain tumor behavior. A series of *in silico* studies on simulating a GBM response to radiotherapy, considering vasculature and oxygen supply, has been conducted [22,58,59]. While in [59] tumor cells were considered individually, in the follow-up studies [22,58], in an effort to overcome the extensive computational demand, cells were clustered into dynamic equivalence classes based on the mean cell cycle phase durations (G1, S, G2, and M, see [41] for a review); that is, tumor response to radiotherapy was investigated on each cluster instead of on each individual cell. Moreover, for performing patient-specific *in silico* experiments as a means of chemotherapeutic treatment optimization, the same authors recently developed a four-dimensional simulation platform based on magnetic resonance imaging (MRI), histopathologic, and pharmacogenetic data, noting that the model's predictions were in good agreements with clinical practice [57]. Taken together, models from both Deisboeck's and Stamatakos' groups pioneered the integration of *continuum* elements into a discrete framework. To put this in perspective, we will detail a strict continuum approach in the following section.

3.2 Continuum modeling

Using a continuum approach, Cristini and co-workers have established a series of exploratory investigations on mathematical analysis of morphologic stability in growth and invasion of highly malignant gliomas [17,18,25,26,53,56,73]. They propose that tumor tissue dynamics can be simply regulated by two dimensionless parameters: one quantifies the competition between local cell proliferation (contributing to tumor mass growth) and cell adhesion (which tends to minimize the tumor surface area), while the other one represents tumor mass reduction related to cell death. The authors then tested the conditions for morphological stability for an independent set of experiments where the levels of growth factors and glucose were changed over a wide range in order to manipulate GBM cell proliferation and adhesion [25]. Most recently, they further confirmed that morphologic patterns of tumor boundary and infiltrative shapes of invasive tumors predicted by their models were in agreement with clinical histopathology samples of GBM from multiple patients [26]. Figure 2 shows a time-series result of the evolving tumor shape over a course of three months using this model. The authors claimed that their algorithm enabled the prediction of tumor morphology by quantifying the spatial diffusion gradients of cell substrates maintained by heterogeneous cell proliferation and an abnormal, constantly evolving vasculature. These models are based on reaction-diffusion equations (that govern variables such as tumor cell density, neovasculature, nutrient concentration, ECM, and matrix degrading enzymes) of the following generic form:

$$v_t = -\nabla \cdot J + \Gamma_+ - \Gamma_- \tag{2}$$

where v represents one of the evolving variables, J is the flux, Γ_+ and Γ_- are the sources and sinks with respect to variable v (expansion formulas differ according to the variable investigated; see [26] for detail). This group's work showed that a continuum approach is capable of 1) accounting for a variety of invasive morphologies observed in tumors *in vitro, in vivo*, and in patients, 2) predicting different growth and invasion behaviors of tumors by calibrating model parameters, and 3) testing the hypothesized phenomenological relationships of tumor adhesion and proliferation that affect tissue-scale growth and morphology.

Several other groups have also been working on applying a continuum approach to the investigation of brain tumor behaviors. For instance, [61] developed a continuum model that incorporated the effects of heterogeneous brain tissue on diffusion and growth rates of glioma cells in an effort to represent asymmetries of the tumor boundaries. This basic work was then extended to examine the growth and invasion of gliomas in a 3D virtual brain refined from anatomical distributions of grey and white matter [62]. By allowing a motility coefficient to differ depending on the local tissue composition (so that glioma cells migrate more rapidly along white matter than in grey matter), the algorithm predicted sites of potential tumor recurrence to a degree beyond the limits of current medical imaging techniques. Interestingly, as supported by the results of this model, two independent factors, velocity of diametric expansion and initial tumor size at diagnosis, were indeed found to be statistically significant in a recent clinical survey on the prognostic evaluation of patients who harbor a grade II glioma [52]. Based on their previous studies [61,62,65], [63] also investigated the effects of chemotherapy on the spatio-temporal response of gliomas. By comparing the simulation results with MRI data of a glioma patient, it was suggested that differential delivery of the chemotherapeutic agent to the grey and white matter could successfully describe the clinical problems of shrinkage of the lesion in certain areas of the brain with continued growth in others. Another recent continuum model confirmed the effects of repeated immuno-suppression treatment (using different protocols) on the progression of glioma, and mathematically revealed the necessity of repeating such treatment in reducing the risk of recurrence [25]. Furthermore, by combining essential methods of two previous approaches [13,50], [68] were able to capture the spatio-temporal dynamics of drug transport and cell-death in a heterogeneous collection of glioma cells and normal brain cells.

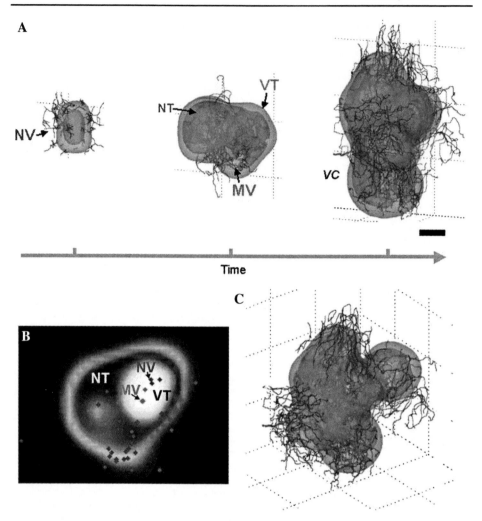

Fig. 2 (a) Time-series of the morphologic features of a growing GBM generated with a 3D model. The morphology is directly influenced by angiogenesis, vasculature maturation, and vessel co-option. The vessels labeled in red are capable of releasing nutrients, e.g., oxygen. (b) Histology-like section of the last frame of the simulation in (a) reveals viable tumor regions (white) surrounding necrotic tissue (dark). The viable region's thickness and extent of necrosis are strongly dependent on the diffusion gradients of oxygen/nutrient in the microenvironment. (c) Another view from the simulation shown in part A. Adapted from [26] with permission

4 Conclusions and perspectives

In recent years, computational cancer research has become a sprawling interdisciplinary field. This is a result of a number of contributing factors. Firstly, in contrast to conventional wet-lab experimental methods, such *in silico* models offer a powerful platform to reproducibly alter parameters and thus investigate their impact on the cancer system studied, at a rapid pace and in a cost-efficient way [42]. Secondly, computational models have demonstrated the ability of providing a useful hypothesis generating tool for refocusing experimental *in vitro* and *in vivo* works [54]. Thirdly, from a practical clinical perspective, computational modeling has already been applied, with some promise, to simulating the impact of chemotherapy, radiotherapy, and drug delivery on brain tumors [19]. Within this *in silico* oncology area, modeling and simulating malignant brain tumors is starting to emerge as a paramount driver for advancing technical developments en route to help addressing important scientific questions.

Demonstrated with examples from the literature, we have reviewed the two major mathematical modeling approaches, discussed their distinct merits and limitations in quantitatively studying brain tumor growth dynamics. In summary: While discrete models perform at the resolution of individual cells, which function independently through a set of behavioral rules that are inspired by biological facts if not fueled with real data, they are limited to a rather small number of cells or constituents. Conversely, continuum models can capture tumor growth at a collective scale that allows monitoring the expansion of a larger cluster of homogeneously behaving cells yet fail to register single cells, genes or proteins. Since both discrete and continuum modeling approaches have their own advantages and shortcomings (Table 1), and because quantifying the relationships between complex cancer phenomena at different scales is highly desirable, we and others have begun to move into the direction of *hybrid* modeling e.g., [4,58,67,72], or more appropriately, towards *hybrid, multi-scale* and *multi-resolution* algorithms as the next stage of cancer modeling in general, and brain tumor modeling in particular. While 'hybrid' refers to the integration of both discrete and continuum techniques, 'multi-resolution' means that cells at distinct topographic regions are treated differently in terms of the modeling approach applied. The overall strategy is clear: achieving discretely high resolution wherever and whenever necessary to maintain (or ideally, improve) the model's overall predictive power while at the same time reducing compute intensity as much as possible to allow for inclusion of sufficiently large datasets, and thus support scalability of the approach to clinically relevant levels. Figure 3 schematically describes the development of a 2D model using this novel strategy. Here, the MRI-demarked hypointense region within the tumor core, often comprised of a large fraction of apoptotic cells if not necrotic tissue, can arguably be described sufficiently as a rather homogenous population, thus at a lower resolution which allows employing a continuum module. Conversely, the highly active, gadolinium enhanced tumor surface supposedly thrives with a genetically and epigenetically heterogeneous population of cells that must at least in part be described discretely, and at the resolution of interconnected signaling pathways, to capture topographic areas that (e.g., with some probability, harbor an aggressive clone that) may impact overall growth patterns in the future. As the tumor grows, these high-resolution regions of interest (ROIs) and, thus the *in silico* modules representing them, will likely have to change dynamically (i.e., in size, number and location) to maintain or, better, improve predictive power

Table 1 Characteristics of discrete, continuum and hybrid brain tumor modeling approaches

Category	Characteristics	References
Discrete	• Autonomous cells, with a set of rules governing their behavior • Capable to investigate tumor dynamics at a single cell level and below • Limited to a comparably smaller scale due to prohibitive computational costs	[39,40,45]
Continuum	• Describing tumor tissue as a continuum medium • Capable to capture larger-scale volumetric tumor dynamics • Computational cost efficiency • Difficult to implement heterogeneous cell-cell and cell-environmental interaction, or molecular level dynamics	[18,26,64]
Hybrid	• Applicable to both small- and large-scale models • Extensive numerical techniques required	[5,6,57,58,72]

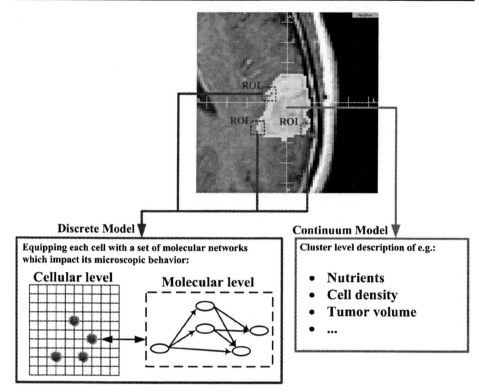

Fig. 3 Schematic illustration of a 2D brain tumor model using a *hybrid*, multi-scale and multi-resolution strategy. "ROI" represents a region of interest which refers to a higher modeling resolution desired and thus discrete-based technique used, versus the larger remaining volume (green) of the tumor tissue that is being modeled with a continuum-based approach. ROIs can be obtained e.g. by using finite elements and other numerical techniques [3, 17, 73]

while training on the patient-*specific* data set. Admittedly, much work needs to be done in this area to tackle the considerable challenges involved that range from data processing in 3D over time (where most computational savings would occur) to automated ROI placement and result driven, dynamic readjustment. However, eventually, such advanced *in silico* oncology approaches should be able to provide, on a clinical level, much needed quantitative insights into the dynamic cross-scale relationships that characterize these and other highly malignant tumors, and thus prove to become an effective and indispensable tool for *personalized systems medicine* in the near future.

Acknowledgements   This work has been supported in part by NIH grant CA 113004 and by the Harvard-MIT (HST) Athinoula A. Martinos Center for Biomedical Imaging and the Department of Radiology at Massachusetts General Hospital. We apologize to all colleagues whose works we could not cite due to space limitations.

References

1. Al-Hajj, M., Clarke, M.F.: Self-renewal and solid tumor stem cells. Oncogene **23**, 7274–7282 (2004)
2. Albeck, J.G., MacBeath, G., White, F.M., Sorger, P.K., Lauffenburger, D.A., Gaudet, S.: Collecting and organizing systematic sets of protein data. Nat. Rev. Mol. Cell. Biol. **7**, 803–812 (2006)

3. Anderson, A.R., Chaplain, M.A.: Continuous and discrete mathematical models of tumor-induced angiogenesis. Bull. Math. Biol. **60**, 857–899 (1998)
4. Anderson, A.R., Weaver, A.M., Cummings, P.T., Quaranta, V.: Tumor morphology and phenotypic evolution driven by selective pressure from the microenvironment. Cell **127**, 905–915 (2006)
5. Athale, C., Mansury, Y., Deisboeck, T.S.: Simulating the impact of a molecular 'decision-process' on cellular phenotype and multicellular patterns in brain tumors. J. Theor. Biol. **233**, 469–481 (2005)
6. Athale, C.A., Deisboeck, T.S.: The effects of EGF-receptor density on multiscale tumor growth patterns. J. Theor. Biol. **238**, 771–779 (2006)
7. Badruddoja, M.A., Black, K.L.: Improving the delivery of therapeutic agents to CNS neoplasms: a clinical review. Front. Biosci. **11**, 1466–1478 (2006)
8. Bailey, A.M., Thorne, B.C., Peirce, S.M.: Multi-cell agent-based simulation of the microvasculature to study the dynamics of circulating inflammatory cell trafficking. Ann. Biomed. Eng. **35**, 916–936 (2007)
9. Ballman, K.V., Buckner, J.C., Brown, P.D., Giannini, C., Flynn, P.J., LaPlant, B.R., Jaeckle, K.A.: The relationship between six-month progression-free survival and 12-month overall survival end points for phase II trials in patients with glioblastoma multiforme. Neuro. Oncol. **9**, 29–38 (2007)
10. Berg, O.G., Paulsson, J., Ehrenberg, M.: Fluctuations and quality of control in biological cells: zero-order ultrasensitivity reinvestigated. Biophys. J. **79**, 1228–1236 (2000)
11. Blume-Jensen, P., Hunter, T.: Oncogenic kinase signalling. Nature **411**, 355–365 (2001)
12. Bonabeau, E.: Agent-based modeling: methods and techniques for simulating human systems. Proc. Natl. Acad. Sci. USA **99**(Suppl 3), 7280–7287 (2002)
13. Burgess, P.K., Kulesa, P.M., Murray, J.D., Alvord, E.C. Jr.: The interaction of growth rates and diffusion coefficients in a three-dimensional mathematical model of gliomas. J. Neuropathol. Exp. Neurol. **56**, 704–713 (1997)
14. Chaplain, M.A., McDougall, S.R., Anderson, A.R.: Mathematical modeling of tumor-induced angiogenesis. Annu. Rev. Biomed. Eng. **8**, 233–257 (2006)
15. Cheng, J.Q., Lindsley, C.W., Cheng, G.Z., Yang, H., Nicosia, S.V.: The Akt/PKB pathway: molecular target for cancer drug discovery. Oncogene **24**, 7482–7492 (2005)
16. Clatz, O., Sermesant, M., Bondiau, P.Y., Delingette, H., Warfield, S.K., Malandain, G., Ayache, N.: Realistic simulation of the 3-D growth of brain tumors in MR images coupling diffusion with biomechanical deformation. IEEE Trans. Med. Imaging **24**, 1334–1346 (2005)
17. Cristini, V., Lowengrub, J., Nie, Q.: Nonlinear simulation of tumor growth. J. Math. Biol. **46**, 191–224 (2003)
18. Cristini, V., Frieboes, H.B., Gatenby, R., Caserta, S., Ferrari, M., Sinek, J.: Morphologic instability and cancer invasion. Clin. Cancer Res. **11**, 6772–6779 (2005)
19. Deisboeck, T.S., Zhang, l., Yoon, J., Costa, J.: *In silico* cancer modeling: Is ready for primetime? Nat. Clin. Pract. Oncol. (in press)
20. Deisboeck, T.S., Berens, M.E., Kansal, A.R., Torquato, S., Stemmer-Rachamimov, A.O., Chiocca, E.A.: Pattern of self-organization in tumour systems: complex growth dynamics in a novel brain tumour spheroid model. Cell Prolif. **34**, 115–134 (2001)
21. Di Ventura, B., Lemerle, C., Michalodimitrakis, K., Serrano, L.: From in vivo to in silico biology and back. Nature **443**, 527–533 (2006)
22. Dionysiou, D.D., Stamatakos, G.S., Uzunoglu, N.K., Nikita, K.S., Marioli, A.: A four-dimensional simulation model of tumour response to radiotherapy in vivo: parametric validation considering radiosensitivity, genetic profile and fractionation. J. Theor. Biol. **230**, 1–20 (2004)
23. Dittmar, T., Husemann, A., Schewe, Y., Nofer, J.R., Niggemann, B., Zanker, K.S., Brandt, B.H.: Induction of cancer cell migration by epidermal growth factor is initiated by specific phosphorylation of tyrosine 1248 of c-erbB-2 receptor via EGFR. FASEB J. **16**, 1823–1825 (2002)
24. Entschladen, F., Drell, T.L.t., Lang, K., Joseph, J., Zaenker, K.S.: Tumour-cell migration, invasion, and metastasis: navigation by neurotransmitters. Lancet Oncol. **5**, 254–258 (2004)
25. Frieboes, H.B., Zheng, X., Sun, C.H., Tromberg, B., Gatenby, R., Cristini, V.: An integrated computational/experimental model of tumor invasion. Cancer Res. **66**, 1597–1604 (2006)
26. Frieboes, H.B., Lowengrub, J.S., Wise, S., Zheng, X., Macklin, P., Bearer, E.L., Cristini, V.: Computer simulation of glioma growth and morphology. Neuroimage **37**(Suppl 1), S59–S70 (2007)
27. Friedl, P., Wolf, K.: Tumour-cell invasion and migration: diversity and escape mechanisms. Nat. Rev. Cancer **3**, 362–374 (2003)
28. Friedman, A., Tian, J.P., Fulci, G., Chiocca, E.A., Wang, J.: Glioma virotherapy: effects of innate immune suppression and increased viral replication capacity. Cancer Res. **66**, 2314–2319 (2006)
29. Gatenby, R.A., Maini, P.K.: Mathematical oncology: cancer summed up. Nature **421**, 321 (2003)
30. Gevertz, J.L., Torquato, S.: Modeling the effects of vasculature evolution on early brain tumor growth. J. Theor. Biol. **243**, 517–531 (2006)

31. Gilbert, D., Fuss, H., Gu, X., Orton, R., Robinson, S., Vyshemirsky, V., Kurth, M.J., Downes, C.S., Dubitzky, W.: Computational methodologies for modelling, analysis and simulation of signalling networks. Brief Bioinform. **7**, 339–353 (2006)
32. Gilbert, N., Bankes, S.: Platforms and methods for agent-based modeling. Proc. Natl. Acad. Sci. USA **99**(Suppl 3), 7197–7198 (2002)
33. Gilhuis, H.J., Bernse, H.J., Jeuken, J.W., Wesselin, P., Sprenger, S.H., Kerstens, H.M., Wiegant, J., Boerman, R.H.: The relationship between genetic aberrations as detected by comparative genomic hybridization and vascularization in glioblastoma xenografts. J. Neurooncol. **51**, 121–127 (2001)
34. Hendrix, M.J., Seftor, E.A., Seftor, R.E., Kasemeier-Kulesa, J., Kulesa, P.M., Postovit, L.M.: Reprogramming metastatic tumour cells with embryonic microenvironments. Nat. Rev. Cancer **7**, 246–255 (2007)
35. Holland, E.C.: Glioblastoma multiforme: the terminator. Proc. Natl. Acad. Sci. USA **97**, 6242–6244 (2000)
36. Jain, R.K., di Tomaso, E., Duda, D.G., Loeffler, J.S., Sorensen, A.G., Batchelor, T.T.: Angiogenesis in brain tumours. Nat. Rev. Neurosci. **8**, 610–622 (2007)
37. Jemal, A., Siegel, R., Ward, E., Murray, T., Xu, J., Thun, M.J.: Cancer statistics 2007. CA Cancer J. Clin. **57**, 43–66 (2007)
38. Kansal, A.R., Torquato, S., Chiocca, E.A., Deisboeck, T.S.: Emergence of a subpopulation in a computational model of tumor growth. J. Theor. Biol. **207**, 431–441 (2000a)
39. Kansal, A.R., Torquato, S., Harsh, G.I., Chiocca, E.A., Deisboeck, T.S.: Simulated brain tumor growth dynamics using a three-dimensional cellular automaton. J. Theor. Biol. **203**, 367–382 (2000b)
40. Kansal, A.R., Torquato, S., Harsh, I.G., Chiocca, E.A., Deisboeck, T.S.: Cellular automaton of idealized brain tumor growth dynamics. Biosystems **55**, 119–127 (2000c)
41. Kastan, M.B., Bartek, J.: Cell-cycle checkpoints and cancer. Nature **432**, 316–323 (2004)
42. Kitano, H.: Computational systems biology. Nature **420**, 206–210 (2002)
43. Lefranc, F., Brotchi, J., Kiss, R.: Possible future issues in the treatment of glioblastomas: special emphasis on cell migration and the resistance of migrating glioblastoma cells to apoptosis. J. Clin. Oncol. **23**, 2411–2422 (2005)
44. Mansury, Y., Deisboeck, T.S.: The impact of "search precision" in an agent-based tumor model. J. Theor. Biol. **224**, 325–337 (2003)
45. Mansury, Y., Kimura, M., Lobo, J., Deisboeck, T.S.: Emerging patterns in tumor systems: simulating the dynamics of multicellular clusters with an agent-based spatial agglomeration model. J. Theor. Biol. **219**, 343–370 (2002)
46. Mellinghoff, I.K., Wang, M.Y., Vivanco, I., Haas-Kogan, D.A., Zhu, S., Dia, E.Q., Lu, K.V., Yoshimoto, K., Huang, J.H., Chute, D.J., Riggs, B.L., Horvath, S., Liau, L.M., Cavenee, W.K., Rao, P.N., Beroukhim, R., Peck, T.C., Lee, J.C., Sellers, W.R., Stokoe, D., Prados, M., Cloughesy, T.F., Sawyers, C.L., Mischel, P.S.: Molecular determinants of the response of glioblastomas to EGFR kinase inhibitors. N. Engl. J. Med. **353**, 2012–2024 (2005)
47. Miners, J.O., Smith, P.A., Sorich, M.J., McKinnon, R.A., Mackenzie, P.I.: Predicting human drug glucuronidation parameters: application of in vitro and in silico modeling approaches. Annu. Rev. Pharmacol. Toxicol. **44**, 1–25 (2004)
48. Mischel, P.S., Cloughesy, T.F.: Targeted molecular therapy of GBM. Brain Pathol. **13**, 52–61 (2003)
49. Mohamed, A., Zacharaki, E.I., Shen, D., Davatzikos, C.: Deformable registration of brain tumor images via a statistical model of tumor-induced deformation. Med. Image Anal. **10**, 752–763 (2006)
50. Morrison, P.F., Laske, D.W., Bobo, H., Oldfield, E.H., Dedrick, R.L.: High-flow microinfusion: tissue penetration and pharmacodynamics. Am. J. Physiol. **266**, R292–R305 (1994)
51. Nathoo, N., Chahlavi, A., Barnett, G.H., Toms, S.A.: Pathobiology of brain metastases. J. Clin. Pathol. **58**, 237–242 (2005)
52. Pallud, J., Mandonnet, E., Duffau, H., Kujas, M., Guillevin, R., Galanaud, D., Taillandier, L., Capelle, L.: Prognostic value of initial magnetic resonance imaging growth rates for World Health Organization grade II gliomas. Ann. Neurol. **60**, 380–383 (2006)
53. Sanga, S., Sinek, J.P., Frieboes, H.B., Ferrari, M., Fruehauf, J.P., Cristini, V.: Mathematical modeling of cancer progression and response to chemotherapy. Expert Rev. Anticancer Ther. **6**, 1361–1376 (2006)
54. Sanga, S., Frieboes, H.B., Zheng, X., Gatenby, R., Bearer, E.L., Cristini, V.: Predictive oncology: a review of multidisciplinary, multiscale in silico modeling linking phenotype, morphology and growth. Neuroimage **37**(Suppl 1), S120–S134 (2007)
55. Schmitz, J., Kansal, A.R., Torquato, S.: A cellular automaton model of brain tumor treatment and resistance. J. Theor. Med. **4**, 223–239 (2002)
56. Sinek, J., Frieboes, H., Zheng, X., Cristini, V.: Two-dimensional chemotherapy simulations demonstrate fundamental transport and tumor response limitations involving nanoparticles. Biomed. Microdevices **6**, 297–309 (2004)

57. Stamatakos, G.S., Antipas, V.P., Uzunoglu, N.K.: A spatiotemporal, patient individualized simulation model of solid tumor response to chemotherapy in vivo: the paradigm of glioblastoma multiforme treated by temozolomide. IEEE Trans. Biomed. Eng. **53**, 1467–1477 (2006a)
58. Stamatakos, G.S., Antipas, V.P., Uzunoglu, N.K., Dale, R.G.: A four-dimensional computer simulation model of the in vivo response to radiotherapy of glioblastoma multiforme: studies on the effect of clonogenic cell density. Br. J. Radiol. **79**, 389–400 (2006b)
59. Stamatakos, G.S., Zacharaki, E.I., Makropoulou, M.I., Mouravliansky, N.A., Marsh, A., Nikita, K.S., Uzunoglu, N.K.: Modeling tumor growth and irradiation response in vitro–a combination of high-performance computing and web-based technologies including VRML visualization. IEEE Trans. Inf. Technol. Biomed. **5**, 279–289 (2001)
60. Stupp, R., Hegi, M.E., van den Bent, M.J., Mason, W.P., Weller, M., Mirimanoff, R.O., Cairncross, J.G.: Changing paradigms–an update on the multidisciplinary management of malignant glioma. Oncologist **11**, 165–180 (2006)
61. Swanson, K.R., Alvord, E.C. Jr., Murray, J.D.: A quantitative model for differential motility of gliomas in grey and white matter. Cell Prolif. **33**, 317–329 (2000)
62. Swanson, K.R., Alvord, E.C. Jr., Murray, J.D.: Virtual brain tumours (gliomas) enhance the reality of medical imaging and highlight inadequacies of current therapy. Br. J. Cancer **86**, 14–18 (2002a)
63. Swanson, K.R., Alvord, E.C. Jr., Murray, J.D.: Quantifying efficacy of chemotherapy of brain tumors with homogeneous and heterogeneous drug delivery. Acta Biotheor. **50**, 223–237 (2002)
64. Swanson, K.R., Bridge, C., Murray, J.D., Alvord, E.C. Jr.: Virtual and real brain tumors: using mathematical modeling to quantify glioma growth and invasion. J. Neurol. Sci. **216**, 1–10 (2003)
65. Tracqui, P., Cruywagen, G.C., Woodward, D.E., Bartoo, G.T., Murray, J.D., Alvord, E.C. Jr.: A mathematical model of glioma growth: the effect of chemotherapy on spatio-temporal growth. Cell Prolif. **28**, 17–31 (1995)
66. Walker, D.C., Hill, G., Wood, S.M., Smallwood, R.H., Southgate, J.: Agent-based computational modeling of wounded epithelial cell monolayers. IEEE Trans. Nanobioscience **3**, 153–163 (2004)
67. Wang, Z., Zhang, L., Sagotsky, J., Deisboeck, T.S.: Simulating non-small cell lung cancer with a multiscale agent-based model. Theor. Biol. Med. Model **4**, 50 (2007)
68. Wein, L.M., Wu, J.T., Ianculescu, A.G., Puri, R.K.: A mathematical model of the impact of infused targeted cytotoxic agents on brain tumours: implications for detection, design and delivery. Cell Prolif. **35**, 343–361 (2002)
69. Wessels, J.T., Busse, A.C., Mahrt, J., Dullin, C., Grabbe, E., Mueller, G.A.: In vivo imaging in experimental preclinical tumor research—a review. Cytometry A **71**, 542–549 (2007)
70. Wishart, D.S., Yang, R., Arndt, D., Tang, P., Cruz, J.: Dynamic cellular automata: an alternative approach to cellular simulation. In Silico Biol. **5**, 139–161 (2005)
71. Wolfram, S.: A New Kind of Science. Wolfram Media, Champaign, IL (2002)
72. Zhang, L., Athale, C.A., Deisboeck, T.S.: Development of a three-dimensional multiscale agent-based tumor model: simulating gene-protein interaction profiles, cell phenotypes and multicellular patterns in brain cancer. J. Theor. Biol. **244**, 96–107 (2007)
73. Zheng, X., Wise, S.M., Cristini, V.: Nonlinear simulation of tumor necrosis, neo-vascularization and tissue invasion via an adaptive finite-element/level-set method. Bull. Math. Biol. **67**, 211–259 (2005)

Editorial Policy

1. Volumes in the following three categories will be published in LNCSE:

i)   Research monographs
ii) Lecture and seminar notes
iii) Conference proceedings

Those considering a book which might be suitable for the series are strongly advised to contact the publisher or the series editors at an early stage.

2. Categories i) and ii). These categories will be emphasized by Lecture Notes in Computational *Science and Engineering.* **Submissions by interdisciplinary teams of authors are encouraged.** The goal is to report new developments–quickly, informally, and in a way that will make them accessible to non-specialists. In the evaluation of submissions timeliness of the work is an important criterion. Texts should be well-rounded, well-written and reasonably self-contained. In most cases the work will contain results of others as well as those of the author(s). In each case the author(s) should provide sufficient motivation, examples, and applications. In this respect, Ph.D. theses will usually be deemed unsuitable for the Lecture Notes series. Proposals for volumes in these categories should be submitted either to one of the series editors or to Springer–Verlag, Heidelberg, and will be refereed. A provisional judgment on the acceptability of a project can be based on partial information about the work: a detailed outline describing the contents of each chapter, the estimated length, a bibliography, and one or two sample chapters–or a first draft. A final decision whether to accept will rest on an evaluation of the completed work which should include

– at least 100 pages of text;
– a table of contents;
– an informative introduction perhaps with some historical remarks which should be
  accessible to readers unfamiliar with the topic treated;
– a subject index.

3. Category iii). Conference proceedings will be considered for publication provided that they are both of exceptional interest and devoted to a single topic. One (or more) expert participants will act as the scientific editor(s) of the volume. They select the papers which are suitable for inclusion and have them individually refereed as for a journal. Papers not closely related to the central topic are to be excluded. Organizers should contact Lecture Notes in Computational Science and Engineering at the planning stage.

In exceptional cases some other multi-author-volumes may be considered in this category.

4. Format. Only works in English are considered. They should be submitted in camera-ready form according to Springer-Verlag's specifications.
Electronic material can be included if appropriate. Please contact the publisher.
Technical instructions and/or LaTeX macros are available via http://www.springer.com/authors/book+authors?SGWID=0-154102-12-417900-0. The macros can also be sent on request.

General Remarks

Lecture Notes are printed by photo-offset from the master-copy delivered in camera-ready form by the authors. For this purpose Springer-Verlag provides technical instructions for the preparation of manuscripts. See also *Editorial Policy*.

Careful preparation of manuscripts will help keep production time short and ensure a satisfactory appearance of the finished book.

The following terms and conditions hold:

Categories i), ii), and iii):

Authors receive 50 free copies of their book. No royalty is paid. Commitment to publish is made by letter of intent rather than by signing a formal contract. Springer- Verlag secures the copyright for each volume.

For conference proceedings, editors receive a total of 50 free copies of their volume for distribution to the contributing authors.

All categories:

Authors are entitled to purchase further copies of their book and other Springer mathematics books for their personal use, at a discount of 33.3% directly from Springer-Verlag.

Addresses:

Timothy J. Barth
NASA Ames Research Center
NAS Division
Moffett Field, CA 94035, USA
e-mail: barth@nas.nasa.gov

Michael Griebel
Institut für Numerische Simulation
der Universität Bonn
Wegelerstr. 6
53115 Bonn, Germany
e-mail: griebel@ins.uni-bonn.de

David E. Keyes
Department of Applied Physics
and Applied Mathematics
Columbia University
200 S.W. Mudd Building
500 W. 120th Street
New York, NY 10027, USA
e-mail: david.keyes@columbia.edu

Risto M. Nieminen
Laboratory of Physics
Helsinki University of Technology
02150 Espoo, Finland
e-mail: rni@fyslab.hut.fi

Dirk Roose
Department of Computer Science
Katholieke Universiteit Leuven
Celestijnenlaan 200A
3001 Leuven-Heverlee, Belgium
e-mail: dirk.roose@cs.kuleuven.ac.be

Tamar Schlick
Department of Chemistry
Courant Institute of Mathematical
Sciences
New York University
and Howard Hughes Medical Institute
251 Mercer Street
New York, NY 10012, USA
e-mail: schlick@nyu.edu

Mathematics Editor at Springer:
Martin Peters
Springer-Verlag
Mathematics Editorial IV
Tiergartenstrasse 17
D-69121 Heidelberg, Germany
Tel.: *49 (6221) 487-8409
Fax: *49 (6221) 487-8355
e-mail: martin.peters@springer.com

Lecture Notes
in Computational Science
and Engineering

23. L.F. Pavarino, A. Toselli (eds.), *Recent Developments in Domain Decomposition Methods.*

24. T. Schlick, H.H. Gan (eds.), *Computational Methods for Macromolecules: Challenges and Applications.*

25. T.J. Barth, H. Deconinck (eds.), *Error Estimation and Adaptive Discretization Methods in Computational Fluid Dynamics.*

26. M. Griebel, M.A. Schweitzer (eds.), *Meshfree Methods for Partial Differential Equations.*

27. S. Müller, *Adaptive Multiscale Schemes for Conservation Laws.*

28. C. Carstensen, S. Funken, W. Hackbusch, R.H.W. Hoppe, P. Monk (eds.), *Computational Electromagnetics.*

29. M.A. Schweitzer, *A Parallel Multilevel Partition of Unity Method for Elliptic Partial Differential Equations.*

30. T. Biegler, O. Ghattas, M. Heinkenschloss, B. van Bloemen Waanders (eds.), *Large-Scale PDE-Constrained Optimization.*

31. M. Ainsworth, P. Davies, D. Duncan, P. Martin, B. Rynne (eds.), *Topics in Computational Wave Propagation.* Direct and Inverse Problems.

32. H. Emmerich, B. Nestler, M. Schreckenberg (eds.), *Interface and Transport Dynamics.* Computational Modelling.

33. H.P. Langtangen, A. Tveito (eds.), *Advanced Topics in Computational Partial Differential Equations.* Numerical Methods and Diffpack Programming.

34. V. John, *Large Eddy Simulation of Turbulent Incompressible Flows.* Analytical and Numerical Results for a Class of LES Models.

35. E. Bänsch (ed.), *Challenges in Scientific Computing - CISC 2002.*

36. B.N. Khoromskij, G.Wittum, *Numerical Solution of Elliptic Differential Equations by Reduction to the Interface.*

37. A. Iske, *Multiresolution Methods in Scattered Data Modelling.*

38. S.-I. Niculescu, K. Gu (eds.), *Advances in Time-Delay Systems.*

39. S. Attinger, P. Koumoutsakos (eds.), *Multiscale Modelling and Simulation.*

40. R. Kornhuber, R. Hoppe, J. Périaux, O. Pironneau, O.Wildlund, J. Xu (eds.), *Domain Decomposition Methods in Science and Engineering.*

41. T. Plewa, T. Linde, V.G. Weirs (eds.), *Adaptive Mesh Refinement–Theory and Applications.*

42. A. Schmidt, K.G. Siebert, *Design of Adaptive Finite Element Software.* The Finite Element Toolbox ALBERTA.

43. M. Griebel, M.A. Schweitzer (eds.), *Meshfree Methods for Partial Differential Equations II.*

44. B. Engquist, P. Lötstedt, O. Runborg (eds.), *Multiscale Methods in Science and Engineering.*

45. P. Benner, V. Mehrmann, D.C. Sorensen (eds.), *Dimension Reduction of Large-Scale Systems.*

46. D. Kressner, *Numerical Methods for General and Structured Eigenvalue Problems.*

47. A. Boriçi, A. Frommer, B. Joó, A. Kennedy, B. Pendleton (eds.), *QCD and Numerical Analysis III.*

48. F. Graziani (ed.), *Computational Methods in Transport.*

49. B. Leimkuhler, C. Chipot, R. Elber, A. Laaksonen, A. Mark, T. Schlick, C. Schütte, R. Skeel (eds.), *New Algorithms for Macromolecular Simulation.*

50. M. Bücker, G. Corliss, P. Hovland, U. Naumann, B. Norris (eds.), *Automatic Differentiation: Applications, Theory, and Implementations.*

For further information on this book, please have a look at our mathematics catalogue at the following URL: www.springer.com/series/3527

Monographs in Computational Science and Engineering

For further information on this book, please have a look at our mathematics catalogue at the following URL: www.springer.com/series/7417

Texts in Computational Science and Engineering

1. H. P. Langtangen, *Computational Partial Differential Equations*. Numerical Methods and Diffpack Programming. 2nd Edition

2. A. Quarteroni, F. Saleri, *Scientific Computing with MATLAB and Octave*. 2nd Edition

3. H. P. Langtangen, *Python Scripting for Computational Science*. 3rd Edition

4. H. Gardner, G. Manduchi, *Design Patterns for e-Science*.

5. M. Griebel, S. Knapek, G. Zumbusch, *Numerical Simulation in Molecular Dynamics*.

For further information on these books please have a look at our mathematics catalogue at the following URL: www.springer.com/series/5151